1992

DYNAMIC AQUARIA

DYNAMIC AQUARIA

Building Living Ecosystems

Walter H. Adey
Marine Systems Laboratory
National Museum of Natural History

Karen Loveland
Office of Telecommunications

Smithsonian Institution
Washington, D.C.

ACADEMIC PRESS, INC.
Harcourt Brace Jovanovich, Publishers
San Diego   New York   Boston   London
Sydney   Tokyo   Toronto

Front cover photo courtesy of BBC Television, Only One Earth.
Series Producer, Peter Firstbrook.

Back cover photos, home ecosystem and Everglades
mesocosm, by Karen Loveland.

Copyright 1991 by ACADEMIC PRESS, INC.
All Rights Reserved.
No part of this publication may be reproduced or transmitted in
any form or by any means, electronic or mechanical, including
photocopy, recording, or any information storage and retrieval
system, without permission in writing from the publisher.

Academic Press, Inc.
San Diego, California 92101

United Kingdom Edition published by
Academic Press Limited
24—28 Oval Road, London NW1 7DX

Library of Congress Cataloging-in-Publication Data

Adey, Walter H.
    Dynamic aquaria : building living ecosystems / Walter H. Adey,
  Karen Loveland,
      p. cm.
    Includes bibliographical references and index.
    ISBN 0-12-043790-2
    1. Aquarium water.   2. Aquariums.   I. Loveland, Karen.
  II. Title.
  SF457.5.A33 1991
  630.3'4--dc20   90-806
                                                CIP

PRINTED IN THE UNITED STATES OF AMERICA
91 92 93 94   9 8 7 6 5 4 3 2 1

CONTENTS

PREFACE

This book is an outgrowth of many years of research aimed at studying the problems of accurately modeling complex living ecosystems, particularly aquatic systems. Over the past decade, much of this work was done at the Smithsonian Institution's Marine Systems Laboratory, where one of us (Adey) directed and worked with other scientists in developing new approaches to a growing field with applications extending broadly across modern society.

Today we can build model ecosystems at many scales that are sufficiently similar to their wild counterparts that they allow the dynamic processes that are vital to the survival of hundreds of species, and the ecosystems as a whole. We can construct these working "machines" with modifications of our engineering expertise and with biological and ecological raw material of the wild. We can tinker with them and conduct sophisticated experiments with them to demonstrate what is vital not only to the individual organisms contained in the systems, but also, and perhaps of greater importance, to the operation of the systems as a whole. For it is our ability to work in community with our biosphere that will ultimately determine the quality of our lives. This book provides scientists and professional and amateur aquarists with the opportunity to further their understanding of how to more closely approximate natural ecosystems in artificial environments. We hope it will increase interest in a growing field that needs a high level of research activity.

With our Smithsonian and National Museum of Natural History base, we inevitably have had a core role in public display and education.

Many of our model ecosystems were built at least partly for this function. This taught us that our broader public still has a biological understanding and conservation interest that is strongly directed toward selected species and the exceptional performance or appearance (in human eyes) of some species. Yet, the primary problem with loss of biodiversity today is rarely direct impact on an individual species. Mostly, it is habitat destruction and ecosystem disruption. Only when a broad public appreciates the surrounding world as a collection of functioning ecosystems with hundreds of species of plants, animals, and microbes that support human populations in many ways can we hope to make progress in limiting the degradation of our biosphere. No longer do aquarists have to present individual organisms as captives in totally artificial environments. Functioning ecosystems can be "exhibited" to a public that by and large craves to understand how to be ecologically oriented.

Museum, zoological park and public aquaria are extremely important elements in this ecological educational equation. However, with more than 20 million aquarium hobbyists in the United States alone, the development of hobbist orientation to ecosystems rather than individual species must be a crucial element to public understanding of our biosphere.

The book is divided into four broad sections each containing four to seven chapters. Most chapters begin with a review of the subject matter relative to the bigger picture of ecology, ecosystems, and the earth's biosphere as a whole. Part of our appreciation of the complexities of smaller ecosystems comes from understanding the more universal context in which all ecosystems operate. The remainder of each chapter deals with the building of microcosms and mesocosms of ecosystems for research or public display. Where appropriate, each chapter closes with examples directed to the unique aspects of small home aquarium systems.

Part I discusses the physical environment, elements of which at the ecosystem level have often been greatly misunderstood by the scientist and ignored by aquarists and hobbyists. We discuss our new understanding about the shapes, material, and construction of the envelope that will hold various size aquaria; the temperature, water composition, and motion; solar energy; and the substrate, or rock, mud, and sand that make up the floor of the system and in part provide for all-critical geological storage.

In Part II on the biochemical environment we discuss the mechanisms of gas and nutrient exchange as well as the management of animal wastes in small models. We particularly examine "ecosystem metabolism" contrasting the functions of plants with animals needed for the successful operation of these dynamic ecosystems. We introduce a new means for biochemical management of model ecosystems, using controlled communities of algae, to achieve the simulation of larger volumes of open water and where appropriate, export to other communities or geological storage.

The biological structure section (Part III) stresses the importance of the role of diversity in creating stability in the system. While theoreticians argue as to the why and how of this equation, those of us who work daily with ecosystem models know that greater diversity generally means greater stability. We also introduce a core concept of ecology to aquarium science—the food web. We discuss some of the problems that arise with the compression of community structure and food web dynamics into limited space.

In Part IV we present various case studies of microcosms, mesocosms, and aquaria, including the coral reef mesocosm of the Great Barrier Reef Marine Park in Australia, the Chesapeake Bay and Estuary wetlands, and the Everglades model at the Smithsonian. We also discuss our experience in modeling salt, brackish, and fresh water ecosystems at the aquarium sizes of 70–130 gallons.

Finally, in Part V, we present a series of principles for establishing and operating living ecosystems. This is where the real scientific learning process begins, in reducing our endeavors to core concepts each one of which we strive to better understand in the framework of the ecological function of the natural world. This is also where we truly begin to understand how we human beings can become a vital part of our ecosystems.

As we near the twenty-first century, and our technological horizons continue to expand, we are confronted with new considerations involving the vital interdependency of water and life. As the human race heads off on its greatest adventure yet, the exploration and colonization of space, being able to understand and recreate the earth's ecosystems beyond the confines of our planet is probably a precondition to our ultimate success. The information presented here is perhaps only a step in the realization of this larger dream, but we believe it is a vital link. Artificial habitats in which we choose, for efficiency, only those few elements that we feel comfortable with in an engineering sense, will not likely provide long term, stable life support. For we must look at ecosystems, including those we create, as essential to higher life.

Finally, this book is dedicated to the aquarists of zoological parks and public aquaria around the world. Your efforts in continuing to bring an ever deeper understanding of ecosystems and our biosphere to the broader public can be crucial to the ultimate survival of the human race on this planet.

Walter Adey
Karen Loveland

ACKNOWLEDGMENTS

This book was conceived within the first few years of the development of the process and basic engineering of algal turf scrubbing. Many times the writing was set aside for a few more years when the further opportunity to try the process with another ecosystem or water quality management or algal production system presented itself. We are therefore indebted to an extraordinary number of individuals and organizations who have helped us bring the book and the model ecosystems it represents to fruition.

First and foremost none of the background research would have been possible without the extraordinary devotion of Walter H. Adey's associates, postdocs, students, and employees. Within the space allotted we can only give credit to a few of the shining lights, among them Susan Brawley, Tim Goertemiller, Jill Johnson, Silvana Campello, Mike Brittsan, Danielle Lucid, John Hackney, Regas Santas, Jim Norris, Matt Finn, Chris Luckett, Donald Spoon, Adam Milton, Janice Bryum, Sue Lutz, and Lynn Ellington. Our sincerest thanks to all of you and to the even larger unnamed group.

Both of us work at an institution justifiably known for the professional freedom it provides. Many of our administrators, especially directors Porter Kier and Paul Johnson, frequently smoothed a path strewn with other obligations so that we could pursue the endeavors that led up to the book itself. Likewise, it was a pleasure to be associated with the Great Barrier Reef Marine Park Authority, and we will always be grateful for the encouragement of its chairman Graeme Kelleher.

For the production of the text and its multitudinous illustrations, we are indebted for the extensive help of artists Charlotte Roland and Alice Jane Lippson and computer graphic specialists Regas Santas, Nina Ahuja, Steven Schloss, and Bradley Tesh. Photographers Nick Caloyianis and Clarita Berger worked tirelessly to achieve needed angles, lighting, and quality results and Mary Ellen McCaffrey, John Steiner, and Louie Thomas of the Smithsonian's Office of Printing and Photographic Services provided high-quality copy work. For administrative follow through, we could not have asked for more than the superb detail work of Addie Fialk and Sarah Hennessey.

The staff of Academic Press have been so pleasant and supportive that our writing and producing has been a continuing pleasure rather than a task. In particular Jeff Holtmeier, Chuck Arthur, Barbara Heiman, and Barbara Williams were always there helping even when it seemed like we were forever doomed to be late.

Microcosms and mesocosms are expensive to construct, to operate, and to acquire data from. Over the years the Smithsonian Institution, the National Science Foundation, the NOAA Marine Sanctuary Program, and Space Biospheres Ventures have all funded aspects of work presented in this book. We are indebted for their support and hope that in return we have assisted with support of their respective endeavors.

Last, but not least, our respective far flung families have shown only encouragement and patience for a task that consumed our lives. In particular, Nathene Loveland helped us edit for an unfamiliar audience, gave us a sun room to be forever wet, and then held our hands even when sailing was to be preferred.

DYNAMIC AQUARIA

CHAPTER 1

INTRODUCTION

On the surface, water may not appear to be a topic worthy of much discussion. Water is—well, just water. However, on planet Earth there is an tremendous amount of this liquid, and it comprises one of the major spheres that make up our natural world: the hydrosphere (mostly ocean), along with the lithosphere (the rocks and the minerals of the earth's crust), the biosphere (organisms), and the atmosphere (the earth's gaseous envelope). All these concentric envelopes interact with one another. Today most scientists believe that our atmosphere derives its oxygen-rich and low carbon dioxide characteristics directly from plant photosynthesis, and that the hydrosphere, biosphere, and atmosphere are even more inextricably and uniquely intertwined. Indeed, even much of the solid earth's crust was formed or influenced by the activity of water and organisms.

In addition, chemically, water is a most unusual material. By accepted rules, under normal pressures one would expect this ubiquitous compound to exist only as a solid or as a gas, depending on temperature. However, due largely to the peculiar polarization of individual water molecules and the tendency of this compound to form a "semicrystalline" liquid at moderate temperatures, water appears in its most familiar liquid form over a relatively wide temperature range. At the same time, it becomes a "universal" solvent. Almost every chemical element that occurs in the earth's crust dissolves in water, ultimately finding its way into the sea. Here, in the earth's great mixing basin, the potential for the interaction of a wide variety of chemical elements, and the almost infinite number of possible chemical combinations, is significantly increased. The ocean, the ultimate collector of everything washed off the exposed lithosphere and rained out of the atmosphere, becomes very much a chemical soup. One might expect a potential for something very unusual indeed from this infinite number of possible chemical combinations.

Water also has one of the highest capacities of any compound for storing and exchanging heat, and it has great surface tension. Thus, this almost miraculous material is a basic stabilizing element, resisting temperature variations, and tending to become "cellular" by forming internal and external membrane-like surfaces. Given the presence of an additional common chemical element such as carbon, with its great ability to combine with many other elements, the occurrence of liquid water virtually preordains the potential for a chemical complexity that is orders of magnitude above that provided by the basic physical structure of the universe.

Biogenic compounds as complex as amino acids have been identified in interstellar dust and gas clouds, and these may be involved in the collapse that leads to stars and potential solar systems. Nevertheless, without the existence of liquid water, which requires highly specialized conditions (e.g., the right distance from a star or sun, the right planet size, and perhaps even the recycling action of plate tectonics), life cannot develop and continue to evolve (National Research Council, 1990). Wherever liquid water is to be found in the universe in any abundance, life, with its unique ability to store information in a genetic code and to use an energy source for ordering, building, and changing everything with which it comes into contact, is surely to be found. Perhaps somewhere in the vastness of space even more sophisticated life than what we know here on earth occurs.

Billions of years before plants and animals were able to live on land, the first living material formed, evolved, and became ever more complex, as sea organisms. Eventually, along ocean shores and lagoons, perhaps up the numerous rivers, lakes, and wet flats leading to the sea, plants began their first tenuous land existence. Later there followed specially adapted fish capable of crawling on land. As they gradually made the transition to living in an atmosphere rather than floating in the hydrosphere, and as they evolved ways to preserve the crucial water in their bodies from evaporation, plants and animals began to conquer the land. Yet 90% of the time during which life has been on earth has been spent only in the water. And most life that has left the watery environment must conserve and limit its loss of water or die.

Water is the largest component of living matter. While many land plants have developed a form of "suspended" life as nearly dry seeds, at the appropriate time these embryos spring to life as germinating seedlings, but only if water is available. Every animal begins life surrounded by an envelope of fluid. Even land animals, when young, lie in a watery tissue or fluid, protected from the drying air by an encasing shell or by the mother's body. Some scientists believe that the blood that courses through our veins and those of all higher animals is, with some modification, chemically speaking, the equivalent of the ancient sea water that simpler animals used to bring oxygen and food to their cells.

While human adaptation to the terrestrial environment is so complete that living in or even on the ocean has required considerable development of the intellect, we remain fascinated with water and the sea, and constantly strive to truly return to our ancestral home with a new command of that environment. Our engineering accomplishments in this regard are impressive.

The beauty and mystery of lakes, rivers, and oceans have inspired poets and naturalists alike to describe them. But to capture them! That has proven more elusive. For centuries, man has attempted to "bring it back alive"—"it" being a few animals species or occasionally a plant of striking beauty or character. But re-creating the aquatic world, especially the salt-water realm, that would keep those extraordinary creatures alive has often proven difficult.

Communities of organisms residing in freshwater environments are generally less complex than those of the more ancient oceans, making a successful freshwater aquarium considerably easier to achieve. Equally important, those plants and animals are adapted to major and rapid changes in their aquatic piece of the terrestrial world, and therefore are more adaptable to less-than-perfect aquarium environments. Even so, freshwater tanks are a step away from disaster for the unwary who would try to make nature in the small container more than it is in that small piece of the river or lake from which it came, or have failed to provide sufficient elements of wild habitat.

Most home hobbyists and scientists have suffered frustration and often failure in their efforts to establish and maintain marine aquaria. "Witchcraft mixed with a little science" is the way in which one well-known professional aquarist described the difficulties of keeping marine life alive in captivity (Spotte, 1979). An earlier colleague defined the problem as due to the "instability of sea water and its organic constituents, when confined in aquaria or circulatory systems, and to the characteristic inability of marine organisms to adjust to changes in their environment." He listed the necessary components of a proper environment as "a chemically inert water system, a low ratio of animal life to volume of water, the control of bacteria, and the elimination of metabolic waste products" through "aeration, filtration, storage in the dark, and treatment with alkalizers, ultraviolet light and antibiotics" (Atz, 1964).

The concept of the "balanced aquarium" comes to mind as the goal to achieve. Yet, it is curious that the very same concept as applied to wild nature is hardly accepted and has been extensively debated in this century (Egerton, 1973). More recently, "balanced nature" at the species and population level has been replaced not only by a concept of balance but by one of increasing ecological stability with time (extreme physical factors excluded). However, the balance we now find is at the biosphere level (Lovelock, 1979, 1986). In this book we seek the balanced mesocosm, microcosm, or aquarium. However, the balance is at the ecosystem level. Populations may or may not fluctuate in the long term, and because of the relatively small ecosystem sizes involved, human intervention in cropping or replacement of individual organisms is occasionally required to provide the population stability that in the wild is provided by large area.

Over the past few decades, scientists in a variety of laboratories around the world have been making significant advances in keeping marine, estuarine, and aquatic organisms in aquaria-like simulations of real marine environments—model ecosystems, or microcosms. Some of these become quite large, and when they exceed 5000–10,000 gallons in water volume, they are sometimes called mesocosms. There is no sharp line between the microcosm and the aquarium. Perhaps it is best to draw the line at the point where the desire for strict ecosystem simulation is relaxed because of size, and more artificial methods of population or species maintenance, especially human intervention, are undertaken. Living ecosystems in captivity or enclosures have been repeatedly used to answer scientific questions. We will not attempt to develop an historical review but will simply refer to several general works that can lead to the scattered literature for those interested.

In 1984, Eugene Odum called for more mesocosm research, citing overspecialization and fragmentation in ecological research. This paper had been preceded and followed by several compilations of research on aquatic models (Grice and Reeve, 1982), marine and aquatic microcosms (Giesy, 1980), and marine microcosms (Adey, 1987). In the United States, the Marine Ecosystems Research Laboratory at the University of Rhode Island has concentrated on problems of estuarine ecology, especially the fates of pollutants, using a series of cylindrical water towers and a slow flow-through from Narragansett Bay (e.g., Perez et al., 1977; Nixon et al., 1984). More recently, attempts have been made to standardize microcosms as general testing systems for toxic compounds (Gearing, 1989; Taub, 1989). H. T. Odum and R. J. Beyers (1991) review the development and present status of microcosms in scientific research.

Many of these efforts have led to notable scientific insights perhaps unattainable from theoretical or field studies. Although it is not marine or aquatic, but is so exceptional in its accomplishment, we particularly cite microcosm research by Van Voris et al. (1980) on an old field community. In this case, the first successful demonstration of an old and very controversial ecological problem, stability versus complexity, was carried out using microcosm techniques. In some cases new organisms of considerable scientific importance have been recognized for the first time in microcosms (e.g., Brawley and Sears, 1981). Indeed, the simplest multicellular animal, *Trichoplax adhaerens*, was first discovered in a marine aquarium and has subsequently become the sole member of a subkingdom of the Kingdom Animalia (Barnes, 1987; Parker, 1982).

In an aquarium context, we have built on the work of Eng (1961) and Risely (1971) as well as many others whose "natural systems" recognize the benefits of a wide diversity of organisms, especially microorganisms, in the marine aquarium. Carlson (1987) undertook a historical review of those aquarists who have sought to develop aquaria capable of keeping stony corals. However, we also draw on the more recent European approaches in freshwater aquaria of including a rich assemblage of plants. We have also attempted to gain a deeper understanding of the processes involved to discover the principles affecting balance in the aquarium as compared to the wild and to formulate the physical, engineering, and biological elements that would lead to successful ecosystem establishment in captivity. Most of the systems we describe, aquatic, estuarine, and marine, are a departure from traditional aquaria in which many important environmental characteristics are omitted and limited numbers of organisms are utilized. Our focus in this book is not on organisms per se, but rather on ecosystem simulation.

There are three basic philosophies of approach used in the management of water quality in closed aquaria and aquatic models. One is abiological, using chemical methods such as ozonation, and physical methods such as physical filtration, protein skimming, and ultraviolet radiation to offset the effects of a basically poor water quality. These methods are almost always used with the second, more generalized technology of bacteriological filtration, which is employed in various forms and has been used in virtually all closed systems of the past 30 years.

The bacteriological (or biological) filter is a device of almost infinite variety used to maximize surfaces with bacterial cultures in close contact with flowing water of the system being managed. The purpose is threefold: (1) the trapping and breakdown of organic particulates; (2) the degradation of the universal waste products urea and highly toxic ammonia to less toxic nitrite and thence to least toxic nitrate; and (3) more recently, either in special anaerobic chambers or in open aerated trickle systems, the denitrification of nitrate nitrogen to atmospheric gas. Either separately or in conjunction with the above systems, oxygen input into the aquarium and carbon dioxide release from the aquarium are maximized to support not only the organisms being maintained, but the essential respiration activity of the bacteria. The intense respiration of the bacteria in these filters releases considerable carbon dioxide, which tends to dangerously acidify the culture. Thus, buffering with calcium carbonate of a wide variety of forms must be used. Moe (1989) provides an excellent summary of modern aquarium methods. In most cases, these methods are sufficient to maintain many organisms. However, rarely do they achieve the quality of unpolluted wild waters. In most cases, the highest quality aquarium water would not meet national minimum standards for wild waters.

The proponents of the basic principles of bacteriological filtration assume that microbes have been the dominant force controlling water quality in the wild. However, this is likely incorrect, since far more organic material is stored in soils and geological sediments than exists in the biosphere. In addition, the earth's atmosphere is rich in oxygen, and prior to human involvement was very poor in carbon dioxide. Plants and algae have created far more organic matter than microbes have degraded, with a concomitant production of oxygen and removal of carbon dioxide from the biosphere. Thus, plants have been and (until humans started burning coal and oil) remain the dominant force controlling earth's chemistry and particularly the needs of higher animals. It is a general tendency of humans to assume that lack of raw materials to maximize production is a basic need that must be managed. In this case, it has been assumed that the primary requirement is rapid breakdown of all organics to basic mineral elements (carbon, nitrogen, phosphorus, sulfur, silica, etc.). We disagree with this concept. While productivity is sometimes limited by the lack of "nutrients," this is not necessarily a bad condition. Excess nutrients usually result in unstable (bloom) conditions in the wild. Farming and aquaculture almost invariably add nutrients so as to drive productivity of a single organism. However, the result is either unstable or semistable, requiring continuous careful management to avoid a variety of "crash" scenarios. Biospheric and ultimately ecosystem stability must lie not in the rapid breakdown of organics but rather in emphasis on their storage as either plant biomass or geological materials. *Stability in the biosphere, in most wild ecosystems, and in microcosms, mesocosms, and aquaria must lie in competition for scarce resources including carbon and nutrients.*

The third philosophy of approach assumes that the primary desire in model ecosystems and aquaria is stability and an environment that is close to that of the wild community being simulated (or at least that from which most of the organisms were drawn). In this case, health of the organisms as a whole rather than maximum productivity of one or a few species is paramount. This approach uses plants, rather than microbes, as the dominant biological controlling element. Since plant photosynthesis and production dominate over animal, bacterial, and plant respiration in this process, nutrients are minimal, controlling and stabilizing, and oxygen is high, often above saturation. In addition, pH is also typically high. It is assumed in systems using plant control of model ecosystem parameters that microbe activity is normal for the ecosystem and that no special approaches are required to increase such activity. Depending on the ecosystem being simulated, the "locking up" of organic reactants (carbon and nutrients) by sedimentation processes may also be a critical element.

Over the past decade we have designed, constructed, and operated microcosm, mesocosm, and aquarium ecosystems in settings as diverse as our home, the laboratories of the Marine Systems Laboratory of the

Smithsonian Institution, and the Great Barrier Reef Marine Park Authority in Townsville, Australia. They have ranged in size from 30 to 750,000 gallons. Some have simulated tropical reefs and lagoons, others subarctic rocky and muddy shore environments, and finally estuaries and rivers. Most have contained hundreds of species. We are now designing a one-million-gallon tropical complex of aquatic ecosystems for a space biosphere, "Biosphere II," in Tucson, Arizona.

The techniques described in this book have enabled us to place complex marine and aquatic microcosms in stable operation in the laboratory as well as the public aquarium. However, many of these systems have also been housed in homemade tanks of plywood, glass, and fiberglass— built on a limited budget, using readily available equipment and materials. More recently, we have developed a series of standard home-size ecological aquaria, 30–150 gallons, some of which are described below.

These approaches have been monitored by sophisticated electronic equipment, and some of the results are described herein. However, extensive instrumentation is not required to construct and operate ecological models once a basic understanding of the system has been acquired. Questions can generally be answered with a few simple measurements or observational techniques. Our home aquaria were purposely developed using the tools available to the home aquarist.

Improved aquarium maintenance and performance through modern techniques can help reduce the enormous losses of organisms in the commercial aquarium trade. The suffering of the animals is deplorable, and there exists the very real possibility that intensive collection will deplete the environment and upset the balance of natural communities. While large numbers of plants and animals may die in the wild during environmental extremes, human impacts in general are becoming severe enough to shift the delicate survival balance negatively for many species and even for ecosystems. For recreation and education purposes, we cannot accept subjecting organisms to stressful conditions beyond their normal environmental range. Even for research purposes, it is crucial that scientists be far more sensitive to the health of the organisms involved and to the potential negative impacts of collecting. The concentration on tropical "reef systems" should cease, and only when the accomplished aquarist has achieved considerable success with less endangered (and equally exciting) ecosystems should a reef community be approached. By adopting ecosystem techniques, distributors, dealers, and hobbyists can maintain functioning systems and reduce losses dramatically. Indeed, experimental ecosystems and their organisms can be maintained separate from wild ecosystems and endangered organisms can be nurtured for return to the wild. Zoological parks have strongly entered into this arena in recent decades, and now public aquaria, with sufficient financial and scientific expertise, can do likewise. Many freshwater fish have been bred in aquaria, and in the past decade increasing numbers of marine species of fish are being added to the list. Because of our success in breeding hundreds of species of aquatic invertebrates and plants in our ecosystem tanks, the prognosis for greatly reducing wild collecting is encouraging.

In the following chapters, we plot an orderly approach to planning and developing microcosms and aquaria. In Part I we first discuss methods of creating the physical environment. Part II treats the chemical environment, which is inseparable from the organisms as physiological elements. The core of the book, Part III, deals with the organisms themselves. They are presented in a food chain or food web format, since the ultimate difficulty in synthetic ecosystems is scaling, a basic problem that is expressed in the way in which food chains are handled. In Part IV we discuss a wide variety of synthetic ecosystems that we have built, or been heavily involved in, ranging from large mesocosms to home aquaria. Finally, in Part V, as a summary, we attempt to express the major elements of the book as a series of steps or principles of synthetic ecology.

We have attempted to direct this book to the professional aquarist and the advanced hobbyist. Hopefully, it will also be of value both for scientists wishing to undertake mesocosm development and for the less advanced but more dedicated hobbyist. There is an extraordinarily large literature for home aquarium management. Much of it is weak and simply a rehash of traditional methodology. A few modern treatments are quite good, in general, though still largely wedded to traditional methods of water treatment. Several of these, particularly Moe, 1989; Hunnam, 1981; Mills, 1986; and Riehl, 1987, are listed among the references given below. This was an old and almost fossilized field that has come alive with new ideas in the past 5–10 years. Some of the best ideas are still in the magazine literature such as *Freshwater and Marine Aquarium* (FAMA) and *Today's Aquarist*. The scientific literature for marine and aquatic mesocosms is likewise extensive, though rather limited in scope. In recent decades, the scientific work has tended toward using ecosystem models for testing the fates of toxic compounds. Several references are listed that can lead to the broader literature. We have specifically tried to avoid extensive literature citation that would reduce the readability of the text. Where material of other authors is used it is cited. When we feel that a particular matter of interest is sufficiently controversial, literature citation is also provided.

Our understanding of the evolutionary relationships of living organisms has been growing by leaps and bounds in the last twenty-five years. To provide some uniformity in a rapidly changing field, we have tried to order all of our taxonomic materials in accordance with Sybil Parker's (1983) *Synopsis and Classification of Living Organisms*. For the reader without a biological background or whose courses in biology predate 1980, we strongly recommend reading *Five Kingdoms* by Margulis and Schwartz (1988).

References

Adey, W. 1987. Marine microcosms. In *Restoration Ecology*. W. Jordan, M. Gilpin, and J. Aber (Eds.). Cambridge University Press, Cambridge.

Atz, J. 1964. Some principles and practices of water management for marine aquariums. In Sea-Water Systems for Experiental Aquariums. J. Clark and R. Clark (Eds.). *Res. Rep.* **63:** 3–16, Bur. Sport Fish. Wildl.

Barnes, R. 1987. *Invertebrate Zoology*, 5th Ed. Saunders, Philadelphia.

Brawley, S., and Sears, J. 1981. Smithsoniella gen. nov., a possible evolutionary link between the multicellular and siphonous habits in the Ulvophyceae, Chlorophyta. *Am. J. Bot.* **69**(9): 1450–1461.

Carlson, B. 1987. Aquarium systems for living corals. *Int. Zool. Yb.* **26:** 1–9.

Egerton, F. 1973. Changing concepts of the balance of nature. *Q. Rev. Biol.* **48:**322–350.

Eng, L. C. 1961. Nature's system of keeping marine fishes. *Trop. Fish. Hobbyist*, Feb. 1961: 23–30.

Gearing, J. 1989. The role of aquatic microcosms in ecotoxicologic research as illustrated by large marine systems. Chapter 15 In S. Levine, M. Harwell, J. Kelly, and K. Kimball (Eds.). *Ecotoxicology: Problems and Approaches*.

Giesy, J. (Ed.). 1980. Microcosms in Ecological Research. U.S. Tech. Inf. Center, U.S. Dept. of Energy, Symp. Ser. 52 (Conf.-781101).

Grice, G., and Reeve, M. (Eds.). 1982. *Marine Mesocosms: Biological and Chemical Research in Experimental Ecosystems*. Springer-Verlag, New York.

Hunnam, P. 1981. *The Living Aquarium*. Ward Lock Ltd., London.

Lovelock, J. 1979. *Gaia, a New Look at Life*. Oxford University Press.

Lovelock, J. 1986. *The Ages of Gaia*. Norton and Co., New York.

Margulis, L., and Schwartz, K. 1988. *The Five Kingdoms*. Freeman, New York.

Mills, D. 1986. *You and Your Aquarium*. Knopf, New York.

Moe, M. 1989. *The Marine Aquarium Reference*. Green Turtle Publications, Plantation, Florida.

National Research Council. 1990. *The Search for Life's Origins. Progress and futue directions in planetary biology and chemical evolution*. National Academy Press, Washington, D.C.

Nixon, S., Pilson, M., Oviatt,C., Donaghay, P., Sullivan, B., Seitzinger, S., Rudnick, D., and Frithsen, J. 1984. Eutrophication of a coastal marine ecosystem—An experimental study using the MERL microcosms. In *Flows of Energy and Materials in Marine Ecosystems*. M. Fasham (Ed.). Plenum Publishing, New York.

Odum, E. 1984. The mesocosm. *Bioscience* **34:** 558–562.

Odum, H. T., and Beyers, R. J. 1991. *Microcosms and Mesocosms in Scientific Research*. Springer-Verlag.

Parker, S. (Ed.). 1982. *Synopsis and Classification of Living Organisms*. McGraw Hill, New York. 2 Vols.

Perez, K., Morrison, G., Lackie, N., Oviatt, C., Nixon, S., Buckley, B., and Heltsche, J. 1977. The importance of physical and biotic scaling to the experimental simulation of a coastal marine ecosystem. *Helgol. wiss. Meeres.* **30:** 144–162.

Riehl, R. 1987. *Aquarium Atlas*. Baensch. Melle, West Germany.

Risely, R. A. 1971. Tropical Marine Aquaria: the Natural System.

Spotte, S. 1979. *Seawater Aquariums*. Wiley, New York.

Taub, F. 1989. Standardized aquatic microcosms. *Environ. Sci. Technol.* **23:** 1064–1066.

Van Voris, P., O'Neill, R., Emanual, W., and Schugart, Jr., H. 1980. Functional complexity and stability. *Ecology* **61:** 1352–1360.

PART I

Physical Environment

Thus, to miniaturize an ecosystem, to put it into a small space for pleasure, observation, education, or research, one is immediately faced with a major consideration—how to scale the miniature so that it can still function as a reasonable facsimile of the wild ecosystem. When the model differs from its wild analog, preferably one knows what the differences are. This question is so intimately related to the entire problem of how we affect our wild environments and how we restore them, as well as how we construct our aquaria and terraria, that it requires all the backup discussion of the entire book to approach it properly. This chapter deals primarily with shape and mechanics—how to achieve the physical enclosure once one has decided what the size will be. In the last chapter (Chapter 25), we will return to summarize the questions of how to achieve a functioning ecosystem of some complexity in limited space.

Shape of the Model Ecosystem

There is little biological reason for the traditional box-like aquarium shape. It results primarily from mechanical and esthetic convenience, that is, ease of construction and placement in a room or laboratory. For many scientists and aquarists the ease of purchase and setting up of a ready-made tank outweighs all other factors when a water-based ecosystem is desired. However, to go to the other extreme, if one wished to simulate an accurate planktonic ecosystem in microcosm or mesocosm, the presence of tank walls that would support benthic or bottom communities or would allow excessive lateral daylight or night light would be undesirable. In that case, a weakly translucent cylindrical tank to minimize attachment surface for a given volume, with a continuous, rotating, wiping mechanism, to keep that surface free of settlement, would be a possibility. Such a tank would have to be lighted from the top in such a way that the portion of the photic zone desired would have its appropriate light range over depth in the tank.

For the hobbyist, aquarist, or scientist who wishes to construct a model ecosystem, the materials are now available to do so in any shape. In most cases it is important to set aside the box convention and think first of the ecosystem that one wishes to model. Only after determining the ideal shape for simulation of the desired system should one become concerned about esthetics, viewing, and construction problems. Those concerns may then result in a modification of the ideal shape to a variant of the traditional form.

The modern aquarium tank is typically a box-shaped glass or plastic container with a height of about one-half the length and a width of about one-quarter of the length. A recent tendency is to reduce tank width for improved viewing, or to develop unusual shapes for purely esthetic reasons. Such steps should be carefully considered on an ecological basis, especially with regard to surface area open to the atmosphere. An old aquarist rule of thumb relates total length of fish in an aquarium to the open surface area of that aquarium. For example, Dick Mills (1986) suggests 12 square inches of tank open surface to an inch of fish for tropical freshwater fish, 30 square inches per inch for cold freshwater fish, and 48 square inches per inch for tropical marine fish.

The need for simulating a higher air surface contact with the water had led to one of the oldest practices of aquarium science, namely, the bubbling of air in the water. In this case the air–water surface in the bubbles themselves can be very large. Movement of the water in the tank itself, using pumps, also meets this requirement as well as increasing equally essential attached organism–water contact. Many more recent variants include the now almost standard "trickle" filter used in "reef systems." All of these have disadvantages that we shall discuss. However, most crucial for shallow water systems, which cover virtually all modeling and aquarium endeavors, these alternate methods for achieving air-to-water contact do not take into account the need for light for plant photosynthesis. With artificial lighting, a source concentrated in bulbs, it is difficult to get enough light into an ecosystem tank. Reducing surface area usually increases this problem. Water-to-air surface is important. It should not be reduced without a good reason.

As far as actual construction is concerned, the development of silicone cement and seal several decades ago made possible the reliable cementing of glass to glass without the necessity of using a separate steel or wood frame (see e.g., Hunnam, 1981). Since the frames were often a source of problems due to corrosion of the steel and rotting of the wood, the use of glass and silicone cement (and now also varieties of acrylic plastic and acrylic solvent) was a major step in low-cost aquarium construction and maintenance. For tanks larger than several hundred gallons there are other possibilities, including epoxy-treated steel frames with glass walls, molded fiberglass, plywood fiberglass, reinforced concrete, and cement block construction. In cases where these materials are opaque, assuming that through-the-wall viewing is required, ports or windows of glass or plastic must be installed.

To illustrate the interaction of both space and construction concerns, we will first discuss several specific cases that we have dealt with. We will then discuss construction materials in greater depth.

A Coral Reef and Lagoon of Steel, Ply-Fiberglass, and Glass

The 3000-gallon reef tank shown in Figure 1 is an example of an epoxy-coated steel frame with a plywood–fiberglass bottom and ends and glass viewing walls. Its shape was chosen to fit the system being modeled.

Figure 1 A 3000-gallon, coral reef microcosm (reef and lagoon) under construction. The ends and bottoms of the two tanks are fabricated of two layers of glass cloth and polyester-coated, $\frac{3}{4}$-inch marine plywood epoxied together. Main frames are welded, $\frac{1}{4}$-inch steel, double-coated with epoxy paint. The two frames are welded together with steel strips or angles (see end to right). All glass was triple-laminated of tempered sheets, $1\frac{1}{2}$-inch total on the deep side, $\frac{3}{4}$-inch total on the shallow side and on the lagoon. The reef and lagoon tanks are connected by 4-inch pvc pipes. This tank was dismantled after 11 years service due to the initiation (without failure) of glass delamination.

Since this reef was a prototype for a public exhibit, it was also desirable to show the typical shallow, sandy lagoon on one side of the reef in contrast to the deeper water on the "ocean" side. A simple box shape could have been used here for the entire ecosystem pair. However, it would have been necessary to handle a considerable weight of reef and sand structure to build the reef from the bottom of the box. A solid calcium carbonate reef with a sandy lagoon to the depth of the tank would weigh more than twice the weight of water alone. This would have greatly increased the cost of the entire exhibit. The contents of a 200-gallon tank of traditional shape containing only water would weigh about 1600 lbs. With half its volume consisting of a porous carbonate reef and lagoon sand, the tank contents could exceed 3000 lbs in weight. Use of the "stepped" shape shown in Figure 1 for a 200-gallon tank would reduce the weight for the same surface area tank to about the original 1600 lbs. If weight is not a serious consideration, such as in smaller tanks or on a thick, stable concrete floor, the greater the amount of calcium carbonate included in the system, the more stable the ecosystem itself is likely to be, and the richer the included reef community. On the other hand, the tank structure must be strengthened to take such a load even though most of it lies on the floor of the tank.

In constructing enclosed ecosystems it is sometimes desirable to separate parts of those systems by some distance to parallel an important ecological relationship or gradient that exists in the wild. Often, in a model, the space to carry out that relationship is not available. The small, glass lagoon tank attached to the reef tank shown in Figure 1 is a simple example of what could be an extended concatenation of tanks used to produce various parts of ecosystems or adjacent ecosystems in limited space. For example, in this case, the physical distance from coral reef to open lagoon is extremely critical in the wild (see Chapter 21). In this model, this distance is simulated by the constriction of a pipe connection. In Chapters 21–23 we discuss the biochemical and ecological functions of numerous concatenated tanks in a variety of mesocosm systems.

A Maine Coast of Molded Fiberglass and Glass

Figure 2 is another example of a simple "shore type" shape for an ecosystem that is very different from that of a reef. While superficially similar to the coral reef tanks, these units were built from self-supporting molded fiberglass with the same type of glass viewing walls. This tank has been in operation for 6 years as compared to 11 years for the coral reef tank. The only problems of note with both tanks have been with the glass itself.

The basic spatial need and the weight problems are similar in both reef and Maine coast tanks. Unlike the reef, the cold-water shoreline is characterized by granite and metamorphic rocks that are about twice as heavy as reef calcium carbonate. Also, if one is attempting to closely simulate the wild Maine coast rocky shore ecosystem, it would not be desirable to make the substructure of the shore as porous as that of a reef. In a rectangular tank, the requirement for rock could result in such a massive weight as to require special support and tank engineering. We have further reduced weight problems in this case by using pumice stone coated with polyester resin as a sublayer in the rock filling. However, since pumice floats, it can be difficult to manage. If not properly weighted and tied down it can break free and create considerable damage, perhaps

Figure 2 Photograph of 2500-gallon rocky shore Maine coast system including an attached mud flat/marsh. This tank was constructed of fiberglass by laying successive layers of glass cloth with sprayed resin on a mold surface, much as a fiberglass boat would be constructed. The glass is a 1½-inch tempered/laminated sheet of three glass layers. Struck with a large workman's tool, the outer sheet of one of the large glass panels massively fractured but without total sheet failure. The addition of an internal acrylic sheet to this panel solved the immediate problem of total replacement.

even breaking the tank glass during filling of the tank with water. In this case, as in the reef lagoon, the shallow muddy bay is simulated with a smaller elevated tank, partially to avoid many feet of heavy mud in the primary tank. This also was done partially to provide the effect of distance between the two communities. In some cases it might even be desirable to simulate both open ocean and bay in the same box-like tank. However, the area required and the weight that would have to be supported would be quite large. For a large system on solid ground, this might be a cheaper and easier approach. However, enough distance, length of tank, or rock baffle would have to be present to keep wave energy from entering the bay.

An Estuary of Ply-Fiberglass

In a Chesapeake estuarine system (Color Plate 1), a quite different approach to system shape has been used. The Chesapeake Bay is characterized by extensive shallows, with a wide salinity range geographically. A mesocosm thus requires a "flatter shaped," multiple system through which salt- and freshwater mobile organisms, fish and invertebrates as well as plankton and plant spores and seeds, can migrate from one salinity and bottom type to another, much as they do in any wild estuarine ecosystem. The separate salinity sections in this case are connected by computer-controlled gate valves. The mode of operation is discussed in detail in Chapter 23. Being low and flat, the entire construction in this case is that of prefiberglassed, $\frac{3}{4}$-inch exterior plywood. The joints in this tank were faired with soft wood angles and "glassed" over with fiberglass strips. The whole interior of the tank was coated with a heavy, nontoxic epoxy paint. The only concession to strength beyond ply-fiberglass was a double ring of 3-inch steel angle to provide an extra safety factor for the large weight of sediment that the system holds. This tank has functioned with only very minor leaks for 5 years.

An Estuary of Reinforced Concrete

Color Plate 2 shows an Everglades mesocosm in which the same basic salinity gradient and flow problems of Chesapeake Bay were encountered. However, on an Everglades shore, salinity ranges widely both in time and over short distances. Between-unit salinity control is not as important as in the Chesapeake system, and weir rather than valve separations were sufficient. In this case basic tank construction (a reinforced concrete base with reinforced concrete block walls) was entirely different. Standard cement block with steel insert reinforcing between the blocks provided insurance against block or cement fracturing. This latter approach is that of permanent construction where weight is not a critical consideration. In this case the block walls and concrete base were coated with a butyl rubber preparation both to protect the block and steel reinforcing from the water and to seal the steel reinforcing from potential water contamination.

General Considerations

The combination of a simple or multiple-stepped tank, simulating sloping coastal or lake bottoms, reefs, or other communities, with the addition of extra tanks for lagoons, bays, tide pools, or refugia, provides great flexibility where close simulation of whole shore communities is desired. If a less faithful model or a small part of a wild community is desired (e.g., a single coral head), then a traditional box-shaped tank with a single columnar or back drop of coral limestone substrate will suffice (Color Plate 3).

It is virtually axiomatic when modeling aquatic ecosystems, whether for scientific, education, or hobby reasons, that system volume be as large as possible (with due consideration to surface area as we mentioned above). Tank volume is primarily a matter of practical consideration, largely determined by how much money and space is available. The weight that can be supported in larger systems can also be a crucial factor. Generally, for reasons that will become apparent in every section of this book, the larger the tank unit, the easier the ecosystem will be to operate, the closer the environmental parameters will be to the wild, and the healthier will be its organisms. We have built and successfully operated 120-gallon coral reef tanks, though we regard that size as a minimum volume for even a very limited tropical reef community. On the other hand, freshwater ecosystems that can better stand the inevitable temperature and water-quality changes and organism interactions in smaller tanks can be handled at sizes of 60 gallons and perhaps less. Because of the nature of this chapter, we have mentioned primarily engineering factors as they relate to ecosystem limitations. In many cases biological scaling factors can become even more crucial. A higher predator requires a relatively large territory on which to feed, or soon it would be one of fewer occupants of a small tank. On the other hand, many herbivores, uncontrolled by predators, will overgraze an ecosystem. Even a single herbivore such as the plant-eating sea cow can be too large for any mesocosm. In microcosms, mesocosms, and aquaria, informed human interaction must play the role of these larger organisms (see Chapters 14, 25).

Having briefly introduced considerations of shape and structure, and deferred many matters for later discussion in depth, we now turn to the materials and methods of construction.

All-Glass Construction

Glass tanks constructed commercially offer many advantages to the hobbyist and even to the scientist and aquarium distributor. Technically a glass is any molten rock that has cooled so rapidly that crystals have not had a chance to form. In the context of this book, it is silica (SiO_2), or uncrystallized, amorphous quartz. In practice, commercial glass is an alloy typically also containing oxides of sodium (Na_2O) and calcium (CaO). This soda and lime glass is the oldest, cheapest, and easiest to fabricate. Many other glasses now exist. For example, borosilicate (borax, sodium borate, plus silica) is the base of chemical and kitchenware glasses such as Pyrex and Kimax. Lead is used in "crystal" glasses. Some commercial glasses for specialized uses do not even contain silica (see e.g., Bansal and Doremus, 1986; Doremus, 1973).

Soda–lime glass (simply glass in the rest of this discussion) is quite transparent and very hard. Very few minerals, such as diamond and carborundum, are harder or more resistant to weathering and most chemicals. Theoretically glass is stronger than steel even in tensile strength. Also, glass components (silica, calcium, and sodium) in micro quantities are not only nontoxic, they are needed micronutrients for almost any ecosystem. Most of these points are well known and are the primary reasons for the traditional aquarium use of glass. It is the theoretical qualification in the strength characteristics of glass that is puzzling and sometimes leads to the choice of other materials. Glass is brittle. It shows little plastic deformation or give before fracture. A very carefully drawn rod of glass can have extraordinarily high strength and, theoretically, could replace steel in rigid steel cable. However, it flaws easily and cracks can start at flaws, greatly reducing strength. Also, even though glass has great weathering resistance, it does in fact react slowly with water (some SiO_2 groups being replaced by SiOH). This effect is particularly seen at flaw points on the glass surface from which crack development is accelerated.

Glass is relatively cheap. Silicon is one of the more abundant elements of the earth's crust. As long as weight is not a significant factor (and this is likely true in aquarium situations because of the proportionally large weight of water), 10 to 1 safety factors placed on normal working strengths will easily solve the indeterminant strength characteristics. Figure 3 shows glass thickness to dimension characteristics currently in use in the commercial aquarium field. This applies to all glass tanks or glass in steel or concrete structures. In 20 years of working with glass in aquaria, we have seen only a single glass panel failure short of obvious physical damage. Generally, in all glass tanks it is the sealant that begins to fail with minor leakage long before there are glass problems. In very sensitive situations, safety glass or tempered glass can be used as an extra safety factor, although safety glass eventually delaminates and tempered or "heat-strengthened" glass can fail explosively. We have used heat-strengthened, laminated glass successfully. With three $\frac{1}{2}$-inch panels and two 0.030-inch plastic laminate layers little visual distortion occurs and cost is minimal.

One interesting side feature of glass, if used as a cover or in greenhouse situations, is that glass alloys tend to reduce the transmission of ultraviolet radiation. If one wishes to transmit ultraviolet (UV), pure silica or "quartz" glass should be used or artificial UV may have to be provided. As with many environmental factors, the presence of UV is often required by many ecosystems but excess can be damaging. It is because of the latter that the required presence at a low to moderate level is often ignored.

When glass tanks are to be constructed in the home and laboratory, the hobbyist and aquarist should in most cases have the glass cut commercially. It is very difficult to cut the requisite straight and smooth surfaces without proper equipment and training, particularly with the thicker dimensions. Whether cut professionally, in the laboratory, or by the hobbyist, edges should be heat-polished or fine-sanded to reduce sites for crack initiation. With some care, high-quality, inexpensive glass tanks can

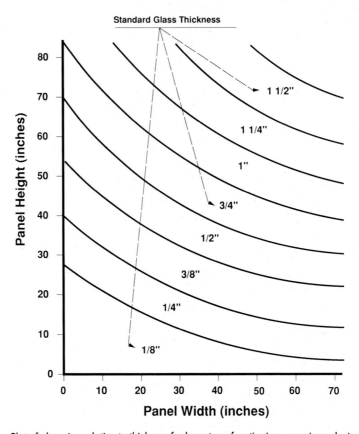

Figure 3 Plot of glass size relative to thickness for long-term functioning aquaria and microcosms.

be constructed at home. At the hobbyist and aquarist level where relatively small sizes are in question (300 gallons and less), Hunnam's (1981) treatment of tank construction is excellent and is repeated here with minor changes (Fig. 4). For larger sizes all-glass tanks are generally impractical.

Acrylic Construction

The primary competitor with glass for small aquarium construction or for ports in larger tanks is acrylic sheet (Plexiglas, Lucite, etc.). Acrylics are made of synthetic hydrocarbons and can be colorless and transparent (see e.g., Levy and DuBois, 1984). Like glass, their strengths in actual practice are significantly less than the theoretical values. The strength of acrylic is generally taken to be somewhat less than glass. Where weight is a factor,

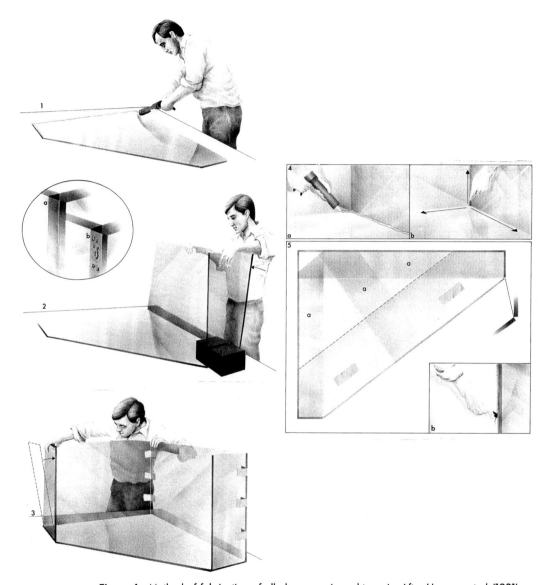

Figure 4 Method of fabrication of all glass aquaria and terraria. After Hunnam *et al.* (1981).

acrylic is about 40% the density of glass, a feature readily apparent to anyone who has carried glass and plastic aquaria. Using the same care as for silicone cement attachment of glass, acrylic-soluble cements can be used to fuse acrylic sheets together. While newer versions of acrylics can be harder than the older sheets, acrylic tanks and ports are subject to intensive scratching by organisms with shells or teeth. Also, in all eco-systems algal, bacterial, and protozoan growths (aufwuchs) are inevitable (and desirable) on available surfaces. If visibility is to be maintained, the glass or acrylic surfaces must be regularly cleaned either by organisms or by hand. Care also has to be taken in construction of complex structures to allow for the considerable flexibility of acrylic sheets. All in all, our preference remains with glass, though we recognize that in public viewing areas, where absolute safety is demanded, and where unplanned impact from a hard object, such as a tool or a bottle, is possible, acrylics have much to offer. Nevertheless, in the past we have chosen safety glass (multilayered glass with thin acrylic sheets molded between) in these situations. This laminated glass product provides the advantages of both materials in aquaria and microcosm construction.

Fiberglass-Reinforced Plastics

In the mid-size enclosure range of 300–10,000 gallons, composite plastic tanks become preferable to all-glass or acrylic units. The base material for these tanks is polyester resin, usually with glass cloth or mat reinforcing. Polyester resins have good strength, toughness, and chemical resistance characteristics. Most important, polyesters adhere well to other materials, particularly glass and wood. The composite material of polyester resin and glass cloth or matting has a flexural strength of nearly 10 times the practical strength of glass and can exceed that of magnesium alloys. Equally important, the polyesters are catalyzed resins, and setting time as a function of temperature and catalyst quantity (which is on the order of drops per gallon) is relatively uncritical. As in all plastics, avoiding contamination, especially by oils and greases, in mixing and application is critical. Although almost any technique will "work," strength and longevity will be severely compromised by any but a technique of scrupulous cleanliness and reasonable adherence to time, temperature, and catalyst quantity tables. Fiberglass unfortunately is not transparent, and therefore glass or acrylic panels must be used where side viewing is required.

Ply-Fiberglass Construction

The construction of fiberglass tanks requires a mold on which to lay the materials (Fig. 5). For one-time units and units of complex shape, molding procedures can be expensive and time consuming. In the 300–10,000 gallon range, plywood can effectively provide a permanent mold for a triple composite of polyester resin, glass cloth, and plywood, as well as serving as a strong and reliable filler. In this technique, cleanliness of materials and application is essential, and all corners should be "faired" with wood angle stripping and carefully fiberglassed, preferably with glass mat. Where wood-to-wood contact is made, epoxy glues should be used to provide full

Figure 5 Fiberglass tank nearing completion. The tank has been constructed over a mold coated with mold release compound. The mold can be re-used more or less indefinitely. This tank is being constructed for aquaculture. However, with strengthening, ports can be cut out and glass installed for side viewing.

bonding. Epoxy putty can also provide for "fairing," sealing, and strengthening of corners, though in quantity it can be expensive. Only waterproof (exterior) plywoods should be used, and for the best construction marine plywood will provide superior results. Where lesser grades of plywood are used, all visible voids should be filled, preferably with an epoxy filler compound. A major qualification to this type of construction is that the polyester is subject to breakdown, chipping, and delamination in the long term when exposed to ultraviolet light. For installations planned to last more than several years, where constant exposure to unfiltered sunlight is present, other techniques should be used. Even where metal halide and mercury vapor lamps are employed, protection from ultraviolet must be provided for fiberglassed surfaces.

In ply-fiberglass construction, epoxy resins can be used instead of polyester resins. Epoxies are stronger, less brittle, and have a generally greater resistance to chemicals than polyesters. In some cases epoxy–glass composites can approach steel in strength. Unfortunately, they are also more expensive and much more sensitive to mixing proportions and temperature at the time of mixing and setting.

In any type of fiberglass construction, side viewing must be accomplished through glass or acrylic ports. Silicone cement is an ideal material for mounting the glass to the fiberglass (or steel or concrete) frame. One must be certain of a continuous bead of silicone when sealing; also, glass strength can be contingent upon a sufficient quantity of silicone being used (Fig. 6). Some silicone cements or seals contain toxic compounds; care must be taken to avoid these materials.

It is difficult to overstate the virtues of glass–resin (polyester and epoxy) composites. Their chemical inertness and strength make them

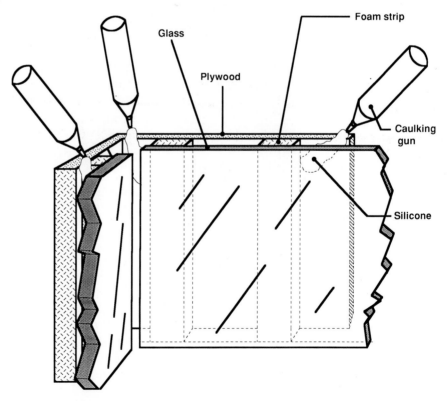

Figure 6 The use of foam strips to assure a full thickness of silicone seal when caulking glass onto fiberglass, steel, or concrete tank main frames.

almost miracle materials for microcosm and mesocosm construction (and many other uses). Even in the smaller aquarium ecosystems where complex shapes are required, these materials have considerable value. For readers not familiar with fiberglassing techniques, or those wishing to update their expertise, we recommend Wiley (1986). Equally important with ease of handling, most fiberglass materials lack the corrosive problems and polluting potential of metals.

Reinforced Concrete Construction

Where weight and space are not factors, and tank size exceeds 10,000 gallons, reinforced cement block or concrete becomes the material of choice. Particularly in larger dimensions, these structures should be carefully designed by experienced designers. However, it is essential to convey the special problems of water, salt water, and ecosystems to the responsible engineers. An engineer will easily appreciate the corrosion problems, but it is only with great difficulty that the sensitivity of ecosystems to contamination by metals can be conveyed. Also, concrete is porous and block can gradually disintegrate when constantly submerged. The inside surfaces of these construction materials must be coated with an impervious layer. Since this is the same problem in reverse as the leaking of building foundations in wet areas, engineers and contractors will understand the sealing problems. However, the final internal coatings should be epoxy or butyl rubber to prevent any water contamination.

There is a natural tendency today for environmentally oriented aquarists to turn away from plastics. Indeed, plastics such as Styrofoam, improperly disposed of, can create serious environmental problems. However, these are organic materials and their leachates are minimum and organically degradable. They certainly do not provide the problems to model ecosystems that concentrated and leached metals do.

Home Aquarium Notes

Although the most typical home aquarium construction is all glass, except in room divider or middle-of-the-room situations or where major backlighting is needed, four sides of glass are not often required or even desired. Maintenance of required environmental conditions, particularly light, darkness, and temperature, are more difficult to achieve in an all-glass tank. While the light situation is more or less obvious, it might not be quite so apparent that it would be much easier to maintain a constant temperature—where temperature constancy is a crucial aspect of the wild environment—by avoiding all-glass construction. Ply-fiberglass, particularly, is an excellent insulator, whereas glass is rather poor in this respect. In establishing a microcosm aquarium, especially when accomplishing the construction at home, one should carefully consider whether four glass sides are really needed. An aquarium with one or two glass sides, with the remaining walls properly constructed fiberglass or ply-fiberglass, is cheaper, stronger, and easier to operate than the more traditional glass tank.

The aquarist should also remember that water is heavy. One cubic foot of salt water weighs about 65 lbs, and fresh water is only a little lighter. A 200-gallon tank will weigh about 1600 lbs. Also, the pressure present at the bottom of the tank is approximately 0.5 psi for every foot of depth. For the individual installing a home aquarium, these simple calculations are desirable simply to place the project in proper perspective and to appreciate the need for strength both for the tank and for the underlying floor.

Glass strength is crucial to tank integrity. If one is not using a commercial tank, care should be taken to insure that thickness is adequate for a given size. The graphs given in Figure 6 can be used to determine plate glass or acrylic thickness. This diagram is based on a variety of successful commercial and one-time construction units. When calculating direct from glass or plastic strength and using acceptable deflections, a safety factor of at least 5 : 1 is desirable for a static system. If a wave machine or other dynamic loading and potentially fatiguing system is present, the safety factor should be increased to 10–15 : 1. Glass edges should be smooth to avoid sites for crack initiation. When setting the large sizes of glass, it is desirable to be certain of a sufficient thickness of silicone. For large systems where failure could be disastrous in many ways, it is desirable to use safety glass. We are personally aware of at least two cases where operating mesocosms have been struck with workman's tools and only safety glass (laminated panels) has averted system disaster and perhaps personal injury. Glass is an extraordinary material that is very hard, clear, and virtually inert relative to aquarium and mesocosm uses. However, it has special characteristics, particularly brittleness, a tendency to fail from damaged edges and serious surface scratches and dents, as well as being subject to long-term fatigue.

Plastics are now available that can compete with glass for aquaria. Unfortunately, standard acrylic scratches too easily for most uses. Special but commercially available hard-surfaced plastics that will serve for most uses are now available. They do tend to be rather expensive as compared to glass or acrylic. In joining plastic to plastic, cements that dissolve the surface of the plastics being cemented, rather than silicone, should be used.

Plumbing of Aquaria and Microcosms

If care is taken in tank construction, there is little need for concern about disastrous loss of water and loss of the entire ecosystem. Beyond massive tank failure (which is a remote possibility in any tank), the most likely way to lose one's microcosm is through external plumbing failure. This can be avoided by withdrawing water from the system only by suction placed relatively high in the tank; if a model system requires drawing from lower down to provide adequate circulation, then a small hole placed high in the suction line should be used to "break suction" with any fall in water level. Thus, if a plumbing leak is present only a small part of the water in the system will be removed before pumping ceases. Also, it is desirable where possible to place all plumbing directly above the tank so that any leakage in the entire system simply drains back to the main tank. Plumbing and the virtues of another "miracle" plastic, PVC (polyvinyl chloride), are discussed in some detail in Chapter 5.

Electrical Considerations

Wild ecosystems are driven or powered primarily by solar energy, although tides can also be important. In the wild, the solar energy is also converted to waves and currents. In most microcosms and mesocosms, artificial energy is likely to be required to circulate and heat or cool any enclosed ecosystem. In many cases, even light will be electrically supplied. Depending on the size of the system, sufficient power must be allowed and, for safety, ground fault breakers should be installed. These are now available as wall-mounted receptacles and are readily mounted in a room lacking such devices.

Toxic Elements and Compounds

Many chemical elements and compounds are toxic to life. Some of these are only mildly poisonous and are often required by many organisms as elements in small quantities and only become toxic in excess. Others are always toxic and only concentration determines effect. Many organisms have evolved the ability to produce poisonous organic compounds, primarily as a means of defense or to facilitate food capture.

Humans have been highly successful at learning the chemical possibilities of the earth, both inorganic and organic. Some of the toxic productions of humans are solely for industrial purposes, but in many cases have been allowed to leak into the wild environment. Other elements and compounds humanity has developed specifically to kill undesirable organisms. Unfortunately, many chemicals that are directed toward specific pest species work their way through ecosystems and become major controlling elements of biological function. DDT was a classic case, and *Silent Spring* (Carson, 1962) presents a grim reality that could have been. While DDT was caught in time and the processes were reversed, many other similar situations are creeping up on us. Chlorine as used for water "purification" in swimming pools, sewer outflows, and drinking water supplies deserves particularly close attention. We may yet destroy ourselves and our higher animal and plant associates through chemistry intended for "better living" if we do not rapidly learn more care in this regard.

The aquarist operates mini ecosystems, which because of their small size are particularly susceptible to contamination by external and internal pollutants. Every chapter in Part I of this book could usefully describe the pitfalls of toxic contaminants. In Chapter 4, we discuss the potential problems that can derive from the water source. Although less likely, atmospheric contamination is also a possibility. Here, we will briefly discuss potential materials problems that would apply equally to the envelope, plumbing, heating and cooling, tide creation, light supply, and substrate. These involve mostly structural elements and thus primarily metals contamination, though many other possibilities exist.

Glass, acrylics, epoxies, polyesters, polypropylenes, polyethylenes, nylons, Teflon, and silicones, among others, are structural materials in common aquarium construction use. *When properly cured* these materials are generally inert, nonbiodegradable, and nontoxic. In some cases fungicides might be added to the materials in use, and these should be guarded against. On the other hand, many metals easily find their way into construction processes and must be avoided. Except for perhaps lead, mercury, cadmium, chromium, nickel, and silver, many metals find micronutrient uses in organic processes (e.g., iron, zinc, and copper). Nevertheless, in abundance even these can provide severe problems.

Copper is one of the most insidious of metal problems for the aquarist. In ionic form it is placed in municipal water systems to kill algae; it is also abundant around human and aquarium situations as copper wiring and piping. Stainless steel, in addition to iron (which is probably the most acceptable of metals, often a micronutrient), has alloy metals such as chromium and nickel. While possibly acceptable in freshwater use, stainless steels should never be used around saltwater ecosystems. No matter what the manufacturer says, a stainless has yet to be developed that will not leach nickel, chromium, and other alloy metals into the water column. Titanium may be acceptable, especially in cooling systems, though the final environmental word has not been received on this metal.

Zinc as "galvanizing" is often used to coat iron and steel to reduce corrosion. It is particularly toxic when dissolved and should be strictly avoided. Galvanized and stainless steel will be recommended by engineers to solve structural problems where corrosion is a potential difficulty; however they should not be allowed into aquatic life processes in abundance.

Many plants take up metals and incorporate them into their structure. Some geological prospecting can be carried out by looking at the plants that either concentrate the element of interest or show characteristic response. Algal turfs also take up many metals preferentially, and the use of this technology to manage water quality can be valuable in this respect (see Chapter 12).

A book on the external toxicity problems of model ecosystems is badly needed. We are not able here to provide the needed depth of information on this important area. We simply suggest that by following the above guidelines, by having one's water sources checked, and by assiduous attention to potential contamination from structures and equipment, these problems can largely be avoided.

References

Bansal, N., and Doremus, R. 1986. *Handbook of Glass Properties*. Academic Press, Orlando, Florida.

Brady, G. S., and Clauser, H. 1986. *Materials Handbook*. McGraw-Hill, New York.

Carson, R. 1962. *Silent Spring*. New York. Fawcett Crest.

Doremus, R. 1973. *Glass Science*. Wiley, New York.

Hunnam, P., Milne, A., and Stebbing, P. 1981. *The Living Aquarium*. AB Nordbok, Göteborg, Sweden.

Levy, S., and DuBois, J. H. 1984. *Plastics Product Design Engineering Handbook*. Chapman and Hall, New York.

Mills, D. 1986. *You and Your Aquarium*. Knopf, New York.

Wiley, J. 1986. *Working with Fiberglass, Techniques and Projects*. Tab Books. Blue Ridge Summit, Pennsylvania.

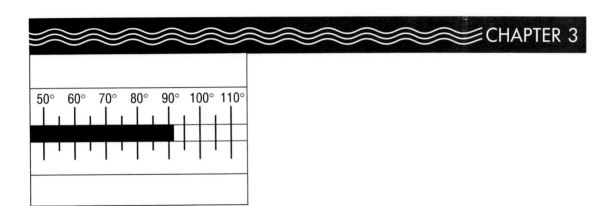

CHAPTER 3

TEMPERATURE
Control of Heating and Cooling

The average temperature of the universe is about −235°C. While stars range from 3000 to over 20,000°C, any point in space unable to collect solar radiation, like the shadow of the planet Mercury, is close to absolute zero (−273°C). The surfaces of most of our own sun's planets and our moon are still rather extreme, generally ranging below −100°C to above +100°C. On the other hand, the extreme temperature range on the surface of the earth, outside of the direct effects of volcanic activity, is a mere −50°C (−68°F) to about +50°C (122°F). This is just about the range that life as we know it can tolerate.

Practically speaking, even on Earth, full life processes are limited from about −2°C to about 40°C and most individual species require a far smaller temperature range to carry out their normal life functions and life cycles. The freezing point of water (0°C in pure fresh water and about −1.8°C in salt water) is a point of special difficulty for many plants and animals, as ice crystals can be very destructive to cells and tissues. Even so, many organisms have developed protective chemicals that function much like antifreeze. This at least allows for a period of hibernation or inactivity at temperatures below freezing.

Abundant water is a major element in limiting the temperature range on Earth. However, to a large extent this extremely small temperature range also results from the activities of organisms over billions of years. The gradual uptake and storage of carbon dioxide (CO_2) as well as the release of oxygen (O_2) by plants, over geologic time, has created an atmosphere that greatly limits temperature extremes on planet Earth. The vegetated surface itself limits temperature variation because of its large water content and indirectly through the control of water movement.

With respect to temperature, our planet has become a large environmentally perfect aquarium and terrarium relative to human activities. If we insist on dominating the whole earth, farming millions of square miles and creating endless buildings, highways and paved malls, grassing our suburbs and limiting organisms cohabiting with us to only those that directly provide benefit or survive in our technical world, we will probably lose temperature control and could create a habitat in which we cannot live.

Temperature and Biomes

Most people are aware that the major biomes (terrestrial life zones of characteristic terrestrial plants and animals) change radically with temperature. From the polar north southward, the polar south northward, and downward from the tops of high cold mountains, the tundra, taiga (conifer forests), and hardwood forests form circumpolar or circummountain bands. Around the warm equatorial and subequatorial bands of the earth, water supply rather than temperature becomes the primary factor determining biome type. These bands, generalized for all continents, are shown in Figure 1, and, in a little different format, (without the role of precipitation), for just North America, appear as the gardener's measure of plant hardiness (Fig. 2).

Although perhaps more muted (as is temperature change), underwater the same kinds and ranges of life zones or biomes occur. Once again these biomes are partly determined by temperature, though substrate becomes crucial and rainfall only indirectly of interest. It has been possible to demonstrate that in the coastal waters of the world ocean, biogeographic zones result from large areas having discreet long-term temperature patterns that differ from those of adjacent areas (Briggs, 1974). There are some very basic biological and ecological differences between cold and warm biogeographic regions. On land, these are obviously related to the need to survive both low temperatures and reduced light levels, accompanied by lower plant production. Generally these latter needs are accomplished by largely shutting down or hibernating to varying degrees and in a wide variety of ways.

Physiological Factors

Underwater the temperature extremes are much smaller, but some very important temperature-controlled factors are operating in addition to

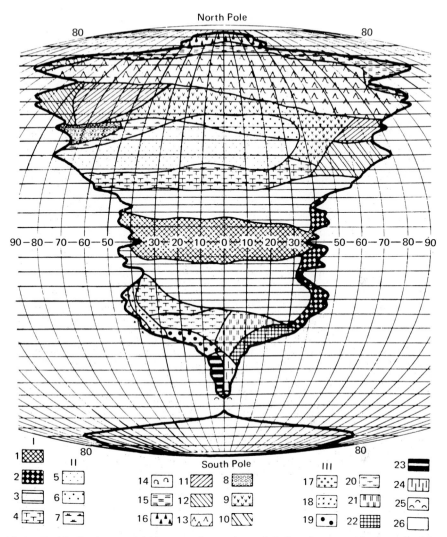

Figure 1 Idealized terrestrial biomes or large areas of similar climate, mostly determined by temperature and to a less extent rainfall. I. *Tropical zones:* (1) Equatorial rain forest; (2) tropical rain forest with trade wind, orographic rain; (3) tropical-deciduous forest (and moist savannas); (4) tropical thornbush (and dry savannas). II. *Extratropical zones of the Northern Hemisphere:* (5) hot desert; (6) cold inland desert; (7) semidesert or steppe; (8) sclerophyllous woodland with winter rain; (9) steppe with cold winters; (10) warm-temperate forest; (11) deciduous forest; (12) oceanic forest; (13) boreal coniferous forest; (14) subarctic birth corest; (15) tundra; (16) cold desrt. III. *Extratropical zones of the Southern Hemisphere:* (17) coastal desert; (18) fog desert; (19) sclerophyllous woodland with winter rain; (20) semidesert; (21) subtropical grassland; (22) warm-temperate rain forest; (23) cold-temperate forest; (24) semidesert with cushion plants, or steppes; (25) subantarctic tussock grassland; (26) inland ice of the Antarctic. After Walter (1979).

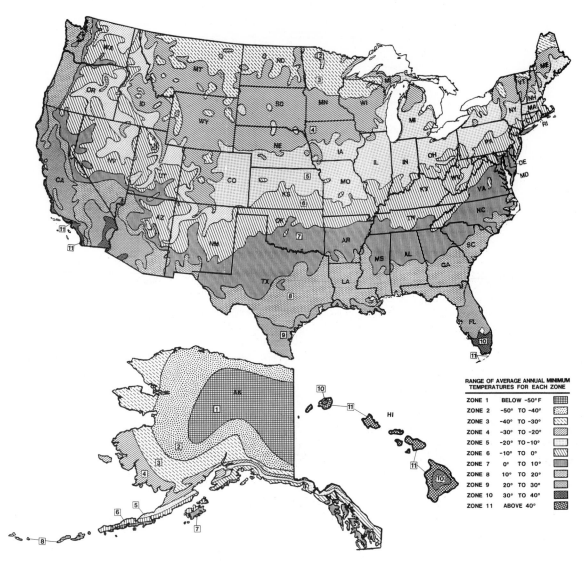

Figure 2 "Hardiness" zones for North America, based largely on minimum winter temperatures. After U.S. National Arboretum.

direct temperature effects. For example, oxygen solubility, and thus the amount of oxygen available to the gills of an aquatic animal, is quite temperature-dependent (low at tropical temperatures and nearly twice as high near 0°C). Tropical and temperate zones have existed for billions of years on the earth. Arctic and subarctic zones, on the other hand, have come and gone. Only a few million years has been available to evolve new organisms and cold-water mechanisms and ecology in the most recent glacial cycle (Pleistocene or Quaternary). This basic limiting factor along with the strong seasonal cyclicity leads one to expect that colder regions would have fewer species and thus a less complicated ecology than the tropics.

The most important factor controlling the limitation of most organisms to relatively narrow temperature or biogeographic zones is a very basic chemical limitation. Rates of chemical reactions, including those characteristic of organic processes, normally are very much a function of temperature, and the thousands of different reactions that occur in any organism are mutually tuned to a limited temperature range. Chemists and physiologists say that Q_{10} ranges from 2 to 3, meaning that any given chemical reaction doubles or triples its rate for every 10°C rise of temperature. Since many mutually dependent chemical reactions in an organism are not likely to have the same Q_{10} values, relatively small temperature changes can quickly unbalance an organism's critical chemistry. Thus, organisms are at a constant risk of poor health and death when the temperature exceeds the ranges, too high or too low, to which they have become genetically and environmentally adapted.

There are two generally applicable subrules that apply to the physiological temperature extreme capabilities of organisms. Most organisms can acclimatize considerably given the time to do so. This rule applies to both individuals (within a time frame of days to months) and populations (many years). Thus, a subarctic clam that might quickly expire given a temperature of 15°C in March would find that same temperature quite optimal when naturally derived in August. Another rule of thumb is that individual organisms, as well as whole populations, are typically nearer their lethal point at the high-temperature end of the scale of survivorship than the colder end (Fig. 3). Temperatures several degrees above the normal range of a population would likely be more dangerous than temperatures several degrees below the normal range.

Beyond concern for the extreme ranges and the short-term survivorship of organisms and their populations, there is also a need to consider normal life cycles and whether or not temperature determines breeding (as it often does) (Fig. 4) and whether feeding patterns (and the availability of feed) and migration are also determined by temperature.

Water has the highest specific heat, and thus the highest ability to store heat, of any naturally occurring compound except ammonia. This very special characteristic of H_2O has considerable bearing on the unique temperature/weather characteristics of our planet and on its ability to support life. It is also important with regard to the effects that bodies of water exert on their surroundings and on organisms both in the water and in the terrestrial environment. This important characteristic of water is also crucial in most modeling of living systems, as we shall discuss below.

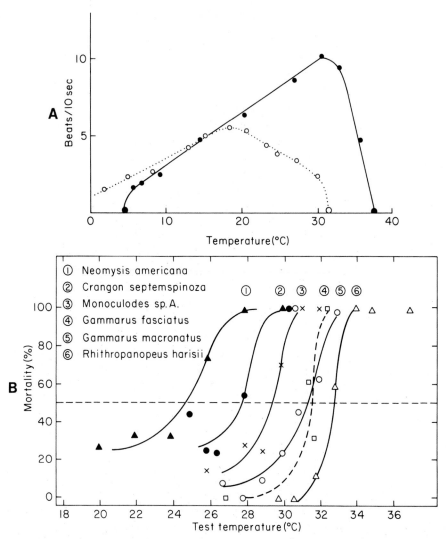

Figure 3　Activity and mortality of marine/aquatic organisms is much more sharply tied to higher rather than lower temperatures. (A) Intertidal barnacle activity (upper intertidal, solid line; lower intertidal, dashed line). (B) Mortality rates of a number of marine crustaceans as they approach their upper temperature limits. After Levinton (1982). Reprinted by permission of Prentice Hall, Englewood Cliffs, New Jersey.

The Temperature Characteristics of Lakes and Oceans

If they are large enough, lakes can have a considerable ameliorating effect on local terrestrial climate (see for example, Fig. 2 relative to the Great Lakes of North America). Lake climate itself in temperate and colder waters is always considerably more moderate than the surrounding ter-

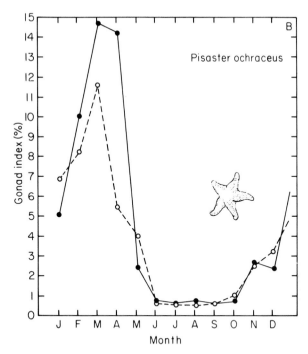

Figure 4 Gonad development in the starfish *Pisaster ochraceus* from the intertidal of northern California. Note that while some seasonal reproductive cycles are tied to day length, this animal reaches its peak reproduction shortly after average temperatures reach their minimum for the year [1955 (•)–1956(○)]]. After Levinton (1982). Reprinted by permission of Prentice Hall, Englewood Cliffs, New Jersey.

restrial climate (Fig. 5). Because the maximum density of water (including the ice phase) lies at about 4°C, cold-climate lakes in the winter have relatively warm deep water (the hypolimnion) while the surface is frozen and the upper few meters lie near 0°C. During the summer, lake surface waters warm considerably and, barring strong winds or other factors, become stable and stratified. The surface layers (or epilimnion) become nearly as warm as the average monthly terrestrial air temperatures, while the bottom water temperature typically ranges from 6 to 15°C. The temperature change from shallow to deep water is often sharp and is called the thermocline. Because the hypolimnion is often relatively isolated, in lakes excessively rich in organic material, bottom waters can become anaerobic in summer if they lie beyond the maximum penetration of light. Tropical lakes, on the other hand, while showing the same basic tendency to stratify, may have a surface to bottom temperature difference of only a few degrees centigrade. Most lakes "overturn," developing uniform temperatures and mixing from top to bottom once or twice per year. Because of their flowing and mixing nature and a much greater contact with the atmosphere, rivers show a wider temperature range than lakes but, at least

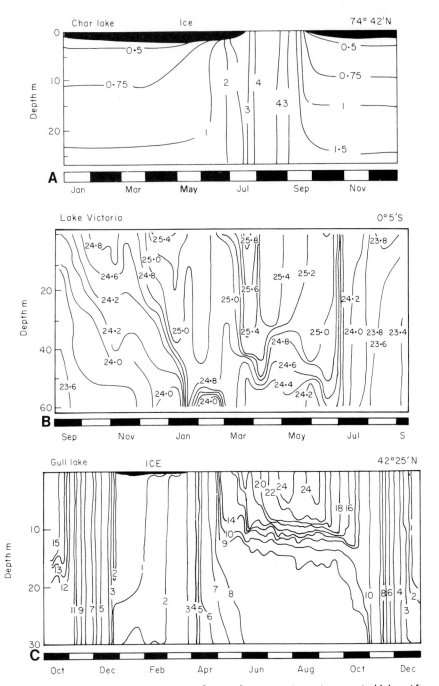

Figure 5 Yearly temperature (°C) ranges surface to deep water in arctic to tropical lakes. After Hutchinson (1957). Reprinted by permission of John Wiley & Sons, Inc. (A) Arctic; (B) tropical; (C) temperate, continental; (D) temperate, maritime.

(*Figure continues*)

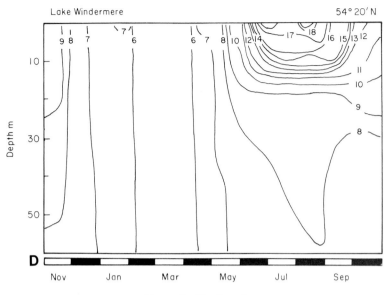

Figure 5 *(Continued)*

in colder climes, will still be considerably more stable than the terrestrial environment.

Even more so, the marine environment varies relatively little in temperature, and marine organisms have evolved to be able to withstand only relatively small changes in ambient temperature. The maximum temperature range throughout the world ocean, outside of very restricted salt ponds or tidal pools, is about 32°C (58°F), as opposed to about 100°C (180°F) in the land environments. The yearly ranges in any one locality, even coastal localities, are much more restricted, ranging from as high as 20°C (36°F) in some temperate or subarctic coastal areas to as little as 3–5°C (5.5–9°F) in some tropical or boreal coastal zones. Generally speaking, the daily and weekly changes at a given depth in coastal and ocean waters are to be measured in no more than tenths of a degree centigrade (Thurman and Webber, 1984).

Marine Biogeography

Coastal marine ecosystems have characteristic yearly temperature ranges, usually following a more or less sinusoidal pattern with a peak and a minimum following solar peak and minimum by 1–2 months. Marine temperature patterns are related to ocean currents, the orientation of coastlines, and the relative "continentality" of the coast, as well as to

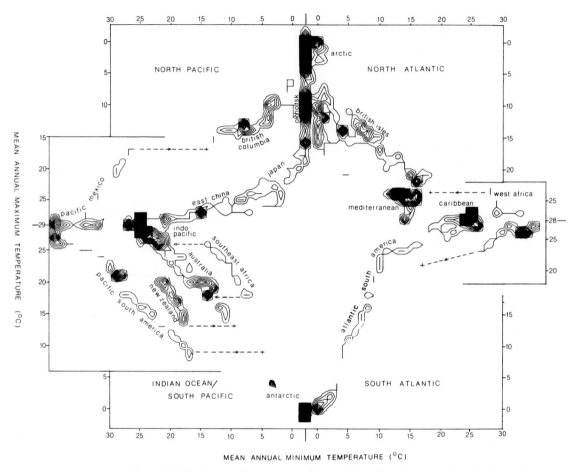

Figure 6 Distribution of temperature characteristics of rocky ocean shore in the world ocean. Each contour represents one nautical mile square (one minute of latitude; a nautical mile = 6080 feet or 1870 m). Note that overlapping coast patterns (e.g., Indo-Pacific and tropical East Pacific; New Zealand and Australia) are separated by the amounts shown to reduce confusing overlap. This diagram shows that large areas of coast line occur under certain temperature regimes. For example hundreds of nautical miles of coast (in the Mediterranean) have a temperature regime of 24–26°C in summer and 10–16°C in winter; on the other hand there is virtually no rocky shore that is 20–24°C in summer and 6–10°C in winter. From W. Adey and R. Steneck (unpublished data).

latitude. Figure 6 shows miles of coastland for the world ocean. There are obviously large areas of coast having certain temperature characteristics and very little coast at other temperatures. If similar plots of temperature for the cold or glacial periods of the last several million years are combined with Figure 6 to show the long-term (3 million years) temperature patterns of coastal areas, effective biogeographic patterns can be determined (Fig. 7). On a map, the data of Figure 7 can be plotted areally (Fig.

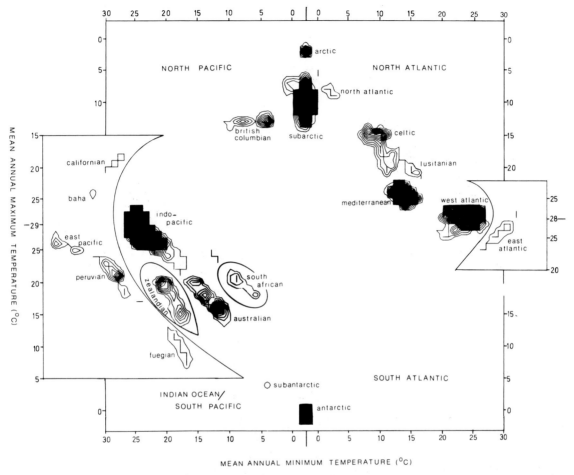

Figure 7 Distribution of average temperature characteristics of rocky ocean shore in the world ocean for the past 3 million years (Pleistocene). This diagram is derived by obtaining the glacial coastal temperature range (after Climap Project Members, 1976) and multiplying (for each degree square) times existing coastal temperature range (Fig. 6). Basically it shows evolutionary time scale (3 million years) coastal temperature patterns.

8), and these biogeographic provinces, as they can be called, developed to provide the basic temperature framework within which large numbers of organisms function and within which the aquarist is also likely to operate (see also Briggs, 1974; Pielou, 1979).

Most organisms in the biosphere are "cold" or "cold-blooded," poikilothermic in scientific terminology. Unlike birds and mammals (homeotherms), which, with adequate energy or food supply, can handle a wide range of temperatures, invertebrates, fish, amphibians, reptiles, and plants can do little or nothing to control their temperature. They function

temperature in large mesocosm systems are less readily available, but still obtainable. Generally one should search for units designed for use in corrosive liquids. If expense is not critical, we recommend temperature control with the mercury-electrode thermometer thermostats. These are obtainable at any scientific supply house; they are accurate to within a few tenths of a degree centigrade and rarely fail. Short of being physically damaged, these units last for many years and justify the investment. When such precision control thermometers are used, actual heating can be obtained using the standard immersion heaters with their controls set slightly above the desired temperature. Multiple units can be set up to obtain a virtual fail-safe system with cost as the limiting factor.

Immersion heaters that are operating for extensive periods in marine or estuarine tanks should be placed in a part of the system with moving water. Not only is heat exchange reduced if the heater is placed in quiet water, but the potential for precipitation formation on the heater walls is present. Such precipitation further reduces the efficiency of the heater and can potentially alter water composition.

For aquaria of less than several hundred gallons, standard aquarium heaters can be used. On the other hand, all standard aquarium heaters that we have used have potential thermostat or switch jamming problems. For small water volumes this could be disastrous. This potential problem can be lessened by using multiple small heaters (e.g., 50–60 W each), rather than a single 200 to 300-W heater, since multiple failure is not likely.

Mesocosm or aquarium systems operated in a habitat designed for human comfort may not need to be heated, particularly when extensive artificial tank and scrubber lighting is utilized, such as on a coral reef. The coral reef system at the Smithsonian Institution has never needed heating even though winter temperatures in the museum are maintained at 20°C (68°F). A 3000-gallon system will, however, maintain considerable stability against normal indoor day-to-night and day-to-day temperature variation of perhaps 2–3°C. A 70-gallon tank in the same environment would have to be heated (or cooled) in most cases.

The same coral reef system at the Smithsonian generally required cooling in the spring before the air conditioning in the building began to operate at its standard level of 24°C (75°F). After that point some cooling was required only on the hottest summer days when building temperatures approach 28°C (80°F). In recent years, the aging Museum heating-ventilating system has wandered widely, creating serious problems for this model reef. In a home air-conditioned environment, where room temperatures can be maintained at a desired level, summer cooling for tropical temperate systems is probably not required. On the other hand, cold-water systems in the home and elsewhere and larger commercial or public mesocosms of tropical or temperate character will likely require cooling.

Generally, cooling of water is considerably more difficult than heating. Commercial units are readily available, but one must use Teflon- or quartz-coated emersion coils to avoid metal contamination, particularly in sea water. Titanium-tubed heat exchangers are now available, but are quite expensive. Often the simplest approach is to build a glass-cored heat exchanger (Fig. 9). If brine (or fresh water if temperatures are not too low) rather than a toxic compound is used as the heat-exchanging liquid, the danger of slight leaks or inadvertent spills into the living system can be reduced or avoided.

Heating and cooling requires energy, and regardless of the direction of heat flow, insulation is generally desirable. As discussed in Chapter 1, if a tank is to be more than a few degrees warmer or colder than the room in which it is placed, an all-glass tank is undesirable, since glass is such a poor insulator. While a double-walled arrangement for all-glass tanks, like "storm windows," is possible, a fiberglass or ply-fiberglass tank with relatively small insert windows (or for a very large tank, insulated concrete

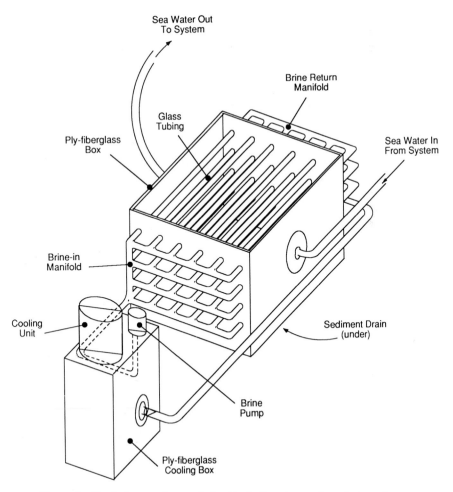

Figure 9 Simple heat exchanger built with ply-fiberglass walls and glass tubing.

block) is more desirable. Insulated tanks not only save energy, they make it easier to maintain a uniform temperature environment within the system. In cold-water systems, properly designed insulation can avoid the condensation that can make viewing difficult and literally destroy tanks and floors. In situations with long pipe runs outside the tank environment, the pipes themselves should be insulated. Neoprene foam is readily available for this purpose. In cold-water tanks it is usually not desirable to keep the room within a few degrees of microcosm temperature. In this case, it is desirable to close the top of the system to reduce cooling required and also to reduce humidity in the room itself. When this must be done, care should be taken not to drastically reduce light levels when full light levels are desired. Ideally a thin quartz sheet that is frequently cleaned would maximize radiation transfer. Practically, a thin glass sheet, frequently cleaned, will generally achieve the needed result.

It is possible to calculate heat loss (or gain) using the following equation:

$$H = \frac{AU(t_o - t_i)}{T}$$

where H = heat flow in Btu per hour, A = surface area (ft.²), U = coefficient of heat transmission, t_o = outside temperature, t_i = inside temperature (°F), and T = thickness of material (inch).

TABLE 1

Heat Transfer (Thermal Conductivity) Coefficients for Common Materials Used In Microcosm and Aquarium Construction[a]

Material	Thermal conductivity (Btu)(inch)/hr/ft²/°F)	Relative scale
Air (still)	0.192	Insulator
Polyurethane foam (10 lbs/ft³)	0.24–0.36	Insulator
Pulp board (wood composite)	0.39	Insulator
Pine wood	0.96	Insulator
PVC (polyvinyl chloride)	0.87–1.45	Intermediate
Epoxy resin	0.87–1.74	Intermediate
Polyester resin	1.2–1.8	Intermediate
Acrylic resin (general purpose)	1.44	Intermediate
Fiberglass (polyester resin)	1.2–2.04	Intermediate
Silicone (RTV)	1.45–2.18	Intermediate
Polyethylene (high density)	2.28	Intermediate
Concrete	6	Conductor
Glass (borosilicate)	7.08	Conductor
Steel	334	Strong conductor
Aluminum	1500	Strong conductor

[a]After Rothburt (1985) and Fink and Beaty (1978).

Heat transfer coefficients for common materials are given in Table 1. Generally, Btu/h = 14.3 (cal/s); Btu/h/12,000 = tons (refrigeration); and Btu/h × 3.41 = watts (heating).

Although it is possible to roughly calculate the heat loss or gain by carefully treating all surfaces through which heat might be transferred, a real situation is so complicated, particularly when water motion is added, that it is best to calculate an approximate worst-case situation, allow an acceptable safety factor, and carry out a test on the selected equipment.

For small home aquarium systems, Moe (1989) has a particularly good section on temperature management. Spotte (1979) discusses larger systems.

References

Briggs, J. 1974. *Marine Zoogeography*. McGraw-Hill, New York.

Climap Project Members. 1976. The surface of the Ice-Age earth. *Science* **191**: 1131.

Constidine, D., and Ross, S. 1964. *Handbook of Applied Instrumentation*. McGraw-Hill, New York.

Fink, D., and Beaty, H. W. 1978. *Standard Handbook for Electrical Engineers*. 11th Ed. McGraw-Hill, New York.

Hutchinson, G. E. 1957. *A Treatise on Limnology*. Vol. 1. John Wiley and Sons, New York.

Levinton, J. 1982. *Marine Ecology*. Prentice Hall, Englewood Cliffs, New Jersey.

Moe, M. 1989. *The Marine Aquarium*. Green Turtle Publications, Plantation, Florida.

Pielou, E. 1979. *Biogeography*. John Wiley and Sons, New York.

Rothbart, H. 1985. *Mechanical Design and Systems Handbook*. 2nd Ed. McGraw-Hill, New York.

Spotte, S. 1979. *Seawater Aquariums*. John Wiley and Sons, New York.

Thurman, H., and Webber, H. 1984. *Marine Biology*. Merrill, Columbus, Ohio.

Walter, H. 1979. *Vegetation of the Earth, and Ecological Systems of the Geo-biosphere*. Springer-Verlag. New York.

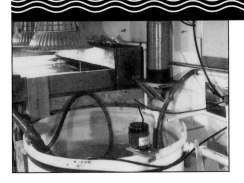

CHAPTER 4

WATER COMPOSITION
The Management of Salinity, Hardness, and Evaporation

As terrestrial organisms bathed in a rather constant atmosphere, mostly of nitrogen (78%), oxygen (21%), and carbon dioxide (350 ppm, 0.035%), the chemical composition of our envelope as it varies from time to time and place to place has had little affect on our behavior and health. Prior to the industrial revolution, only individuals sensitive to the pollen output of some plants had reason to object to atmospheric quality. Even in the last century, only in polluted, "industrial" environments, with toxic additions, did change of atmospheric chemistry affect us or our fellow plants and animals. Times are changing. Some cities, where stagnant air masses develop under some meterological conditions, have "killer smogs" that are largely the result of automobile exhausts. Likewise progressive societies have begun to restrict smoking public places. Also it would appear that man has now begun to significantly change the earth's atmosphere as a whole. CO_2 has increased 30% in the last century. Ozone, while sometimes excessive in cities, has begun to decrease in the upper atmosphere where it functions as an ultraviolet shield. Nevertheless, it is not yet a serious part of our consciousness to be controlled by the composition of our air envelope. For most aquatic organisms, the situation is very different.

Water is the key element in all life, and for terrestrial organisms water loss can be the key factor that decides community composition or even existence. Because of the crucial importance of water loss, land plants and animals have mostly developed shields in the form of largely impervious skin and waxy coatings. A wide variety of controlled ways to prevent water loss, and yet exchange atmospheric gases, have evolved (e.g., mouths and trachea to lungs in vertebrates; spiracles to trachea in insects; stomates to spongy mesophyll in higher plants). Even so, except for carbon dioxide (CO_2) abundance or absence in a closed greenhouse environment, or perhaps the concentration of water vapor in a rain forest or a hot desert, changing atmospheric composition remains with little significant effect or control on terrestrial organisms. As for our body wastes other than CO_2, when crowded we have generally sought a convenient body of water to flush them away. Thus, it is easy for us to forget that for marine, estuarine, and freshwater organisms, the picture is indeed different.

The aqueous medium or hydrosphere as a whole is highly variable in its chemistry over the face of the earth. Even the ocean, the most constant of waters, varies considerably in some aspects of its chemical composition as coasts and rivers are approached. At the same time, aquatic organisms are much less chemically removed from the water medium than their terrestrial counterparts are from the atmosphere. In the course of organic evolution cell walls were undoubtedly first evolved, several billion years ago, because they provided an advantage of self control to many organic processes that had previously had to compete in the larger oceanic soup. Likewise, the first closure of the vascular or blood system some one billion years ago took the previously open sea water that used to carry food and oxygen to each cell and isolated it, as a transporting medium, in blood vessels and body cavities. While this isolating device provided more control and allowed the development of larger organisms, to obtain oxygen from water, gills, devices basically arranged to bring water in as close a contact as possible to the flowing blood, had to be developed. Even more recently, in a geologic sense, perhaps 400 million years ago, some fish entering fresh waters evolved kidney glomeruli. These were necessary to constantly pump out the fresh water that now continuously flowed into the blood through the gills, once the salt balance between inside and outside changed. While kidneys have gone on to become much more complex and varied structures, the problem of sensitivity to the aqueous medium remains. Many algae and small invertebrate animals remain extremely sensitive to changes in the chemical composition of water. Others such as anadromous fish (salmon, for example) have adapted to manage enormous changes of salinity, and yet remain sensitive to acidity changes. It is unfortunate to note that one of the more routine human additions to the atmosphere, industrial stack emissions that result in acid rain, have devastated fish populations in the streams and lakes of thousands of square miles of Scandinavia and northeastern North America. . In the establishment and operation of microcosms, mesocosms, and aquaria, an understanding of the basic nature of the chemical composition of water and its effects on organisms is essential.

Water Structure and Characteristics

Water is a most extraordinary compound. Because of its unusual molecular structure (Fig. 1), it has a number of unexpected chemical characteristics that are generally favorable to life on earth (Table 1). Its change of density with temperature (most dense at 3.98°C) is crucial for freshwater ecosystems to function in cold climates. If it were more normal, more dense below zero, all but the smallest bodies of water in colder climates would freeze to the bottom, and only the surface would melt in summer. Perhaps even more important to life as we know it is the dissolving power of water. Allowing virtually every naturally occurring element and most compounds into its semi-open liquid structure, water is the universal soup in which an incredible number of chemical interactions become possible.

In this chapter we will discuss what can be called the conservative properties of water chemistry, basically those that are not strongly changed by the effects of the activities of organisms. These are the chemical features (salinity and hardness) that are primarily environmental and specifically geological and meteorological in origin. The properties of water that are strongly affected by the activity of organisms, the interactive properties, we will discuss in Chapters 9–11.

Ocean Salinity

The salinity of the open oceans has a rather uniform chemical composition in both time and space (Tables 2 and 3) that ranges from about 34 to 37 ppt (parts per thousand), or 3.4–3.7% by weight. Sodium (Na^+ at 30.61%) and chlorine (Cl^- at 55.04%) make up over 85% of this salt, and only four additional ions—sulfate (SO_4^{2-} at 7.68%), magnesium (Mg^{2+} at 3.69%), calcium (Ca^{2+} at 1.16%), and potassium (K^+ at 1.10%)—bring the composition to over 99%. When sea water is diluted by normal fresh waters, along an ocean coast or in estuaries, the proportional chemical composition remains essentially unchanged, though the total quantity of salts drops with dilution. This rather uniform composition over time (at least on the order of thousands of millennia) results from the constant leaching of terrestrial rocks by rainfall and streams into the enormous volume of the world ocean. Evaporation, precipitation, and river supply certainly affect seawater salt composition, and for that reason the major ocean and sea surface waters vary slightly in salinity. Also, variation can occur on a geologic time scale. For example, it is known that epicontinental or enclosed seas have often dried up, leaving large salt beds. In

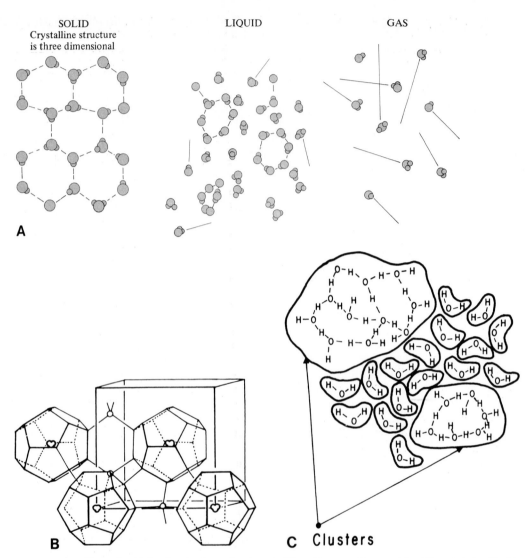

Figure 1 Molecular structure of water at different temperatures and in different states. Liquid water is a "semisolid" because of the asymmetric and charged nature of the molecule. It is this "mixed" state that gives rise to most of its "miraculous" qualities (see Table 1). (A) After Thurman and Webber. Copyright © 1984 by Scott, Foresman and Company. Reprinted by permission of HarperCollins Publishers. (B) Pauling's self clathrate water model. (C) Frank and Wen's flickering clusters model, B and C after Horne (1969). Reprinted by permission of John Wiley & Sons, Inc.

the case of the Mediterranean in the late Miocene epoch, 5–6 million years ago, repeated isolation and complete evaporation probably affected salt composition of the world ocean by several parts per thousand. (This in turn may have affected world climate by increasing sea ice formation.)

TABLE 1

Unusual Physical Properties of Water and Their Importance in Biological/Ecological Systems[a]

Property	Comparison with other substances	Importance in physical–biological environment
Heat capacity	Highest of all solids and liquids except liquid NH_3	Prevents extreme ranges in temperature Heat transfer by water movements is very large Tends to maintain uniform body temperatures
Latent heat of fusion	Highest except NH_3	Thermostatic effect at freezing point owing to absorption or release of latent heat
Latent heat of evaporation	Highest of all substances	Large latent heat of evaporation extremely important in heat and water transfer of atmosphere
Thermal expansion	Temperature of maximum density decreases with increasing salinity. For pure water it is at 4°C.	Fresh water and dilute sea water have their maximum density at temperatures above the freezing point. This property plays an important part in controlling temperature distribution and vertical circulation in lakes.
Surface tension	Highest of all liquids	Important in physiology of the cell Controls certain surface phenomena and drop formation and behavior
Dissolving power	In general dissolves more substances and in greater quantities than any other liquid	Obvious implications in both physical and biological phenomena
Dielectric constant	Pure water has the highest of all liquids	Of utmost importance in behavior of inorganic dissolved substances because of resulting high dissociation
Electrolytic dissociation	Very small	A neutral substance, yet contains both H^+ and OH^- ions
Transparency	Relatively great	Absorption of radiant energy is large in infrared and ultraviolet. In visible portion of energy spectrum there is relatively little selective absorption, hence is "colorless." Characteristic absorption important in physical and biological phenomena.
Conduction of heat	Highest of all liquids	Although important on small scale, as in living cells, the molecular processes are far outweighed by eddy conduction.

[a]From Sverdrup *et al.* (1942). Reprinted by permission of Prentice Hall, Englewood Cliffs, New Jersey.

TABLE 2

Elements Present in Sea Water[a]

Element	mg/kg Cl = 19.00 °/oo	mg-atoms/l Cl = 19.00 °/oo	Atomic weight (1940)	1/atomic weight
Chlorine	18980	548.30	35.457	0.02820
Sodium	10561	470.15	22.997	0.04348
Magnesium	1272	53.57	24.32	0.04112
Sulphur	884	28.24	32.06	0.03119
Calcium	400	10.24	40.08	0.02495
Potassium	380	9.96	39.096	0.02558
Bromine	65	0.83	79.916	0.01251
Carbon	28	2.34	12.01	0.08326
Strontium	13	0.15	87.63	0.01141
Boron	4.6	0.43	10.82	0.09242
Silicon	0.02–4.0	0.0007–.14	28.06	0.03564
Fluorine	1.4	0.07	19.00	0.05263
Nitrogen (ionic)	0.01–0.7	0.001–0.05	14.008	0.07139
Aluminum	0.5	0.02	26.97	0.03708
Rubidium	0.2	0.002	85.48	0.01170
Lithium	0.1	0.014	6.940	0.14409
Phosphorus	0.001–0.10	0.00003–0.003	30.98	0.03228
Barium	0.05	0.0004	137.36	0.00728
Iodine	0.05	0.0004	126.92	0.00788
Arsenic	0.01–0.02	0.00015–0.0003	74.91	0.01335
Iron	0.002–0.02	0.00003–0.0003	55.85	0.01791
Manganese	0.001–0.01	0.00002–0.0002	54.93	0.01820
Copper	0.001–0.01	0.00002–0.0002	63.57	0.01573
Zinc	0.005	0.00008	65.38	0.01530
Lead	0.004	0.00002	207.21	0.00483
Selenium	0.004	0.00005	78.96	0.01266
Cesium	0.002	0.00002	132.91	0.00752
Uranium	0.0015	0.00001	238.07	0.00420
Molybdenum	0.0005	0.000005	95.95	0.01042
Thorium	<0.0005	<0.000002	232.12	0.00431
Cesium	0.0004	0.000003	140.13	0.00714
Silver	0.0003	0.000003	107.880	0.00927
Vanadium	0.0003	0.000006	50.95	0.01963
Lanthanum	0.0003	0.000002	138.92	0.00720
Yttrium	0.0003	0.000003	88.92	0.00125
Nickel	0.0001	0.000002	58.69	0.01704
Scandium	0.00004	0.0000009	45.10	0.02217
Mercury	0.00003	0.0000001	200.61	0.00498
Gold	0.000006	0.00000002	197.2	0.00507
Radium	$0.2–3 \times 10^{-10}$	$0.8–12 \times 10^{-13}$	226.05	0.00442
Cadmium				
Chromium				
Cobalt				
Tin				

[a]Note: Virtually all the gases of the atmosphere are also present, and together these provide a wide variety of organic and inorganic chemical compounds. From Sverdrup et al. (1942). Reprinted by permission of Prentice Hall, Englewood Cliffs, New Jersey.

TABLE 3

Ionic Composition of Sea Water[a]

Element	Chemical species	Molar	$\mu g \, l^{-1}$
H	H_2O	55	1.1×10^8
He	He (gas)	1.7×10^{-9}	6.8×10^{-3}
Li	Li^+	2.6×10^{-5}	180
Be	$BeOH^+$	6.3×10^{-10}	5.6×10^{-3}
B	$B(OH)_3$, $B(OH)_4^-$	4.1×10^{-4}	4440
C	HCO_3^-, CO_3^{2-}, CO_2	2.3×10^{-3}	2.8×10^4
N	N_2, NO_3^-, NO_2^-, NH_4^+	1.07×10^{-2}	1.5×10^5
O	H_2O, O_2	55	8.8×10^8
F	F^-, MgF^+	6.8×10^{-5}	1.3×10^3
Ne	Ne (gas)	7×10^{-9}	1.2×10^{-1}
Na	Na^+	4.68×10^{-1}	10.77×10^6
Mg	Mg^{2+}	5.32×10^{-2}	12.9×10^5
Al	$Al(OH)_4^-$	7.4×10^{-8}	2
Si	$Si(OH)_4$	7.1×10^{-5}	2×10^6
P	HPO_4^{2-}, PO_4^{3-}, $H_2PO_4^-$	2×10^{-6}	60
S	SO_4^{2-}, $NaSO_4^-$	2.82×10^{-2}	9.05×10^5
Cl	Cl^-	5.46×10^{-1}	18.8×10^6
Ar	Ar (gas)	1.1×10^{-7}	4.3
K	K^+	1.02×10^{-2}	3.8×10^5
Ca	Ca^{2+}	1.02×10^{-2}	4.12×10^5
Sc	$Sc(OH)_3$	1.3×10^{-11}	6×10^{-4}
Ti	$Ti(OH)_4$	2×10^{-8}	1
V	$H_2VO_4^-$, HVO_4^{2-}	5×10^{-8}	2.5
Cr	$Cr(OH)_3$, CrO_4^{2-}	5.7×10^{-9}	0.3
Mn	Mn^{2+}, $MnCl^+$	3.6×10^{-9}	0.2
Fe	$Fe(OH)_2^+$, $Fe(OH)_4^-$	3.5×10^{-8}	2
Co	Co^{2+}	8×10^{-10}	0.05
Ni	Ni^{2+}	2.8×10^{-8}	1.7
Cu	$CuCO_3$, $CuOH^+$	8×10^{-9}	0.5
Zn	$ZnOH^+$, Zn^{2+}, $ZnCO_3$	7.6×10^{-8}	4.9
Ga	$Ga(OH)_4^-$	4.3×10^{-10}	0.03
Ge	$Ge(OH)_4$	6.9×10^{-10}	0.05
As	$HAsO_4^{2-}$, $H_2AsO_4^-$	5×10^{-8}	3.7
Se	SeO_3^{2-}	2.5×10^{-9}	0.2
Br	Br^-	8.4×10^{-4}	6.7×10^4
Kr	Kr (gas)	2.4×10^{-9}	0.2
Rb	Rb^+	1.4×10^{-6}	120
Sr	Sr^{2+}	9.1×10^{-5}	8×10^4
Y	$Y(OH)_3$	1.5×10^{-11}	1.3×10^{-3}
Zr	$Zr(OH)_4$	3.3×10^{-10}	3×10^{-2}
Nb		1×10^{-10}	1×10^{-2}
Mo	MoO_4^{2-}	1×10^{-7}	10
Tc			
Ru			
Rh			
Pd			
Ag	$AgCl_2^-$	4×10^{-10}	0.04
Cd	$CdCl_2$	1×10^{-9}	0.1
In	$In(OH)_2^-$	0.8×10^{-12}	1×10^{-4}
Sn	$SnO(OH)_3^-$	8.4×10^{-11}	1×10^{-2}

(Table continues)

TABLE 3

(Continued)

Element	Chemical species	Molar	$\mu g\ 1^{-1}$
Sb	$Sb(OH)_6^-$	2×10^{-9}	0.24
Te	$HTeO_3^-$		
I	$IO_3^-,\ I^-$	5×10^{-7}	60
Xe	Xe (gas)	3.8×10^{-10}	5×10^{-2}
Cs	Cs^+	3×10^{-9}	0.4
Ba	Ba^{2+}	1.5×10^{-7}	2
La	$La(OH)_3$	2×10^{-11}	3×10^{-3}
Ce	$Ce(OH)_3$	1×10^{-10}	1×10^{-3}
Pr	$Pr(OH)_3$	4×10^{-12}	6×10^{-4}
Nd	$Nd(OH)_3$	1.9×10^{-11}	3×10^{-3}
Pm	$Pm(OH)_3$		
Sm	$Sm(OH)_3$	3×10^{-12}	0.5×10^{-4}
Eu	$Eu(OH)_3$	9×10^{-13}	0.1×10^{-4}
Gd	$Gd(OH)_3$	4×10^{-12}	7×10^{-4}
Tb	$Tb(OH)_3$	9×10^{-13}	1×10^{-4}
Dy	$Dy(OH)_3$	6×10^{-12}	9×10^{-4}
Ho	$Ho(OH)_3$	1×10^{-12}	2×10^{-4}
Er	$Er(OH)_3$	4×10^{-12}	8×10^{-4}
T m	$Tm(OH)_3$	8×10^{-13}	2×10^{-4}
Yb	$Yb(OH)_3$	5×10^{-12}	8×10^{-4}
Lu	$Lu(OH)_3$	9×10^{-13}	2×10^{-4}
Hf		4×10^{-11}	7×10^{-3}
Ta		1×10^{-11}	2×10^{-3}
W	WO_4^{2-}	5×10^{-10}	0.1
Os	ReO_4^-	2×10^{-11}	4×10^{-3}
Ir			
Pt			
Au	$AuCl_2^-$	2×10^{-11}	4×10^{-3}
Hg	$HgCl_4^{2-},\ HgCl_2$	1.5×10^{-10}	3×10^{-2}
Tl	Tl^+	5×10^{-11}	1×10^{-2}
Pb	$PbCO_3,\ Pb(CO_3)_2^{2-}$	2×10^{-10}	3×10^{-2}
Bi	$BiO^+,\ Bi(OH)_2^+$	1×10^{-10}	2×10^{-2}
Po	$PoO_3^{2-}\ PoO(OH)_2(?)$		
At			
Rn	Rn (gas)	2.7×10^{-21}	6×10^{-13}
Fr			
Ra	Ra^{2+}	3×10^{-16}	7×10^{-8}
Ac			
Th	$Th(OH)_4$	4×10^{-11}	1×10^{-2}
Pa		2×10^{-16}	5×10^{-8}
U	$UO_2(CO_3)_2^{4-}$	1.4×10^{-8}	3.2

[a]From Spotte (1979). Reprinted by permission of John Wiley & Sons, Inc.

Fresh waters, on the other hand, vary widely in salinity (here called hardness) from those of virtually pure water to, for example, the Dead Sea (226 ppt) and the North American Great Salt Lake (203 ppt), with salinities far above that of the ocean and chemical compositions that vary considerably from that in the ocean. On the average, rain water has a salinity of about 0.008 ppt, and is slightly enriched in calcium and sulfate, though near sea coasts a considerable increase in chlorine can be found (Hutchinson, 1957). River waters average 0.1–0.16 ppt, though those draining predominantly igneous rock areas are typically below 0.05 ppt. Rivers usually do not extend above 0.2 ppt, barring human effects or the localized leaching of salt beds. The River Jordan reaches 7.7 ppt, for example.

Lakes with no outflow, that is, endorheic lakes, can have very high salinities (see above), and the chemical composition generally varies greatly from that of the sea and from one lake to another (Table 4).

Hardness of Fresh Waters

Water hardness is handled a little differently from salinity in that the unit of measure is the degree in some countries. One German degree of hardness equals 10 mg/l or 10 ppm of calcium and magnesium oxide or typically 17.8 ppm of $CaCO_3$ in the United States. This is not at all equivalent to salinity in that a relatively low salinity water rich in calcium and magnesium could have a relatively high degree of hardness. Nevertheless, the mean river of Hutchinson (1957) with a salinity of 0.13 ppt would have hardness of about 1.9° i.e., very soft). The standard terminology relating degrees of hardness to descriptive elements is given in Table 5.

Unfortunately, hardness through calcium is tied to the very nonconservative carbonate system, as measured by pH. The pH is very much affected by organism respiration and photosynthesis and is treated at length in Chapter 9. In practice, the two can be handled separately as long as one remembers that there can be a connection. Hard waters usually also have a high pH.

Within the range of the standard hardness test kit the wild freshwater community that one wishes to simulate can be determined. If one is creating a synthetic system, the best manuals describing standard aquaria fish usually provide optimum numbers for each species. Many fish species can adapt to a wide range of hardness, particularly if it is changed gradually. Generally, with freshwater systems, as long as evaporation is not excessive and the hardness of replacement water is not too high, occasional small volume changes (a gallon a week for a 70-gallon system) are sufficient to prevent salt buildup and the effective development of an endorheic or salt lake in the aquarium environment. As discussed below, algal scrubbers will tend to adjust disproportionate elements as long as the system is not allowed to become too unbalanced. The difficulties arise

TABLE 4

Chemical Composition of Selected Lakes Dominated by Different Anions (%)[a]

	Na	K	Mg	Ca	CO₃	SO₄	Cl	SiO₂	(AlFe)₂O₃	Salinity (mg/kg)
Chloride waters										
Bear River										
Upper, Wyoming	4.49		6.86	23.69	52.68	5.76	2.68	3.84	—	185
Lower, Utah	20.54		4.76	10.12	21.53	8.16	32.36	—	2.53	637
Great Salt Lake, Utah	33.17	1.66	2.76	0.17	0.09	6.68	55.48	—	—	203,490
Jordan at Jericho	18.11	1.14	4.88	10.67	13.11	7.22	41.47	1.95	1.45	7,700
Dead Sea	11.14	2.42	13.62	4.37	Trace	0.28	66.37[b]	Trace	—	226,000
Sulfate waters										
Montreal Lake, Saskatchewan	4.9	2.3	10.8	16.8	56.5	1.8	2.5	3.9	0.5	150.5
Redberry Lake, Saskatchewan	12.0	0.85	12.3	0.56	2.58	70.5	1.1	0.03	0.07	12,898
Little Manitou Lake	16.8	1.0	10.9	0.48	0.47	48.4	21.8	0.019	0.21	106,851
Carbonate waters										
Silvies River, Oregon	10.42	2.45	3.13	12.88	34.76	7.35	2.88	25.13	0.08	163
Malheur Lake, Oregon	24.17	5.58	4.13	5.58	44.63	7.64	4.55	2.89	Trace	484
Pelican Lake, Oregon	29.25	3.58	2.62	2.27	30.87	22.09	7.97	1.21	0.02	1,983
Bluejoint Lake, Oregon	37.70	2.62	0.63	0.53	38.68	5.67	13.85	0.55	0.02	3,640
Moses Lake, Washington	19.86		7.25	8.41	51.56	2.87	3.88	5.06	1.11	2.966

[a]After Hutchinson (1957). Reprinted by permission of John Wiley & Sons, Inc.
[b]And 1.78% Br.

TABLE 5

Hardness Scales for Fresh Waters[a]

dGH = dKH + PH
GH   = total hardness (dGH=German total hardness)
KH   = carbonate hardness (dKH=German carbonate hardness)
PH   = permanent hardness

Carbonate hardness based on bicarbonate is not permanent because the CO_2 can be driven off by boiling. The remaining calcium and magnesium by definition determines permanent heardness.

Total hardness can be illustrated as follows:
DEGREES OF HARDNESS

0–4° dGH=very soft	12–18° dGH=fairly hard
4–8° dGH=soft	18–30° dGH=hard
8–12° dGH=medium hard	over 30° dGH=very hard

One degree of hardness = 10 mg/l of CaO or MgO

[a]Partly after Riehl and Baensch (1987).

in fresh- or saltwater microcosms or aquaria when the fresh water being used to replace evaporated water is excessive in iron, sulfur, or magnesium.

We have worked with a hard water on the Atlantic Coastal Plain that is rich in both iron and sulfur. Even though a standard water softener is used on the well water, the taps and sinks all show excessive iron staining, and water out of the tap has a sulfurous odor particularly in dry years. In any case, no aquarium problems were encountered for about a year. However, after several dry years, pH rose to over 9 in some tanks and iron deposits began building up on plants and on the algal turf scrubber itself. Being quite low in nitrogen and phosphorus, and with a moderate fish and invertebrate load, the ionic load from an evaporative replacement rate of about 1 gallon per day exceeded the ability of scrubbers to remove the iron and sulfur. A second cichlid tank with the same basic input problem but an excessively heavy fish load showed high turbidities, very strong algal growth in the scrubber, and no obvious iron or sulfur problems. The problem in the first tank was solved by increasing fish loading. In recent years, we have successfully used replacement water "softened" by algal scubbing.

The character of well waters varies widely, and it would be impossible to cover the problems of microcosm and mesocosm water supplies for all cases. In general, if a standard water softener does not solve the problems, we have found that a separate scrubber-managed water reservoir adjusted for nutrient levels and pH can remove almost any contaminant including heavy metals. Of course, one can resort to distilling water or

TABLE 6

Concentrations of Selected, Analyzed Chemical Elements, Ions, and Compounds from a Variety of City Water Supplies[a]

Added in treatment—At least in part	Element, ion, or compound	Parts per million, mg/l						
		Boston	Chicago	Dallas	Baltimore	Los Angeles	San Francisco	Wash. D.C.
*	Aluminum	?	0.45–0.5	?	0.08–0.16	N.D.–0.2	0.02–0.08	0.042–0.2
	Arsenic	< 0.005	< 0.005	< 0.005	< 0.005	N.D.–0.02	N.D.	0.000–0.001
	Barium	< 0.1	< 0.05	< 0.05	0.021–0.03	N.D.	N.D.–0.01	0.028–0.07
	Cadmium	< 0.002	< 0.001	< 0.001	0.001	N.D.	N.D.	0.0–0.00
*	Chlorine-free	?	?	2.12–2.29	?		0.4–0.7	1.6–2.1
*	Chloride	12–44	10.9–11.1	15–27	18–22	18–127	2–19	15.6–29.0
	Chrominum	< 0.005	< 0.003	< 0.01	< 0.001	N.D.	N.D.–0.001	0.0–0.006
*	Copper	0.01–0.08	< 0.003	< 0.01–0.02	0.004–0.006	N.D.	N.D.–0.015	0.001–0.110
*	Flouride	0.06–0.98	0.92–0.93	0.77–0.78	0.9–0.97	0.1–0.7	0.3–1.2	0.93–1.14
*	Iron	0.02–0.16	< 0.010	0.01–0.01	0.02	N.D.–0.07	N.D.–0.06	0.001–0.07
	Lead	< 0.002	< 0.003–0.010	< 0.005	< 0.001	N.D.	N.D.–0.001	0.0–0.001
	Manganese	0.01–0.05	0.001–0.002	< 0.005	0.01	N.D.	N.D.–0.01	0.001–0.004
	Magnesium	.5–2.4	12.3–12.5	3–5	4.4–6.2	5.7–27	?	6–11
	Mercury	< 0.001	< 0.0005	< 0.001	< 0.0005	N.D.	N.D.	0.0000–0.0003

Nickel	?	< 0.003	< 0.01–0.01	?	?	?	0–0.003
* Nitrate(asN)[b]	?	0.25–0.26	0.22–0.45	1.7	N.D.	0.04–0.1	1.29–2.67
* Phosphate (as P)[b]	?	0.010–0.018	0.03–0.09	0.01	.02–.07	0.002–0.017	00–0.14
Potassium	0.9–2.3	1.5	3.7–4.8	2.2–2.6	3.6–4.8	0.2–0.8	2.11–3.27
Selenium	< 0.005	< 0.001	< 0.001–0.001	< 0.005	N.D.	N.D.	0.0–0.002
Silver	< 0.005	> 0.001	< 0.01	< 0.005	N.D.	N.D.	0.0
* Sodium	7.2–26.3	5.8–6.0	11.0–28.5	7.1–8.3	37–86	1.0–15.5	4.8–15.6
Strontium	?	0.119–0.142	0.10–0.25	?	?	?	0.093–0.24
* Sulphate	8.3–14	26.4–27.5	31–50	13.6–15	28–232	1.6–3.6	22.1–48.9
Zinc	< 0.02	0.004–0.005	< 0.01–0.02	0.028–0.038	N.D.	N.D.–0.010	0.0–0.005
Total Trihalom ethanes	?	?	0.0227	41	13–72	0.063–0.075	26–137
Endrin (pesticide)	?	?	N.D.	< 0.04	N.D.	N.D.	N.D.
2.4-D (pesticide)	?	?	N.D.	< 0.05	N.D.	N.D.	N.D.

[a] Sources: Indicated metropolitan water authorities. Note: tap waters can have additional or increased contamination levels. N.D., not determined.
[b] Reactive.

running it through a reverse osmosis purifier before use. This is discussed further below.

In general, city or town waters can be a much more serious problem. While rarely are city tap waters saline or hard as compared to average natural waters (Table 6, Los Angeles extreme, for example), the salinity of an aquarium or microcosm can become that of an endorheic, or closed basin lake, if topping up for evaporation continues for some time and water exchange is not carried out. However, the primary problem with city tap water is the additives used to control human pathogens, to adjust the taste, to control algae in artifically eutrophic reservoirs, or to reduce corrosion in pipes (Table 7). The addition of chlorine is well known and is the primary characteristic that renders tap waters objectionable to humans. It is added to kill pathogens, but the chlorine also would kill most of the animals that one would wish to maintain in an ecosystem. Fortunately, being in the gaseous state it is more or less easily removed by bubbling or allowing the water to stand for several days.

Since chlorine in combination with organic compounds occurring

TABLE 7	
Chemicals Typically Used in the Treatment of City Water Supplies[a]	
Chemical	Purpose
Aluminum as Aluminum Sulphate $(Al_2(SO_4)_3\ 18\ H_2O)$	Clarification
Ammonium Hydroxide $(NH\ OH)$	Taste and odor control
Carbon, activated (C)	Taste and odor control
Chlorine or Chloramines (Cl_2) $(NH_2Cl, NHCl_2, NCl_3)$	Sterilization
Copper Sulphate $(Cu\ SO_4)$	Algal reduction
Florosilicic Acid $(H_2Si\ F_6)$	Reduction of dental decay
Ferric Chloride $(Fe\ Cl_3)$	Clarification
Hexametaphosphate $(PO_3)_6$	Reduce corrosion in metal pipes
Lime $(CaO, Ca(OH)_2)$	pH adjustment
Sodium Chlorite $(Na\ ClO_2)$	Control of tastes, odors, and algae

[a]Source: metropolitan water authorities as given in Table 6.

naturally forms compounds (trihalomethanes; see Table 6) that have been shown to be carcinogens, some cities have started to add chloramines instead of chlorine to control human pathogens. Unfortunately chloramines are more toxic than chlorine to fish and invertebrates and are more slowly lost to the atmosphere than chlorine. If water to be used for a model ecosystem cannot be bubbled for at least a week or allowed to stand for several weeks to a month, then a filter with fresh, activated carbon can be used to recirculate the tap water (in a reservoir separate from the aquarium or microcosm). Several hours to a day of filtering is required to assure removal of the chloramines. A brochure describing the dangers of chlorine and chloramines and methods of removal is available from Public Information, Metropolitan Water District, Box 54153, Los Angeles, California 90054.

Copper sulphate is added to many drinking water systems that include reservoirs to prevent the excessive algal growth that sometimes results from artificially high nutrient levels. Unfortunately the copper is also toxic to the algae that is necessary to maintain virtually all natural, aquatic ecosystems. While additional chemical treatments are available to remove all of these contaminants, we prefer physical (reverse osmosis) and plant production (algal scrubbing) methods for cleaning tap waters to levels acceptable for ecosystem management. Scrubbers utilized in this way are typically dominated by blue-green algae, as these are the only algal species that can withstand water direct from reverse osmosis and/or ativated carbon prescrubbing of city tap waters.

Finally, a major problem of many city tap waters is that they are eutrophic, i.e., excessive particularly in dissolved nitrogen and phosphorus. Washington, D.C. tap water typically carries 50–200 μM (0.7–2.8 ppm) concentrations of nitrogen as NO_3^{2-}. However, in the Potomac River, dissolved nitrogen concentrations in excess of 25 μM (0.35 ppm) are regarded by ecologists as being at bloom levels. Even these levels of nitrogen (at least at the lower end) might not be so bad if phosphorus as PO_4^{3-} were in short supply. Phosphorus in the dissolved state as phosphate often limits productivity in high quality (oligotrophic) mountain lakes. However, concentrations of about 0.02 ppm phosphorous as PO_4^{3-} are also at bloom levels and these are approached or exceeded in many tap waters.

Water and Model Ecosystems

Considering the above discussion, with what does one fill the tank? We recommend total use of natural waters unless logistics and transportation costs dictate otherwise. Of course, filling a South American stream tank with water from South America is possible for very few aquarists. However, the use of local unpolluted stream water with its microorganisms intact is still better than most tap waters. Likewise, it is better to use natural sea water as long as it too is not polluted or excessively diluted, as from an inner bay or estuary. This is a major difference between our approaches to aquaria and mesocosm management and the recommendations of more traditional aquarium manuals. However, we note that Moe (1989) agrees to the extent that natural sea water carries many essential organic and inorganic trace elements that are not in a sea salt mixture. He notes however, that after 2–3 months it may not make much difference. That is also why a small water change (about 1% per month—but not the standard 10–20% requirement to remove nutrients) is necessary, as we discuss below.

We have a bias toward natural waters, particularly natural sea water, that transcends easy accessibility and the expense of "sea salts." We believe in introducing the appropriate microorganisms and microchemical constituents early in the development of a system. Fear of introducing disease or lethal compounds into a microcosm system from unpolluted natural waters is mostly unwarranted. When substrate and organisms from a natural environment are introduced, some microcomponents will be brought in with them. They cannot be avoided, unless of course one wishes to select all organisms, including protozoans and microbes, individually (a virtually impossible task) and then to pass them through sublethal or other baths hopefully selective for the desired organisms. The sterilization approaches of modern medicine and some algal and bacterial culture methods are not applicable to the synthesis of ecosystems. Thus, the use of "sea salts" or distilled water to achieve or maintain sterility is generally pointless. If one wishes to include all or most of the elements of a natural community, it is difficult to totally avoid introducing disease vectors. Quarantine of fish, especially those purchased, and particularly those from fresh water, is essential to avoid occasional serious disease problems. On the other hand, if the organisms are in a healthy environment and the ecosystem is managed properly, disease will be minimal. We discuss these matters further in Chapter 20.

Once a microcosm, based on the principles discussed in this book, is established, water changes are minimal and aimed primarily at preventing "salt drift" in sea waters and the change in the proportional salt content caused by evaporative accumulation in fresh waters. In short, very minor changes should be made to avoid drift in conservative elements that would create a "Great Salt Lake" or endorheic basin. As we discuss in Chapters 9–12, nonconservative elements should be maintained dynamically by balanced systems loading. In the Smithsonian's Marine Systems Laboratory, we change only about 0.05% per day (about 1 pint of water in a 100-gallon tank). Since this level of water exchange is so critical, an example involving heavy metals illustrates the point. Average levels of zinc in city tap waters (see Table 6) are about 0.01mg/l. Given a 130-gallon aquarium requiring 10 gallons per week evaporative top-up and a one gallon per week exchange rate, zinc concentrations in the tank would be expected to build up to a level of about 0.1 mg/l (assuming no export by algal scrubbing or other means). After ten years of operation, zinc concentrations in the Smithsonian Coral Reef lie at 0.07 mg/l. This compares to 0.005 mg/l in sea water and a maximum acceptable drinking water level of 0.5 mg/l (Lorch, 1987). The same author reports 0.1 mg/l as being acceptable for human haemodialysis make-up. The latter is perhaps closer to an acceptable value for an ecosystem. While the Washington, D.C. tap water data taken from Table 6 shows rather low levels for zinc, copper ranges from 0.001 to 0.110 mg/l. After ten years of equivalent exchange, the concentration of copper in the Smithsonian Coral Reef lies at about 0.05 mg/l, suggesting either a very high level of sequestering in sediments, or more like a high level of removal of copper by the algal turf scrubbers. In summary, it seems clear that the exchange level suggested (1–2% per month), preferably on a daily or weekly basis, is acceptable for contamination levels equivalent to those shown in Table 6. The exchange rate could be adjusted one way or the other in accordance with contamination levels in local water supplies. This simple analysis also points out the critical need to avoid metals contamination from construction and plumbing materials in closed or semi-closed systems.

If sea water or other natural water in any quantity is difficult to obtain, an initial filling with fresh water and sea salts or aged tap water alone followed up by use of natural sea water or stream or lake water for changes is acceptable. Exclusive use of sea salts for marine aquaria will certainly "work"; however, in our judgment it is simply another element of many reducing the accuracy of the system. Also, because of the relatively high level of nutrients in most commercial salts, scrubbing levels will have to be generally higher (see Chapter 12).

Algal Scrubbing and Water Composition

In both our mesocosm and aquarium endeavors, we have emphasized the use of plants, often algae, to manage water quality. As we discuss in many places in this book, the key to ecosystem management is stability achieved by locking nutrients up in biomass rather than by using bacterial filtration to rapidly reduce all nonliving organics and organism excretions to freely available elements and ions. The algal scrubber is the principle technology in this management system and is discussed in depth in Chapter 12. The nonconservative or interactive elements, particularly carbon, oxygen, nitrogen, and phosphorus, are discussed in Chapters 9–11. Here, we are principally concerned with potential effects on the conservative elements.

Algal scrubbing is used primarily to maintain a balance, to simulate the effects of the larger body of low-animal biomass water that balances out the requirements of high-biomass systems. This might be smoothing out the differences between day and night in small model ecosystems or between seasons in large mesocosms. If there is no import to a system, because it is large enough and rich enough in photosynthetic plants to provide sufficient energy to the community of organisms maintained, then export is not required. The algae removed by scrubbing as a daily or seasonal requirement are dried and eventually returned. Thus, there is little chance for the removal of critical elements from the water medium. However, when the aquarist is running a balanced system with significant input (usually dried or live food) and export, either because the model has excessive biomass and is being driven hard or because the input and export simulate similar features in a wild ecosystem, then inbalances are possible. On the other hand, particularly when biomass export is involved and nitrogen and phosphorus input as feed or fertilizers is employed, micronutrient depletion usually results from the harvest of a monoculture. The algal turfs of scrubbers are communities, typically with many species from most algal divisions and thus tend to be self-balancing.

Algae and other aquatic plants synthesize a wide variety of organic compounds including many vitamins (Ragan, 1981). It is unlikely in diverse communities of fifty to several hundred species of plants that contain many herbivores, even in relatively small models, that lack of synthesized organic compounds would hinder ecosystem development. On the other hand, many algae have requirements for inorganic micronutrients, such as iron, calcium, manganese, molybdenum, boron, cobalt, copper, and iodine (Weissner, 1962) that could theoretically provide such limitations. Moe (1989) discusses removal of trace elements particularly by ozonation and protein skimmers, but also through uptake by algae. Iodine, particularly needed by crustaceans for molting, was cited as a characteristic problem. We refer to a general pattern of crustacean success in our systems (Chapters 21–24), where extensive algal scrubbing has been used for many years. For aquaria, we particularly note over three years of continuous molting in the Rare Reef Lobster. In addition, after 10 years of continuous operation that never included a water change other than a standard 2% per month, the Smithsonian Coral Reef contained concentrations of iron, manganese, silica, copper, and strontium that remained very close to or slightly above concentrations in Delaware coastal waters.

For a research laboratory, the limiting or potential limiting of micronutrients in ecosystem models could be followed and adjusted by monitoring. For the hobbyist or ecologist, this may be impractical, and a few simple rules that are in effect what happens in the wild will avoid problems: (1) do not run import and export highly out of balance (see Chapter 12); (2) use a wide variety of living and dried foods; (3) in the long term insist on small water changes (except for transient puddles, endorheic basins, or the infinite ocean, constant "new water" exchange is a feature of wild ecosystems); and (4) where calcification is an important feature of a model ecosystem, be certain that calcium carbonate, especially as aragonite, is readily available (see also Chapters 8 and 10).

Marine Microcosms and Aquaria

An attempt to faithfully re-create a marine system must include monitoring of salinity levels, and knowledge of daily, weekly, and monthly salinity variations in the wild. Normal ranges of salinity for most nonestuarine coastal situations are 28–34 ppt for colder waters (with a variation of no more than several percent a day) and 34–37 ppt for tropical oceanic situations (varying no more than 0.05 ppt daily). (The idiosyncracies of brackish water aquaria are discussed in Chapter 23.)

Evaporation is an everyday, ongoing problem with all open tanks, especially those with trickle filters, wave generators, aerators, and intense lighting. In the 3000-gallon coral reef system at the Marine Systems Laboratory a loss of 20–30 gallons a day to evaporation is normal, potentially resulting in a salinity increase of about 0.05 ppt. Thus 20–30 gallons is replaced daily with fresh water, and special sensing is employed because the drop in the water level is only 1–2 mm. If measuring by eye, a drop of 5 mm would be a good benchmark for topping off: salinity would vary about ±0.1ppt. We use topping up for small tanks and cold-water or brackish systems. But for controlling salinity in tropical reef tanks, precision leveling devices now available from most aquarium stores and laboratory supply houses should be used.

The hydrometer measures salinity by specific gravity (or density) and is the least expensive and most trouble-free device of this type. A conversion is required (Fig. 2) since temperature is also an important parameter relating salinity to specific gravity. However, even the larger hydrometers are difficult to read with a consistent effective accuracy of better than ±0.5 ppt. A refractometer costs a little more and is easier to use. However, it provides about the same accuracy. Both devices require very careful observation and conversion for effective use on reef systems. For other coastal communities and estuaries, either device is quite adequate.

We check the salinity of our reef microcosms twice daily with or a hydrometer. Salinity is maintained at between 35.5 and 36.5 ppt for up to a month at a time by adjusting for positive or negative drift with sea water or distilled water. The sensitive level control device shown in Figure 3 is easily constructed. It keeps daily and weekly fluctuations to a minimum We use a capacitance-level controller on the sight glass. Attached to a pump, it automatically replaces about 50 ml when that amount has evaporated. The infrared sensors, now easily available, also work quite well. The sensors themselves need to be wiped off every few weeks to avoid failure.

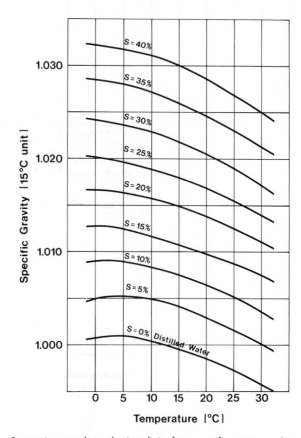

Figure 2 Conversion graph to obtain salinity from specific gravity and temperature.

Failure of a water delivery system can be just as disastrous as the jamming of a heater. A check of the tank daily can prevent the problem from becoming crucial. However, continuous pumping of fresh water for hours can destroy a sensitive marine microcosm. We use a manually filled freshwater reservoir for replacement water. It contains the maximum amount of water that can safely be pumped in between checks.

Water Quality of Top-Up Water

For coral reef systems, high-grade distilled water is recommended for replacement water because contaminants continually placed in the tank are not removed by evaporation. Algal scrubbers certainly help, depending upon the contaminant. However, poor input water quality is just another element tending to degrade a model ecosystem.

The inexpensive water treatment system illustrated in Figure 4 will

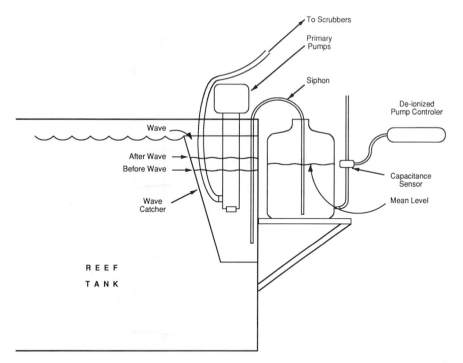

Figure 3 Level control device used on Smithsonian coral reef microcosm.

be quite adequate in most localities. As typically used to supply the Smithsonian Chesapeake mesocosm fresh water input (river and rain water), an incoming dissolved nitrogen concentration in the tap water of 150–200 μMN (as $NO_2^- + NO_3^-$), is reduced to about 15–25 μMN (0.2–0.4 ppm) by the laboratory grade milli-R.O. system. The algal scrubber, connected to the reservoir, further reduce dissolved nitrogen to about 0.2–2.0 μMN, depending on use rate. The entire reduction could have been accomplished with scrubbers, though a larger area would have been required. (See also discussion accompanying chapter 23, Figure 7.) Some municipal water systems routinely use a copper salt to attack algal growth. This would be disastrous to a mesocosm if present in any quantity, as proper algal growth would be difficult to sustain, and the copper would be generally toxic to many organisms. In cases where the drinking water is severely contaminated, reverse osmosis or reagent-grade deionizer treatment may be called for. High-quality freshwater replacement supplies pose no problem for many coastal microcosm systems, but oceanic ecosystems and coral reef ecosystems in particular are adapted to unusually pure sea water. Most city tap waters are not suitable for direct long-term evaporative water replacement to ecosystem models.

The problem of very acid freshwater top-up is solved with a simple

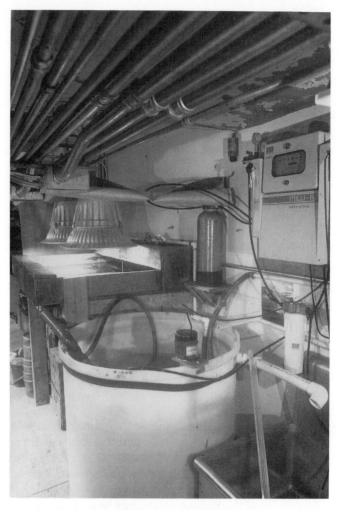

Figure 4 Tap-water cleaning system consisting of a laboratory grade reverse osmosis unit, a reservoir, and an algal scrubber.

device that trickles water through crushed carbonate. Washed *Halimeda* or oölitic sand is especially effective. Most sea-water systems operate at pH levels of 8.00 or higher. The range for typical coral reef sea water is 8.15 to 8.30 depending on the time of day. Tap waters that are highly acid can interfere with calcification or shell-building, which are enhanced at a high pH (see Chapter 10).

With the evaporation of water comes a change in the tank's salt composition. We counteract this in our 3000-gallon reef system by replacing about a gallon a day—about 1.5% a month. Most traditional aquarium manuals suggest a normal replacement of 20% a month—much higher if problems develop. In our 10 years of experience with these microcosm systems we have never needed to conduct a massive water change. Peculiar effects can result from mechanical damage to a toxic organism (see Chapter 21). In such a situation we prefer to remove fish for 24 hours rather than change the water, and the aquarist with extensive and expensive collections may wish to keep an emergency holding and perhaps quarantine tank in addition to the main display tank.

Hardness in fresh water is usually much less of a problem than salinity maintenance for salt waters. Nevertheless, in special cases it can be crucial. Hardness testers can be bought at most aquarium stores or scientific supply houses. If one is simulating a local stream or lake, determining the hardness desired is relatively easy. For more exotic systems it must be remembered that great variety occurs in the wild. It would be easy to assume, for example, that all South American fish are adapted to soft, low-pH waters and all African fish to hard, high-pH waters. That is not the case, and one should refer to a good "dictionary" of freshwater fish for the appropriate data, particularly if breeding is desired (Riehl and Baensch, 1987).

If one has acid, low-pH water and desires the opposite extreme, it is simple enough to place carbonate sand and rock in the tank, or to trickle incoming water through carbonate as described above. If a specific high ion content (such as magnesium) is desired, chemical addition is possible. A more difficult problem is a hard, high-pH source water, when the opposite extreme is desired. Most manuals suggest using sphagnum moss to acidify. In our experience this works only when extremely small decreases in pH are needed. One can add a balanced nitrogen and phosphorus supply using a calculated ratio of nitric and phosphoric acid to give a $5:1$ to $10:1$ ratio of $N:P$ and then scrub the reservoir water with an algal turf. However, quality measurement capability for nitrogen and phosphorus is required to carry this out successfully.

Coarse measurement of the conservative elements of make-up, top up, and model ecosystem waters can be carried out with kits available at better aquarium stores. Depending on the elements, the concentrations involved and the sensitivity of the ecosystems being worked, this may be adequate. However, in many cases a realistic understanding of the water chemistry requires precision measurement, most efficiently obtained either by professional analyses or with high-quality water analysis kits (Hach, 1990). For those aquarists desiring to carry out their own analyses, we suggest the following references: Fresenius et al., 1988; Strickland and Parsons, 1968.

References

Fresenius, W., Quentin, K., and Schneider, W. (Eds.) 1988. *Water Analysis*. Springer-Verlag. Berlin.
Hach Company. 1990. *Products for Analysis*. Loveland Colorado.

Horne, R. 1969. *Marine chemistry*. John Wiley, New York.

Hutchinson, G. E. 1957. *A Treatise on Limnology*. Vol. I. John Wiley, New York.

Lorch, W. 1987. *Handbook of Water Purification*. Wiley, New York.

Moe, M. 1989. *The Marine Aquarium Reference*. Green Turtle Publications, Plantation, Florida.

Ragan, M. 1981. Chemical constituents of seaweeds. In *The Biology of Seaweeds*. C. Lubban and M. Wynne (Eds.). University of California Press, Berkeley, California.

Riehl, R., and Baensch, J. 1987. Mergus-Verlag, H. A. Baensch.

Spotte, S. 1979. *Seawater Aquariums*. Wiley-Interscience, New York.

Strickland, J. and Parsons, T. 1968. *A Practical Handbook of Seawater Analysis*. Bull. Fish. Res. Bd. Canada *167*: 1–311.

Sverdrup, H., Johnson, M., and Fleming, R. 1942. *The Oceans, Their Physics, Chemistry and General Biology*. Prentice-Hall, Englewood Cliffs, New Jersey.

Thurman, H., and Webber, H. 1984. *Marine Biology*. Merrill, Columbus, Ohio.

Weissner, W. 1962. Inorganic micro nutrients. In *Physiology and Biochemistry of Algae*. R. Lewin (Ed.). Academic Press, New York.

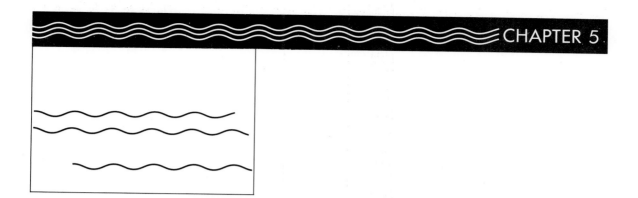

CHAPTER 5

WATER MOTION
Waves, Currents, and Plumbing

Natural waters, to varying degrees, are very much chemical soups, as discussed in the preceding chapter. The possibilities for chemical interactions on a large scale are greater in water than in any other environment on the earth. They are probably greater than anywhere else in the universe, except where water might be present. On the other hand, simple diffusion of an element or an ion through water can be very slow. While water allows most elements and many compounds and gases into its loose chemical structure, often as weakly charged ions, the rate of mixing and interacting can be limited in the short term by the rate of diffusion.

Organisms are rarely static. The basic unstable nature of individual life requires a constant flow of energy and materials to keep the biological and then ecological processes operating. Whether it is the stirring of water in the home or laboratory, the wind blowing across a lake, waves crashing on a rocky shore, or tides coursing in and out of a bay, all significantly contribute to ecological and ecosystem processes. These physical energy inputs force the mixing required to varying degrees by living organisms to carry out an active life. Scientists have been examining the direct effects of waves and currents on physical variables such as oxygen and temperature and directly on individual species for a long time. But only recently has it begun to be apparent that these forms of physical energy have a major impact on primary ecosystem processes such as photosynthesis and respiration (Adey and Hackney, 1988; Nixon, 1988; Leigh *et al.*, 1987).

Figure 1 shows the contrast between the muddy bottom infaunal biomass, with depth for fresh and marine waters. Nixon (1988) relates the approximate one-half order of magnitude difference between marine and fresh waters to the driving effects of tide and wind (waves and current). Note that the one lake that approaches marine situations in benthic biomass is the large Lake Michigan.

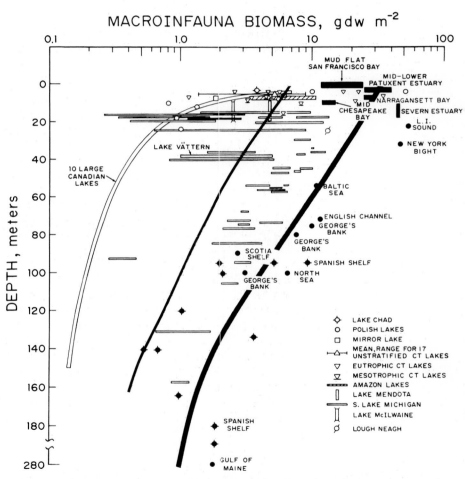

Figure 1　Comparison of the biomass of soft bottom infauna in marine, estuarine, and freshwater environments. After Nixon (1988), with mean curves fitted by eye. Weights are shell-free, ash-free dry weights. The wider curve to the right represents salt and brackish water. The narrow curve to the left represents fresh water.

All forms of mixing accelerate chemical and therefore biological processes up to a certain saturation point beyond which other factors (light intensity, temperature, chemical concentration, etc.) begin to be limiting. Thus, waves, currents, and tides (Chapter 6) should not be omitted from the ecosystem model any more than from the wild community. They need not be, and often cannot be, of the great force and energy that they sometimes are in the wild. However, wild levels during storms are often far above saturation levels in terms of their effects on ecological process. On the other hand, storm or even weekly or monthly high waves or currents may actually determine community structure in many cases. If these high-energy communities are to be simulated, at least in a localized way within a model system, then steps need to be taken to provide strong point sources of energy.

Currents in Fresh Waters

Everyone is aware of the currents that occur in rivers and would not think it unusual that, in a river or stream, currents are critical ecologically. That currents also occur in lakes and are of great importance is not as widely recognized (Figs. 2 and 3). A well-known example of the critical effects of mixing occurs in lakes and estuaries the world over. Typically lakes are mixed by wind and, in colder climates, by autumn and spring overturns. Overturns result from surface cooling (or heating in early spring) until the surface waters become heavier than deeper waters. Under some conditions, particularly in warmer and less windy seasons, or in the tropics

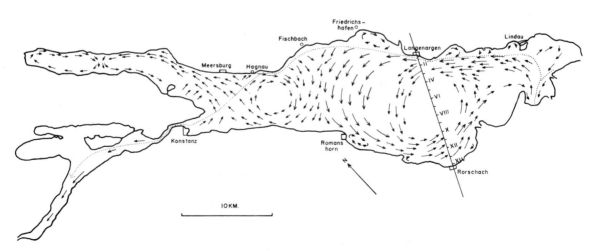

Figure 2 Surface current patterns in Lake Constance, Switzerland. Average rates in shallow water are 10–20 cm/s and reach 27 cm/s at some localities. After Hutchinson (1957). Reprinted by permission of John Wiley & Sons, Inc.

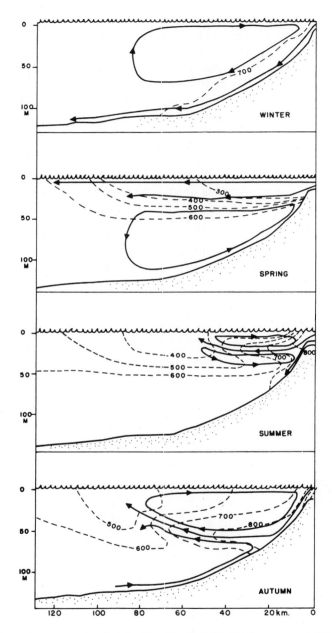

Figure 3 Vertical current patterns in Lake Mead on the Colorado River, seasonally. The combination of incoming salts in the river (to 0.7 ppt) and temperature changes of lake and river seasonally produce the currents as well as the basic distribution patterns of the salts. ←, current; −−700−− = 0.7 ppt. After Hutchinson (1957). Reprinted by permission of John Wiley & Sons, Inc.

where lakes can be warm at the surface year round, the surface waters become warm, light, and strongly stratified. When this happens, mixing with deep water stops. If the bottom sediments are even moderately rich in accumulated organic material the bacteria, worms, and clams in the mud can use up all of the available oxygen and leave the bottom sediments and even some of the bottom water virtually dead (Fig. 4). Lake Tanganyika in East Africa is a well-known example of such a lake, with about 90% of its total volume being permanently devoid of oxygen.

It is also not unusual for even quite large lakes and estuaries to become unnaturally anaerobic due to the nutrifying effects of human activity. Chesapeake Bay in eastern North America, one of the largest estuaries in the world, is just such an example. A deoxygenated layer has recently begun to appear in deeper waters during the summer. Unfortunately in these cases, since phosphorus does not diffuse easily through oxygenated sediment, slight deoxygenation releases stored phosphorus in large quantities. This triggers more surface productivity and a "snowballing" of deoxygenation. Many lakes in populated or heavily farmed areas have reached a similar state.

Currents in the Marine Realm

In open, shallow-water marine environments constant and relatively rapid turnover of living biomass is the rule, longevity the exception. Although

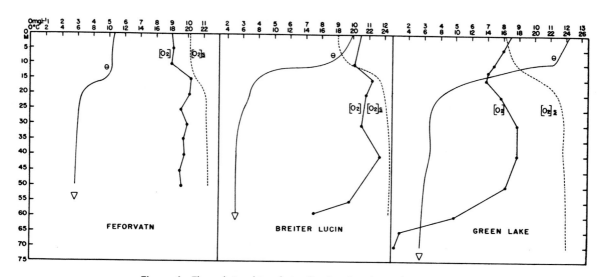

Figure 4 The relationship of stratification (as shown by temperature curves) and oxygen in different types of lakes (Feforvan—biologically sterile; Breiter Lucin—intermediate; Green Lake—highly productive). [O_2] = oxygen observed; [O_2]$_s$ = saturation level at temperature; Θ = temperature; ∇ = bottom. From Hutchinson (1957). Reprinted by permission of John Wiley & Sons, Inc.

organic-rich sediments equivalent to marshes on land can form in deep
water and in protected lagoons and bays, and these organics can become
geologically stored for hundreds of thousands or millions of years,
the short-term "in situ" storage of biomass as in terrestrial forests and
swamps does not readily occur in the open marine or aquatic realms. A
rapid exchange of gases (oxygen, carbon dioxide), metabolites (food,
nutrients, nitrogenous wastes), and salts and minerals between aquatic
organisms and their watery environment is essential. Mixing in high-
energy surface waters is critical to accomplishing the rapid turnover characteristic of these areas. On the other hand, these low-biomass, high-
water-quality zones remain in that state in large measure because of export
and storage of organics elsewhere, whether in deeper water or in lagoons
and bays.

The world ocean is characterized by major surface currents (Fig. 5).
These currents are created primarily by prevailing winds, modified by
Coriolis (earth rotational) and density forces as well as by land geography.
Generally moderate in rate, ocean currents however can reach several
knots (for example, the southern Gulf Stream). Ocean currents basically
provide mixing and heat transfer for the entire ocean. They also interact
with the earth's atmosphere and are major factors in determining weather

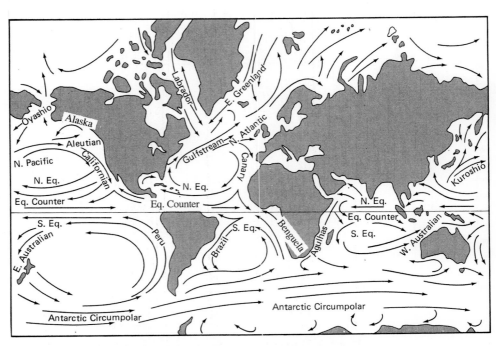

Figure 5 Surface circulation of the world ocean. After Kennett (1982).

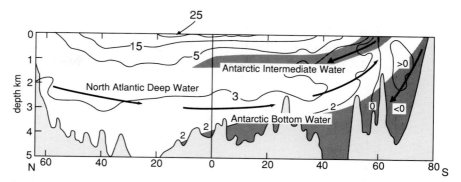

Figure 6 Primary vertical circulation in the Atlantic Ocean. Temperature in °C. After Thurman and Webber (1984). Copyright © 1984 by Scott, Foresman and Company. Reprinted by permission of HarperCollins Publishers.

patterns. Below several hundred meters, circulation generally becomes slower, being driven primarily by density as established by salinity and temperature (Fig. 6). As in lakes, the limited circulation in deep water results in oxygen depletion (Fig. 7). However, in this case because of the enormous depths involved, the organic material raining from the surface

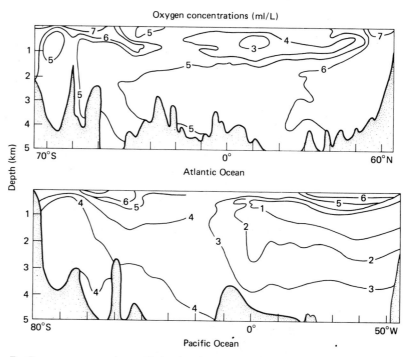

Figure 7 Oxygen concentrations with depth in the Atlantic and Pacific Oceans (ml oxygen per l). After Kennett (1982).

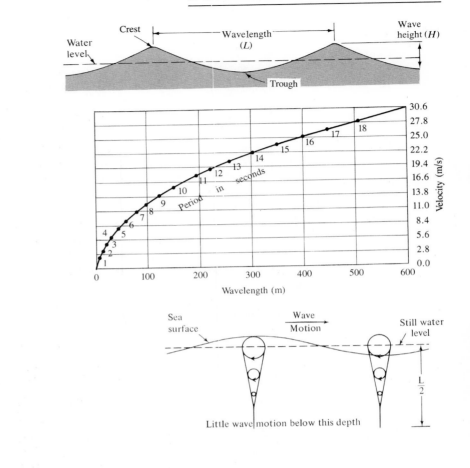

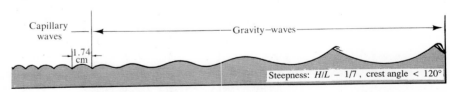

Figure 8 Motion of water particles in wind-driven waves. The circles describe the extent of orbital movement of individual water particles. After Thurman and Webber (1984). Copyright © 1984 by Scott, Foresman and Company. Reprinted by permission of HarperCollins Publishers.

is largely broken down within the upper 1000 m. The oxygen minimum occurs in the 500–1500 m range, and oxygen concentrations generally rise in even deeper water. The deep and bottom water is generally very cold and relatively oxygen-rich, being derived through very slow-moving bottom currents from the surface water of the Arctic and Antarctic.

Wave Action

We have very briefly discussed the larger-scale circulation of lakes and oceans. While this circulation is in large measure driven by the wind, especially in the ocean, wave action also more directly and strongly mixes the surface layers of water. Figure 8 shows how a wave, whether a wind wave or swell, disturbs and mixes surface waters by its motion. Waves driven for several hours or more by any wind over 12–15 knots and for a distance of over several miles can have large effects on the shore and the biological and ecological processes of that shore. Prediction curves for wave height and period relative to wind velocity and fetch are given in Figure 9. While wave period can be simulated easily in an aquarium or mesocosm, wave height, except in the largest mesocosm (see Chapter 21, Australian reef mesocosm), would not likely be fully matched. Consideration of the wild environment in this respect will at least lead the scientist and aquarist to attempt a partial match and to consider the effects of reduced wave energy.

In surface waters of the ocean, a well-developed circulation is crucial to ecological function. For example, a rich shallow-water reef, with its

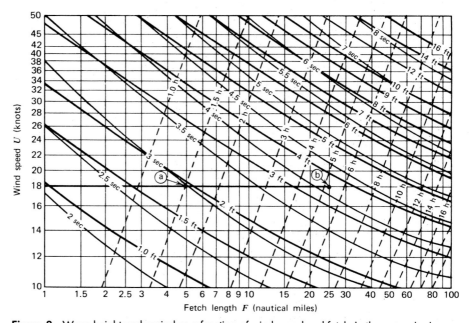

Figure 9 Wave height and period as a function of wind speed and fetch. In the example shown, an 18 knot wind blowing over a 5 nautical mile fetch will produce a 1.9 foot sea of almost a 3 second period. It will take about 1.5 hours to build to that height. For a 25 mile fetch, the sea will build to 3.6 feet, but will require over five hours to fully develop. After McCormick (1981).

extraordinary respiration and primary productivity, resembles a city in that it requires continuous exchange with its surroundings—in this case the adjacent ocean and lagoon. With a greater concentration of plants and animals than it can support within itself, the wild coral reef experiences rapid oxygen depletion of the strongly inflowing, oxygen-rich ocean water during the night (Fig. 10).

Water Motion and Model Ecosystems

Rarely has sufficient physical energy been provided for mixing in microcosm and aquarium situations. Moe (1989) discusses this problem relative to the hobbyist and describes how the equipment is now becoming available to allow achievement of this ecosystem requirement. The preceding brief review of wave and current energy in aquatic environments was undertaken to stress this need and to provide the framework in which to discuss its use.

The role of wave surge as an important mixing element is easily seen in a reef microcosm where a well-defined flow and surge are present (Fig. 11). Removal of the surge component alone markedly reduces oxygen production and therefore photosynthesis (Fig. 12). On quiet nights in a high-biomass reef, low current and surge can result in low oxygen concentrations. A well-developed reef community would quickly expire in the stagnant conditions of a nonflowing, unaerated, closed-system aquarium because of the oxygen draw-down and the carbon dioxide and ammonia pollution, which occurs rapidly with any appreciable animal load.

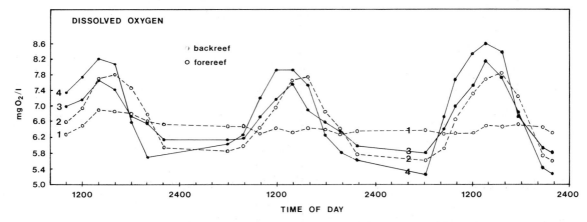

Figure 10 Oxygen reduction at night in water driven over a Caribbean coral reef by the trade-wind sea and swell. 1, Front of forereef; 2, back of forereef; 3, front of backreef; 4, back of backreef. After Adey and Steneck (1985).

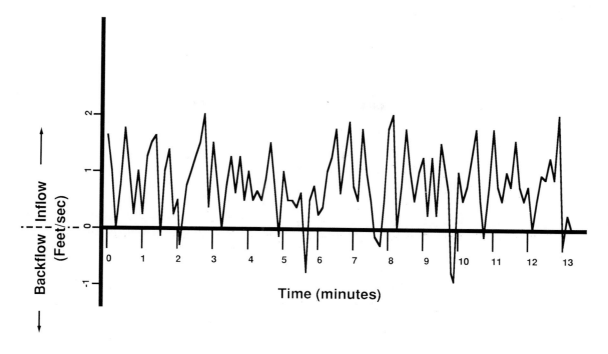

Figure 11 Wave surge and current in the Smithsonian coral reef model. See Chapter 21, Figure 1.

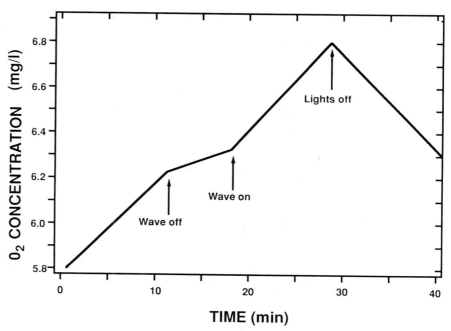

Figure 12 Reduction of oxygen production with reduced wave action in coral reef microcosm algal scrubber (Adey and Hackney, 1989).

Traditional aquaria often employ bubblers or aerators to achieve water motion with its resulting atmospheric exchange. Although a bubbler can induce some water movement when set up as an air lift, it is relatively inefficient at creating currents and cannot create a surge. Filter devices usually have associated drive pumps. However, they are rarely used to establish current or waves, although that possibility sometimes exists. In this case even a pump of very high rating could provide little flow if the filter unit itself is partially clogged. Newer pumping devices often include "power heads" and are capable of producing locally high current zones (though see our discussion on the effects of centrifugal pumps, below). We prefer more natural approaches for most microcosm systems. In shallow coastal and shore communities, marine or fresh water, on which this book focuses, water motion occurs as a result of simulated tides, currents, wave action, and its resulting surge. Tides are treated separately in Chapter 5.

Many devices have been constructed to create waves in tanks for experimental purposes, including push boards and large pistons. Most recently a pneumatic device has been extensively used for creating waves in swimming pools, and a version of this device has been used on at least

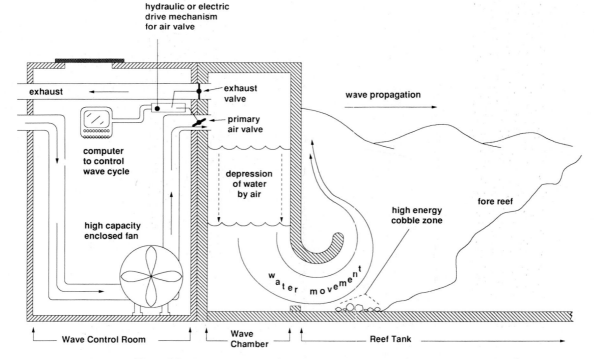

Figure 13 Pneumatic wave drive system used in Australian Great Barrier Reef aquarium at Townsville.

one mesocosm (Fig. 13). For most mesocosms and aquaria the simple dump bucket (Fig. 14) is the most reasonable approach in terms of cost and maintenance. If the dump rate is matched to the natural frequency of the tank (determined by "rocking" or seiching the tank water mass by hand), a high rate of surge can be created. Often, however, it is desirable to slightly mismatch wave dump with natural frequency to avoid waves that become excessive. If diaphragm pumps are used (see below), a surge or seiching motion can also be created by the periodic suction.

Besides being an important mixing agent, wave machines can be highly effective aerators under the right wave conditions where those are normally achieved in the wild. In the Smithsonian Caribbean reef tank, for

Figure 14 Photographs of dump-bucket wave maker. Note that these devices are highly sensitive to weight shift and easily become unbalanced with algal growth. Proper setting of the hinge axis and the use of friction-minimum bearings that will not corrode are essential. We have found that Teflon against Teflon is the only foolproof method, although Teflon against other plastics is usually acceptable. We have also used a newer plastic, Nylatron, as the axis against a Teflon bearing.

(*Figure continues*)

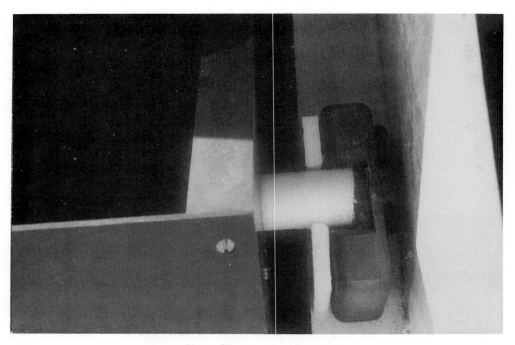

Figure 14   (*Continued*)

example, where the night biological load is about 11 g O_2/h, oxygen concentrations do not fall below 6 mg/l—a higher-than-average night-time
level for Caribbean reef flats, one that would normally be attained only
with trade winds exceeding 15 knots (see Chapter 10 for details).

Moderate waves created by the techniques described in this chapter
and in Chapters 21–24 are normally sufficient to maintain community
structure on the scale of days and perhaps weeks. However, many ecological communities, especially those of rocky shore and reef environments
are partly structured by occasional strong wave action. Small epiphytic
algae, particularly blue greens, are strongly affected by strong wave action.
It is often necessary in models of these systems, especially in the smallest
microcosms and aquaria, to create unusual turbulence either by hand or
with apparatus designed for this purpose.

Generally, we have used two arrangements for wave and current flow
across open-shore communities in microcosm:

1. For coral reefs a one-way inward current flow is generated across
the community. Return flow resulting from wave action is collected
"shoreward" of the community itself in the lagoon on the opposite side of
wave generation and wave input (Fig. 15). In this approach, the pumps

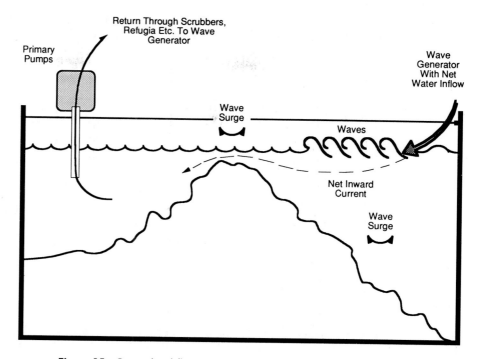

Figure 15 Generalized flow patterns on a reef-type shore configuration.

that constantly recirculate water from the lagoon to the seaward end of the tank are the equivalent of both the trade-wind sea and the reef pass out of which accumulated water flows.

2. For sandy-beach and rocky-shore simulations, the erratic "seaward" return flow is back "seaward" under the wave and across the shore community (Fig. 16). Pumps not only drive the wave generator but also effectively simulate rip zones on a beach or rocky shore.

Pumps

Among the wide variety of pumps available for moving water in microcosms, mesocosms, and aquaria, the most commonly used is the centrifugal pump. These can be inexpensive, reliable, and quiet, and an enormous choice exists to fit almost any system design. Submersible centrifugal pumps are readily available and can be quieter and esthetically more desirable than those designed for operation in air as they can be hidden within a tank. However with plumbing failure they also have the potential for pumping a tank virtually dry unless the intake is placed very high. Submersible pumps also tend to raise system temperatures, and

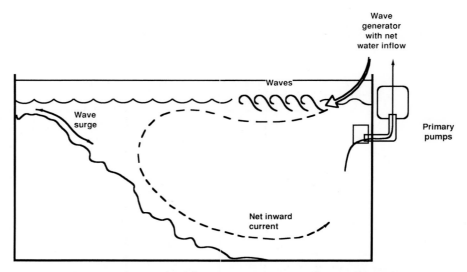

Figure 16 Generalized flow pattern on an open shore configuration.

some varieties have the potential for leaking oil. The latter should be strictly avoided. As with all pumps destined for use in model ecosystems, if any internal pump parts are metal, they should be made of the highest quality stainless steel. It is best to have no metal, even stainless, exposed to salt water.

The unfortunate difficulty with most centrifugal pumps is that their internal turbulence, pressure and shear kill many plankters and the swimming or floating reproductive states of plants and animals. We have been able to demonstrate a greater than 90% mortality of large zooplankters, such as *Artemia salina*, on passing through such a pump. While additional experimentation is needed on the effects of centrifugal pumps on a broader spectrum of plankton, other choices are clearly needed to continue to improve the veracity of model ecosystems.

We have experimented with alternatives to the ubiquitous centrifugal pumps. Among the most likely possibilities are relatively large diameter diaphragm pumps that contain "flapper valves" and alternate relatively slowly from a slight suction to a slight pressure. Some compressed-air-driven types for transferring corrosive slurries in the chemical and other industries work quite well and are easily available. However, they are expensive. We have designed and extensively used the bilge pump conversion shown in Figure 17 on many small aquarium systems (see Chapters 21 and 23). While these are a little noisy, relatively expensive as "one-of-a-kind" designs, and the diaphragms have a tendency to slowly crack after several months continuous use, a little engineering and manufacture for a mass market could certainly solve these problems.

Figure 17 Thirty-rpm, 5-gpm diaphram pumps as installed in a 130-gallon scrubber-operated reef aquarium (see Chapter 21, Figs. 9 and 10). The dark area to the right is a refugium. The scrubber unit is off the photo to the right (see Color Plate 5). Photo by Nick Caloyianis.

Similar vacuum-driven pumps without diaphragms, that simply suck water to the desired height with a timed or sensed release, additionally offer some promise.

We have also experimented with slow-moving piston pumps. These function quite well, although characterized by moderate expense and maintenance problems. The latter also could likely be easily solved by engineering and mass manufacture.

An interesting aspect of diaphragm and piston pumps is their ability to create surge as well as flow. We have used this feature in several of our aquarium ecosystems described below, and considerable further development is likely.

Archimedes screw pumps are ancient devices used centuries ago, particularly in irrigation. Today, they still merit usage in sewage plants where a simple, virtually uncloggable pump to move large quantities of water at minimal height is desired. As yet, these pumps have not been commercially developed in small sizes. However, they are a very promising possibility for both mesocosms and home ecosystem aquaria. We have successfully used a screw pump on our Everglades mesocosm, though not without some maintenance problems (Color Plate 4).

Pumping rates vary from system to system. At the Smithsonian, the coral reef system overturns approximately every 60 min, and the Maine rocky-shore, mud-flat tank overturns approximately every 45 min. The small, home 120-gallon reef described in Chapter 25 overturns every 20 min. The 750,000-gallon reef system at Townsville, Australia (Chapter 21), is completely pumped around every 2.9 h (Jones, 1988). There is no way to arrive at an ideal rate without taking oxygen measurements in the fully functioning system (see Chapter 10) and comparing actual currents and surge, depending on design, with the wild analog. The mean flow rate across the St. Croix analog coral reef is 10 cm/sec, three times as high as that in the Smithsonian model at 3 cm/sec. On the other hand, turnover time (ocean water replacement) on the wild reef averages about 6h as compared to 1h in the model mentioned above. Thus, these related parameters are both within the right order of magnitude, but displaced in opposite directions because of critical size restrictions in the model. A secondary and related parameter, water quality in the model as measured by dissolved oxygen concentration is remarkably close to that in the wild (see Chapter 10, Figure 9). This kind of give and take matching is crucial to the modeling process. In general, warmer and smaller microcosms and aquaria should turn over more frequently than larger, colder systems. Oxygen solubility in cold water is almost twice that in tropical waters (see Chapter 10). Therefore draw-down for a given biological load is less critical in cold water. Also, smaller systems will usually be more overloaded than larger ones whether research or pleasure or educational viewing is planned. Thus, greater circulation in proportion to volume is desired.

Many factors determine actual pumping rate as compared to the rate given on the pump or pumps. Besides the pump rating, actual pumping rate is a function of both suction and delivery heights, length of flow, and the size of the pipe, as well as any restrictions to flow. Traditional filtering devices can greatly reduce actual output, as well as destroying plankton. It is best not to use such filters in model aquatic ecosystems.

In our discussions we refer to actual flow. In general, if algal scrubbers are used to control water quality, a minimum rule-of-thumb flow rate over the scrubbers is 5 gallons per minute per 6 square feet of scrubber surface. If this rate is not acceptable to achieve the currents desired, then a separate recycling pump can be used, either on the primary tank or on the scrubbers. A simple device that combines scrubbing with current, wave action, and wave surge is shown in Color Plate 5. This "dump scrubber" is particularly effective when minimum space is available in the small home system and the desirability of mounting all apparatus directly on top of the tank is a foremost consideration.

It is best to split up the pumping among as many small pump units as economically and physically feasible, thereby lessening the threat to the whole system when a pump breaks down, as it inevitably does. There should always be back-up pumps readily available.

Plumbing

Plumbing a system calls for considerable care because blockage and/or major water loss is usually disastrous both for the mesocosm and for the facility in which it is housed. Also, plumbing becomes another substrate for organisms. Excessive growth will reduce or block flow and increase ecosystem respiration. Here we simply offer some tips and suggestions.

We use PVC piping or Tygon and similar tubing. Where lighting is intense, fouling is less in dark PVC; we have had no indication that dark PVC is slightly toxic, as has been suggested.

• Opaque piping or tubing rather than transparent tubing is recommended for lighted areas to prevent algal fouling, which can quickly block lines. Considerable fouling by a wide variety of organisms has developed on much of our piping. Because internal fouling is inevitable, a "pig" or cleaner should be able to pass through each line, or pipe should be cleared or replaced every few years.

• Avoid bends greater than 45° whenever possible, especially those 90° or greater. Sharp angles can greatly reduce pumping efficiency, particularly if the run is long or the head great.

• Whenever possible, place all intake lines including pump lines at shallow depths to avoid pumping tanks dangerously low in the event of a leak while the system is unattended. If proper circulation requires intake placement deep in a tank, use a siphon break (Fig. 18). In a well-designed system, loss of pumping and wave action over a single night probably will not be critical; total water loss probably will be fatal.

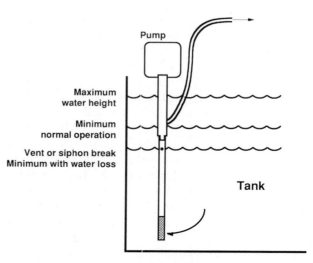

Figure 18 Simple suction-breaking device to avoid disastrous water loss in the event of a plumbing failure.

• Practice joining fittings and pipes together, testing for leakage and strength of PVC junctions before doing the final construction. Great care is necessary with every joint.

• Use "hose barbs" wherever possible for connecting hose and hose clamps to PVC pipe. Take care in attaching the hose clamps, and replace them regularly because they quickly corrode, especially in a saltwater environment. Under water use only plastic hose clamps, and in most cases use more than one clamp.

• Finally, whenever possible, design a system so that failure anywhere will result in water flowing by gravity back to the main system as the primary reservoir. No matter how carefully a system is constructed, sooner or later a break or blockage occurs in the plumbing and water loss and spillage occur. However, in a design that allows all water to return to the main tank in the event of failure, make sure the main tank has sufficient capacity to handle the entire system volume.

References

Adey, W., and Hackney, J. 1989. Harvest production of coral reef algal turfs. In *The Biology, Ecology and Mariculture of Mithrax spinosissimus Based on Cultured Algal Turfs.* W. Adey (Ed.). Mariculture Institute, Washington, D.C.

Adey, W., and Steneck, R. 1985. Highly productive eastern Caribbean reefs: Synergistic effects of biological, chemical, physical and geological factors. In *The Ecology of Coral Reefs.* M. Reaka (Ed.), pp. 163–187. NOAA Symposium Series of Underwater Research. Vol. 3. Washington, D.C.

Hutchinson, G. E. 1957. *A Treatise on Limnology*. Vol. 1. *Geography, Physics and Chemistry*. John Wiley, New York.

Jones, M. 1988. The Great Barrier Reef Aquarium—A Matter of Scale. Northern Reg. Eug. Conf. (Australia), Townsville, June 10–13.

Kennett, J. 1982. *Marine Geology*. Prentice Hall, Englewood Cliffs, New Jersey.

Leigh, E., Paine, R., Quinn, J., and Suchanek, T. 1987. Wave energy and intertidal productivity. *Proc. Natl. Acad. Sci. USA* **84:** 1314–1318.

McCormick, M. 1981. *Ocean Wave Energy Conversion*. John Wiley and Sons, New York.

Moe, M. 1989. *The Marine Aquarium Reference*. Green Turtle Publications, Plantation, Florida.

Nixon, S. 1988. Physical energy inputs and the comparative ecology of lake and marine ecosystems. *Limnol. Oceanogr.* **33:** 1005–1025.

Thurman, H., and Webber, H. 1984. *Marine Biology*. Merrill. Columbus, Ohio.

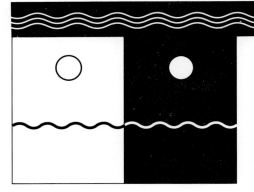

CHAPTER 6

TIDES
Simulating the Effects of Sun and Moon

The ebb and flow of tides is one of the most fascinating aspects of the sea. At an early age we learn of the gravitational effects of the sun and moon on the earth's oceans. Tides are perhaps the most visible signs of how the two orbs work in tandem, their gravity pulling on us, affecting our planet, its oceans, and its inhabitants in many ways. Fascinating and visceral as tides and lunar effects may be, why are they important to the ecologist bent on modeling and to the aquarist?

Even though lakes show seasonal or meteorological changes in level, tidal effects are virtually absent and ecologically lakes seem to function quite well, at least when away from human influence. Indeed, on some ocean coasts tides are small enough to be more or less negligible. However, as we pointed out in the previous chapter, bottom biomass in lakes is generally less than in the sea, probably in large part due to tidal effects (Chapter 5, see Fig. 1). Many ecological communities (salt marshes and rocky intertidals, for example) depend entirely on tides, and it has been demonstrated that the mixing effects of tides provide an energy subsidy to ecosystem function (Fig. 1). In addition, many organisms key important elements of their life cycles, particularly reproduction, to the tides and to the moon.

As an example of the complex ways in which tides and moon, together and separately, can affect an organism (and therefore a community),

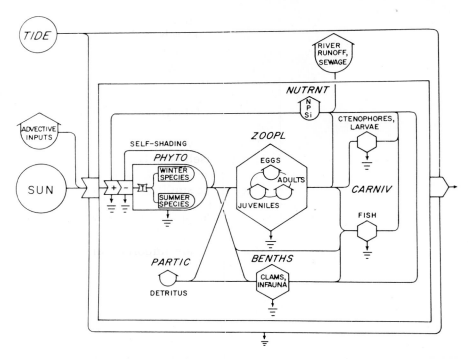

Figure 1 Energy flow diagram developed for the Narragansett Bay ecosystem. Note that tide has been given a major controlling role both on solar input and on export from the bay. The boxed form of tidal attachment also indicates tidal effects driving all levels of the ecosystem. After Kremer and Nixon (1978).

we cite the case of the Caribbean/West Indian magpie shell, *Cittarium pica.* This large rocky intertidal and upper subtidal snail, characteristic of exposed wave-beaten shores, achieves some gastronomic use in the Caribbean region, and in some areas, such as Puerto Rico, it is virtually a national dish. Three to five days after the new moon (i.e., during the darkest nights of the month), the males and females crawl into exposed tide pools at high tide. When tide level lowers sufficiently to isolate the pool, the snails start releasing eggs and sperm into the water. It is not known whether elevated temperatures, hydrostatic pressure, lack of wave action, or some other factor keys the animal to knowing that the pool is isolated. In any case, fertilization takes place in the pool and the developing larvae are washed into the ocean as the rising tide floods the pools. This complex and multikeyed reproductive pattern offers many advantages. To an animal living on a wave-beaten shore where eggs and sperm released into the water would have a difficult time coming together before being washed out to sea, it provides a quite precise way to securely concentrate large numbers of eggs and sperm. Also, the tide pool situation is one that is difficult for many fish to occupy. Wrasses, for example, would otherwise eat the eggs as they are released. Thus, the tide pool provides relative freedom from fish predation and it seems likely that in part this also has resulted in the evolution of this curious pattern.

A similar reproductive keying to spring-higher high tides on sandy California beaches is practiced by a fish, *Leuresthes tenuis*, the grunion. Thurman and Webber (1984) describe this process and its relationships to tide in some detail. To the aquarist wishing to model ecosystems and curious about keying by tides this is a story worth reading.

The intertidal itself is truly the interface between the terrestrial and the marine, and this area is the most easily accessible of marine environments. Many scientific studies have been carried out in the intertidal and some excellent books describe the zone. One of the finest is *Between Pacific Tides* (Ricketts *et al.*, 1985), first written in 1939 and now in its fifth edition. Particularly the rocky intertidal and salt marshes, restricted to the intertidal zone, are important and fascinating subjects for model ecosystems.

Ocean Tides

Average ocean tides measure 2–3 feet from high to low water (0.5–1.0 meters), but they can range from virtually nothing to 50 feet (15 meters), depending on special geographic features (see Fig. 2). Generally, any body of water has a natural frequency of oscillation as a standing wave. This can be demonstrated in an aquarium by "pushing down" on the water on one side in a pulsing manner until the entire mass starts "rocking." When the natural frequency of a body of water is close to that of the sun/moon-created tidal wave, tide heights can become much larger than those in the open ocean. The English Channel, the Gulf of Maine, and the semi-enclosed waters north of Australia are well known for their extraordinary tides.

Tides are typically semidiurnal, with two high points and two low points a day, each one about 6 hours and 10 minutes apart and more or less equal in height. However, diurnal tides of one high and one low daily, or combinations, mixed tides, somewhere in between, are not uncommon (Fig. 3). The Florida Everglades coast, for which we discuss modeling in depth in Chapter 22, has a nearly perfect mixed diurnal/semidiurnal tide. In this case, the Atlantic Ocean has a semidiurnal tide, the Gulf of Mexico, a diurnal tide. The southwest Florida coast is where they meet.

Every 2 weeks when the sun and moon are on the same side of the earth, and also when they are on opposite sides, the earth has large "spring" tides. Roughly 7 days after a spring tide, when the sun and moon form right angles to the earth, "neap" tides occur. These are a third to a quarter smaller (Fig. 4).

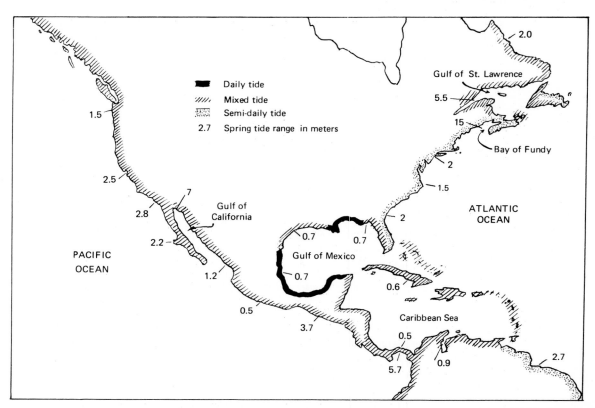

Figure 2   Distribution of tide heights around North America. After Thurman and Webber (1984). Copyright © 1984 by Scott, Foresman and Company. Reprinted by permission of HarperCollins Publishers.

The Intertidal

The intertidal band of bottom or shore between high and low tides is a difficult region for living organisms to occupy. Relatively few plants and animals have adapted to this zone. It is alternately cooled and wetted with salt water, sometimes frozen in the winter, dried and baked by the sun in the summer, and sometimes flooded with totally fresh water during heavy rains at low tide. Depending on tidal range and slope of the shoreline, it can be quite narrow or many miles wide. It is neither marine nor terrestrial. On rocky shores it is dominated by relatives of primarily marine groups, on marshes and swamps mainly by terrestrial-derived species. Environmentally difficult as the intertidal can be, for those organisms able to withstand its rigors successfully, competition and predation are generally reduced and for the plants both maximum light and a very reliable water supply can be achieved.

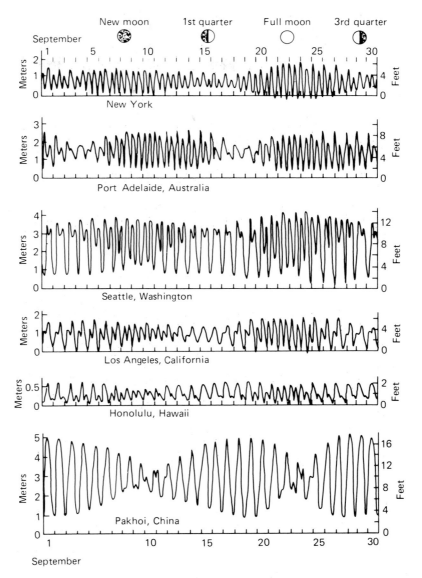

Figure 3 A variety of tidal curves from ports scattered around the earth, September, 1958. After Gross (1982).

Though the intertidal zone is not a favorite of aquarists, it is often a strikingly beautiful region that can tolerate extreme conditions and therefore is relatively easy to maintain in microcosm. The intertidal itself is often strongly subzoned, and such zonation can be related to patterns of tidal form and height. Typically, worldwide, the basic pattern of zonation

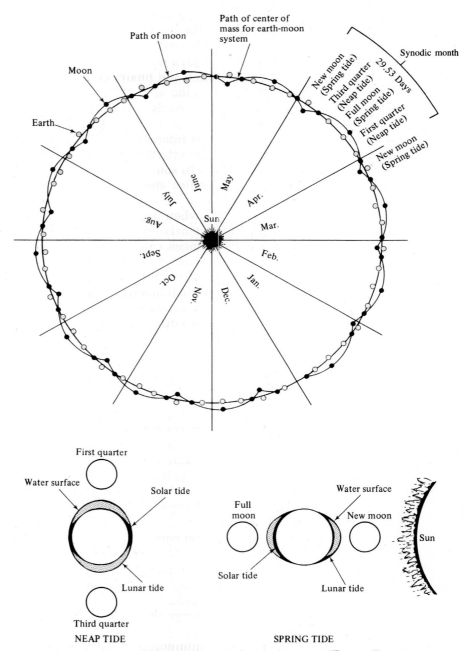

Figure 4   Earth–moon rotation and their relationship to the tides. After Thurman and Webber (1984). Copyright © 1984 by Scott, Foresman and Company. Reprinted by permission of Harper-Collins Publishers.

from top to bottom in the rocky intertidal is (1) a black band of blue green algae and lichens (in the spray zone or just at or above the highest regular tides); (2) a periwinkle snail or littorinid snail zone (roughly the upper half of the tide zone); (3) a white and very rough barnacle zone (the lower half of the regular tide); and finally (4) a mussel zone between neap and spring low tides. Specific areas have even more characteristic communities, such as the rockweeds (brown algae) and Irish moss (red algae) of the North Atlantic (see Chapter 22) (Figs. 5 and 6).

Some of the most interesting marine and estuarine plants and animals richly occupy the intertidal zone and form unique and highly productive communities. In some areas salt marshes and mangrove swamps cover many square miles and provide extensive habitats for intertidal organisms, insects, and birds. These too are strongly zoned to the tide (Color Plate 6), even when beyond the limit of salt water (Odum et al., 1984 and Fig. 7). Extensive mud flats, often found in protected areas having extreme tide ranges, can become rich reservoirs of organic particulates. These organic-rich, muddy bottoms are largely derived by wash-in from more exposed areas and from streams and rivers. Tidal flats are occupied by a host of small invertebrates, including many clams prized as human food. On rocky coasts, tidal plant species form communities unlike any other in the sea or on the land.

Tides and the Model Ecosystem

While a few research systems built to investigate salt marshes in ecological models have been constructed and some more progressive public aquaria show the effects of tides in appropriate tanks, tidal models are virtually nonexistent, and in Moe's (1989) excellent modern treatment tides are not even mentioned. This situation is peculiar since accomplishing a rather accurate intertidal simulation is relatively simple. (See Fig. 8 for standard intertidal zone terminology.) Many Smithsonian microcosm and mesocosm systems have included tides and intertidal zones, and we have achieved tides that control community structure in home aquaria as small as 70 gallons (see Chapter 23).

Whether tides should be considered a factor in a microcosm depends on the marine community being simulated. None of the Marine Systems Laboratory coral reef tanks has had a tidal element. While in some places of very high wave energy the intertidal algal ridge is one of the most fascinating communities on Caribbean reefs, for most reefs in this region tides are relatively unimportant. On the other hand, on the Australian Great Barrier Reef, tides of 3–10 feet (1–3 meters) are an extremely important ecological element. The Great Barrier Reef Marine Park reef aquarium in Townsville, Australia, included a moderate tide range (see Chapter 21).

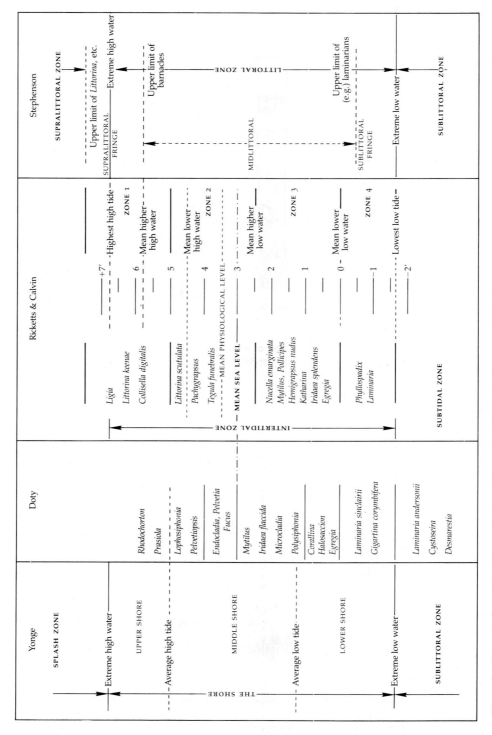

Figure 5 The vertical zonation of organisms occurring on intertidal rocky shores on the Pacific Coast. Several authors as noted, after Ricketts *et al.* (1985).

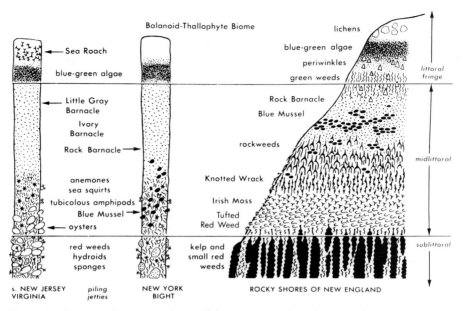

Figure 6 The vertical zonation of intertidal organisms on hard bottoms of the Atlantic Coast. Modified after Gosner (1978). Reprinted by permission of Houghton Mifflin Co.

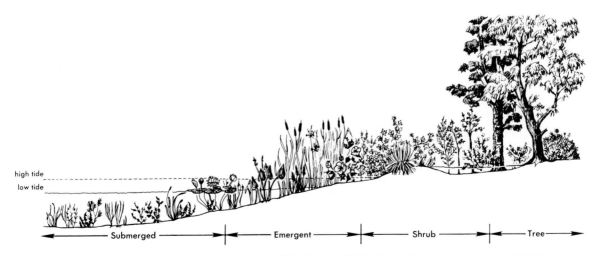

Figure 7 Zonation in a Chesapeake Bay tidal fresh marsh. From Lippson *et al.* (1979).

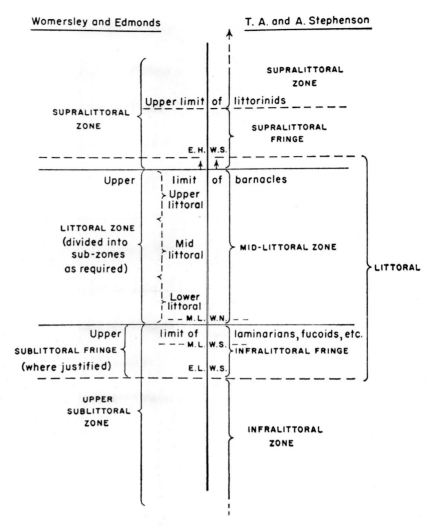

Figure 8 Diagram showing the variety of terminology used for describing tide ranges. From Doty, 1957.

In our simulations of the rocky Maine coast and Chesapeake Bay and the Florida Everglades, tides are too important to ignore.

Developing a tidal system in a closed microcosm is in part a matter of time-regulating higher and lower water levels. This is most easily accomplished by temporarily storing water at times of low tide in a separate reservoir (Fig. 9). If it is desirable to save space, two separate reservoir systems or two separate parts of the same tank can be operated on alternate tide cycles. In our Florida Everglades mesocosm, the estuary served as the tidal reservoir for the Gulf of Mexico portion of the system. Pumps

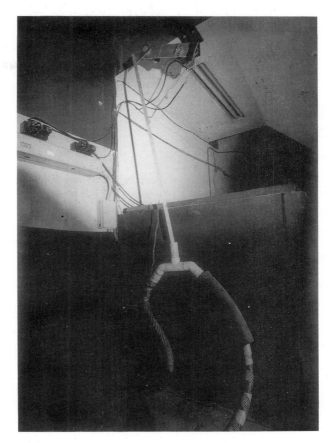

Figure 9 Photograph of tidal controller attached to a tidal reservoir on the Maine coast microcosm. The center motor rotates approximately twice a day to create the semidiurnal component. The outer motor rotates once every two weeks to create neaps and springs.

first pumping in one direction (to the reservoir) and then in the other direction (back to the main tank) can also be used; however, unless a complicated multiple pump arrangement is employed, the result is a sawtooth tidal pattern rather than the sine-type curve of the wild. The sawtooth curve reduces by about 25% the total times of high and low tide.

We have designed a water-level tidal management system in the Marine Systems Laboratory based on interval timer control of stepping motors (Figure 10 and Color Plate 7). This method creates pure semidiurnal tides and provides for biweekly springs and neaps as well as for the 50-minute daily timing advance to "follow the moon." The approach is generally trouble-free and inexpensive. A similar drive was described by Cripe in 1980. The system can easily be adapted to a diurnal tidal cycle simply by changing the timing on the primary stepping motor drive. Stepping

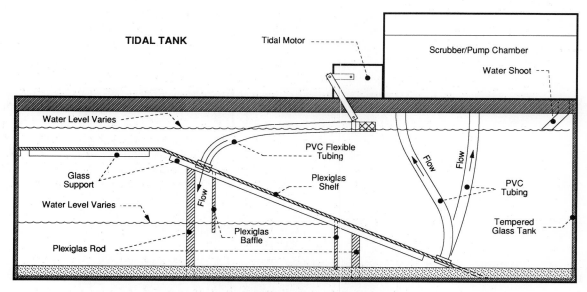

TIDAL TANK

Figure 10   A 130-gallon Chesapeake tank showing tidal/refugium (lower left).

motor control can be modified to simulate virtually any tidal cycle, even a mixed diurnal/ semidiurnal tide.

On the larger Smithsonian mesocosms we have used Superior Electric Company S550 stepping motors with 256:1, P5 planetary reduction gears, which provide ample available torque (125 inch-lbs) with a rotation rate of 0.027 r.p.m. We set pairs of interval timers at 3.6 seconds every minute for the central (semidiurnal) tide motor and 3.1 seconds every 1800 seconds for the outer (biweekly) tide motor. On smaller, 70-gallon tanks, we have used less expensive low-torque stepping motors for simple semidiurnal tidal control (see Chapter 22). The springs/neaps cycle is set manually in this case.

An irregularly shaped tide can result if the reservoir is box-shaped and when the intertidal in the main tank has a low slope causing the volume of water to change with water level. In the example shown in Fig. 10, the distance between high-water springs and neaps (supralittoral) is much less than between low-water springs and neaps (infralittoral). A purer wave form and more proportionate tidal zones can be achieved by using an appropriate reservoir shape (Chapter 23, Fig. 21). As in the example just shown, the tidal reservoir can also serve as a fine-sediment set tling trap and refugium for the numerous worms and small crustaceans that characterize deeper water mud bottoms. As we discuss throughout this book, refugia and settling traps are critical elements of model ecosystems that allow the simulation of much larger wild environments.

Mud flats and rocky shores are relatively easy to establish and manage in an intertidal microcosm. Some care should be taken to stock organisms at their proper zonal level, and this can be quite difficult when a large vertical tidal zone in the wild analog is reduced to a relatively narrow zone in the model. However, community adjustments in the microcosm generally develop slowly over many months without detrimental effects. Rocky intertidals in particular are adjusted in the wild to wave action, with given zonal bands being higher with increased wave action (Fig. 11). Also, rocky intertidals are notoriously patchy, with the effects of settlement, wave action, local terrain, and local predation varying widely. These effects have been favorite subjects for studying the dynamics of community structure (see Levinton, 1982). In an aquarium or small mesocosm, the available shore can become a single patch changing from time to time. The process can be altered and directed by the "disturbance" of the aquarist. Algal mussel and barnacle zonation on the rocky intertidal of an aquarium system, with the grazers and predators that occupy the same environment, can be an exciting model to manage.

Marsh communities are considerably more sensitive to tidal levels than rocky intertidals. In microcosms of small or microtidal ranges (less than 2 feet in the wild), it is important to place the sod surface of marsh grasses at the equivalent tidal level so that total exposure (and sub-

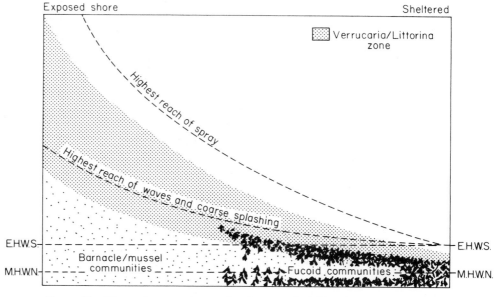

Figure 11 Effects of wave action on the elevation of tidal bands. E.H.W.S, extreme high water springs; M.H.W.N., mean high water neaps. After Levinton (1982). Reprinted by permission of Prentice Hall, Englewood Cliffs, New Jersey.

mergence) times in the microcosm are the same as the in wild. Marsh tidal ranges are always much less than the total tide, generally occupying only the upper half of the tide range.

It is not yet known how much tides can be compressed in a marsh microcosm and still maintain the character of the simulated marshes. However, if submergence time can be kept to that in the wild and drainage is adequate, anything over 6 inches should be acceptable. Some emergent aquatics such as the *Spartina* species may die back when transplanted, especially in the fall. The sod, however, is likely to regrow emergent parts from rhizomes in the spring. These relationships are discussed in depth for specific systems that we have modeled in Chapters 22 and 23.

References

Cripe, C. 1980. Apparatus to produce tidal fluctuations in a simulated salt marsh system. *Progressive Fish Culturist* **42**: 32–33.

Doty, M. 1957. Rocky Intertidal Surfaces. In *Treatise on Marine Ecology and Paleoecology*. Vol. 1. *Geol. Soc. Am. Mem.* **67**: 1–1296.

Gosner, K. 1978. *A Field Guide to the Atlantic Seashore*. Petersen Field Guide Series, Houghton Mifflin, Boston.

Gross, G. 1982. *Oceanography, A View of the Earth*. Prentice Hall, Englewood Cliffs, New Jersey.

Kremer, J., and Nixon, S. 1978. *A Coastal Marine Ecosystem Simulation and Analysis*. Springer-Verlag, Berlin.

Levinton, J. 1982. *Marine Ecology*. Prentice Hall, Englewood Cliffs, New Jersey.

Lippson, A. J., Haire, M. S., Holland, A. F., Jacobs, F., Jensen, J., Moran-Johnson, R. L., Polgar, T. T., and Richkus, W. A. 1979. *Environmental Atlas of the Potomac Estuary*. Martin Marietta Corp., Baltimore.

Moe, M. 1989. *The Marine Aquarium Reference*. Green Turtle Publications, Plantation, Florida.

Odum, W., Smith, T., Hoover, J., and McIvor, C. 1984. The ecology of tidal freshwater marshes of the United States East Coast: A community profile. U.S. Fish & Wildlife Service OBS-83/17.

Ricketts, E., Calvin, J., Hedgpeth, J., and Phillips, D. 1985. *Between Pacific Tides*. 5th Ed. Stanford University Press, Stanford, California.

Thurman, H., and Webber, H. 1984. *Marine Biology*. Merrill Publishing Co., Columbus, Ohio.

CHAPTER 7

THE INPUT OF SOLAR ENERGY
Lighting Requirements

Solar energy is the ultimate driving force in virtually all existing terrestrial and marine ecosystems. Although there is considerable debate among scientists who study the earliest forms of life on earth (and those were in the ocean), most would agree that early biological energy sources were chemical. The early atmosphere and ocean, rich in carbon dioxide, ammonia, and perhaps hydrogen sulfide and without oxygen or ozone, was bombarded with short-wave radiation that created a wide variety of physically synthesized organic compounds. The earliest "organisms" evolved from chance combinations of these compounds, perhaps with the involvement of the intense energy of lightning strikes. Some of these developing "creatures" likely made direct use of simpler energy-rich compounds much as many present-day heterotrophic bacteria are able to do. This pattern of energy supply went on for perhaps a billion years or more and probably could have been effective only in the absence of oxygen.

Photosynthesis and Its Origin

About $3\frac{1}{2}$ billion years ago, some bacteria developed the ability to utilize sunlight directly and to split a variety of abundantly available noncarbon compounds, many with hydrogen and an available electron. The same or similar bacteria still exist though they are rather rare, requiring an anaerobic environment in conjunction with abundant solar energy, a somewhat unusual situation on today's Earth (Rheinheimer, 1985).

Finally, about a billion years later, the blue-green algae (often treated today as bacteria since they lack a well-defined nucleus) developed the ability to use the most abundant compound available in the earth's oceans (water) in this same basic process of energy supply (Fig. 1). Photosynthesis eventually became a highly sophisticated chemical process of two basic steps, the first requiring light and using chlorophyll and associated pigments to capture light energy and to split water to provide electrons and hydrogen ions (and incidently release oxygen). The second step can take place in the dark. It uses the energy carried by excited electrons from the first step and the hydrogen ions to build simple sugars ($C_6H_{12}O_6$ or $6CH_2O$) from carbon dioxide, thereby providing the basic chemical energy supply for the plants and the animals that eat them (Fig. 2). The great variety of photosynthesis in the modern-day Plant Kingdom probably has

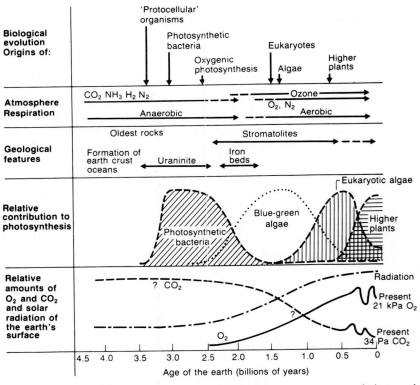

Figure 1 Evolution with time of the major groups of organisms, the process of photosynthesis, and the composition of the atmosphere. After Lawlor (1987).

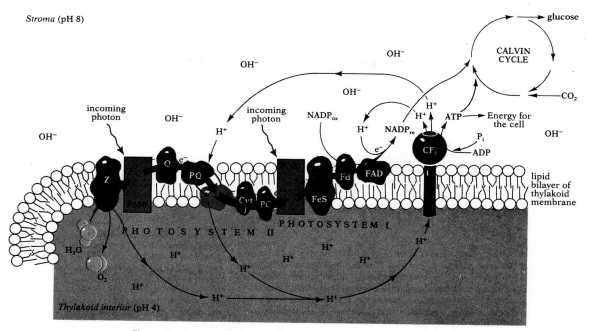

Figure 2 Process of photosynthesis and the sequence and location of light-sensitive phases on the membranes of plant chloroplasts. After Keeton and Gould (1986).

evolved from this invention of blue-green algae some 2 billion years ago. The evolution of photosynthesis was one of the major advances in the development of life. The great complexity now found in the earth's biosphere, the very character of its atmosphere, and the oxygen that allows cellular respiration (a rapid means of removing energy from food) to occur are to be credited primarily to this event.

Today it is still true that in the oceans and hot springs some biological sources are chemoautotrophic and probably not even indirectly derived from solar energy. A prime example, discovered and popularized in the press in the early 1980s, is provided by the thermal vent communities and their ecosystems. along mid-ocean ridges. These ecosystems are based on energy supplied from the earth. However, interesting as these systems are, and as interesting as they would be to bring into a microcosm, it is sunlight that drives the vast majority of ecosystems likely to be of interest to the ecologist and the aquarist. In environments such as deep soft bottoms where no light is present, the ecosystem's driving energy is obtained through the rain of organic matter from shallow water or possibly through turbidity or other currents. Also, a tropical jungle stream survives largely through the breakdown of leaves that fall into it or are made available to the community during floods of the adjacent forest floor. In simulating such systems one would perhaps wish to "feed" them with appropriate organic particulates or leaf litter rather than primarily with light through a second model ecosystem. However, many of the most interesting and affordable marine and fresh water microcosms, mesocosms, and aquaria that are likely to be attempted at this time are shallow-water systems where light is the most critical factor to be considered.

Finally, just as the basic character of our atmosphere derives from plant photosynthesis, so the ultimate "purity" of our surface waters also derives directly from plant photosynthesis. If we wish to simulate wild environments to any degree like their unpolluted analogs, we must either use photosynthesis as the major controlling element in our models or use high-quality wild waters as rapid replacement. Of course we have the technology, assuming we can afford to use it, to chemically and physically "create" high-quality waters of whatever character we desire. However, we will never be able to chemically and physically undo the damage we have caused to natural waters in our biosphere, for lack of affordable energy supply and economic reasons, among others. Therefore, in our models, we must learn to use photosynthesis, nature's way of providing not only energy but the requisite water quality, in the hope that this will eventually translate to our relationship to the earth's biosphere. We will explain this view repeatedly throughout this book.

Light and Ecosystems

Even if an aquarist were to be primarily concerned with keeping a few interesting animals as a hobby or for conservation or breeding purposes and were not at all interested in an ecosystem for its own sake, it is important to remember that a "balanced aquarium" and healthy animals depend upon the appropriate use of natural recycling processes to maintain the stability of water chemistry in an enclosed system. Many substances produced by animals and bacteria as products of their metabolism, including some compounds that are toxic to the very animals that produced them, are very much needed by plants. The plants, in their turn, produce substances (food, oxygen, vitamins) required by animals. In this exchange, animal wastes that would be harmful if allowed to accumulate are converted by plants into beneficial forms. If the interchange between plants and animals is balanced so that all the waste products are utilized or otherwise appropriately removed from the aquarium system, a stability of water condition is maintained much as it is in healthy wild communities. Also, complex ecosystems have a built-in stability that can be used by the aquarist. Just as in farming, productivity may be higher with a monoculture directed to the specific economic needs of the farmer. However, a monoculture is inherently unstable, just waiting for the wrong weather, the wrong insect, or a flock of birds at the wrong time to dramatically topple months of work.

The requirement for, and the control by, sunlight in shallow-water ecosystems goes far beyond the energy requirements of the plants in the system. Some elements of solar radiation are even harmful, particularly ultraviolet radiation. Just as some races of human beings have become pigmented and more or less protected from the radiation found in their ancestral habitat and other members of the human species have lost or only weakly evolved such protection [since ultraviolet (UV) is needed in their solar-radiation-poor environment], many plants and animals can be ranged on a similar spectrum of tolerance to UV. Even beyond tolerance or intolerance for portions of the solar spectrum, many plants and animals use light of all or part of the spectrum to time their lives to feeding, reproduction, defense, and virtually all of the functions of organisms. Of course, vision, or at least light sensitivity in eye-like organs, forms one of the major sensing devices of animals throughout the animal kingdom. Also, just as it is known that certain land plants, including many house-plant varieties, use day length to determine flowering time, so many marine plants and animals also key to the length of the daylight. Other animals use the polarization of sunlight to navigate, and still others use narrow portions of the solar spectrum to locate food.

Finally, many shallow-water animals have developed symbiotic relationships to algae living in their tissues much as lichens have done on land. Large groups, including most corals and their gorgonian and anemone cousins as well as some clams and snails, require sunlight for their symbiotic associates. In general, marine and aquatic communities show a strong depth zonation that is mostly dependent on the amount and quality of light that reaches the individual members of that community. These relationships must be carefully considered in the management of healthy communities in microcosms, mesocosms, and aquaria, and cannot be ignored even if animals alone are to be kept in captivity.

Under some conditions, such as tropical aquaria in the tropics and subarctic aquaria in a subarctic region, natural light can be used to operate synthetic ecosystems. However, often there is a considerable price to pay in heating or cooling where this is done. A greenhouse environment can be a difficult one in which to obtain enough light that includes some ultraviolet radiation without excessive heating or cooling. Generally, in enclosed aquatic and marine ecosystems, light must be supplied artificially, and the greatest attention must be given to the design of this element of the system. Most traditional aquarium endeavors have treated light primarily as an esthetic factor, of interest to the viewer and only secondarily of interest to the aquarium itself. Indeed, in most aquarium work prior to the present decade, light was regarded as a nuisance or a danger that could start blooms in unnaturally nutrient-rich water. This concept is rapidly changing, particularly among the tropical invertebrate specialists. However, aquarium technology, as presently practiced, usually does not employ lighting that is strong enough or of the proper color balance to supply the needs of a plant community, yet capable of balancing the animal activity in the tank. The optimal growth of plants and animals from shallow-water marine communities requires appropriate conditions of intensity (brightness), period (day length), and spectrum (color), conditions that vary with the species being grown and the native environment from which they come.

Solar Radiation and Water

Any atom or body, including the sun, radiates electromagnetic energy in accordance with its temperature. Beginning at about 800 K (degrees centigrade plus 273), visible light in the red end of the spectrum begins to be radiated. As the temperature increases, emitted wavelengths (and the amount of energy radiated) increase. An incandescent bulb has a color temperature of 2800 K and is quite red; a cool-white fluorescent bulb radiates at about 4000 K and is strongest in the red orange. Daylight fluorescent bulbs and the sun radiate at about 6000 K.

Visible radiation is, unfortunately, measured in many ways. In this book we will use an energy measure, micro-einsteins/m²/s (mmol). For rough comparison, 2000 $\mu E/m^2/s$ is approximately 100,000 lux or 10,000 foot-candles.

In the middle of a bright sunny day, early summer at high latitudes and year-round in the tropics, approximately 2000 $\mu E/m^2/s$ of visible or photosynthetically active solar radiation (PAR) is received at the earth's surface. Depending on the clarity of the water in question, this radiation can extend to great depth or be limited to a few meters (Fig. 3 and Table 1). The total solar radiation received per day over the earth is shown by month for various latitudes in Figure 4. The yearly total is shown in Figure 5. Generally, higher latitudes have greater seasonal variability in incoming light. While progressively north and south of latitude 40° total yearly radiation is considerably reduced, in a broad range of mid latitudes total incoming radiation is mostly a function of cloudiness or atmospheric clarity.

Light incoming to the earth's outer atmosphere has roughly the spectrum shown in Figure 6. By the time it passes through the atmosphere and reaches the sea surface through clear sky or clouds it looks more like the second or third curves. Shaded areas without significant local reflection and those under green foliage are the fourth and fifth curves shown in Figure 6. Further, attenuation in lakes or the sea provides a color balance that is even more removed from the original spectrum. The resulting attenuation at different wave lengths varies greatly depending upon depth and water character. A family of curves for light distribution with depth for open ocean, clear coastal water, and a lake are shown in Figure 7. In particular cases, the unique absorption characteristic of a particular coastal lake or river water to be modeled might also have to be considered.

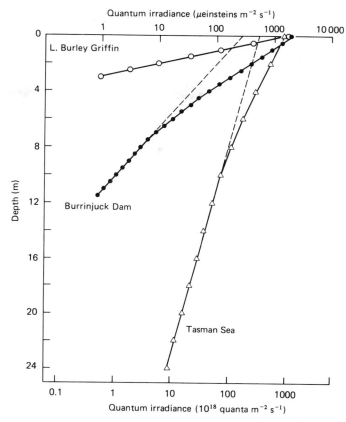

Figure 3 Attenuation of photosynthetically active radiation (PAR) in different types of water. After Kirk (1983).

TABLE 1
Attenuation Coefficients for PAR for a Variety of Fresh and Marine Water[a]

Water body	$K_d(PAR)$ (m^{-1})
I. Oceanic waters	
Sargasso Sea	0.03
Gulf Stream, off Bahamas	0.08
Pacific Ocean, 100 km off Mexico	0.11
II. Coastal and estuarine waters	
Bjornafjord, Norway	0.15
Gulf of California	0.17
Tasman Sea, Australian coast	0.18
Port Hacking estuary, Australia	0.37
Clyde R. estuary, Australia	0.71
	(Table continues)

TABLE 1

(Continued)

Water body	$K_d(PAR)$ (m^{-1})
III. Inland waters	
North America	
Crater Lake, Oregon	0.06
L. Ontario, Canada	0.15–0.58
San Vicente reservoir, California	0.64
Lake Minnetonka, Minnesota	0.7–2.8
McConaughy reservoir, Nebraska	1.6 (av.)
Yankee Hill reservoir, Nebraska	2.5 (av.)
Pawnee reservoir, Nebraska	2.9 (av.)
Europe	
Esthwaite Water, England	0.8–1.6
Loch Croispol, Scotland	0.59
Loch Uanagan, Scotland	2.35
Sea of Galilee	0.5
Sea of Galilee (*Peridinium* bloom)	3.3
Africa	
Lake Simbi, Kenya	3.0–12.3
Lake Tanganyika	0.16 ± 0.02
Australia	
(a) Southern Tablelands	
Corin Dam	0.87
Lake Ginninderra	1.46 ± 0.68
(3-year range)	0.84–2.74
Burrinjuck Dam	1.65 ± 0.81
(6-year range)	0.71–3.71
L. Burley Griffin	2.81 ± 1.45
(6-year range)	0.86–6.93
L. George	15.1 ± 9.3
(5-year range)	5.7–24.9
(b) Northern Territory (Magela Creek billabongs)	
Mudginberri	1.24
Leichardt	1.68
Georgetown	8.50
(c) Tasmania (lakes)	
Perry	0.21
Ladies Tarn	0.41
Risdon Brook	0.58
Barrington	1.13
Gordon	2.16

[a]After Kirk (1983).

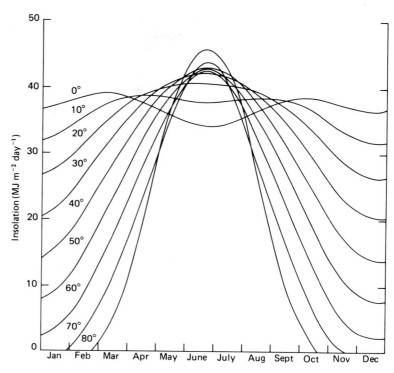

Figure 4 Change in incoming solar radiation at the top of atmosphere calculated seasonally and for different latitudes. After Kirk (1983).

In general, pure water absorbs longer wavelengths above 550–600 nm and thus, given enough depth, will appear green or ultimately blue. Suspended particulates and dissolved substances absorb the shorter wavelengths, and typically as one passes from open ocean to coast, to estuary and then river and lake, more of the wavelengths below 500 nm and eventually even 600 nm are absorbed. Thus, depending on the extent of matter in the water, one passes from blue, green, yellow, to red or brown in making such a transect. This is also an indication of the kind of light that is available at depth. However, it must be remembered that some mountain lakes or streams can be as clear as open ocean water.

Light Absorption by Water Plants

Plants, including algae, utilize a wide variety of accessory energy-absorbing pigments in addition to the all-important chlorophyll. Each of these pigments requires specific wavelengths to be photosynthetically efficient. Plants are generally classified into evolutionary groupings, most easily

Figure 5 Mean annual insolation (solar radiation) on the surface of the earth (calories/cm^2/yr × 10^3). Note that the greatest yearly totals occur in the horse latitudes rather than in the tropics proper. However, the seasonality is considerably greater in the horse latitudes, with higher levels in summer and considerably lower levels in mid winter (see also Fig. 4). After Gates (1980).

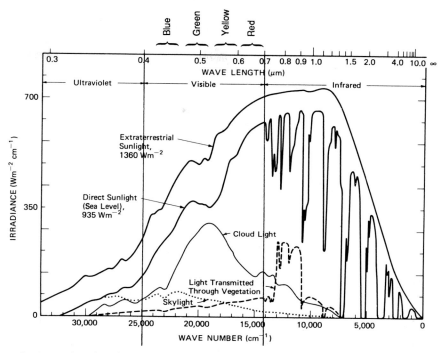

Figure 6 Spectral quality of incoming solar radiation, light reaching the earth's surface (through clear and clouded sky), and light transmitted through terrestrial vegetation. After Gates (1980).

designated by color and therefore by their photosynthetically active pigments. The green plants include the flowering plants such as marsh grasses and all submerged aquatic vegetation as well as algae including *Ulva* (sea lettuce) and *Enteromorpha*. The brown algae, with virtually no aquatic or terrestrial representatives, include large marine plants such as *Laminaria* (kelp) and *Sargassum*. The red algae, mostly marine with a few species in fast-running streams, include *Chondrus* (sea moss) and *Rhodymenia* (dulce). The blue-green algae are filamentous or unicellular plants that form slimy crusts, ribbons, or cushions of varying hues (green, red, black, etc.) that are common in most lighted wild environments as well as in many fresh- and saltwater aquariums. Finally, a number of widespread algal groups particularly important as plankton include goldenbrown algae (diatoms) and yellow-brown algae (dinoflagellates).

The chlorophylls and the accessory pigments used by higher plants and marine and aquatic algae absorb light for photosynthesis at particular wave lengths or groups of wave lengths (see Fig. 8). These absorption peaks tend to be in the blue-green and the far-red wavelengths. However, the actual absorbance by plants and the action spectra (or photosynthetic activity) of those plants, including the major phyla of marine algae (Figs. 9 and 10), are more smoothed or spread out. Thus, even if the spectrum of a

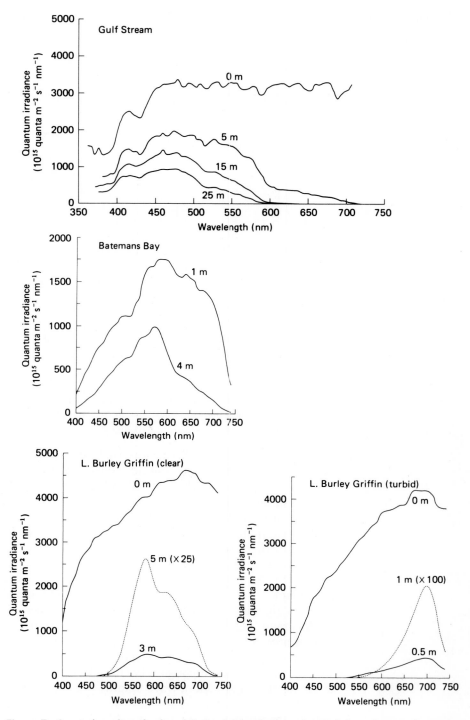

Figure 7 Spectral quality of solar radiation having been transmitted through natural waters (marine and fresh) of varying character. After Kirk (1983).

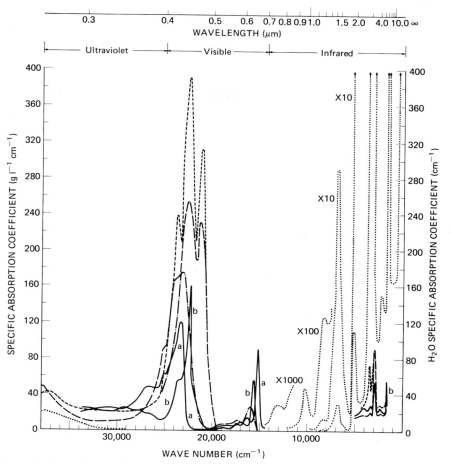

Figure 8 Spectral absorption of solar radiation by higher plant pigments. (— chlorophyll; —·— protochlorophyll; — — α-carotene; -------- lutein; ········ liquid water. After Gates (1980).

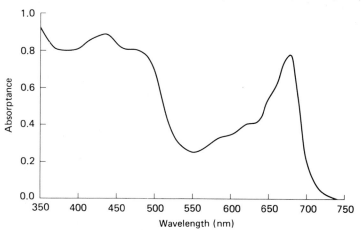

Figure 9 Spectral absorption of the freshwater macrophyte (flowering plant) *Vallisneria spiralis*. After Kirk (1983).

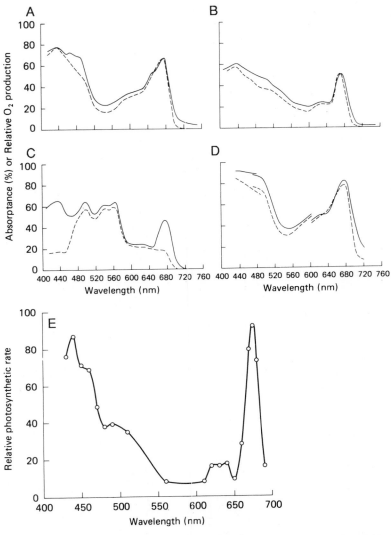

Figure 10 Action (————) and absorption (—) spectra of photosynthesis in various marine algae. While the two curves usually coincide and light absorbed is utilized in photosynthesis, that is not necessarily the case, particularly at the ends of the spectrum. (A) *Ulva* (green); (B) *Coilodesme* (brown); (C) *Delesseria* (red); (D) *Chlorella* (green); (E) *Skeletonema* (diatom). After Kirk (1983).

planned photosynthetic light source were "spiked" rather than continuous, one would probably be closer to natural conditions by attempting to match the "envelope" of naturally available light rather than any specific light requirements of chlorophyll. Even more so, the summation of the individual requirements of the entire plant community covers a broad range of the solar spectrum. Perhaps, if one were interested in growing specific plants and maximizing the production of those plants for the least light energy used, then a light source that omitted light in the red or green-blue might be the most efficient way to proceed. However, in an ecosystem and where stability of function and not maximum productivity is desired, this is not warranted. Remembering that light is also used for many organism activities in addition to photosynthesis, omission of yellow light might even alter ecosystem function.

Light Intensity and Plants

In addition to color balance or spectrum of solar radiation, a major concern in photosynthesis is the total intensity of the radiation. Many plants including marine and aquatic plants are highly adaptive to available light. Plants of a given species grown at high light intensity will perform more poorly when given low light than plants raised at low intensity. It is a characteristic of many algae that with a major change in solar radiation the plants will die or die back and develop new growth adapted to the new light levels.

Terrestrial plants from normal sunny habitats typically show a pattern of photosynthesis with available light similar to that for the bean (Figs. 11 and 12). At cool temperatures, leaves of this plant show a more or

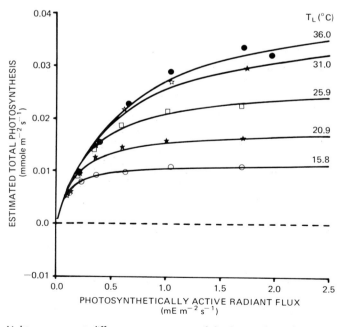

Figure 11 Light response at different temperatures of the bean *Phaseolus vulgaris,* a tropical plant characteristic of sunny sites. At lower temperatures photosynthesis is limited by molecular and chemical processes controlled by temperature, rather than by light itself. After Gates (1980).

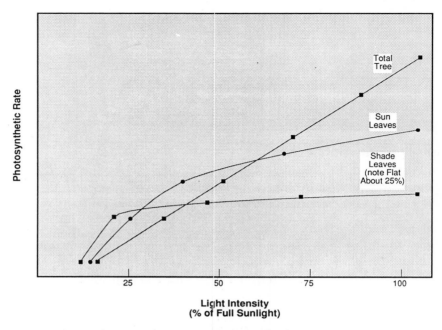

Figure 12 Photosynthesis rate for two terrestrial, woody plant species—individual leaves as compared to the entire trees. After Oliver and Larson, 1990.

less direct and strong increase of photosynthesis up to about one-third of full tropical sunlight. Above that level to the strongest natural light possible, photosynthesis continues to increase but at a slower rate. On the other hand, at high midday leaf temperatures photosynthesis considerably increases up to the most intense solar light possible on earth. Most physiological studies of plants are carried out on small specimens or pieces of those plants. As shown in Figure 12, there is a major difference between the photosynthetic performance of whole plants as compared to their parts. Here, the relationship between light and photosynthesis for entire trees is almost direct to the full intensity of sunlight. As we discuss below, there is reason to suspect that this is even more universal for whole plant communities.

An often quoted pattern of underwater photosynthesis is that for phytoplankton, as in Figure 13. Here, photosynthesis peaks at about one-quarter of surface light, and at higher intensities it is actually inhibited. On the other hand, benthic plants, both flowering plants and algae, do not show an obvious inhibition (Fig. 14).

One might conclude from these data that while many land plants, given no limitation in other factors (e.g., carbon dioxide or temperature), can use all of the possible solar radiation available for photosynthesis,

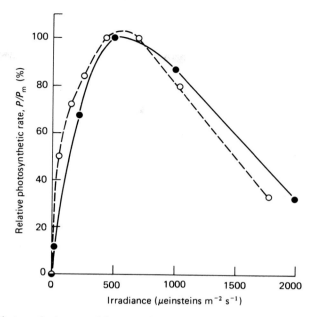

Figure 13 Photosynthetic rates of Sargasso Sea (•) and Lake Windemere (○) phytoplankton as a function of light levels (laboratory experiments). After Kirk (1983).

marine and freshwater plants saturate or even lose production at levels of one-quarter to one-half of available surface radiation. However, if one looks at production by entire benthic plant communities in the wild, almost invariably primary production (or water purification if one is thinking in those terms) is limited seasonally and daily by the available solar radiation (Fig. 15–18). Figure 15 illustrates the relationships particularly well. While 50% of production is achieved when 500 mE/m²/s is reached (a few hours after dawn), full production is only achieved at noon, at light intensities close to full tropical sunlight. Below maximum intensity (and down to about 500 mE/m²/s) every 10% reduction in intensity causes a 5% reduction in production. Likewise in a cold-water kelp community (Fig. 16), light and temperature are not in phase. Yet yearly photosynthesis closely follows available light independent of temperature. Other factors may also limit plant production. Plankton production in higher latitudes, in lakes and oceans, for example, typically rises sharply in the spring, as the light returns and no nutrient limitation exists. There may be dips in production in the mid summer, but this is usually due to nutrient limitation or grazing by zooplankton. Phytoplankton production and more rarely benthic plant production can be limited by ultraviolet radiation near the surface. This limitation is minimal, however, and likely more than made up by increased plant production in deeper water, particularly in benthic communities. All of this is not meant to suggest that

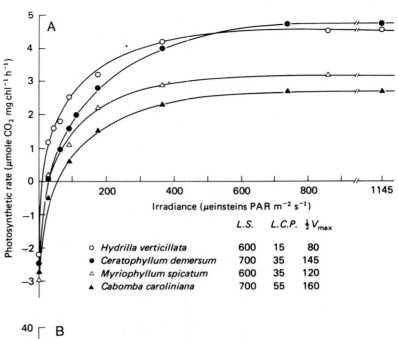

	L.S.	L.C.P.	$\frac{1}{2}V_{max}$
○ *Hydrilla verticillata*	600	15	80
● *Ceratophyllum demersum*	700	35	145
△ *Myriophyllum spicatum*	600	35	120
▲ *Cabomba caroliniana*	700	55	160

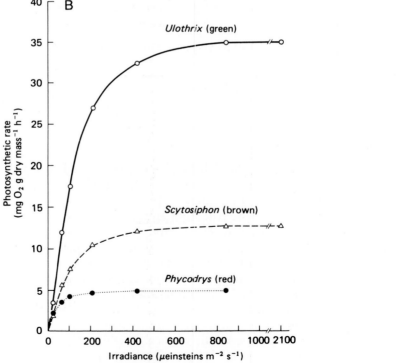

Figure 14 Photosynthetic rates of (A) various submerged macrophytes (flowering plants) and (B) algae as a function of light levels. After Kirk (1983).

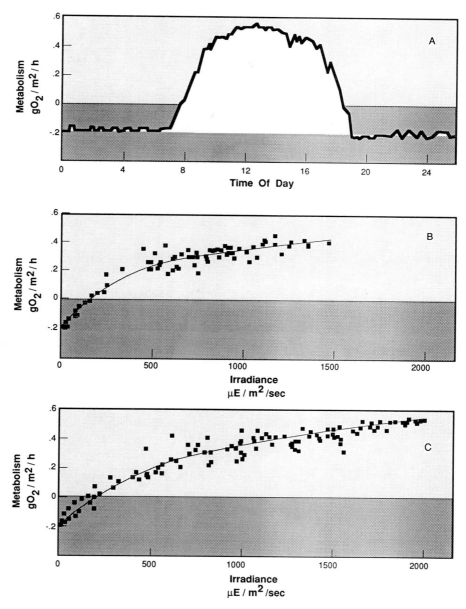

Figure 15 Photosynthesis (day) and respiration (night) as measured on a Caribbean coral reef (Panama): (A) oxygen exchange vs. time; (B) oxygen exchange vs. light *in situ* on reef; (C) oxygen exchange vs. light for an *in situ* microcosm of the reef. After Griffith *et al.* (1987).

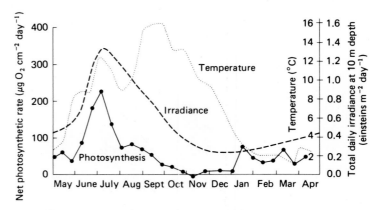

Figure 16 Photosynthetic rates of kelp as related to light levels (at depth of growth) throughout a 1-year cycle. After Kirk (1983).

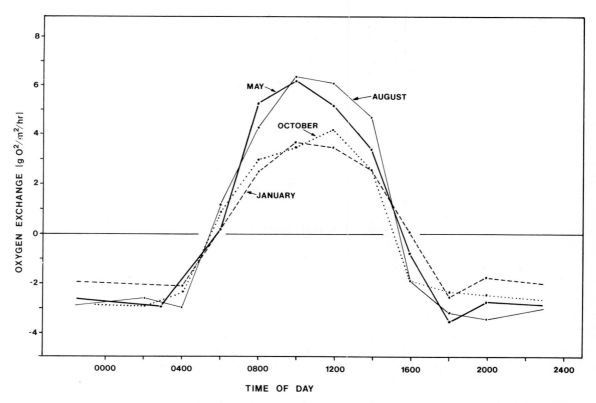

Figure 17 Diurnal oxygen exchange of a tropical coral reef community by season. After Adey and Steneck (1985).

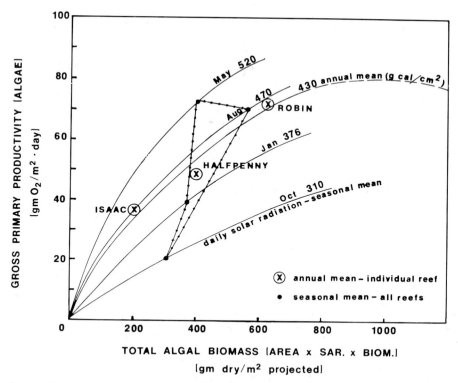

Figure 18 Gross primary productivity of a shallow-water tropical coral reef community as a function of light, season, available surface, and biomass. After Adey and Steneck (1985).

some deep-water corals and other invertebrates might not be damaged by ultraviolet. The aquarist should take these factors into account when building a model ecosystem. Full photosynthesis and primary production, high water quality, and shading or deeper-water simulation can certainly all be attained with careful design.

Light and Model Ecosystems

Until the 1980s, because of the rule of thumb that aquatic plants only used a fraction of available sunlight, model ecosystems tended to be operated with relatively low light levels. In the aquarium world, light was largely considered to be solely for the viewer and perhaps for the activities of fish. Particularly in marine tanks, if algae were present, they were primarily the encrusting blue-greens that could manage the low light levels (and low water quality). Since these algae were often black (and slimy), even such a minimum presence was not desired.

Moe (1989) documents this history and provides an excellent review of the changes in attitude among aquarium hobbyists, particularly those specializing in "reef" tanks. Nevertheless, the attachment of aquarists to the absorption spectra of chlorophyll rather than that of whole communities remains. Also, the basic view that moderate levels of light (e.g., in the range 100–400 $\mu E/m^2/s$) are sufficient for reef simulations is still the rule. Certainly low light is reasonable for keeping deeper-water corals. On the other hand, most *Acropora*, *Millepora*, and *Porites* species, among many others, grow most rapidly where they are receiving greater than 1000 $\mu E/m^2/s$ through the middle of the day (Adey, 1978). Algal turf communities in open lagoon culture reach peak production at mid-day light intensities of about 1500 $\mu E/m^2/s$ (Adey and Hackney, 1989).

When dealing with light simulation in microcosms, mesocosms, and aquaria, our rule of thumb is to try to match both light spectrum and light intensity as closely as possible to the wild ecosystem being considered. In general, if desired for reasons of economics, light intensity might be reduced by 20–30% of that in the wild model without major effects, but greater reductions would seriously compromise the system biologically and ecologically. Likewise if there were considerable benefits to be derived, one might omit some of the green/yellow part of the light spectrum, but it is likely that some part of the ecosystem would likewise be compromised.

A great number of artificial light sources are now available for use on model ecosystems that produce a diverse range of intensity and spectra (Color Plate 8). Tungsten-filament lamps produce light useful to plants. However, a tungsten lamp produces little green and blue light, and it is in the blue-green end of the spectrum that aquatic plants carry out a major part of their photosynthesis. They are also relatively inefficient in terms of PAR produced as compared to heat. On the other hand, the gas lamps—fluorescent, mercury vapor, and metal halide—produce a wide variety of light spectra, much of it within the photosynthetically useful range (Color Plate 9). In addition, they are quite efficient in terms of power usage. Many spectral types of fluorescent lamps are available and, with appropriate mixing of color types, the spectrum of incoming sunlight to a marine or aquatic community can be reproduced. The intensity of the standard fluorescents is, however, relatively low and, although they are strong enough to supply the light requirements of the deepest-water or heavy-shade communities, they are not generally suitable for most sun-dependent ecosystems, especially those from shallow tropical areas.

HO (high output) and VHO (very high output) fluorescent lamps can be found in most of the spectral types of standard fluorescents. For the same size these lamps put out approximately twice and four times as much radiation respectively. (Output is roughly proportional to wattage.) For most small systems (less than 200 gallons), especially freshwater and cold-water coastal microcosms, these lamps are quite suitable. For coral reef or similar brightly lit communities it is possible to use VHO lamps, particularly if the tank is relatively shallow. However, the light levels of the shallowest coral reefs cannot be effectively produced by this means.

For large (deeper) microcosms and mesocosms, and even larger coral reef aquaria, only the high-intensity discharge lamps, which can be obtained from 250 to 1000 W, can provide sufficient light to simulate shallow water intensities. Mercury vapor lamps have sufficient intensity to match any natural lighting. However, the light produced is too blue for conversion to most natural spectra (Fig. 19). They can be used as supplemental lighting with metal halides to simulate oceanic conditions. For larger systems, of all commonly available light sources, metal halide lamps such as the Sylvania metalarc or General Electric multivapor have been found to provide the best combination of spectrum and intensity for simulating natural sunlight (Color Plate 9).

In ecosystem models, the intensity and period for which artificial lighting is used can be varied to suit conditions in the native environment of the community being supported. For instance, the microcosms holding the coral reef communities (see Chapter 21) are modeled after areas in the tropical Caribbean between 10 and 25° north latitude. Day length at those localities (photoperiod) ranges from about 11 to 13 h over the year, and incoming light intensity at noon, just beneath the water surface, measures about 1800 $\mu E/m^2/s$ in summer and 1400 $\mu E/m^2/s$ in winter. The "cold tank" (Chapter 22) represents the Maine coast, 2000 miles to the north, where photoperiod and light intensity change significantly with the season. In summer the sunlight is almost as strong as that of the tropics and the day is even longer. However, in the winter, light intensity drops dramatically to about 800 $\mu E/m^2$ just below the water surface at noon, and there are only 6–7 daylight hours. Lighting for these microcosms was planned with these conditions in mind. Diagrams of the arrangements that reproduces them are shown in Chapters 21 and 22. Clear 1000-W lamps are used to light the tropical tanks, the brighter lamps over the deeper areas. Although output directly beneath individual bulbs exceeds the intensity of natural sunlight in the tropics, the light level decreases rapidly with distance from the center of concentration. The physical size of the light units prevents there being more than one bulb for every 3 square feet of tank surface; thus the mean light intensity per unit surface is lower than that found in the tropics. Peak intensity is, however, stretched out over a larger period each day. On the other hand, lower intensity in midday results in a rapid dwindling of light as it penetrates the water and increases the effective depth of the microcosm. The light at the bottom of the reef tanks at 6 feet is equal to that measured from 40 to 60 feet in the natural reef environment. The result is that the microcosm contained in this tank is essentially a scale model, relative to light, and includes 40 to 60 feet of reef profile (relative to light) in only 6 feet of water. Many opportunities and problems are presented by this compression of depth zones.

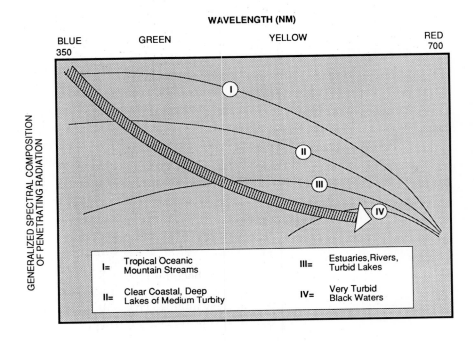

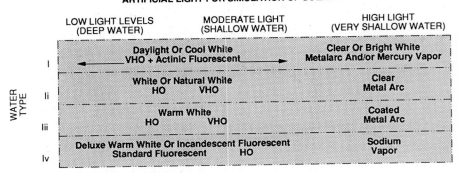

Figure 19 *Generalized spectral characteristics of natural water bodies and the artificial light typically required to provide underwater light in aquarium and mesocosm models of those water bodies.*

The cold-water and temperate microcosms, as we have arranged them, are also lighted at least partially by metal halide lamps. The bottom in the Maine tank slopes steeply, and the turbidity is moderate. The community represents a cross section that reaches from the intertidal nearly to the limit of the photic zone. While summer radiation in the Maine intertidal can be close to that received by an exposed reef, or algal ridge, in the tropics, even during times of strongest sunlight, the energy that reaches the subtidal organisms in the coastal Gulf of Maine is much less than in the clear water of the tropics. Turbidity is caused by runoff from the land, tidal stirring of sediments, and the bloom of planktonic organisms responding to high nutrients and the seasonal increase in light. A light measurement equivalent to a depth of about 50–60 feet is registered at the bottom of the Maine tank, while the intertidal receives light levels close to those found in the wild. The lights are raised and the daylight period shortened to simulate winter intensities.

In the Smithsonian Chesapeake estuarine mesocosm, the shallow and emergent marshes required maximum intensity, and we used four 1000-W metal halides, at a height of about 3 feet above the soil surface, for about 20 square feet of marsh area. It is interesting to note that while a radiation of 1200–1800 $\mu E/m^2/s$ is characteristic of the upper third of the marsh plants, many will grow well into the high-intensity cone of light, exceeding 4000 $\mu E/m^2/s$, with apparent healthy color ceasing only when temperatures finally become too high. On the deeper end of the Chesapeake tank, eight 160-W VHO lamps are used to simulate reduced light in the highly turbid bay analog. Thus, for the marsh, a maximum equivalent intensity of 200 W per ft^2 is used whereas in deeper water levels drop to 32 W per ft^2.

The period of illumination of each mesocosm, microcosm, and aquarium that we have constructed, the "day length," is controlled automatically by timers that open and close the light circuits according to a preset program. However, full intensity is not delivered to the microcosms for the entire period of illumination, but changes gradually to imitate the periods of dawn and dusk. This is accomplished by lighting or extinguishing parts of lights in sequence over a period lasting up to 2 h between each light and dark cycle. This allows the tank inhabitants a transition period between day and night activities. Most marine organisms are sensitive to light changes. Some animals are nocturnal, functioning only during dark periods while others are active in the daytime. In some cases, these transitions are striking. Parrot fish, for instance, rest in a secreted bag of mucous in the dark hours and feed continuously during the day. As possible, a reasonable twilight or changeover time is desirable.

The total light energy available to a microcosm is a function of period as well as intensity, the length of time as well as the brightness. To a limited extent, a deficiency of light intensity can be compensated for by lengthening the period, but probably some loss of accuracy in simulation results. Many plant and animal responses are related to day length, especially in populations from higher latitudes where seasonal variation is significant. Often growth and reproductive cycles can be closely connected with light period.

The same approaches that we have described above have been

put to use in developing a variety of small aquaria, coral reef, estuarine, and freshwater systems (see later chapters). The depth of the average home tank is at most half that of the mesocosm and microcosm systems described here and in most cases HO and VHO fluorescents can be used. This greatly reduces heating problems for small tanks. The guidelines, presented in Figure 19, may be followed in choosing the correct type of lamps for an aquarium, or actual light measurements can be taken.

Light and Physiological Considerations

Reprocessing of animal wastes and oxygen production occurs as a natural part of algal metabolism in well-lighted situations. However, a sufficient biomass of plant material must be maintained to provide the production of plant food needed to accommodate the full requirements of plant-eating animal populations. It is possible to balance this interaction within a single tank, but usually this requires that the number of animals be limited to their average abundance in the natural environment, especially if herbivorous species are included that might deplete the algae. When the lights go out at night the plants stop producing oxygen and, if the animal population is large, the oxygen supply dissolved in the tank water can be greatly reduced before morning. Many shallow marine and aquatic environments have relatively dense populations. However, in the wild, during the night, the constant flow of water from less populated areas of the open ocean or open lake or river supplements the supply of oxygen and removes wastes.

This situation of one ecosystem supplying the needs of another is common in the wild and can be adopted to preserve the night-time balance in an aquarium. Rather than using a large reservoir to replenish the water at night, a method of insuring continuous water conditioning by plants can be utilized. Our microcosm systems are connected to separate units that are reserved for the cultivation of a specialized community of algae. These units are lighted at night when the microcosms are in darkness, and they supply a constant flow of oxygenated, decontaminated water. This process (algal turf scrubbing) and the mechanisms built to support it are described fully in Chapter 12.

Light measurement is an important part of microcosm management. It is also one of the most difficult of physical factors to quantify. A number of instruments are available for sensing intensity, though they can be expensive. Examination of spectral characteristics, while not difficult, requires even more expensive instrumentation. However, using the guidelines given above, a photographic light meter based on the diagrams can provide approximate information.

References

Adey, W. 1978. Coral reef morphogenesis: A multidimensional model. *Science* **202:** 831–837.

Adey, W., and Hackney, J. 1985. Harvest production of coral reef algal turfs. In *The Biology, Ecology and Mariculture of Mithrax spinosissimus Utilizing Cultured Algal Turfs*. W. Adey (Ed.). Mariculture Institute. Washington, D.C.

Adey, W., and Steneck, R. 1985. Highly productive eastern Caribbean reefs: Synergistic effects of biological, chemical, physical and geological factors. In *The Ecology of Coral Reefs*. M. Reaka (Ed.). NOAA Symp. Ser. Underwater Research, Vol. 3, Washington, D.C.

Gates, D. M. 1980. *Biophysical Ecology*. Springer-Verlag, Berlin.

Griffith, P., Cubit, J., Adey, W., and Norris, J. 1987. Computer automated flow respirometry: Metabolism measurements on a Caribbean reef flat and in a microcosm. *Limnol. Oceanogr.* **32:** 442–451.

GTE/Sylvania. 1987. Color Is How You Light It. Catalog. 10pp. Sylvnaia Lighting Center. Danvers, Massachusetts.

Keeton, W. T., and Gould, J. L. 1986. *Biological Science*. 4th Ed. Norton Co., New York.

Kirk, J. T. O. 1983. *Light and Photosynthesis in Aquatic Ecosystems*. Cambridge University Press, Cambridge.

Lawlor, D. W. 1987. *Photosynthesis: Metabolism, Control and Physiology*. Longman Science and Technology, New York.

Moe, M. 1989. *The Marine Aquarium Reference*. Green Turtle Publications, Plantation, Florida.

Oliver, C., and Larson, B. 1990. *Forest Stand Dynamics*. McGraw-Hill, New York.

Rheinheimer, G. 1985. *Aquatic Microbiology*. John Wiley & Sons, New York.

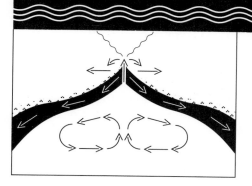

CHAPTER 8

SUBSTRATE
The Management of Rock, Mud, and Sand

The biosphere and each of its biomes and ecosystems are so complex that each scientist will try to narrow the range of his or her interests as much as possible. Most biological science is reductionist, treating a specific species or genus, the interaction of a single ecological factor, or the food chain or web by itself. The very act of simplifying may in one sense be necessary to achieve the concrete answer desired. On the other hand, because of the extensive interaction of factors in the real world, the relevance of experimentally achieved ecological answers to that real world is sometimes in doubt, often in major ways.

Likewise, aquarists traditionally have tended to ignore substrate and to reduce it to a noninteracting element. In the fish tanks of past decades, "clean" relatively inert gravel and undergravel filters have provided environments unlike all but the most specialized natural situations. Except for gravel bottoms in relatively unproductive "hard rock" mountain streams, or sandy beaches, with sandstone composition, rarely is the substrate chemically neutral. Also, in most wild aquatic and marine environments, soft substrates are rich organic reservoirs that harbor a myriad of important invertebrates and microbes and support a rich plant growth. Limestone substrates control water chemistry, and reef rocks with their myriad organisms, spaces, and peculiar chemistry determine the very character of the organisms growing on the surface of the reef. The interest in including so-called "live rock" in "coral reef" aquaria in recent years is the beginning of a tendency to replace tank sterilization approaches with a real ecology. Likewise, the more recent addition of "trickle trays" with calcium carbonate pebbles shows a developing interest in the carbonate cycle and pH control. Conversely, acid, black water streams are most likely in granite or sandstone areas where the natural acidity of the rain and tannic acid from the forest litter will not be neutralized. It behooves the aquarist to use hard rocks and silica sand in a black water system.

The geological world is less remote from the object of his intentions than the aquarist might think, and the lithosphere warrants a brief background discussion before proceeding with a more detailed discussion of substrates for microcosms, mesocosms, and aquaria.

The Solid Earth and Life

As we discussed earlier, the earth as viewed from space, with approximately 70% of its surface covered with water, would be most appropriately called "planet water." The size of the earth, its distance from the sun, and its moderate level of internal energy production have allowed it to retain a massive volume of water (in the liquid state), which more than any other single factor has led to the development of life. If the hard surface of the earth, the lithosphere, were smooth (i.e., nearly spherical), then the solid earth would be covered uniformly with approximately 2600 m (8000 feet) of ocean. While life would probably have developed and evolved under such a regime, without a more active inclusion of the lithosphere in the atmosphere–hydrosphere interaction, it would certainly be far less diverse. Probably it would still be very primitive, as it was several billion years ago.

Our planet still has considerable radioactively derived heat in its hot, liquid or semiliquid interior. The hard crust of the earth caps a slowly churning cauldron, and this has given rise to a rather remarkable evolution of the surficial crust that has been a major catalyst to the evolution of life and its ecology. The "boiling up" of this cauldron has gradually concentrated relatively light minerals particularly rich in silica, aluminum, and potassium in scattered "floating mounds" called continents, which are raised above the general heavier calcium-, magnesium-, and iron-rich crust of the earth (Fig. 1). The oceans occupying primarily the deeper parts of this two-level surface (Figs. 2 and 3) have a mean depth of 3730 m, while the continents are slightly raised to an average height of 870 m. Further, as the relatively new concept of plate tectonics (continental drift) has explained, the "boiling up" of the earth's cauldron is constantly splitting the crust, changing the shape of both continents and oceans (Fig. 4). When pieces of the crust driven by the cauldron collide, earthquakes, mountain chains, and sometimes very active volcanoes are created and ocean levels are changed.

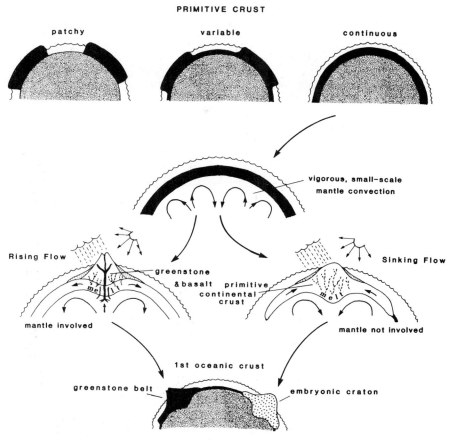

Figure 1 Hypothetical process of early formation of the earth's heavy oceanic crust and light, "floating" continents. After van Andel (1985).

Thus, there has been throughout much of the earth's history, and is today, a very active relationship between the earth's lithosphere, with its minerals and elements, and the hydrosphere, the home of life. This relationship is strengthened by the continuous formation of limestone and other sediments in which living or recently dead organisms play a crucial role. Sedimentary rocks or structures include those that are formed largely by the skeletons or even the organic matter of dead organisms. The earth's crust is a melange of rocks created by tectonic and volcanic activities and those that include or have resulted from the activities of living organisms.

Crucial additions to an already active rock, water, and atmosphere interface, changing mostly on the scale of many millions of years, are major alterations of overall earth climate on the scale of tens to hundreds of thousands of years. Partly due to slight cyclical changes in solar radiation and partly due to the change of the positions of continents and sea

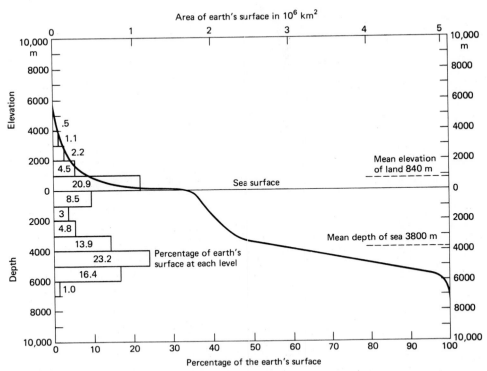

Figure 2 Hypsographic curve. The distribution of the amount of earth surface at different elevations. After Kennett (1982).

bottom relative to ocean currents, cooling of the poles results in large-scale glaciation. These "continental" glaciers produce both radical alterations in rock weathering (by ice) and sea-level changes. Even in mid latitudes, sea-level changes resulting from the locking of water in ice caps and the bending of the earth's crust as weight distribution of ice and water changes, result in marked alterations of both coastal areas and, as the gradient changes, the lakes and streams that drain into them. In addition, while some lakes and rivers can result from tectonic effects—for example, the African Rift Valleys (Fig. 5) and the Andean lakes—the majority of lakes and drainage in high latitudes results from either the scouring or the mounding-up of sediment by glaciers (Fig. 6). The Great Lakes in North America are an example of this process. Elsewhere, a variety of local factors such as volcanism, beach drift along the shore, rivers that change their course, and of course human activities result in lake and estuary formation. Stream formation results from the runoff of rain working against tectonism or major earth movements and the relative resistance of the underlying rocks to stream erosion.

Geological scale processes have done more than establish water/rock and air/rock boundaries and provide an attachment for organisms that

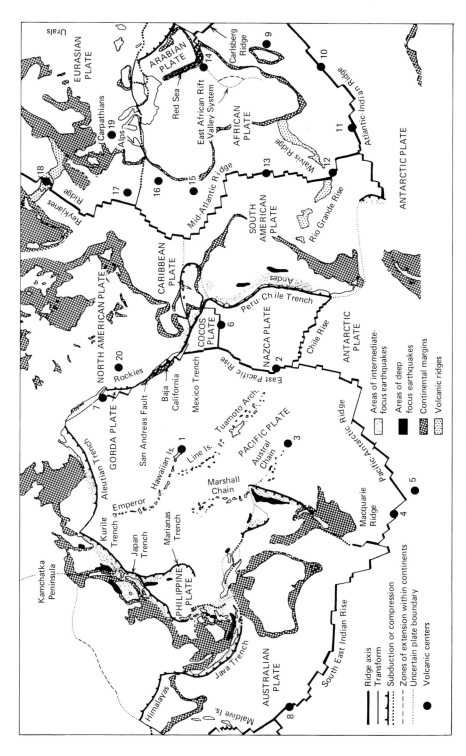

Figure 3 Worldwide plate system showing the placement of "raised" continents and oceanic crust with its deep ocean. Continental margins are shallow coastal areas, presently submerged but belonging to continents and are shown cross-hatched. The ocean-ridges form over hot up-flowing areas in the underlying mantle and are sites of new ocean crust formation. Ocean crust slides away from the ridges and sinks and melts in the subduction zones. Volcanic hot spots are mostly on ridges or subduction zones, though a few (like the island of Hawaii) are isolated beneath either ocean crust or continents. After Gross (1982).

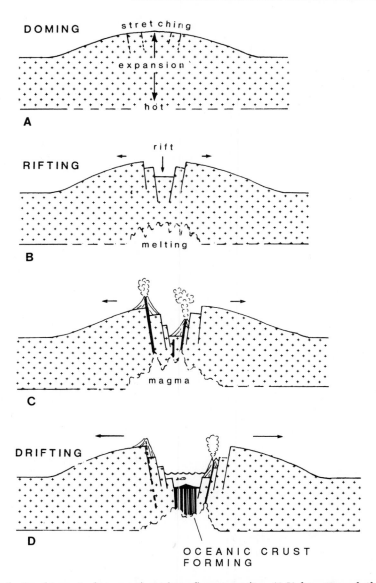

Figure 4 Development of oceans through seafloor spreading: (A,B) formation of rifts; (C) rift valley with lakes stage (see Fig. 5); (D) "Red Sea" stage; (E) young ocean stage. After van Andel (1985).

(Figure continues)

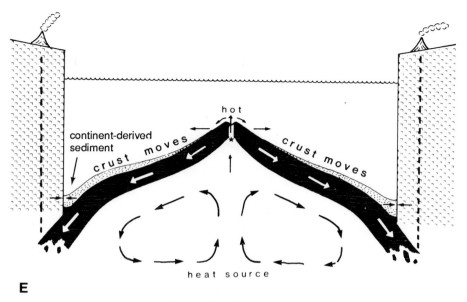

hot

continent-derived
sediment

crust moves

crust moves

heat source

E

Figure 4   (*Continued*)

benefit greatly from the movement of water past the rock. Rocks break down, largely due to a weathering process that includes water, organisms, and atmosphere, providing a continuous supply of essential chemical elements. However, the breakdown also produces rock and mineral fragments or sediments (and soils) that provide a different kind of home for organisms and their organic and inorganic productions. These sediments progressively move to lower levels due to gravity and assisted by water. Eventually they collect in basins of various sizes and elevations; in the process of becoming sedimentary rocks, the sediments sequester a very major part of the biomass that has moved and formed with them. All of these materials, including the organic matter, are likely to surface again, from geological storage, but that may be thousands to many millions of years later.

As we have briefly discussed, rock and water interaction has been in the past a major element in the development and evolution of life, and that relationship remains today at all scales. A major part of the biosphere is attached to or imbedded in mineral substrate, whether rock, mud or sand. Benthic life at the hydrosphere/lithosphere interface mostly operates with both high production and high biomass. In large part because of the remoteness of this connection, shallow water, mid-ocean ecosystems operate at very low productivities and biomass. Whether we refer to biogeological cycling or the ancient dictum "dust thou art and to dust thou shall return" in our modeling efforts, we must be aware of the

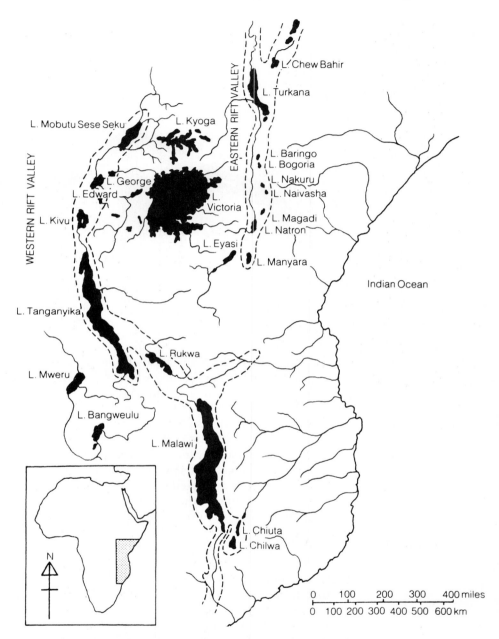

Figure 5 Formation of large continental lakes by rifting of the earth's crust. After Burgis and Morris (1987).

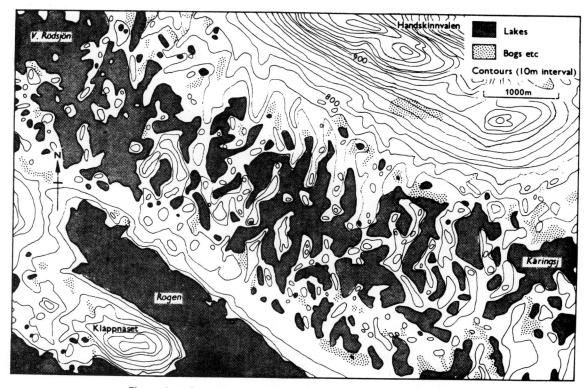

Figure 6 Lake and bog complex in Sweden formed by glaciers. After Davis (1983).

relationships. The connections of the biosphere to the atmosphere, hydrosphere, and lithosphere are essential and ecological systems have evolved attuned, not just to their existence, but also to the magnitude of raw materials in each segment and to the rate of exchanges between them.

The Solid Earth and the Model Ecosystem

In general, the problems of the physical handling of rock substrate in an aquarium or model environment are well known. Dick Mills (1986) discusses the preparation, placement, and management of various rock substrate for the traditional glass aquarium. The potential for damaging tank structure, whether glass, fiberglass, or even coated cement block, is real and should be carefully guarded against. Also, rock orientation and dimension relative to water volume can greatly modify water movement. We would strongly emphasize obtaining general knowledge of the minerologic type of rock that occurs in the wild analog, and particularly of the spatial role that rock plays in the community. Surfaces are extremely important to most biological and ecological processes. Rock substrate is often crucial in structuring a community in that it supplies a secure base for those organisms able to tightly attach. In addition, the spaces of cracks or voids, sometimes abundantly available, provide greatly increased surface for attachment and spaces for hiding. The aquarist should try to match the configuration or spatial heterogeneity of the wild analog as closely as possible. Of course, one does not add rocks (or gravel) that are iron-rich, lead-rich, arsenic-rich, etc., particularly in a small system. It is important to ask what is the rock substrate in contact with the ecosystem to be modeled in the wild: for example, carbonate rock for a reef, marine, or any hard-water system, or a silica-rich hard rock for a black or other soft-water ecosystem. Buffering and the carbon dioxide/pH cycle in relationship to carbonate use are discussed in depth in Chapter 10. However, the sediments, in themselves, are a more critical concern, and that is what we will concentrate on here.

Generally, whether dealing with a fresh- or a saltwater environment, there is a strong relationship between current and wave energy and the coarseness or fineness of the bottom sediments (Fig. 7). In microcosms, mesocosms, and aquaria, sediments should be used that are the same as in the systems being modeled, and in no cases that we can reasonably think of should an undergravel filter be used. Strong wave action or currents, whether on exposed lake or ocean coasts or in a mountain stream, give rise to exposed bedrock or large cobbles or boulders providing a bare (or usually algal-colonized) rock. In an aquarium or mesocosm environment, if these surfaces are to be preserved as in the wild, equivalent or at least sufficient wave or current action must be present. Otherwise the surface

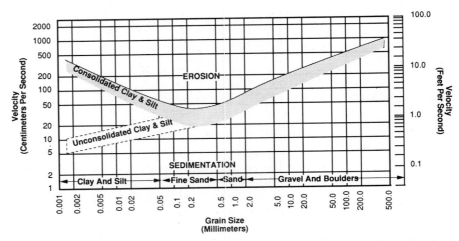

Figure 7 Relationship between the movement of sediment and the grain size of the sediment as a function of water velocity. After Davis (1983).

will become coated with fine sediment and will fail to recreate the wild ecosystem. There is a very major difference between a bare rock bottom to which organisms must attach (or bore as in softer rocks) and over which water must have considerable motion, and a sediment bottom.

In Chapters 22–25, we discuss a variety of ecosystems in mesocosms and aquaria with and without bare rock and coral rock substrate. It is best to discuss these on an individual basis, since the very layout of this substrate often provides the primary habitat for the ecosystems. This chapter mostly deals with finer substrate, and although there is considerable overlap with Chapters 15 and 19, the generalities will be discussed here.

Sediments and Model Ecosystems

As quieter water is approached whether in widened areas or billabongs (ox bow lakes) in a stream, in a lake small enough to prevent large waves, or in a bay or coastal lagoon along a sandy coast, the sediment becomes progressively finer from gravel, to sand and silt, to a soupy, silty-clay mud. Coarse sands or gravels are perhaps the most difficult benthic environments for organisms to adapt to, and within sand and gravel habitats there are relatively few species. To remain sand, the bottom must stay in motion and therefore special adaptations are required by any organisms that will inhabit such bottoms. A few larger animals, such as the *Donax* clam and the mole crab, have developed rapid burrowing techniques (Fig. 8). Otherwise, organisms must be small (less than 0.5 mm) and worm-like, so that the sand grains appear large, to rapidly burrow. These are the relatively poorly known meiofauna (Higgins and Thiel, 1988) (Fig. 9). On the other hand, even bacterial numbers tend to be limited in sand and gravel since their organic substrates are often "washed out" (Table 1). Also, it should be remembered that in model construction, sandy shores have a rather long profile in the energy regime required to keep them sandy. There is little use in trying to sandwich a sandy beach between a dune and a wave-broken sandy bottom within a few meters. It does not work in the wild, as many coastal landowners have found to their chagrin, and it does not work in an aquarium or mesocosm (Fig. 10) (see also Chapters 23 and 24). With difficulty, sandy beaches can be simulated in mesocosm and microcosms. In aquaria, it is extremely difficult, unless it is the only community included.

As we discuss in Chapter 13 in depth, the break between high-energy shores with their rock or mobile sand substrate and a quieter mud or sandy-mud bottom is ecologically great. We treat them as separate biomes, the highest community level differentiation. Generally algae occupy the highly disturbed but stable shores and higher plants (marsh plants and submerged aquatics) dominate the finer, less energetic shores. Even in a lake this differentiation is apparent, large lakes lacking reed beds on the

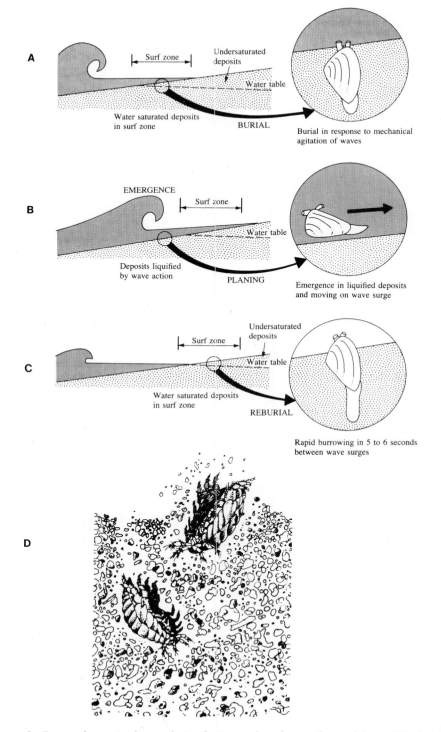

Figure 8 Two moderate-sized invertebrates from a sandy surf zone. The sand digger (*Neohaustorius*) (D) and the clam *Donax* (A–C) are among the few macrofauna to adapt to the sandy surf zone. After Thurman and Webber (1984). Copyright © 1984 by Scott, Foresman and Company. Reprinted by permission of HarperCollins Publishers.

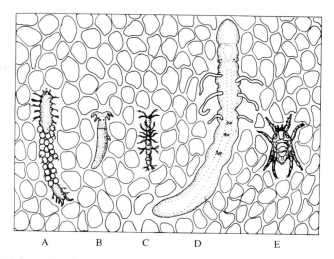

Figure 9 Meiofauna (small invertebrates, less than 1 mm and greater than 42 μm) in relation to fine sand grains. Three phyla are represented: (A) polychaete worm; (B) mollusk; (C) arthropod; (D) polychaete worm; (E) arthropod mite. After Thurman and Webber (1984). Copyright © 1984 by Scott, Foresman and Company. Reprinted by permission of HarperCollins Publishers.

shores (except in protected coves) and very small lakes often being continuously rimmed with emergent aquatic flowering plants. In the ocean, this relationship is also apparent. The outer shores, depending on sediment supply, range from rock with more or less abundant macroalgae to bare sand. The only plants from the high-water line seaward are algae. On the other hand, in protected bays, or behind reefs and barrier islands, mud bottoms prevail and the marsh communities and their flowering plants dominate the landscape.

Finer sediments, sandy-silt to silty clay mud, typically have a very rich fauna, usually richer than a rock or boulder surface, though probably not richer than a coral reef. On a typical rocky shore, there are dozens of

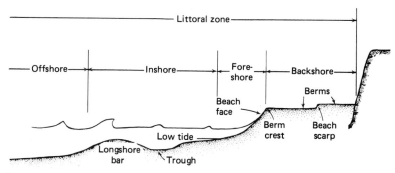

Figure 10 Generalized characteristics of sandy beaches and of wave action on those beaches. After Kennett (1982).

TABLE 1			
Numbers of Bacteria in Sediments of Differing Grain Size[a]			
	Grain size (μm)	Water content (%)	Bacteria ($\times 10^{-3}$ g^{-1})
Sand	50–1000	33	22
Silt	5–50	56	78
Clay	1–5	82	390
Colloidal sediment	<1	>98	1510

[a]After Rheinheimer (1985).

common species of algae, barnacles, snails, and small crustaceans in what appears to be an extremely rich rocky intertidal and subtidal flora and fauna. On the other hand, in the very uniform and often vacant-appearing muds on the bottom of adjacent bays, hundreds of species of worms, amphipods, and clams, to mention a few, are largely hidden beneath the surface. An examination of numbers of macroinvertebrates occurring in the soft sediments of 13 estuaries and bays scattered around the world yielded figures from 722 to 30,000 individuals per m^2, the mean number being 7400 individuals/m^2 (Maurer et al., 1978). Some of the life habitats of these organisms are shown in Fig. 11. It is to be noted that the communities of such bottoms can also change radically with time and location over several miles. Sometimes they can be quite patchy even on a local scale (Fig. 12). Basically the same kinds of organisms occupy muddy

A

Figure 11 Selected infauna from muddy marine bottoms. (A) left to right *Macoma* clam, cockle, polychaete worm, *Mya* (soft shell clam), polychaete worm and snails; (B) *Arenicola,* a polychaete worm—the food is organic particulates in the sediments, water moved through is used to receive oxygen; (C) *Chaetopterus* polychaete worm—a feeder on detritus particles suspended in the water column; (D) *Amphitrite,* a polychaete worm that feeds on surface deposits of organic particulates. After Thurman and Webber (1984). Copyright © 1984 by Scott, Foresman and Company. Reprinted by permission of HarperCollins Publishers.

(*Figure continues*)

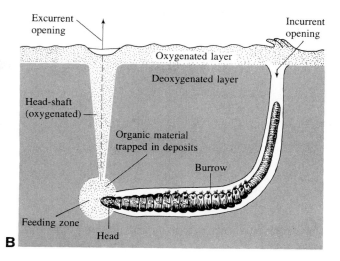

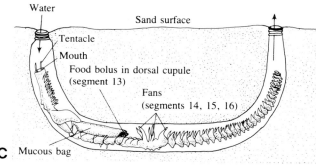

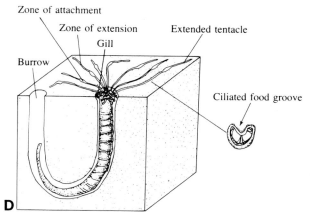

Figure 11   (Continued)

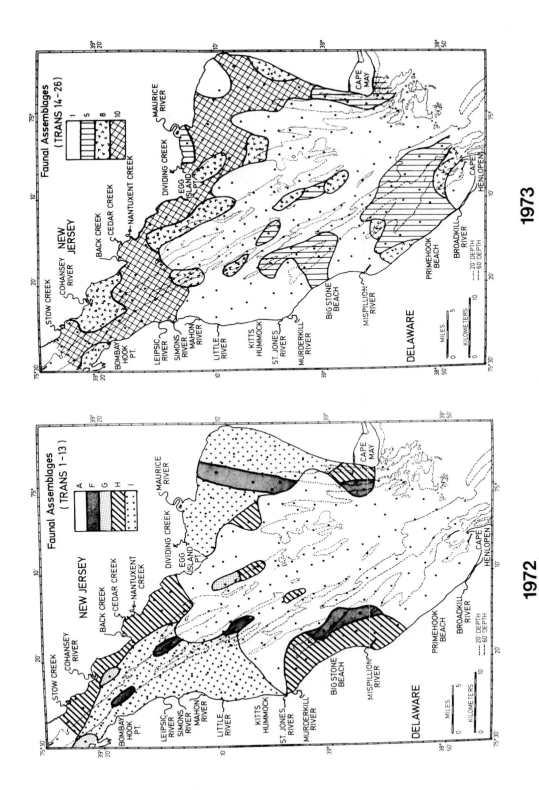

Figure 12 Results of the analysis of the soft bottom macroinvertebrates (greater than 1 μm) communities from Delaware Bay on 2 successive years (1972, 1973). Although most of the same species occur in the 2 years, none of the "community" groupings is obviously the same. This analysis includes 169 species: 40.8% annelid worms; 28.9% arthropods; 17.8% molluscs; 7.1% bryozoans and 5.4% miscellaneous phyla. By feeding types the species could be grouped as 45% deposit feeders; 24.8% suspension feeders; 18.3% carnivores; 10.7% omnivores; and 2.2% miscellaneous. After Maurer et al. (1978).

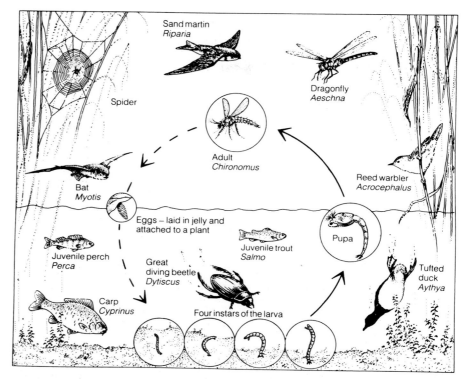

Figure 13 Life cycle and predators of mayflies. The flying stage is reproductive only and lasts for quite a short time. The larval stages of many species are burrowers in soft sediments of lakes, ponds and streams. After Burgis and Morris (1987).

bottoms in fresh waters. However, here the insects, almost absent from salt waters, particularly larval forms (Fig. 13), become extremely abundant, while the dominant worms are oligochaetes (earthworm relatives) rather than polychaetes.

Muddy bottoms are typically collection areas for large quantities of fine organic material from "producer ecosystems" such as coral reefs, rocky shores, or mud flats, or from terrestrial sources such as forests and fields. This abundant food source gives rise to the rich diversity of species and feeding types that typically occupy the mud or sandy mud bottom. In spite of the very active reworking of the mud by the infauna or by fish, crabs, and even diving birds seeking food, the abundant bacterial activity on the organic particulates provides for strong oxygen utilization. Anaerobic conditions typically exist close to the surface, and hydrogen sulfide production occurs within a few centimeters (Fig. 14). This sediment chemistry is very different from most soils in the terrestrial environment and provides a very strong control on the structure and function of the biological community.

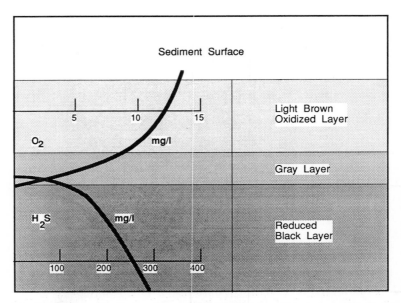

Figure 14 Distribution with depth of oxygen and hydrogen sulfide in a muddy bottom. This profile typically occurs over a few to at most a few tens of centimeters from the mud/water surface. After Levinton (1982). Reprinted by permission of Prentice Hall, Englewood Cliffs, New Jersey.

In this brief discussion we have concentrated on the animals of benthic soft bottoms. Although very much limited in area of occurrence, many plants, particularly higher plants, occupy muddy and sandy-mud bottoms in the shallowest, usually subtidal, zones. Although relatively few species of flowering plant genera and species occur in the sea (Fig. 15), these can occupy very large areas of shallow, well-lighted bottoms. And although difficult for model systems, except for the largest mesocosms, in the tropics, mangroves form very extensive swamps in muddy intertidal zones (see Chapter 24). In fresh waters, on the other hand (see Chapter 15), large numbers of submerged aquatic flowering plants tend to dominate over the algae on shallow sediment bottoms (Table 2).

Fine sediment bottoms in the wild are typically very active, and rich communities that are important components of their ecosystems and should also be equivalently important elements in mesocosms and aquaria. If enough light is present, algae and flowering plants (particularly in fresh water) are able to photosynthesize and to directly provide new production to grazers or to detritus in the community. Algae derive all of their nutrient supply directly from the water, as in any environment. Many of the higher plants, however, extract their nutrient needs from the sediments (as on land). This provides an important link between buried organic materials and their nutrients and the overlying water. Equally important, the algae, and, more important, the flowering plants act to trap

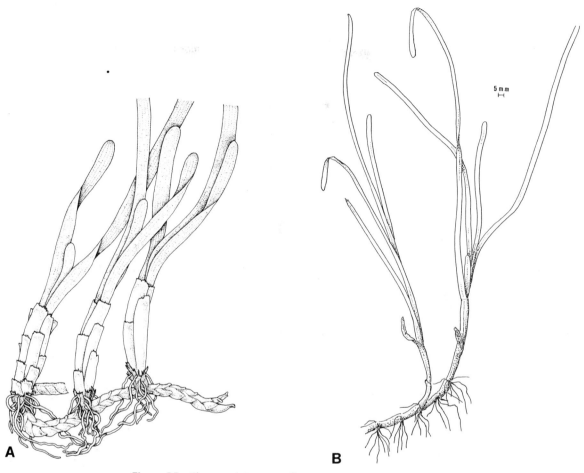

Figure 15 Characteristic marine flowering plants from muddy sand and muddy bottoms: (A) the tropical *Thalassia testudinum* (turtle grass), a dominant of many reef lagoons; (B) the cold-water Northern Hemisphere *Zostera marina* (eel grass). Note that neither species is a true grass. After Dawes (1981). Reprinted by permission of John Wiley & Sons, Inc.

and thereby increase sedimentation. Marshes are among the best examples of land-building by sediment trapping. Submerged "grass" beds also are important in this respect.

Most crucial, whether shallow or deeper, fine sediment bottoms are generally receivers of abundant organic detritus—plant fragments, dead animals and animal parts, and animal feces (Fig. 16). Much of this "waste" organic material tends to be infractile to higher animals. On the other hand, it is easily utilized and fully broken down to carbon dioxide, water nutrients, and minerals by a host of bacteria. Most of the bacteria are, in

TABLE 2

Plants Dominating Sediment Bottoms in Temperate Lakes and Their Patterns of Change with Time (Succession)[a]

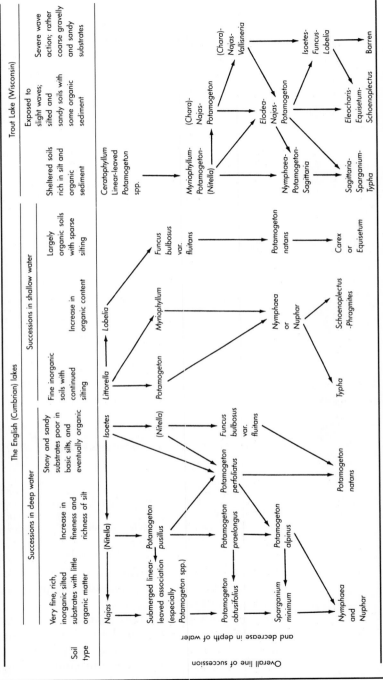

[a] In all the above successions, only the dominant(s) of each community is indicated. Algal dominants are enclosed in brackets. After Sculthorpe (1985).

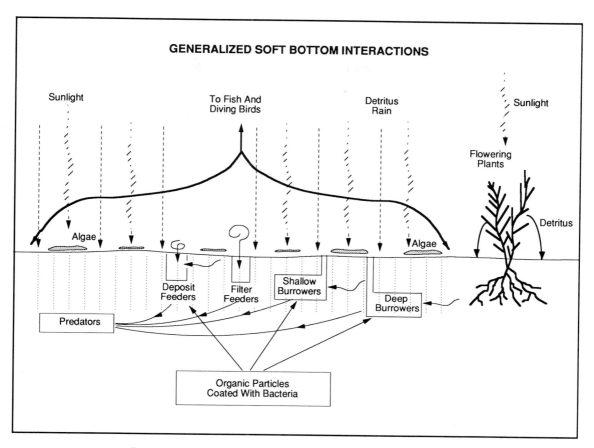

Figure 16 Generalized feeding patterns in a typical shallow-water, soft-bottom community.

turn, utilizable as food by many inhabitants of the sediment bottom, particularly clams, worms, and insect larvae. Even though the sediment tends to be anaerobic at or immediately below the surface, many organisms are able to work the rich organic load by drawing in oxygenated water from above the mud. In the wild, rarely is all of the organic storage in the sediments recycled in the present time frame. Some of it, when geological conditions are right, becomes the oil, coal, and organic-rich shales of the future.

Traditionally, aquarium practice has been to avoid the natural detrital processes and to keep bacteria in filters that to some extent act like a very reduced bottom community. However, in a filter the variety and capability of the bacteria are limited. Few or no animals are present to eat the bacteria and, in turn, the fish do not have a rich, invertebrate bottom in which to browse. Thus, aquarium procedures in the past have tended to short-circuit the natural cycling processes. This results in loss of valuable energy to the many larger members of the community. Microcosms, mesocosms, and aquaria that by design do not have a fine sediment community should have a separate sediment trap that periodically can be partially drained of sediment. Particularly if it is desired to drive a system faster than normal for scaling reasons, or if import and export are desired because of the size and coverage of the model ecosystem, organic sediments can be used as an export tool.

In the wild, nutrients are exchanged between sediments and water column in important ways. Detritus to bacteria to worms to fish is an important pathway to recycle nitrogen and phosphorus. Another pathway is through the root hairs, rhizomes, and up into the leaves of higher plants, to be eaten by fish or snails or become more detritus. Some nitrogen in the anaerobic sediments is denitrified to a gaseous form and lost to the atmosphere. Phosphorus, as long as the sediment surface is aerobic, tends to remain locked in the sediments, since iron oxides under these conditions link up with and trap phosphorus. When the normally subsurface zero-oxygen levels extend into the water column, stored phosphorus tends to be released. Sulfur also is utilized by bacteria and becomes the very odiferous hydrogen sulfide that one associates with anaerobic conditions. In nutrient-poor lakes, phosphorus tends to become the limiting nutrient because it is locked in sediments by high oxygen levels. In richer lakes and estuaries with large amounts of sediment in contact with the waters, phosphorus tends to be released in abundance and is responsible for algal blooms if nitrogen is available. In the open ocean where fine organic sediments are deep and to a large extent out of reach of the shallow water column, nitrogen and phosphorus tend to be closer in importance with the final limitation usually going to nitrogen.

For the aquarist, fine sediment bottoms should not be ignored. Given their full reign, with the proper environment and biota available, they can be important buffering communities that will provide great stability to an ecosystem simulation.

Geological Storage

Bioturbation (or bottom disturbance) by animals and the rooting and shooting activities of flowering plants in shallow water are processes that continuously return the energy and chemical elements of organic sediment to the water column. However, in many cases sedimentation is rapid enough to bury organic materials out of reach of living processes for geological time. This could be thousands of years for lakes or freshwater environments or millions of years for marine situations. On the earth, a significant part of the plant primary production that has occurred in the past 500 million years has been stored as coal, oil, gas, and oil shales.

Limestone production likewise is indirectly related to the photosynthetic process. As we shall discuss further in Chapter 9, virtually all carbon on the earth's surface and in its crust has been cycled through organisms, yet it has been estimated that greater than 1600 times as much organic carbon is buried in the earth's crust as exists in the biosphere proper. Most of this buried carbon was derived from the process of photosynthesis and was removed, as carbon dioxide, from the atmosphere. The process also resulted in the evolution of oxygen into the atmosphere. This fact, along with the high oxygen and preindustrial low carbon in the atmosphere, is the primary argument for considerable net positive primary production over respiration.

Most human management of the earth's organic resources, including aquarium management, is philosophically based on rapid recycling and readily available raw materials (nutrients). In more recent history of the earth's biosphere, organic storage and great limitation of available nutrients has been the rule. This is undoubtedly a major factor providing long-term stability to our biosphere. Likewise in modern aquarium management, it behooves us to keep our nutrients locked in biomass, either active or stored. In general, depending on the system being modeled, plant production should dominate over microbe breakdown.

In addition to algal scrubbing (or other plant removal) to simulate model communities characterized by organic sediment burial, a storing or exporting sediment trap is necessary. There are many ways to do this, and several units are described in Chapters 21–25. Sometimes, when export, in the time frame of the model planned, is not desired, and bottom disturbance is not excessive, the basin of the tank itself becomes the sedimentation trap and organic storage facility. Separate settling traps, within the model plumbing, and with tap-off values, can also be used as refugia (see Chapter 13).

References

Burgis, M., and Morris, P. 1987. *The Natural History of Lakes.* Cambridge University Press, Cambridge.

Davis, R. 1983. *Depositional Systems, A Genetic Approach to Sedimentary Geology.* Prentice Hall, Englewood Cliffs, New Jersey.

Dawes, C. 1981. *Marine Botany.* John Wiley & Sons, New York.

Gross, M. Gran. 1982. *Oceanography: A View of the Earth.* Prentice Hall, Englewood Cliffs, New Jersey.

Higgins, R., and Thiel, H. 1988. *Introduction to the Study of the Meiofauna.* Smithsonian Institution Press, Washington, D.C.

Kennett, J. 1982. *Marine Geology.* Prentice Hall, Englewood Cliffs, New Jersey.

Levinton, J. 1982. *Marine Ecology.* Prentice Hall, Englewood Cliffs, New Jersey.

Maurer, D., Watling, L., Kinner, P., Leethem, W., and Wethe, C. 1978. Benthic invertebrate assemblages of Delaware Bay. *Marine Biol.* **45:** 65–78.

Mills, D. 1986. *You and Your Aquarium.* Alfred Knopf, New York.

Rheinheimer, G. 1985. *Aquatic Microbiology*, 3rd Ed. John Wiley, New York.

Sculthorpe, C. 1985. *The Biology of Vascular Plants*. 1985 reprint of 1967 ed. Koeltz Scientific Books, Konigstein.

Thurman, H., and Webber, H. 1984. *Marine Biology*. Merrill Publishing, Columbus, Ohio.

van Andel, T. 1985. *New Views of an Old Planet*. Cambridge University Press, Cambridge.

PART II

Biochemical Environment

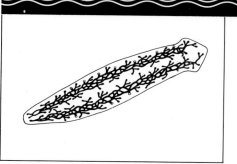

CHAPTER 9

METABOLISM
Respiration, Photosynthesis, and Biological Loading

This chapter is intended to provide a basic understanding of essential biogeochemical processes and cycling as they would apply to the modeling of ecosystems. The practical management of biological metabolism in synthetic systems, microcosms, mesocosms, and aquaria will be discussed in Chapters 10–12.

When one is maintaining a human being or almost any terrestrial mammal or bird, the "hotel" requirements seem relatively simple, at least on the surface. Unfortunately, it is this apparent simplicity that has placed the human race in its present increasingly difficult environmental situation. As long as there were relatively few of us, and our tools and access to the earth's great storehouse of fossil energy were limited, we took for granted many essentials that Mother Nature provided. No longer can we assume that good-quality water and atmosphere will automatically be available, and that the products of our daily activities can simply be discarded for "processing" by nature. With our increasing numbers and intense energy use, we are massively changing our planet. Unless we can quickly learn how to carry out fusion, as a boundless and pollution-free energy source, and how to manufacture good water and atmosphere on a global scale, we will have to reduce our numbers and learn to live with Mother Nature.

The problem of maintaining aquaria, microcosms, and mesocosms for the hobbyist and scientist is similar on a small scale to the global environmental problem for the human race. We can operate an aquaculture for one or a few species, and assuming nature can provide us with a good supply of water we need to worry only about the temperature, salinity, light, space, and food needs of those few species. But what about good water quality? That is hard to get from most taps and many shores these days, and even if the quality is reasonably high, can we afford to treat the tap like an endless stream? Many professional production aquaculturists are increasingly finding themselves in great difficulties because of the pollution of natural waters that their intense and massive culture is producing. In addition, in the model ecosystems we not only want organisms to grow fast, as if we were going to eat them, like a herd of cows in a feed lot or chickens in a coop, but also we want them to behave normally, to feel good. This is the context in which we have passed more than 99% of our evolution, one in which thousands of species around us behaved "normally" in a mutual environment or biosphere.

Water quality is even more important to an aquatic organism than it is to us. The entire bodies of underwater animals are immersed, including the gills and, effectively, their internal vascular or blood transfer systems. Aquatic and marine ecosystems functioned quite well for hundreds of millions of years before we humans came along. That leads us to ask: "Why not learn from Mother Nature?" Why not operate an ecosystem as it is (or was) in the wild instead of trying to operate it "our way," a method that has clearly not worked on the global scale, and, as many aquarists would attest, often does not work on the aquarium scale.

So far in this book, we have discussed physical and environmental factors, factors that to a large extent modern man immediately understands as part of life in an industrial society. Now we approach life in a biochemical context. We also come to a most appropriate question that should precede our attempts at recreating a living ecosystem based on hundreds of living units. That question, which is rarely tackled by life scientists, is: "What is life?" A typical dictionary definition would be "a quality which distinguishes a plant or animal from the inanimate such as rocks, earth, or water," sometimes followed by "especially characterized by reproduction and growth by accreting materials from the surroundings." To say that something alive is not dead is rather circular reasoning, however, and in any case, many mineral crystals not only reproduce and grow; they often "look alive," at least in the plant sense. Life, of course, is rich in complex molecules of carbon, hydrogen, and oxygen, but that is also true of the plastics that we now routinely manufacture.

The U.S. National Aeronautics and Space Administration also had this problem when they proposed to find out, as their space probes and landers touched down elsewhere in the solar system, whether or not life was present. Dr. James Lovelock, the author of *Gaia* (1979) and *Ages of Gaia* (1988), proposed an answer that is both workable and instructive relative to how we approach both life on earth and our mesocosms.

In a physical context, the science of thermodynamics tells us that energy is always moving to a lower level of intensity or organization. The disorganization of this energy in the form of the bonding and motion of molecules and atoms (entropy) is always increasing. Thus, in Earth's time frame, the universe has been running down like a battery, as matter flies out from the big bang. Some billions of years from now this almost inconceivable collection of matter and energy is heading for death close to absolute zero. Perhaps when it has converted all its energy to the gravitational form, it then collapses to another big bang and a rebirth. Our concern here, however, is the running down process. On the time frame of hours to years rather than a few billion years, life temporarily reverses this apparent senescence. It is capable of collecting small amounts of mostly solar energy to chemically organize and to store that energy and direct it to rather intensive usage. On the scale of the individual, life can literally store and concentrate solar energy, eventually to defy gravity. On the scale of the community, it can store and concentrate energy to be released as heat a few or millions of years later; on the scale of the biosphere, life has been and still is capable of massive alterations of our planet's surface. The oxygen-rich atmosphere that allows a rapid scale of animal life and has in part given us equitable temperatures for several billion years, the very soil that supports so much terrestrial biomass, the organic-rich sediments that become an integral part of earth geology (perhaps as some would have it even allowing continental drift and the essential features of the earth's geology) are all part of the accomplishments of life. These are the unique features of a living earth. They are contrary to what the physical evolution of the solar system would have offered.

The ability to reduce entropy and produce highly unlikely levels of organization might be the modern physicist's answer to the nature of the basic entity that we call life. The biologist's answer is different, but just as important (see Mayr, 1988). Life is uniquely characterized by information—the information encoded in the genes. Several billion years ago life developed the process of respiration. That information has been passed through countless generations of organisms. Every cell in our body has the age-old code for respiration. Many millions of years ago our distant relatives added another level of respiratory complication. Oxygen and carbon dioxide are exchanged in the lungs to the bloodstream, which in turn meets each cell to provide the raw materials for respiration. Every cell in our bodies also "knows" this. Whether a cell acts to "pull its weight" at this level depends on whether or not it is part of this higher-level respiratory pathway.

This may seem like rather heady stuff to the aquarium hobbyist who simply wishes a pleasant-looking aquarium or even the ecologist who wishes to model a small marsh. It is not so distant at all, as we shall see. Life has what we can call a self-organization capability that occurs at many levels. While we humans may well be very bright as animals go, we still do not understand much about how ecosystems work. Yet most of these complex systems preceded us by many millions of years, functioning perfectly well without the benefit of our brains and powers of organization. Throughout this book we suggest that since we know only vaguely how to internally organize and operate ecosystems in a long-term stable fashion, or even in the short term, we should let those ecosystems be "free" to do what they are more than capable of doing. They "know" because they consist of organisms with information that dates back millions of years, encoded in their genes. As many naturalists and environmentalists have suggested over the past centuries, we should set aside our arrogance, our desire to conquer and control everything, and walk hand in hand with Mother Nature. Thus, broadly, but in the context of this book, the modeler or the aquarist should supply the right environment as closely as possible, supply the right genetic material, then sit back and watch or begin the experiment as one's endeavor directs. It is possible, often even necessary, because of the small scale and location of a given mesocosm or aquarium, that the human needs to participate to supply energy, to fill the role of missing larger organisms, or to offset patchiness. However, given the right outside parameters (light, temperature, salinity, nutrients, etc.), the "captured" ecosystem will generally take care of itself. This may be hard for humans who wish to feel that they are totally in control.

With this rather lengthy introduction as a statement of philosophy of approach, let us move on to the nitty-gritty of the chapter. A basic understanding of metabolism in living organisms and how it relates to the chemistry of the environment is required to develop a husbandry capability for captured ecosystems. We will attempt to present the essentials without becoming overly lengthy or complex. The references will provide the interested reader with a more in-depth treatment of the subject.

Metabolism

Life in the whale, the tree, the alga, or the protozoan survives by the same very basic process, the chemical "burning" or oxidizing of organic matter (food) at the cellular level. While it is possible to obtain some energy without oxygen (by fermentation, for example), cellular respiration using oxygen is the most efficient pathway. The food, a carbon, hydrogen, and oxygen complex, sometimes with nitrogen, phosphorus, and sulfur and minute quantities of other elements, can be produced internally or can be taken from another organism. The energy stored in the food is directed to the production of special chemicals that are transferred to specific sites in the cell for release as heat or motion or to the building of structural materials for reproduction or growth.

Generally, plants are producers. Whether algae or vascular plants, through photosynthesis and acquisition of solar energy plants build sugars, starches, and oils, thereby storing the captured energy for building the even more complex structural and chemical compounds needed such as cellulose and proteins. The needs of plants are great: water, carbon dioxide, nitrogen, phosphorus, and many micronutrients. On the average, growing plants are also oxygen producers. Photosynthesis transfers the energy of sunlight to hydrogen obtained by the splitting of water and the release of the oxygen (see Chapter 6). Thus, in an ecological sense, given light, plants are purifiers; they are constantly removing nutrients from their surroundings and adding oxygen.

Animals, on the other hand, eat plants or other animals. When they are no longer growing they return all intake materials in reduced form to their environment. Even while growing, efficiency is low and therefore a large part of ingested materials is also released to the surroundings. As long as the delivery rate of animal wastes to the local environment is no faster than the uptake of the plants that require those "wastes," then a balance is present.

An idealized model of animal versus plant requirements is shown in Figure 1A. If excess plant production occurs, and direct-feeding animals as well as detritus feeders are not available to consume all of plant production, assuming plenty of water, light, and carbon dioxide from the atmosphere, and some nutrient input, then plant material accumulates. Given the right conditions, as in a swamp, this excess organic material could go into geological storage for tens of thousands to millions of years. Under wild conditions, excess animals occur under only unusual conditions that do not last long, usually because the food supply runs out or predators arrive to enjoy the excess. In a culture or man-operated environment, as in a human city, where plants do not or cannot balance animals, and food is artificially introduced, something entirely different happens. Here waste products accumulate (carbon dioxide, as well as nitrogenous, phosphorus, and sulfur-rich compounds). Bacteria happily use the excess waste products. However, bacteria, which in most cases metabolically act like miniature animals, when in excess can also radically alter environmental chemistry. Also, when atmospheric access is slow, in water and muds, for example, oxygen can be used up, creating an inefficient situation that few higher plants and animals can tolerate. This metabolism; out of environmental context or organic pollution, is the primary subject of this chapter.

For those wishing only a superficial understanding of metabolism as it relates to aquaria, the previous discussion may be adequate. The essentials for understanding Chapters 9–11 are in Figure 1. On the other hand, we feel

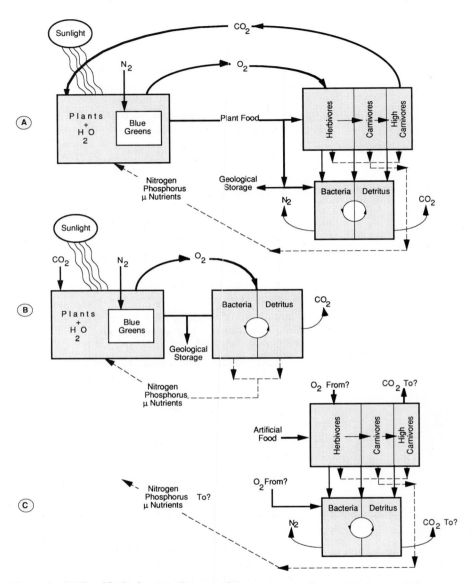

Figure 1 (A) Simplified schematic diagram of the movement of essential compounds by organisms in an idealized closed ecosystem. (B) "Unbalanced" community of plants in which excess plant production is being stored (in sediments, for example). (C) "Unbalanced" community of animals in which food is provided.

that a more in-depth appreciation of the mechanisms of metabolism would be helpful to understanding the problems of synthetic ecology. Since the basics of photosynthesis were treated in Chapter 6, we will begin with the animal side of the cycle, particularly digestion, respiration, and excretion. It is from within these processes that the essential environmental effects are expressed. Following a discussion of bacterial metabolism, we will briefly return to photosynthesis to more fully treat the synthesis of organic compounds. Although the specialized terminology will be kept minimal, a concise biological dictionary such as Curtis, 1985, may help.

Respiration

Animals come in a wide variety of sizes and complexities. They range from unicellular protozoa (now typically placed in a separate kingdom, the Protista), a fraction of a millimeter long, to elephants and whales where tons of complex tissues and multimillions of cells work together. The basic pattern, based on the cell, is the same in all cases. In the more complex animals, individual cells become highly organized parts of tissues and organs specialized in one or a few of the many functions of cells. A generalized animal cell is shown in Figure 2. The reader is referred to a good modern text on biology (e.g., Keeton and Gould, 1986) for a discussion of all cell components. Here we will concentrate on the cytoplasm and the mitochrondria, since it is in these locales that the basic respiration and glycolysis processes take place.

When an animal eats a plant or another animal, a complex digestive process begins. The energy scrubbing is that of the oxidization or "burning" of sugars. Simplistically this is $C_6H_{12}O_6 + 6O_2 \rightarrow 6CO_2 + 6H_2O + 670$ kcal/mole of energy. One glucose plus six oxygen modules gives six carbon dioxide plus six water molecules plus energy. The cell cannot simply burn the sugars. Instead, it uses a small-scale chemical transfer process. Respiration produces adenosine triphosphate (ATP), a nitrogen-rich nucleic acid plus a sugar and attached phosphorus and oxygen. ATP carries away small packets of energy (about 1/100 that of an entire glucose molecule) and delivers that energy throughout the cell for many purposes. Figure 3 shows a very condensed version of the entire process of breakdown of a simple sugar and conversion of its energy to ATP. Carbon dioxide and water are the byproducts. One critical feature that can be seen in this diagram is that some ATP can be produced without oxygen. The amount is small, but this primitive process (glycolysis or fermentation), probably the dominant one in the early eon of life on earth, allows some organisms and some tissues within organisms to live in an anaerobic or partially anaerobic environment.

When an animal feeds on a plant or another animal, it obtains for digestion many more complex organic chemicals than sugars. Proteins

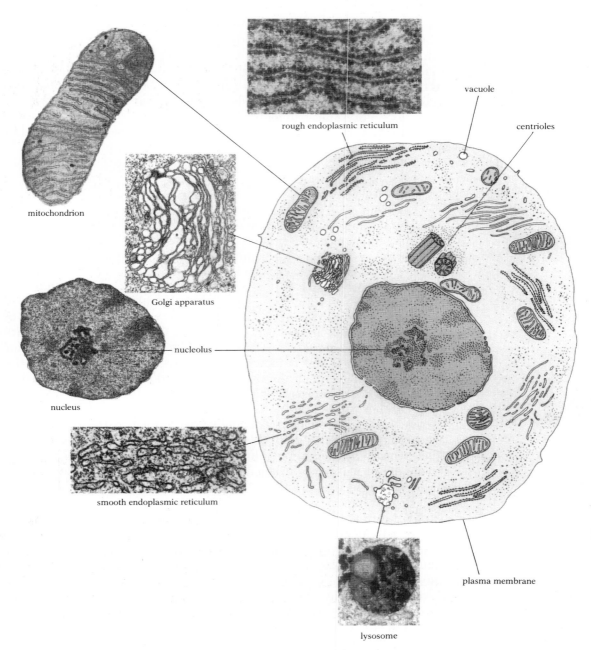

mitochondrion

rough endoplasmic reticulum

vacuole

centrioles

Golgi apparatus

nucleolus

nucleus

smooth endoplasmic reticulum

plasma membrane

lysosome

Figure 2   Idealized animal cell with its multiplicity of internal organelles. After Keeton and Gould (1986).

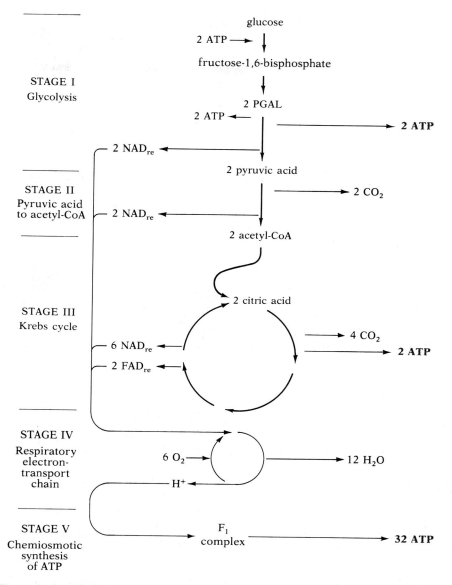

Figure 3 Simplified diagram showing the process of the complete breakdown of glucose, in respiration, to achieve 36 ATP for energy transfer elsewhere in a cell. After Keeton and Gould (1986).

and fats are essential to the animal in many ways, providing vitamins, minerals, and other important compounds. However, much of the protein and fats is also simply digested and respired to produce energy, much as sugars (Fig. 4). Herein lies a crucial element of the effects of animals on their environment: the production of nitrogenous wastes and, to a lesser extent, phosphorus.

In the simplest of animals, protozoans, digestion and excretion can be seen in its basic form (Fig. 5). A small plant or animal is captured by being engulfed and enclosed in a vacuole (phagocytosis). The golgi apparatus through the lysosomes provide digestive enzymes to the food-filled vacuole. After the food is broken down, undigested particles (a diatom silica wall, for example), along with ammonia and phosphoric acid, are transported to the wall and simply excreted to the exterior of the cell by exocytosis. Sugars, amino acids, and nutrients needed are taken into the cytoplasm through the vacuolar membrane.

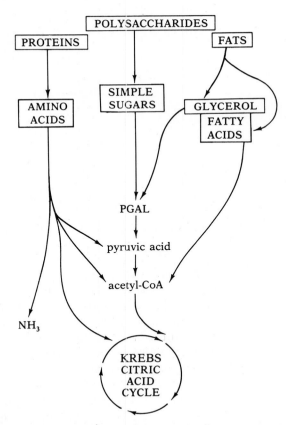

Figure 4 Generalized process of breakdown of proteins and fats to achieve introduction of these compounds into the respiration process. After Keeton and Gould (1986).

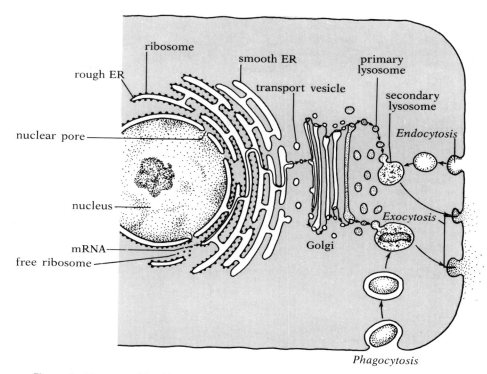

Figure 5 Digestion of food by idealized protozoan cell. After Keeton and Gould (1986).

Excretion

Early in evolutionary history another process entered into the excretory picture. In any waters less than full ocean salinity, outside fresher water continuously moves into the cell by diffusion. The resulting cell dilution is called osmosis, and it often results in a pressurized internal environment. To avoid dilution of the cell to the point of death, some vacuoles called contractile function to collect this water, eventually expelling it to the exterior. A more advanced form of this process, including a specialized cell, the flame cell, is seen in the simple planarian (Fig. 6).

In more complex organisms, water control and excretion have tended to be joined together. In a sense, the unit of the mammalian kidney, the nephron (Fig. 7), is an advanced flame cell. Since in land animals water loss can become crucial, the nephron and the kidney came to reabsorb much water, leaving more concentrated nitrogen compounds as urea. In fish, on the other hand, the primary excretory product remains ammonia and, particularly in fresh water, large quantities of water and ammonia are excreted through the kidneys (Bond, 1979). The kidneys are by no means the only site of excretory activity. The liver, in addition to being a

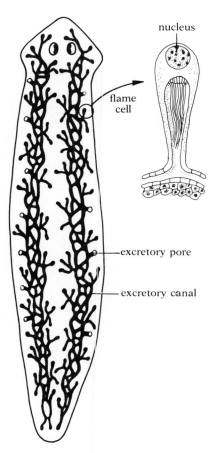

Figure 6   Planaria flame cell. A cell specialized for water regulation—a primitive kidney. After Keeton and Gould (1986).

digestive organ, excretes waste compounds. Also, the digestive system itself excretes, along with the nondigested wastes of the feces, nonabsorbed nitrogen and phosphorus-rich compounds that have been broken down both by the animal itself and by bacteria in the gut. In animals with gills, including fish, much ammonia is excreted through the gills, along with carbon dioxide.

Bacterial Metabolism

Bacteria are neither plants nor animals and today are placed in a separate kingdom called the Monera. Some bacteria (including the very important cyanobacteria or blue-green algae) are capable of photosynthesis, thereby

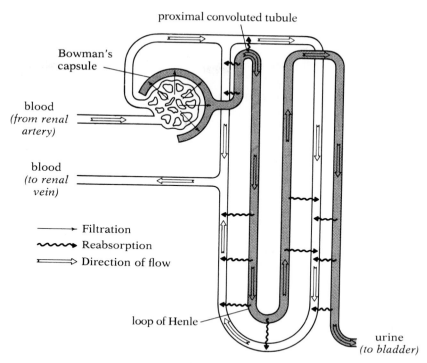

Figure 7 Diagram of the principal cellular element of the human kidney. After Keeton and Gould (1986).

acting like plants. Most bacteria act like animals and break down dead organisms or are predators acting as parasites in living organisms. Bacteria are very simple cells, filaments, and colonies. They lack a nucleus and other cellular organelles enabling them to carry out the complex cellular digestive/excretion process we discussed above. Bacteria excrete their digestive enzymes into their surroundings and absorb through their cell membranes the simple sugars and reduced nitrogenous and phosphorous compounds that they require. Nitrogenous and other wastes develop at least partly external to the cell. In a sense, when bacteria are abundant, the external environment locally becomes like that of the stomach or small intestine. We will discuss this in depth later, but excess dead organic material free in a relatively closed environment becomes like a stomach: acid, rich in carbon dioxide, methane, and ammonia, devoid or nearly devoid of oxygen, and having many enzymes capable of organic breakdown. This is a specialized environment, which in the open world could be the subsurface of an organic-rich mud flat; however, it is not one that many higher organisms can tolerate. It is also an unsatisfactory environment for the open water of most aquaria.

And now we briefly return to Figure 1 and the basic ecological point of this chapter. Animals and most bacteria require complex organic foods, plants and other animals, dead or alive. The principle requirement is simple sugars for energy, and while a small amount of vitamins, amino acids from protein, and many other micronutrients are required, only a small part of the nitrogen, phosphorus, and sulfur in complex compounds is needed. These become organic wastes, some of which are poisonous *in quantity* to many organisms. On the other hand, these wastes are required by living, photosynthesizing, and growing plants to build their tissues. In a balanced system and in Earth ecosystems as a whole organic wastes do not accumulate; they are used as fast as they are produced. In the more-or-less restricted localities where wastes do accumulate, they create a special environment, in which a few specialized organisms are capable of functioning. Gaia has means for storing the components of excess organic production—organic rich sediments and eventually coal and oil, for example—and the storage equivalent can be accomplished in a mesocosm, and, with a little more difficulty, in the aquarium.

Photosynthesis

In Chapter 6, while discussing the biological role of light in both ecosystems and aquaria, we briefly treated the process of plant photosynthesis. There we emphasized the light reactions. Here we complete discussion of the photosynthetic process to provide preparation for the later "practical" chapters.

A typical plant cell is shown in Figure 8. We will concentrate on the photosynthetic process as it takes place in the chloroplasts (Fig. 9). Following the splitting of water in the chloroplast thylakoid, using light energy, the energy is delivered by electron transport to the Calvin (PCR) cycle. This process, which takes place within the chloroplast, provides energy-rich sugars and starch for storage in the cell as starch or for immediate metabolism or transfer to other cells (Fig. 10).

Much as in animal cells, the energy-rich sugars produced by plant cells are respired as needed to produce ATP. The ATP-stored energy derived from photosynthesis is then transferred throughout the cell and used to build walls (cellulose), nucleic acids, proteins, phospholipids on cell membranes, and more chlorophyll, to name a few. All of these are needed as cells grow and divide. These building processes result in a need for nitrogen, phosphorus, sulfur, and other micronutrients, which are primarily taken up as dissolved salts (phosphate, nitrate, sulfate, etc.). Ammonia at least in small concentrations is a preferred source of nitrogen. Often, especially in algae and aquatic plants, higher forms of the most essential requirements, such as urea, can also be taken up. In the aquatic environment removal of these compounds by plants can be crucial, as, in

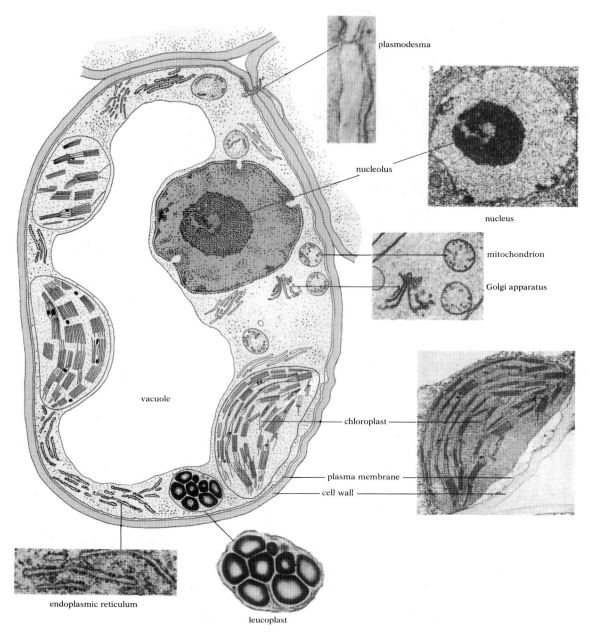

plasmodesma

nucleolus

nucleus

mitochondrion

Golgi apparatus

chloroplast

plasma membrane

cell wall

vacuole

endoplasmic reticulum

leucoplast

Figure 8 Idealized plant cell showing essential similarities and differences with an animal cell (see Fig. 2). After Keeton and Gould (1986).

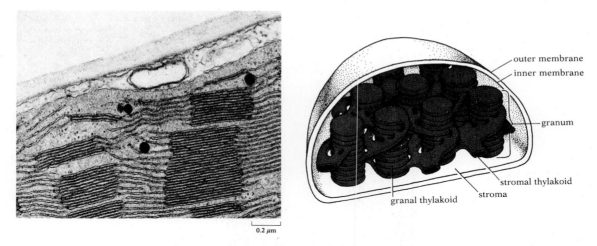

0.2 µm

Figure 9 Electron micrograph and diagrammatic view of a chloroplast structure. After Keeton and Gould (1986).

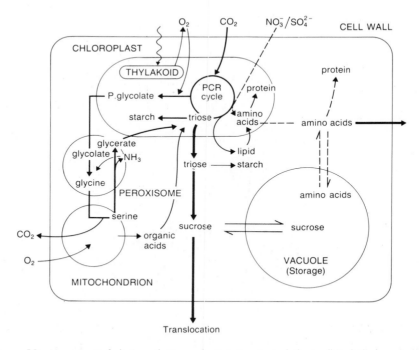

Figure 10 Movement of photosynthetic productions in a typical plant cell, including most marine and aquatic plants (C3). After Lawlor (1987).

abundance, they are toxic to most animals. If these compounds are left to bacteria degradation, the result will be lower water quality, perhaps even an anaerobic environment.

Plant cells also respire, much as animal cells, and at night generally require oxygen and release carbon dioxide. In some cases, both in the wild and in closed ecosystems, the fact that not only animals and bacteria but also plants are respiring in the dark can be crucial to ecosystem function. The practical aspects of this situation are discussed in the next chapter.

Biological Loading

As must be very clear from the above discussion, plants and animals can radically change the chemistry of their surroundings. This becomes very obvious during a red tide, or near an odiferous mud flat, and becomes particularly noticeable when one "walls off" a piece of that environment. When scale modeling a living ecosystem, the buffer effect of the larger surroundings is gone and one loses the normal balances that result from patchiness in the larger ecosystem. Biological loading is the term that we use to describe the effects of the organisms present on the physical and chemical environment. In a simulation system, biological loading can be classified in two general categories: chemical exchange and its requirements, as discussed in this chapter, and behavioral requirements and interactions. While there is some connection between the two, they are largely independent of each other. Behavioral aspects of loading are treated in Chapters 13–20.

Metabolism refers to all the complex of chemical reactions that go on inside a living organism. Respiration refers specifically to the basic energy exchange mechanisms that involve the use of oxygen to "burn" (chemically) the appropriate organic matter to provide energy for all the various elements of organic function. Carbon dioxide is produced by this process, and a typical measure of respiration is O_2 use or CO_2 production. Most important in this context is that the use or production of O_2 and the use or production of CO_2 are the most critical and obvious ways in which marine or aquatic organisms change the chemistry of their surroundings and the most immediate factors to be dealt with in an enclosed system.

If one is operating a closed ecosystem, the respiration problem is basically no different than that faced by the wild community, but several factors have to be kept in mind. For example, a high-biomass (and high-metabolic-rate) rocky-bottom community typically relies on constant movement of water from the overlying or off-lying water mass, often a planktonic community, which normally has much lower loading effects. The community in a tank might very well be functioning exactly as in the wild and yet not have sufficient oxygen to survive a night because of lack of contact with the equivalent larger adjacent body of water. Indeed, this basic situation happens in the wild in a mud-flat environment when the oxygen needs of the community as a whole cannot be met because of normally inadequate oxygen exchange mechanisms for the large metabolic requirement. A mud-flat community is adapted to this situation. A rich rocky-bottom community would often not survive both in the tank, as well as in the wild, when deprived of this need.

There are many other metabolic effects of marine and aquatic organisms that lead to changes, buildups or exhaustion, of elements or compounds in the water in which they live. Chief among these are the variety of compounds that result from nitrogen metabolism, the primary problem in this case being the excretion of ammonia, a toxic compound, and related products that result from the constant breakdown of proteins in animal cells and by some bacteria. In a broadly considered wild marine ecosystem, ammonia is very much needed by the plants, whether phytoplankton, algae, or higher plants that inhabit the community. Thus, it should never be a critically toxic element in either the wild or in a microcosm. However, in both cases potential imbalances exist. On an organic-rich muddy bottom, animal biomass often greatly exceeds plant biomass, which in deeper waters could be totally absent. While bacteria may then take over the role of plants in taking up and utilizing ammonia, the levels of ammonia present are likely to be much higher than in a community frequented by plants. Thus, in aquarium science, when a bacterial filter is used to break down ammonia, not only is more oxygen used, the ammonia concentrations are likely to be considerably higher than they would be in a plant-rich community like a shallow-water coral reef or a rocky shore. Also, the dilution factor plays a major role here again in potentially transferring toxic ammonia from an animal rich site of production to a broad area of dilute but more than sufficient plant production. There are many other ways in which use of a bacterial filter can negatively affect many marine organisms and communities. These are discussed in detail in Chapter 11.

The basic concept under discussion here can be extended to carbonate metabolism, silica metabolism, and all of the potential water chemistry-altering activities of organisms. When one is considering microcosm or mesocosm simulation of a community, the question should be asked: "How does the wild community in question avoid the problems created by metabolic imbalances?" If they are solved within the community on a daily as well as hourly basis, then faithful reproduction of environment and community in microcosm will produce the same result. However, if the problem of metabolic imbalance in a wild situation is solved by interaction with another community, or effectively by dilution, then either the alternate community or the dilution must be supplied or its effects simulated. Another aspect of biological loading is long- or short-term storage of organism tissues or organic materials derived from those tissues. Some entire ecosystems, such as bogs, are accumulating storage biomass. In these situations, some high-level organic compounds that could be available to organisms are not being used for a variety of reasons. Fossil fuels derive from long-dead ecosystems having functioned in this manner. Other biological communities such as reef lagoons, mud flats, and temperate forests are characterized by the accumulation of organic detritus in part of the system (e.g., the soil or lagoonal sediment), where it is gradually used by organisms specialized for the environment. A coral reef system stores very little of its organic production in the reef itself, and there is major loss of plant fragments to lagoons by most reefs. This potential organic loss to a reef could be partially made up by the capturing of zooplankton swept in by currents from the open ocean.

Such mass balances need to be carefully considered in microcosm work. For example, in a coral reef microcosm, if 1 of dried shrimp is fed to the system each day to simulate the planktonic input from the open ocean in the wild, then more than 1 (dry) of algae or other organics must also be removed from the tank. This may not necessarily be carried out daily or even weekly—the import/export schedule is a function of how much imbalance a system will normally take. There are several ways in which this can be done, as we discuss in depth later: (1) weeding of macroalgae, (2) scraping of tank glass and collection of the removed microalgae in a glass wool filter (left in the system for only a short time after scraping), (3) sediment settling traps that include organic particulates (see Chapter 8), (4) removal of larger organisms (fish, invertebrates), or (5) the use of an algal turf scrubber (Chapter 12). Method 2 is quite effective since the quantities of algae removed from the large glass surface are considerable. However, a major part of the algae removed are diatoms, and the net result after periods of several months is silica depletion. While this might not be undesirable in some cases, generally it results in considerably reduced numbers of both planktonic and benthic diatoms.

Biomass accumulation or loss in a microcosm system could be important in itself. For example, gradual buildup of plant or animal tissue could create a situation where oxygen transfer mechanisms, either natural or simulated, are not sufficient. However, as pointed out above for a case involving silica, chemical element or compound loss could also be crucial. As another general example, excess import of organic-rich water as a simulation of organic input will likely gradually raise dissolved nitrogen levels, particularly if it is not possible for buildup of plant biomass to occur (for lack of additional surface or excess grazing, for example).

In the next three chapters we discuss the primary chemical problems of biological loading, respiration and nitrogen metabolism, and the generalized means of simulating the required adjacent ecosystem effects. In Chapters 21–24, in discussing the major types of microcosms and mesocosms on which we have worked, other generally less crucial aspects of biological loading will be considered further.

References

Bond, C. E. 1979. *The Biology of Fishes*. Saunders and Co., Philadelphia.

Curtis, N. 1985. *Longman Illustrated Dictionary of Biology*. Longman York Press, Burnt Mill, Essex, England.

Keeton, W. T., and Gould, J. L. 1986. *Biological Science*. 4th Ed. Norton and Co., New York.

Lawlor, D. W. 1987. *Photosynthesis: Metabolism, Control and Physiology*. Longman Scientific/John Wiley, New York.

Lovelock, J. 1979. *Gaia: A New Look at Life on Earth*. Oxford University Press, Oxford.

Lovelock, J. 1988. *The Ages of Gaia: A Biography of Our Living Earth*. Norton and Co., New York.

Mayr, E. 1988. *Toward a New Philosophy of Biology*. Harvard University Press, Cambridge, Massachusetts.

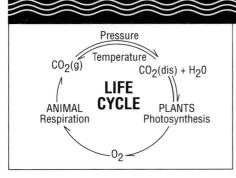

CHAPTER 10

ORGANISMS AND GAS EXCHANGE
Oxygen, Carbon Dioxide, and pH

The metabolism of living organisms affects water chemistry in two basic ways: (1) gas exchange (mostly oxygen and carbon dioxide) and (2) exchange of dissolved nutrients (nitrogen, phosphorus, and a variety of micronutrients). Animals also release undigested food in the form of feces and plants lose or detach parts, which relative to the environment are dead organic materials undergoing further breakdown primarily by microbes. They also excrete organic compounds such as ammonia and urea that will undergo further microbe degradation. All of these are steps that will ultimately use oxygen, release carbon dioxide, and produce nutrients. In this chapter and the next, we will discuss gas and nutrient exchange, respectively. For perspective, we will briefly discuss selected wild aquatic and marine environments followed by examples from a variety of captured ecosystems. Finally, in Chapter 12, we will examine methods of controlling gas and nutrient exchange in microcosms, mesocosms, and aquaria. In these model ecosystems, control or compensation is needed either because of the small size of the system in a day–night cycle, the presence of unnaturally large biomass, or the lack of a compensating larger adjacent body of water.

 Gases from the atmosphere diffuse into and out of any water body. If the water body lacks organisms and organic materials, an equilibrium or saturation will be established that is a function of temperature, salinity, and pressure. Nitrogen is the most abundant gas in the earth's atmosphere. However, while it is abundantly present as a dissolved gas in water, it is largely inert (*as a gas*) and little affected by biological activity. We will discuss nitrogen fixation and denitrification (exchange from the gaseous state to the organic or dissolved state and the reverse) in Chapter 11. Here, our concerns are primarily for the next most abundant atmospheric gases, oxygen and carbon dioxide. The concentration of these gases is radically and constantly altered by organic activity in aquatic and marine environments. The carbon dioxide of the earth's northern hemisphere atmosphere can been seen to vary seasonally due to the activity of plants (about 5 ppm out of 350 ppm). Indeed, so great is the potential for exchange of these gases by biological activity that the earth's atmosphere has been drastically changed (over billions of years) primarily by plants. In addition, a considerable percentage of the rock on the earth's surface is limestone or marble (primarily $CaCO_3$) derived from carbon dioxide through the shell-creating or environment-changing activities of plants and animals. Other rocks, such as shales (derived from muds), can also be very rich in organic carbon derived from the bodies of ancient organisms.

Tables 1 and 2 give the saturation values, that is, concentration at equilibrium, for oxygen and carbon dioxide, respectively, at different temperatures and salinities. These values apply to surface waters and, for "nonliving" waters, they are good reference points. However, they hardly ever occur exactly in nature because of the constant exchange activities of organisms. In aquatic and marine ecosystems oxygen is generally the most important and straightforward of the two gases. Carbon dioxide (CO_2)

TABLE 1								
Saturation Levels of Oxygen Gas Dissolved in Water as a Function of Salinity and Temperature[a]								
Temp. (°C)	Salinity (ppt.)							
	0	5	10	15	20	25	30	35	40
5	14.8	14.4	13.9	13.5	13.0	12.5	12.1	11.6	11.2
10	13.0	12.6	12.2	11.8	11.4	11.0	10.6	10.2	9.8
15	10.3	10.0	9.7	9.4	9.2	8.9	8.6	8.3	8.1
20	9.4	9.1	8.8	8.6	8.4	8.1	7.9	7.6	7.4
25	8.5	8.3	8.0	7.8	7.6	7.4	7.2	6.9	6.7
30	7.8	7.6	7.4	7.2	7.0	6.8	6.6	6.4	6.2

[a]Recalculated from date of Horne (1969). Reprinted by permission of John Wiley & Sons, Inc. Values given as mg/l.

TABLE 2
Solubility of Carbon Dioxide in Water as a Function of Salinity and Temperature[a]

Temp °C	Salinity (ppt.)								
	0	5	10	15	20	25	30	35	40
0	3.39	3.31	3.22	3.15	3.07	2.99	2.90	2.83	2.75
6	2.72	2.65	2.59	2.52	2.46	2.39	2.34	2.28	2.21
12	2.21	2.16	2.11	2.07	2.02	1.97	1.93	1.88	1.83
18	1.83	1.80	1.75	1.72	1.67	1.64	1.60	1.57	1.53
24	1.54	1.51	1.52	1.45	1.42	1.39	1.36	1.33	1.30
30	1.32	1.29	1.27	1.25	1.22	1.20	1.17	1.15	1.13

[a]Recalculated from data of Riley and Skirrow (1965). Values given as g/l. Note that oxygen (in Table 1) is given in mg/l as compared to g/l here. Effectively the solubility of CO_2 is hundreds of times greater than O_2.

reacts with water to form carbonic acid and its ionic forms. It is also involved with both inorganic and organic calcification. We will discuss carbon dioxide second.

Oxygen Exchange

The vast majority of organisms, animals and plants, living on the surface of the earth and within its waters require an oxygenated environment. While plants produce oxygen, often in great overabundance, photosynthesis requires light, which is rarely continuous. Simple plants cannot store oxygen and must remove some from their environment at night (even if they produced a great overabundance during the day). More complex plants, such as many submerged aquatic plants and mangroves, for example, are adapted to living on oxygen-deficient soils or substrates and have spaces within their tissues for the storage and transport of oxygen. Nevertheless, an oxygenated water column or atmosphere is required. Some animals and plants can temporarily use non-oxygen-requiring metabolic pathways to derive energy from food. These pathways can be utilized in low-oxygen environments or to carry out "extra" metabolism (for example, the emergency heavy use of muscles) over and above average oxygen transport capabilities. Such pathways are, however, much less efficient than those using oxygen.

Unicells, or simple filaments, such as bacteria, protozoa, and fungi, will take up oxygen directly from the environment through their cell membranes. Multicellular aquatic animals have evolved a variety of organs called gills to remove oxygen from the water column. They have likewise evolved a number of blood pigments to carry oxygen in the bloodstream (Table 3) so that cells situated deep inside their bodies can also receive oxygen. In general, it is primarily the microbes, bacteria, yeasts, and some fungi, protozoans (protists) and certain parasites that are adapted to the anaerobic environment. Some of these undoubtedly trace their ancestry to the early days of life on earth when oxygen was scarce. A number of alternate-energy, "chemosynthetic" and non-oxygen-using carbohydrate breakdown pathways exist. Some produce compounds like hy-

---

TABLE 3

Respiratory (Oxygen-Carrying) Pigments in the Animal Kingdom[a]

Hemocyanin. Copper-containing protein, carried in solution.
  Mol. wt. = 300,000–9,000,000
  Mollusks: chitons, cephalopods, prosobranch, and pulmonate gastropods: not in lamellibranchs.
  Arthropods: Malacostraca (sole pigment in these): Arachnomorpha: *Limulus, Euscorpius*

Hemerythrin. Iron-containing protein, always in cells, nonporphyrin structure.
  Mol. wt. = 108,000
  Sipunculids: all species examined
  Polychaetes: *Magelona*
  Priapulids: *Halicryptus, Priapulus*
  Brachiopods: *Lingula*

Chlorocruorin. Iron-porphyrin protein, carried in solution.
  Mol. wt. = 2,750,000
  Restricted to four families of Polychaetes:
    Sabellidae, Serpulidae, Chlorhaemidae, Ampharetidae
  Prosthetic group alone has been found in starfishes, *Luidia* and *Astropecten*

Hemoglobin. Most extensively distributed pigment: iron-porphyrin protein: in solution or in cells.
  Mol. wt. = 17,000–3,000,000
  Vertebrates: almost all, except leptocephalus larvae and some Antarctic fishes (*Chaenichtys*, etc.).
  Echinoderms: sea cucumbers
  Mollusks: *Planorbis*, Pismo clam (*Tivella*)
  Arthropods: insects *Chironomus, Gastrophilus*. Crustacea *Daphnia, Artemia*
  Annelids: *Lumbricus. Tubifex, Spirorbis* (some species have hemoglobin, some chlorocruorin, others no blood pigment). *Serpula*, both hemoglobin and chlorocruorin.
  Nematodes: *Ascaris*
  Flatworms: Parasitic trematodes
  Protozoa: *Paramecium, Tetrahymena*
  Plants: Yeast, *Neurospora*, root nodules of leguminous plants (clover, alfalfa)

[a]After Schmidt-Nielsen (1975).

drogen sulfide (H$_2$S) that are quite poisonous to most animals. Many organisms are adapted to aqueous environments in which anaerobic sediments are overlain by oxygen-rich water or the atmosphere. However, in cases where oxygen depletion is permanent and extends up into the water column, along with the concomitant H$_2$S, the environment becomes largely "dead" except for anaerobic microbes. This can happen on a very large scale such as in the Black Sea where deep waters are isolated by the shallow sill at the Bosphorus.

In the surface waters of the ocean, oxygen is generally supersaturated (Fig. 1). While extensive surface exchange (wave action) under conditions of low biomass would not allow a large negative difference from saturation values, either seasonally or diurnally, excess plant biomass and photosynthesis in the well-lighted zones tends to keep ocean surface waters above saturation levels. The ocean is a major original and continuing source of oxygen to the atmosphere. It can be argued that before man evolved and utilized oxygen in the burning of forests and fossil fuels, maximum atmospheric oxygen levels had been attained. The use of fire in hunting by Stone Age men probably already had significantly lowered oxygen concentration and raised that of carbon dioxide. The domination of Australian forests by eucalypts and the abundance of park-like grasslands and sedgelands rather than the previously dominant *Nothofagus* and *Araucaria* was probably created by the aborigines' burning activities (Flood, 1983). Atmospheric oxygen levels prior to man's influence were in part limited by the natural combustion levels of forests. In short, "the world is (or at least was) green." Plants probably have the capability to raise atmospheric oxygen concentrations higher than they are, perhaps to 22%. The potential for subspontaneous combustion of forests on land, the release and oxidation of methane from anaerobic deposits, and perhaps the reduction of carbon dioxide to the point where photosynthesis becomes very slow set the limit to atmospheric oxygen levels (see Lovelock, 1979).

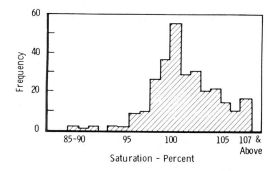

Figure 1 Oxygen saturation levels of the surface waters of the South Atlantic Ocean. After Richards (1965).

In somewhat deeper water (100–700 m) in the open ocean, oxygen levels reach their minimum (Fig. 2). This is below the lighted or photic zone of plant activity. It is also a level at which the rain of plant, animal, and organism waste material from the surface provides relatively high animal and bacterial activity. Finally, in the lower half of the deep ocean, oxygen concentrations return to near-surface values. These deep waters are cold and largely derived by deep currents from arctic and antarctic zones. At their high-latitude sites of origination, saturation levels were initially relatively high. Also, at cold temperatures plant photosynthesis is relatively more efficient than either plant or animal respiration. Open coastal waters are similar to the open ocean, though fluctuations are greater.

Lakes and rivers tend to operate under similar oxygen distributions but, on the average, show a shift to lower oxygen levels. This situation derives from the addition of terrestrial organic matter from the surrounding watershed. There are almost as many oxygen distribution and exchange patterns as there are lakes. For more detail see Chapter 5, Figure 3 and the extensive discussion by Hutchinson (1957). In summary, the surface waters of most lakes are near oxygen saturation levels. Relatively clear, unproductive lakes in which a more ordered organic and oxygen exchange occurs can average at supersaturated levels of oxygen at the surface and at moderate oxygen levels at depth. At the other extreme, in eutrophic or nutrient-rich lakes, variations in oxygen concentration can be very large. Under the right conditions in the spring or summer, such lakes can bloom with intensive plant growth. In these cases, oxygen su-

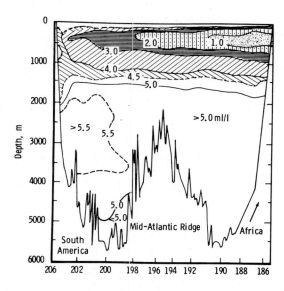

Figure 2   Dissolved oxygen levels in the sea (in ml/l). After Dietrich (1963).

persaturations of 50% or more can eventually be followed by a "crash" in oxygen levels by late summer. Such eutrophic lakes build up biomass to the point where oxygen diffusion at night is not sufficient to prevent anaerobic conditions resulting from the very large oxygen requirements. Extensive fish kills often result. Unlike the oceans, lakes and rivers acquire significant organic loads from the terrestrial environment. Where depths extend well below the photic zone mean oxygen deficits can exist, and a lake will use more oxygen than it produces or can diffuse through its surface. Estuaries are in between lakes and rivers and the ocean with regard to mean oxygen concentration and can be more like one or the other depending on the nature of the organic and nutrient input of the fresh waters and the amount of exchange with the ocean. In lakes, rivers, and estuaries, man's organic and nutrient inputs are crucial to oxygen levels and organic stability. Mesocosms and aquaria are similar.

Photosynthesis and respiration are often summed up as $6CO_2 + 6H_2O +$ energy (light/chemical) $\leftrightharpoons C_6H_{12}O_6 + 6O_2$. Ecologically this implies a balance in that all energy-rich organic compounds created by the photosynthesis of plants are either metabolized by the plants themselves or metabolized by animals through a food chain. This is not the case. Much plant and some animal organic material goes into geological storage, eventually becoming peat, organic-rich shale, coal, gas, and oil. It is estimated that over 1600 times as much organic material and other carbon of organic origin lies in geological storage as presently exists in the biosphere as biomass (Stumm and Morgan, 1981). Also, in oxygen-poor environments, glycolysis and fermentation result in the partial breakdown of plant-produced organics without the use of oxygen. This is how oxygen derived from molecular water has gradually built up to high levels in the atmosphere. Even if we have now reached a balance (independent of man) in oxygen production, and levels are no longer building up in the atmosphere, the excess oxygen required by methane, coal, and oil as they are exhumed on the surface of the earth is not generally recycled by organisms in aquatic environments. Thus, most natural water ecosystems not made eutrophic by man, particularly shallow-water environments, will tend to be supersaturated or at least rich in oxygen. This basic consideration must be given careful thought when modeling water systems. Generally, photosynthesis and the concomitant oxygen-rich water must rule in microcosms, mesocosms, and aquaria as they do in the wild or the organisms involved will be subjected to the same stresses that develop and cause a shift to microbes in wild, poorly aerated environments.

Oxygen and Model Ecosystems

In microcosms and mesocosms where one is attempting to simulate all aspects of a particular environment and ecosystem, presumably one provides enough light and the appropriate plant community to simulate wild levels of photosynthesis. Certainly if diurnal and season oxygen measurements show oxygen levels below those in the natural community one is attempting to simulate then there is a serious problem that should be corrected. Assuming that community structure is more or less correct and photosynthetic plant biomass and animal biomass are properly balanced, a problem of low oxygen levels is likely to be either inadequate light or a failure to simulate water flow from areas of higher oxygen concentration, particularly at night. The first problem was discussed in depth in Chapter 7. We will discuss the solution to the second problem in Chapter 12.

In aquaria, even if the builder and operator are attempting to maximize equivalences between the wild environment and aquarium, scaling and inadequate ratios of water surface to water volume can provide great difficulties relative to oxygen. In the aquarium, where display is a primary function and volume is small, animal biomass is likely to be higher than normal, particularly for the marine environment. Also, artificial feeding is almost invariably provided to an aquarium in excess of wild equivalents. Thus, it is virtually impossible, except for the more-or-less unusual environments normally low in oxygen that one might try to match, to simulate a proper oxygen environment by simple aeration. The newer methods of trickle filtration greatly improve the situation but unfortunately concomitantly prevent the supersaturation levels that are also needed. One could, of course, use bottled medical oxygen, but this approach is expensive and potentially dangerous if improperly used. The answer is to treat the model ecosystem or aquarium like the wild and to use plant photosynthesis in a controlled manner.

In summary, in natural aquatic and marine ecosystems there are communities, particularly deeper-water bottom communities, that normally operate at low oxygen levels. These could be managed in a model by aeration if so desired and if nitrogen balance is not a factor. In shallow water, well-lighted communities of average oxygen levels are likely to be either saturated or supersaturated. To keep the higher organisms, plants, invertebrates, and vertebrates from more-or-less continuous oxygen stress, mechanisms for providing pure oxygen in excess of atmospheric availability are necessary. These are discussed below.

Carbon Dioxide Exchange

The other major gas involved in exchange between organisms and their environment is carbon dioxide. The activity of CO_2 is, in a sense, the inverse of oxygen, the two being primarily cross-exchanged between plants and animals (Fig. 3). However, it is also very different from oxygen in that it reacts chemically with water. Also, CO_2 is intimately bound to calcification or shell and wall formation in organisms, both plants and

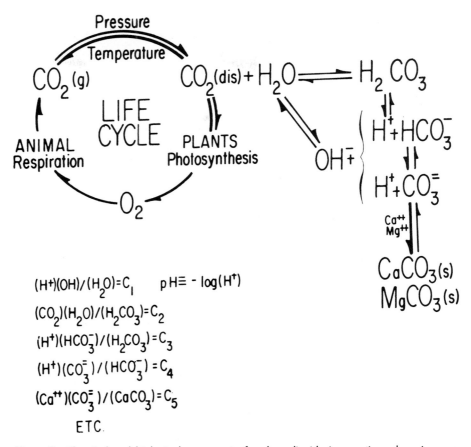

$$(H^+)(OH)/(H_2O)=C_1 \qquad pH \equiv -\log(H^+)$$

$$(CO_2)(H_2O)/(H_2CO_3)=C_2$$

$$(H^+)(HCO_3^-)/(H_2CO_3)=C_3$$

$$(H^+)(CO_3^=)/(HCO_3^-)=C_4$$

$$(Ca^{++})(CO_3^=)/(CaCO_3)=C_5$$

ETC.

Figure 3 Chemical and biological movement of carbon dioxide in aquatic and marine ecosystems. After Horne (1969). Reprinted by permission of John Wiley & Sons, Inc.

animals, and to the chemical precipitation of limestones. Unlike oxygen, relative to respiration, the lack of carbon dioxide in special environments (including greenhouses) may slow down photosynthesis to some degree. However, in the aquatic and marine environment there is usually a nearly infinite supply of this essential compound (as bicarbonate, HCO_3^-) available for all biological activity. On the other hand, excess carbon dioxide will tend to increase the acidity of an environment and will render the calcification requirements of an organism more difficult (Fig. 4).

In the CO_2–$CaCO_3$ system, the oceanic parallel to high photosynthesis, high oxygen in the surface photic zone, can be seen in calcium carbonate saturation (Fig. 5). In tropical oceans, ultimately due to photosynthesis, there is great excess of calcium carbonate in crystalline form as shells. However, in deeper water, carbonate concentration rapidly falls and calcite becomes undersaturated (i.e., shells begin to dissolve as they

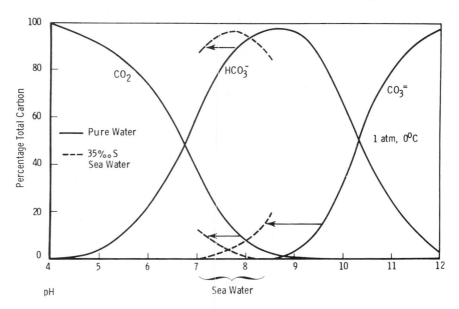

Figure 4 The distribution of dissolved compounds in the carbonate/water reaction chain as a function of pH. Note that sea water is buffered and only occurs over a limited pH range, while pure water can range widely. After Horne (1969). Reprinted by permission of John Wiley & Sons, Inc.

sink). Where the oxygen minimum and the carbon dioxide maximum meet, sea water has only 70% of saturation levels. Finally, below about 4500 m the rate of supply of calcium carbonate from the surface is exceeded by dissolution. Below that depth, carbonate shells are virtually absent from sediments.

Most of the carbon on the surface and in the crust of the earth does not occur in organisms. At any one time, carbon occurs mostly in limestones, in coal and oil, and in other organic rock derivatives. These rocks and fossil fuels were mostly deposited in shallow waters (Fig. 6). The exchange rate, however, primarily through carbon dioxide, is very large. For example, carbon turnover between the atmosphere and organisms occurs on the order of every 20 years. Also, as large as the reservoir of carbon is in the rocks, on a time scale of about 400 million years, geological processes bring the carbon to the surface where, partly through the atmosphere, organisms can again turn over that stored carbon. All of the carbon on the surface and in the continental crust, except for a small amount of new carbon arriving from deep in the earth, mostly through volcanoes, has been cycled through organisms several times. It is to be noted that while all exchange elements in Figure 6 are balanced (as much carbon into each compartment as out), these numbers are hardly more than guesses. At least prior to man's use of fossil fuel, it is likely that the

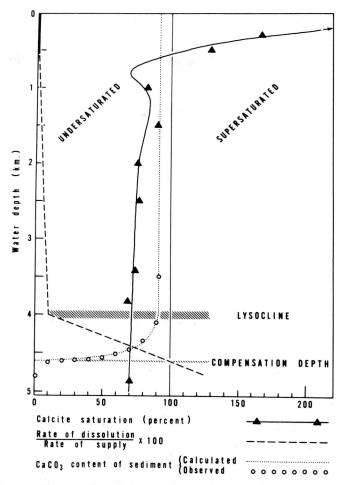

Figure 5 Factors affecting the distribution of calcium carbonate as calcite, with depth, in the equatorial Pacific Ocean. After Kennett (1982).

storage of "organically derived" carbon in the earth's rocks was continuing to grow slowly as the continents enlarged.

The large-scale carbon balance on earth might seem to have little to do with the aquarist. On the contrary, balancing a small ecosystem directly relates to the way in which man as a whole is now manipulating Mother Earth. If environments and ecosystems are to be operated, in model, as they were prehistorically on earth, or where they are minimally affected by man, then the same principles apply. For example, one can attempt to operate a coral reef system with traditional bacterial filtration, low light levels, and external feeding. However, with regard to carbon

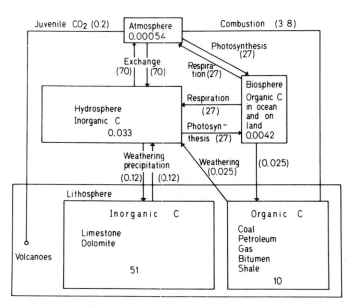

Figure 6 The carbon cycle on earth showing the approximate quantities of carbon and relative exchange between compartments. Mass exchange rates are given as microgeomoles/yr (10^{14} moles/yr). Carbon mass is given as geomoles (10^{20} moles). After Stumm and Morgan, 1981. Reprinted by permission of John Wiley & Sons, Inc.

dioxide, there is no major carbon removal mechanism except slow aeration release to the atmosphere. Excess carbon dioxide will result in low pH and lack of calcification. Such an environment would in some ways resemble a mud flat. However, this would not be acceptable to most reef organisms. Modern trickle systems with a calcium carbonate gravel greatly improve the pH problem; however, only excess carbonate production resulting from photosynthesis can truly simulate the typical growing reef environment.

It is possible, using electronic sensing and glass electrodes, to measure carbon dioxide, bicarbonate, and carbonate concentration in an aqueous medium. Knowledge of any two of these carbonate system components (including pH) defines the state of the entire system. Nevertheless, in practice, measurement of pH and the use of the diagram of Figure 4 is the simplest way to establish the relative status of this complex. While simple paper or chemical tests available at chemistry supply houses or most aquarium stores are adequate for pH measurement in fresh waters, in salt waters very careful examination using electronic sensing is essential. In the latter case, precision test buffers must be employed to calibrate the instrument being used.

Salt water is strongly buffered for pH and generally ranges from 8.00 to 8.40. Because of the ions present, great changes in carbon dioxide are necessary to affect a small change in pH. Nevertheless, most marine organisms are particularly susceptable to small pH changes, and in this respect the environmental status is crucial. It is of course possible to increase the buffering effect of sea water using calcium carbonate chips or substrate within a model system. This is an essential fail-safe feature for any marine microcosm and aquarium, and in small and difficult situations with heavy animal loads or high calcification requirements, the use of sand or silt size aragonite, the least stable form of calcium carbonate, can be invaluable. The shells of some calcareous algae (e.g., *Halimeda*) or crushed coral skeleta can provide the aragonite (when clean of organic material). Oolitic or nearly pure Bahamian aragonitic sand is marketed in bulk in the United States. As was partially demonstrated in Figure 5 and is repeated for emphasis in Figure 7, most shallow marine waters are highly supersaturated for calcium carbonate. In a model system, where benthic substrate and sediments are inevitably of proportionally greater extent than in the wild, accepting subsaturation levels of carbonate by simply maintaining chips in the system to maintain proper levels of the pH/carbonate complex is certainly stressing many organisms. Only by appropriately reducing carbon dioxide production and by removing carbon dioxide through "excess" photosynthesis can one hope to achieve a status of the pH/carbonate complex that is equivalent to most shallow sea waters. This process is discussed in greater depth in Chapter 12.

Rainwater from an unpolluted atmosphere typically has a pH of less than 5.6 (i.e., moderately acid) due to the interaction of atmospheric carbon dioxide with the water while it cycles through the atmosphere. On limestone and related substrates, this natural acidity can be strongly buffered. In some cases, however, particularly on largely unbuffered granite rock basements and with the addition of organic acids from some vegetation, the natural acidity of the rainwater can be increased in the groundwater. Freshwater lakes, ponds, and rivers vary typically from slightly acid (pH 6) to moderately basic (pH 9). Extremes occur from acid volcanic lakes (near pH 1) to soda lakes (at pH 12), though these would be of concern only to a few aquarists. Because of great concern over the polluting affects of acid rain and the resulting environmental acidification, considerable study has been directed to the negative effects on flora and fauna of acidification (Fig. 8). This is a complex and not fully understood subject; the interested reader is referred to Cresser and Edwards (1987).

In the modeling of freshwater systems, the aquarist is seeking to simulate waters of a given pH. Where these are basic waters, above pH 8.0 the same rules apply as we previously discussed for salt waters. On the other hand, if neutral or acid waters are desired, the situation becomes more complex. Certainly if a high-nutrient, low-pH environment is desired and oxygen levels are not a concern, then traditional bacterial filtration is an ideal approach to system management. On the other hand, if a

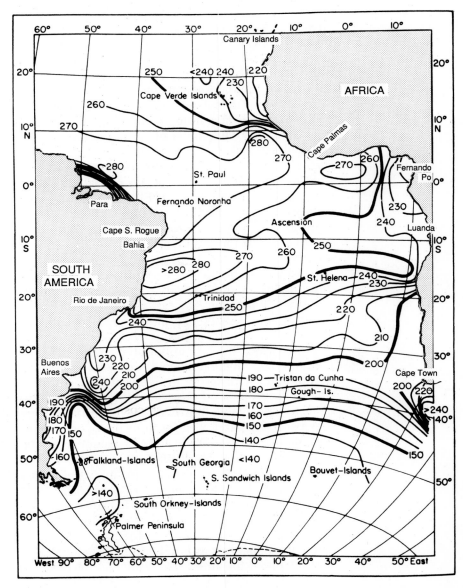

Figure 7 Percent saturation of calcium carbonate in the surface waters of the central and south Atlantic Oceans. After Horne (1969). Reprinted by permission of John Wiley & Sons, Inc.

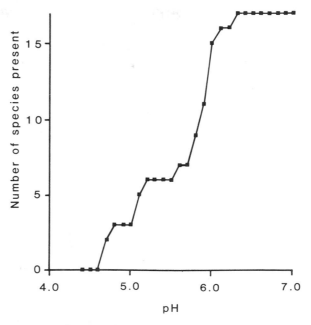

Figure 8 Lower pH tolerance limit of common molluscs and crustaceans found in Norwegian fresh waters. After Cresser and Edwards (1987).

low-nutrient, low-pH (e.g., a black-water stream) environment is needed, the situation can become considerably more difficult.

In mesocosm and microcosm simulations of black-water streams where fish and invertebrate biomass are kept appropriately low and where energy input is derived largely from slowly breaking down leaf litter, minimum mechanical filtration or, preferably, settling traps may be all that is required. For simulating highly acid natural waters, acidification derived from CO_2 bubbling or direct acid addition can be employed. On the other hand, in an aquarium situation, high animal biomass feeding supplementation and therefore high nutrient supply will almost certainly provide a basic problem for a low nutrient, low pH, and high oxygen requirement. As we will discuss later, this environmental pattern can be accomplished with an algal scrubber combined with carbon dioxide bubbling, or by using tannin-rich waters (either natural or artificial) resulting from leaf break down and a sphagnum moss community.

Gas Exchange and Selected Synthetic Ecosystems

The southern bank barrier coral reef on the Caribbean island of St. Croix can be regarded in many ways as a generalized tropical reef. It is mature, well developed geologically and biologically, and has been studied in some detail. Given the low rates of metabolic activity in the offshore water, the water moving onto the reef due to wave and current action typically has an oxygen concentration close to 6.5 mg/l, a level that varies only by a few tenths day and night (Chapter 5, Figure 10). This is essentially the saturation point. As this sea water flows in over the reef, driven by the constant trade winds, it picks up extra oxygen, becoming highly supersaturated during the day because of excess photosynthesis, that is, photosynthesis that exceeds respiration. On the other hand, at night the same water loses oxygen, due to the respiration (without photosynthesis) of both plants and animals in the reef. The effect of extensive wave breaking on the exchange of oxygen is also marked and causes oxygen loss during the day and oxygen gain at night, the net effect being to smooth out metabolic effects on the gas concentrations.

To simulate this typical reef environment in a microcosm system, the aquarist would be trying to obtain a night-to-day oxygen concentration range from 5.5 to 8.5 mg/l. The traditional marine aquarium with its heavy animal oxygen requirement, increased by bacterial action in biological filters, cannot achieve supersaturation. Even if well lighted, it is likely to remain undersaturated for oxygen and may well go below 5 mg/l at night. The lower limit of oxygen for water overlying a reef community is probably typically about 4 mg/l under exceptionally calm conditions. Extensive periods lower than this level in the main mass of overlying water would likely be detrimental to reef plants and animals, especially since oxygen concentration would be considerably lower in the interstices of the reef. Intensive air bubbling carried out in well-kept and reasonably loaded traditional tanks could produce minimum oxygen levels at or above 4–5 mg/l. However, without intense lighting and abundant photosynthetic plants (or an oxygen bottle) it would not be possible to raise oxygen concentrations above 6.5 mg/l and likely not even to that level. The "dry" type trickle filter is excellent for raising oxygen levels from the lower potential values. It too will, however, lose oxygen during the day. A dry trickle filter could be operated attached to the system at night and separated during the day only to overcome this objection.

The oxygen concentration in a Smithsonian coral reef microcosm as compared to the wild reef is shown in Figure 9. As described in Chapter 12, using an algal turf scrubber rather than a bacterial filter, even during the dark hours oxygen levels are stabilized just below saturation levels. During the day oxygen levels in this well-lit reef tank (see Chapter 21) rise well above saturation and in most respects match wild reef values closely.

Using an oxygen meter, one can measure the rate of oxygen uptake in a mesocosm or aquarium system during the dark at near saturation levels to determine if the community respiration load is equivalent to that in the environment one wishes to simulate. It is necessary to make this measurement at or near saturation so as not to have to account for exchange with

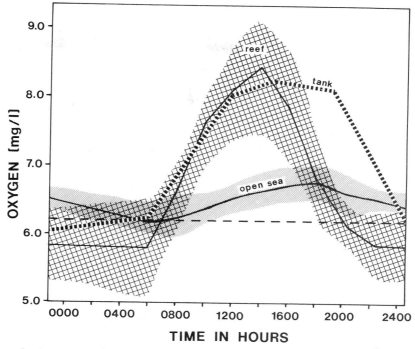

Figure 9 Comparison of oxygen levels in the open sea and on a St. Croix coral reef with levels in a coral reef microcosm (see also Chapter 21). After Adey (1983). ———, Oxygen saturation.

the atmosphere. In one of our 1800-gallon reef tanks, for example, the dark respiration rate was about 2 g O_2/m^2 of reef surface/h or about 24 g O_2/m^2 per day. This is only about one-half of the wild St. Croix reef rates, which presumably means that we could have doubled our animal respiratory load, assuming that wave or other exchange keeps oxygen concentrations at a minimum of about 4 mg/l. Later tanks have had rates of 3–5 g $O_2/m^2/h$, rather closer to rich reef levels. However, it is necessary in this case to be certain that photosynthesis is equivalently elevated. If this is not done, CO_2 levels would rise in the system, depressing pH, reducing carbonate formation, and generally stressing many of the reef organisms. For the hobbyist who does not have access to an oxygen meter, a reasonable estimate of the time of day that oxygen becomes supersaturated can be made by watching the production of oxygen bubbles on algae growing in the system. Typically, these will show up an hour or so after the water becomes supersaturated. A shallow reef aquarium simulation should be supersaturated for 8–10 h per day. Consider a typical 60-gallon tank with a half square meter of reef surface having a dark consumption rate of 3 g O_2/h. If one starts with oxygen concentration at the saturation level (6.5 mg O_2/l), and assumes little or no atmospheric interchange, then all of the oxygen in the system would be used in one hour. The reef would be highly stressed within a half hour. Intense wave action bubbling or trickle filtering can offset this unstable and drastic situation, as mentioned above. However, oxygen is relatively soluble in water and exchanges rapidly with the atmosphere in bubbling or wave action. Carbon dioxide exchanges more slowly and even though sufficient oxygen is being supplied, CO_2 concentration may well rise rapidly in a bubbling system, thereby lowering pH even while oxygen remains at satisfactory levels. Equally important, as discussed in Chapter 11, nitrogenous exhange between plants and animals is not significantly affected by atmospheric exchange. Even if oxygen is artificially kept above 5 mg/l by physical atmospheric exchange methods, that does not mean that water quality is not otherwise rapidly degenerating. It is wise to either have adequate water buffer to carry a system through the night or have a plant compensating system.

Normal sea water ranges from about 8.0 to 8.4 in pH, and in that range most of the inorganic carbon appears as bicarbonate ion with the concentration of CO_2 being very low, on the order of 30 ppm. pH is the most convenient measure of the state of the system, although complete information on the total inorganic carbon present requires knowledge of two components, for example, pH and CO_2. A typical diurnal variation of pH in the reef environment is 8.05–8.20. The numerous organisms within a reef that calcify basically use the dissolved CO_3^{2-} that is present. At pH values less than 8.00, calcification is difficult for organisms to accomplish in that it requires greater energy expenditures, and for some organisms that may not be possible. It seems likely that the primary reason for the development of algal symbioses in reef-building corals is the calcification advantage provided by the uptake of CO_2 in the light.

In many of our older reef tanks, pH reached a minimum of about 7.93 at about 0900 h and a maximum of about 8.17 at 2000 h. Considerable calcification occured, although rates appeared to be somewhat lower than on a natural reef. If the common calcified green algae *Halimeda* can be used as an example of general calcification in the system, calcification occurs primarily from about 1800–2400 h during peak periods of pH.

The critically important variable of photosynthesis, or primary production, is treated in several chapters. However, it is so inextricably linked to oxygen, carbon dioxide, pH, and system loading that its role is repeated again here in the context of a coral reef model. Net primary production (in terms of oxygen) for the St. Croix reef is shown in Figure 10. If oxygen loss due to respiration, which we presume to be roughly constant at all times, is removed from the data, a curve of approximate total photosynthesis is derived. Several important pieces of information can be derived from this diagram. First, total photosynthesis is very high, at approximately 40 g O_2/m²/day. If this is converted to actual plant production (leaving out what the algae themselves use), one finds a rate of about 25 g (dry wt)/m²/day. In the laboratory, measured rates of actual

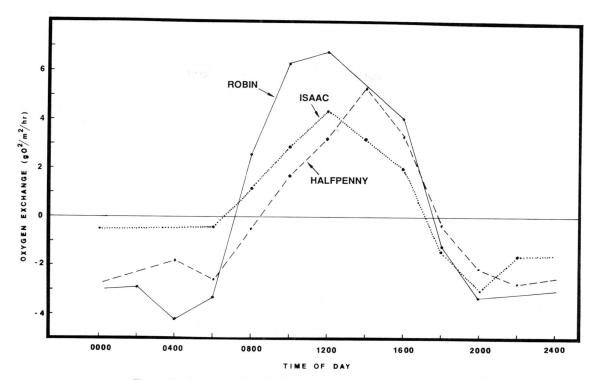

Figure 10 Mean yearly diurnal oxygen exchange for several reef transects on the island of St. Croix. After Adey and Steneck (1985).

algal production consistently achieve 10–20 g (dry wt)/m²/day on flat screens. Since a typical reef has several times the mean surface of a flat screen, the reef value measured by oxygen seems not unreasonable. Also it can be seen that production (photosynthesis) continues to rise during the morning, peaking out around 1200 h when sunlight is at its greatest intensity. In summer, it appears that there is a slight drop by noon, but this is probably at least partly due to saturation of metabolites. Unlike what most scientists have thought with regard to other marine algae, the reef plants can use most of the available sunlight and as a community do not seem to be inhibited by the intense light. The high respiration, and biomass levels, of a shallow-water reef and hopefully a reef microcosm are internally driven by a high level of plant production—one that needs full sunlight to fully perform. Traditional aquaria, even where reef communities are being portrayed, rarely reach one-tenth of full sunlight, and this provides an oxygen and pH stress that is exacerbated by the need to extensively provide supplemental feeds.

In recent years higher light levels have tended to be used for tropical "reef" aquaria. However, they still remain at levels of one-quarter to one-third of wild ecosystems. Since most of the ecosystems that we are likely to place in microcosm rely on internal plant production for most of their energy supply, a careful consideration of light supply to microcosm and aquaria is extremely important (Chapter 7).

We have based our discussion of oxygen and carbon dioxide exchange to this point on simulations of coral reef systems. However, the

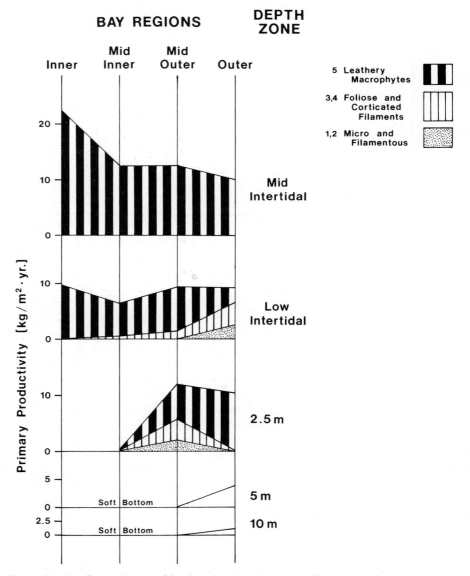

Figure 11 Yearly productivity of benthic algae in (in wet weight) rocky intertidal and subtidal zones in the northern Gulf of Maine. After Adey (1982).

points made can be underscored by looking at other model ecosystems. One tends to think of cold-water northern ecosystems as highly seasonal ones that may reach high levels of activity for short periods in the summer and essentially shut down for the colder, darker part of the yearly cycle. However, examination of primary production in Gulf of Maine rocky-bottom kelp and rockweed communities shows that this is hardly the case (Fig. 11). Photosynthetic efficiency, or available light converted to biological use, on a wave-beaten rocky shore is about the same, in proportion to available light, as on a coral reef. Interestingly enough, on the rocky shore itself, grazing rates are relatively low and thus there is an actual excess of production. In the wild, roughly three-quarters of the plant production is eventually lost from the rocky bottom to beach drift and finally delivered by tides and currents to the inshore mud flats. In short this community is a strong net exporter of biological production, and this must be taken into account when designing such a system. In the Smithsonian Maine Coast system, both a place for drift algae and a small mud flat have been provided (see Chapter 22). Since the available area for export is not sufficient relative to production area in this unit, simple cropping and removal is also used.

In the rocky Maine shore microcosm, oxygen levels average higher than on a coral reef, as a reflection of this system as a net producer and exporter of plant biomass. At 8.5–9.5 mg/l of dissolved O_2 average oxygen concentration remains above saturation for most of the year. This occurs in spite of the fact that considerable wave action is present, continuously, in the model.

References

Adey, W. 1982. A resource assessment of Gouldsboro Bay, Maine. Report to NOAA, Marine Sanctuary Program. No. NA81AA-D-Cz076. 47 pp.

Adey, W. 1983. The microcosm: A new tool for reef research. *Coral Reefs* **1**: 194–201.

Adey, W., and Steneck, R. 1985). Highly productive eastern Caribbean reefs: Synergistic effects of biological, chemical, physical and geological factors. In *The Ecology of Coral Reefs*. M. Reaka (Ed.). NOAA Symp. Ser. on Underwater Research, Vol. 3, Washington, D.C.

Cresser, M., and Edwards, A. 1987. *Acidification of Freshwaters*. Cambridge University Press, Cambridge.

Dietrich, G. 1963. *General Oceanography*. Wiley-Interscience, New York.

Flood, J. 1983. *Archaeology of the Dreamtime*. Collins, Sydney.

Horne, R. A. 1969. *Marine Chemistry*. Wiley and Sons, New York.

Hutchinson, G. E. 1957. *A Treatise on Limnology*, Vol. I. Wiley and Sons, New York.

Kennett, J. 1982. *Marine Geology*. Prentice Hall, Englewood Cliffs, New Jersey.

Levinton, J. 1982. *Marine Ecology*. Prentice Hall, Englewood Cliffs, New Jersey.

Lovelock, J. 1979. *Gaia, A New Look at Life on Earth*. Oxford University Press, Oxford.

Richards, F. 1965. Dissolved Gases, In *Chemical Oceanography*. J. Riley and G. Skirrow (eds.), Academic Press, London.

Schmidt-Nielsen, K. 1975. *Animal Physiology*. Cambridge University Press, Cambridge.

Stumm, W., and Morgan, J. 1981. *Aquatic Chemistry*. John Wiley and Sons, New York.

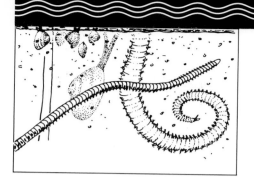

CHAPTER 11

NITROGENOUS WASTES AND NUTRIENTS
Nitrogen, Phosphorus, and the Micronutrients

Typical open ocean sea water contains about 35 parts per thousand (35,000 ppm) by weight of salts and other elements, in addition to the hydrogen and oxygen that make up the remaining 965 parts per thousand. Over 99.6% of the "salt" weight is made up, in order of abundance, by chlorine, sodium, magnesium, sulfur, calcium, and potassium. These elements are conservative. They vary little in their proportions over the entire world ocean and even into most estuaries. Most elements known to man occur in sea water, generally in extremely small quantities. However, there are a number of elements, required by some or all plants, that occur normally at concentrations of a small fraction of a ppm to perhaps as much as 0.2 ppm. The most important of these elements to organisms are nitrogen, phosphorus, and silica.

In fresh waters, the "conservative elements" are generally present in only very small quantities (see Chapter 4), with salt lakes in basins without an outflow being striking exceptions. In salt lakes, the dominant salts vary widely, depending on the rocks present in the drainage basin. However, the same basic nutrients as in the sea tend to be limiting to plant production in the earth's fresh waters.

Nutrients in Natural Waters

A typical tropical open ocean distribution pattern of dissolved (reactive) nitrogen and phosphorus is shown in Figure 1. Surface values of these essential compounds are typically less than 1 μM (0.014 ppm) of nitrogen as nitrite plus nitrate and less than 0.1 μM (0.003 ppm) for phosphate. Below the photic zone, and a little below the oxygen minimum and carbon dioxide maximum, where phytoplankton and zooplankton are being eaten or are breaking down without light for recycling, nutrient levels climb to about 20–30 times surface values. In the deepest waters, which are cold with very slow currents derived from Arctic and Antarctic surface waters, nutrient levels are moderate at all times. Largely inaccessible to plants except at unusual points of current upwelling, reactive nitrogen and phosphorus remain in storage for hundreds to thousands of years in the deep ocean.

In relatively shallow coastal waters the picture is rather different, especially outside the tropics where surface warming in the summer tends to prevent mixing. The yearly cycles of nitrogen, phosphorus, and silica levels for the English Channel are shown in Figure 2. Here, the mixing of tides and winter storms accompanied by low light allows moderate levels of nutrients to build up in winter. Note, however, that in the winter at 5–8 μM for nitrogen as nitrite plus nitrate, they are still well below deep ocean levels. Most aquatic animals excrete ammonia. This very toxic compound is metabolized by some bacteria to the mildly toxic nitrite and finally to

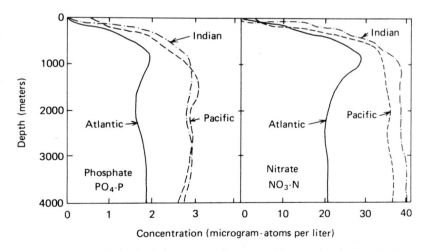

Figure 1 Distribution with depth of phosphate and nitrate in the tropical and subtropical portions of the oceans. Note: μg−at/l = μM; concentrations of N as nitrate with depth are 0.3–0.6 ppm; P as phosphate are 0.06–0.09 ppm. After Sverdrup *et al.* (1942). Reprinted by permission of Prentice Hall, Englewood Cliffs, New Jersey.

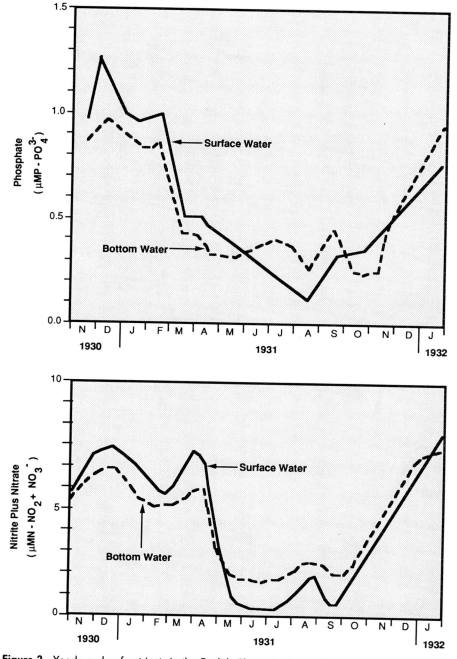

Figure 2 Yearly cycle of nutrients in the English Channel. Note: milligrams per cubic water = μg/l; for nitrate, 100 μg/l ≈ 7 μM; for phosphate, 30 μg = 1 μM; for silicate, 200 μg ≈ 7 μM. After Gross (1982). Reprinted by permission of Prentice Hall, Englewood Cliffs, New Jersey.

the relatively benign nitrate. Both ammonia and nitrite occur at a fraction of the concentration of nitrate in the water columns of most aquatic ecosystems. Algae and bacteria both compete for these compounds as sources of nitrogen and energy, and if it is possible to measure ammonia and nitrite by any but the most sophisticated of chemical analytical processes, that is, if either is over about 0.5 μM (0.0007 ppm), they are probably much too high.

The nutrient picture in fresh waters not subject to intensive human activity is similar to that of the ocean, but different in some critical ways. First, nitrogen has many potential sources of atmospheric bacterial fixation on the land. The equivalent process is not available for phosphorus, and thus phosphorus tends to be limiting for plant production, rather than nitrogen. It is phosphorus that operates on a yearly cycle to limit primary production. Figure 3 shows typical summer patterns on several lakes with low levels of phosphorus at the surface markedly increasing below the thermocline. On the other hand, near the bottom, additional mechanisms quite different from those applicable to nitrogen are active (Fig. 4). Under aerobic conditions, phosphate combines with ferrous iron in the sediments. The precipitate that forms remains locked in those sediments until they approach anaerobic conditions. Thus, in the stratified and stagnant summer conditions, low oxygen reverses this reaction and allows reactive phosphorus to escape into the lower water column. However, being generally below the photic zone except in the smallest lakes, it remains largely unavailable for plant growth.

Silica is a special case in that it is essentially only diatoms that make an ecologically important use of the element. Silica is one step up from carbon in the periodic table of the elements. Thus, chemically it is quite similar to carbon, but is a little over twice as heavy. Forming an enormous number of compounds as minerals in the earth's crust, it is to the lithosphere what carbon is to the biosphere. Some science fiction writers have taken to the concept of life elsewhere in the universe based on silica instead of carbon. Diatoms use silicate to form their cell walls, instead of cellulose or related compounds. They are essentially alone in this use, although a few other groups such as radiolarians and some sponges make silica skeletons, which can be extremely abundant both in the plankton and on the bottom in specialized localities. Thus, while silica (as silicate) is moderately abundant in water environments (about one-tenth of carbon and six times that of reactive nitrogen), it can sometimes be limiting to plant production when that production is dominated by diatoms.

The point that we have been leading up to is that under prehuman conditions, in all aquatic and marine environments except relatively rare naturally eutrophic situations, the low concentrations of nitrogen, phosphorus, and silica were providing considerable restrictions on the level of plant activity and therefore on the level of total biological activity. An upwelling zone with a prevailing offshore wind, a whale washed up on

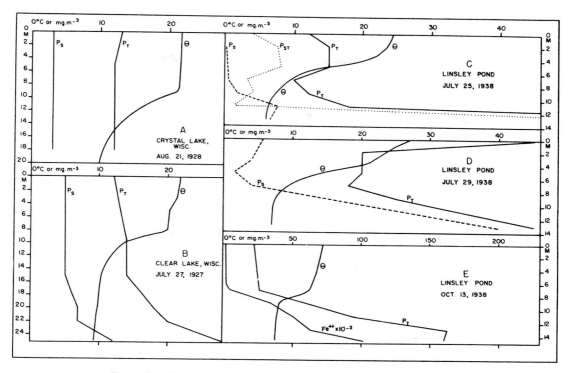

Figure 3 Phosphate concentration in several temperate lakes during the summer. θ, temp. °C; phosphorus: P_T, total; P_S, soluble; P_{ST}, in particulates. After Hutchinson (1957). Reprinted by permission of John Wiley & Sons, Inc.

the shore of a restricted inlet, a marshy basin on a recently burned slope, organic detritus concentrated on a beach or a mud flat by a storm—these were naturally eutrophic or nutrient-rich situations, restricted in space and time. Man's tendency to concentrate a wide variety of organic resources at points deemed most efficient for his activities has led to most organic pollution.

In the last chapter we discussed the basic contrasting roles of plants, animals, and bacteria in nutrient cycling. Plants "seek out" nutrients to build their tissues. Animals eat plants, their products, or plant-eating animals. Finally, bacteria break down all organics, releasing nutrients to soils, sediments, and natural waters. As long as excessive waste organics and thereby nutrients are not concentrated, the only problem is maintaining recycling sufficiently rapid to keep up plant activity. Without man there is no nutrient problem in the biosphere. The few naturally eutrophic water bodies, mostly small ponds or lakes subject to spring flooding, simply provide an element of diversity.

Some heterotrophic bacteria, primarily under anaerobic conditions,

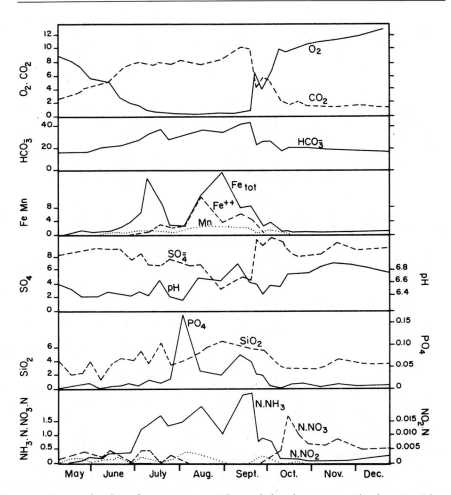

Figure 4 Seasonal cycling of oxygen, iron as Fe²⁺, and phosphate just over the deepest mud, at 14 m, in an English lake. After Hutchinson (1957). Reprinted by permission of John Wiley & Sons, Inc.

utilize nitrate or nitrite and produce the gases nitrous oxide and nitrogen. This process of denitrification removes nitrogen from an aqueous medium or sediments to the atmosphere. In the water column of most lakes and the oceans, this process is negligible. However, in the anaerobic sediments of eutrophic bodies of water (enriched by human activities) denitrification rates can be considerable, in some cases at the levels of primary production (Seitzinger, 1988). On the other hand, in noneutrophic shallow waters, low in nitrogen, denitrification, mostly from underlying sediments, occurs at rates an order of magnititude or more below primary production, particularly when that is achieved largely by macrophytes or attached algae. Below about 15–20 μM (N as $NO_2 + NO_3$) (0.2–0.3 ppm) denitrification is negligible. Below about 1–2 μM (N as $NO_2 + NO_3$) (0.014–0.03 ppm) nitrogen fixation begins to occur (Lucid, 1989; Chapters 20 and 22).

Eutrophication (Nutrification) of Natural Waters

Excess concentration of organics and nutrients leads to a variety of problems in wild ecosystems. Most critical for marine and freshwater environments is ammonia. When this very toxic nitrogenous excretion of most water-living organisms (see Chapter 9) is available in abundance, and sufficient plant or microbe activity is not present to remove it, it can become deadly in restricted areas. A number of bacteria and fungi also perform nitrification, or breakdown of ammonia to nitrite, though they generally require oxygen and release carbon dixoide. Excess organic material without consequent plant return to the stored reactive carbon state also leads to a drawdown of oxygen and eventually anaerobic conditions. The combination of anaerobic conditions in the water column and the often toxic products of anaerobic bacteria (including H_2S) provides environmental conditions that are suitable for very few organisms.

As we will discuss under Bacterial Filtration in Chapter 11, it has become customary for man to avoid the most serious problems of excess concentration of organics (generally sewage) by providing large bacteria-operated industrial plants to break down these organics. A sewage plant releases carbon dioxide to the atmosphere, reducing biological oxygen demand (BOD) in the water effluent. The bacterial biomass that develops (sludge) is hauled away to environments of less concentration (ocean dumping, for example). Most of the nutrients, as nitrites, nitrates, and phosphates, are then dumped into estuaries, lakes, and streams. While this process has avoided the most immediate difficulties of mass sewage—namely, disease transmittal and local low oxygen—when carried out with large human populations it only extends the basic problem a few years. In and around cities we are clearly exceeding the carrying capacities of local environments with current technology.

Many algae and some aquatic plants are adapted to make use of the temporary availability of nutrients in wild ecosystems (the dead whale, for example, or the school of fish). The life strategy of these "opportunists" is extremely rapid growth, given adequate light, temperature, and a large peak of nutrients. Given summer-long or permanently high nutrients, these plants multiply rapidly, choking bodies of water and eventually reducing oxygen, killing themselves as well as many invertebrates and fish at the same time. One does not have great difficulty in understanding why Jim Lovelock in *Gaia* (1979) chose an adulterous union between a phosphorus-gathering bacterium and a blue-green alga to slime-coat the world ocean and kill off themselves and Gaia in the process. Likewise, for many humans, it is the fast-growing algae or the equivalent macrophyte such as *Hydrilla* or water hyacinth that is the problem, not ourselves who have created the unnatural high nutrient situation. On the other hand, it is possible to use the same organisms in a controlled way to reduce eutrophication (see next chapter).

Nutrients and Model Ecosystems

Why are we discussing serious man-made environmental problems in a book on mesocosm and aquarium technology? Because basically the organic pollution problems are the same for the aquarist as they are for human society at large. Only the scale is different, and if as many humans kept ecosystems in terraria and aquaria as now keep dogs, cats, and canaries, we would be a giant step closer to solving our environmental problems on a world scale. It is instructive to compare wild ecosystem nutrients discussed above with those for traditional aquaria. We quote from Spotte (1979): "It is inevitable that seawater aquariums will become eutrophic. . . . inorganic nitrogen levels in aquarium water can be staggering—[e.g.] 50 mg NO_3^- $-N/l$; 165 mg NO_3^- $-N/l$; 309 mg NO_3^- $-N/l$; 70 mg NO_3^- $-N/l$; [with] a 10% partial water change weekly 3 mg PO_4-P/l; 6 mg PO_4-P/l." These values are over 5000 times and 1000 times higher, respectively, than the natural levels discussed earlier in the chapter. We recently had a reliable report for a major U.S. reef aquarium of 450 mg/l for NO_3^- $-N/l$. Few, if any, invertebrates are present in this aquarium. Many of the fish are sickly and the death rate is high.

As we discussed in Chapter 9, animal tissue that has become food for other animals or for bacteria or fungi undergoes a continuous process of breakdown and assimilation. Carbohydrates and most fats in this breakdown process are metabolized using oxygen and result in the release of carbon dioxide. On the other hand, proteins, nucleic acids, and phospholipids (along with other compounds) provide an excess of nitrogen and phosphorus in addition to the basic carbon, hydrogen, and oxygen of organic tissues. The net result is a requirement for the excretion of nitrogen-rich and phosphorus-rich compounds into the ambient water. Chief among these excretory compounds is ammonia, a substance that is quite toxic to most organisms when it occurs in abundance. The chief problem in traditional aquarium management, and in most model ecosystems for that matter, has been the preventing of the "pollution" of the system by these excretory compounds. It is primarily to solve this very basic problem that the traditional biological filter was developed. Most marine aquarium manuals devote one to several chapters to filtration methods, the chief function of which is nitrification or oxidation of ammonia by bacteria living in the filters. It is management of this filtration sytem, which is really a microvariant of the municipal sewage system or the hotel tertiary system, that is the basis of both home and public aquarium operation. Unfortunately, in many cases reliance on the bacterial or "biological" filter is counterproductive in that the solution of the ammonia problem is accompanied by a number of negative side effects that render natural ecosystem management difficult or impossible to achieve.

Marine and freshwater planktonic algae actively photosynthesizing and growing have a continuous and often production-limiting requirement for nitrogen and other nutrients. Approximately one nitrogen atom is needed for every 7 carbon atoms built into plant tissue and one phosphorus atom for every 100 carbon atoms. Benthic algae, on the other hand, are capable of producing at a considerable deficit when nutrients are low (Fig. 5). Likewise many fleshy algae re noted for their "luxury" or excess consumption of nutrients when they are available. Most algae accept ammonia as a nitrogen source and are capable of rapidly taking it up. The nitrification products of ammonia, nitrite, and nitrate, as well as more complex nitrogenous excretory products such as urea, can also be used by many algae. In reef tank ecosystems, with dense algal and higher plant communities, nitrogen salts, measured with autoanalyzers, typically occur at a concentration of less than 1–2 μM. This is equivalent to the situation normally encountered in a wild reef and yet is at great variance with traditional aquarium systems where nitrogen salt levels of over 100 ppm are acceptable. Limiting of sensitive organisms in aquarium systems by ammonia, nitrate, and nitrite, as well as the unstable environment these nutrients create in excess, is a major element in the unstable nature of the traditional aquarium.

Tropical reef communities in particular are characterized by low nutrient levels. Nitrate levels in incoming ocean water in St. Croix reefs are about 0.4 μM (0.006 ppm). More open ocean or equatorial current situations (such as Grand Turk, for example) are even lower, on the order of 0.1 μM. Blue-green algae in reef waters are known to fix gaseous nitrogen much as the bacteria of the root nodules of legumes do in the terrestrial environment. However, these blue-greens are either not present or unable to fix nitrogen in any but the nutrient-minimum environment. At the very least, high nutrient levels cause a shift in algal community structure, a shift that could be very detrimental to an attempt to simulate an ecosystem. It has been well known for several decades that perhaps next to siltation, elevated nutrients (or eutrophication) are one of the primary factors causing degeneration of wild reefs. Equally important, though indirectly so, a serious problem for coral reefs as well as most open shallow-water ecosystems is the relation of the presence of such high nutrient

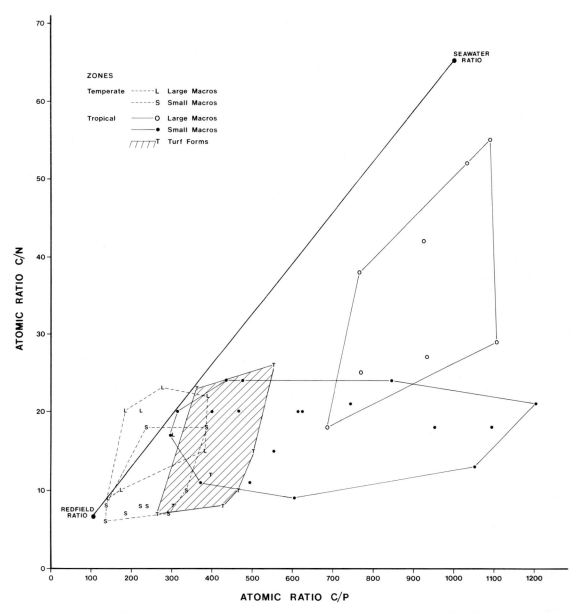

Figure 5 Carbon, nitrogen, and phosphorus ratios in marine plants. Note that temperate algae studied were apparently not nutrient deficient, tropical algal turfs were not nitrogen deficient (due to nitrogen fixation) but were slightly phosphorus deficient, and larger tropical algae were deficient in both nutrients. Productivities as measured by biomass increase were approximately equivalent in all groups. (Adey, 1987).

levels to the growth capabilities of some planktonic algae. Solar radiation or equivalent light levels soon induces disastrous planktonic algal blooms at the nutrient concentrations of traditional aquarium tanks.

Moving on to other communities that have been involved in microcosm or mesocosm simulation, both Maine shore and temperate estuarine systems such as the Chesapeake Bay have relatively high concentrations of dissolved or reactive nitrogen (Figs. 6 and 7). Those naturally rich ecosystems, however, at 5–10 μM $(N - NO_3^-)$, for unpolluted waters and 10–80 μ $(N - NO_3^-)$ for polluted waters,), are still far below the older type aquaria at 5000 μM $(N - NO_3^-)$ or higher. Note that in Figure 7, 25 μM $(NO_2^- + NO_3^-)$ and 13 μM (total phosphorus as PO_4^{3-}) are noted as critical levels. Both systems are also characterized by having both a sediment sink and an open ocean sink for such nutrients. If rocky shore and estuarine communities are operated properly, with considerable nutrient importation, then a "nutrient sink" is required to prevent eutrophication and radical alteration of the biological communities. Most such coastal marine communities are characterized by runoff from the adjacent land. Without the ocean sink to eventually dilute and remove those nutrients, eutrophication would result. Most researchers feel that Chesapeake Bay is at present on a rapid slide to organic death partly because of eutrophication. In this case it is runoff from the fertilization of soils by farmers and

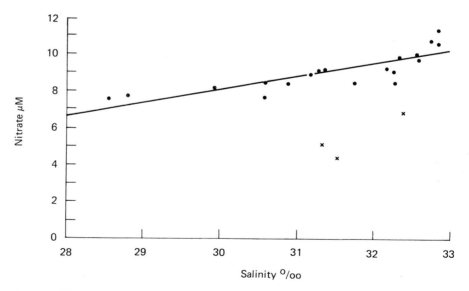

Figure 6 Nitrate concentration as a function of salinity in the Sheepscot estuary (Maine), September 1986. After Fefer and Schettig (1980).

gardners, as well as nutrient-rich outflows from sewage plants. Without a comparable increase of oceanic exchange, the bay is doomed. In a sense, through intense human overloading, we are making the bay into a huge

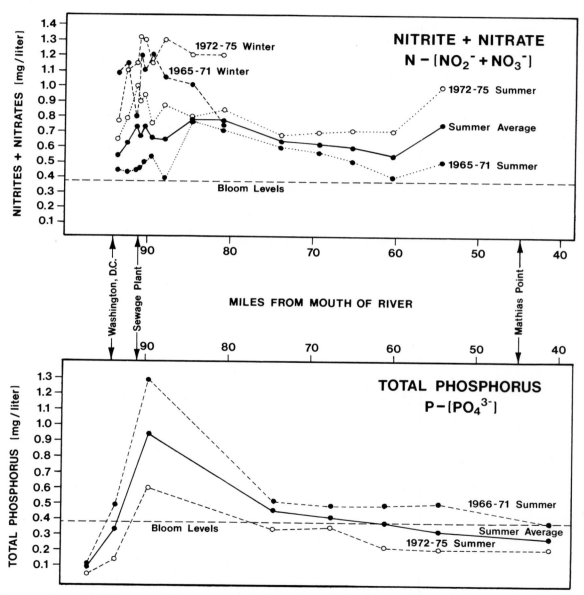

Figure 7 Nitrate and nitrite and total phosphorus concentrations as PO_u on the upper Potomac estuary, summer, 1972–1975. After Lippson *et al.* (1979).

aquarium, which must be handled properly or it will "crash" just as traditional marine aquaria have been prone to do. In the simulations of Chesapeake Bay and the rocky Maine shore described in Chapters 22 ₐ 23, dissolved nitrogen concentrations are normally maintained in the 10 μM range, below presently existing levels in the wild ecosystem, l probably slightly above prehuman levels.

Some aquarists have advocated using the process of denitrificatiₒ in special anaerobic traps to remove excess nitrogen from mesocosm aₗ aquarium environments (see Moe, 1989; also see Chapter 12). Indeed, th process probably occurs to some extent in virtually all closed systems. l eutrophic aquaria with rich sediment bottoms, it will certainly be an important process. However, in sediments and their overlying waters that are not enriched far beyond natural levels, denitrification rates are likely to be below 30–50 μM N/m²/h (Seitzinger, 1988). This is one to two orders of magnitude below levels of the primary production or photosynthetic removal rate of nitrogen, as we will discuss in Chapter 12. In addition, denitrification does not remove a balanced array of nutrients or pollutants as primary production would tend to do.

Summary—Nutrients and Model Ecosystems

The handling of nitrogenous and phosphorus wastes in aquarium simulations of ecosystems can be reduced to several crucial requirements:

1. An animal load that is at least close to that in the wild (an excessive load technically can be managed; however, it makes the system more equivalent to an aquaculture or farming unit and presents numerous balancing problems).

2. A built-in system for rapidly taking up ammonia and other waste compounds. In most aquarium systems this is the bacteria or "biological" filter. In most natural shallow-water ecosystems and in the examples in this book, this is accomplished directly by plants or by export/mixing with an adjacent community (a larger open body of water).

3. A means of ultimately removing added nutrients. In the wild this is partly accomplished by oceanic dilution and sinking to deep water, although burial in sediments and soils acts as a primary sink.

The first requirement, that of an appropriate animal load for the area and volume involved, is discussed in depth in a number of the following chapters. The second requirement can be met by providing the proper plants for the system being worked, and the right environmental factors, light, and water motion enabling those plants to function. These elements are discussed primarily in Chapters 13 and 15. Ion exchange media, including charcoal and newer resins and plastic materials with great porosity and ion adsorbtion characteristics, can be used to reduce nutrients as well as dissolved organics. These media all suffer from difficulties of control and balance in the chemical soup that is a "living" water, especially sea water. Also, as in all filtration methods, plankton and larval stages are trapped and mostly destroyed (see also Chapters 12 and 18). Finally, the effects of an open water sink relative to nutrients could also be provided by plants, given a proper control mechanism. Several mechanisms have now been devised and are discussed at length in the next chapter.

References

Adey, W. 1987. Food production in low-nutrient seas. *Bioscience* **37**: 340–348.

Fefer, S. I., and Schettig, P. 1980. *An Ecological Characterization of Coastal Maine.* Vol. 2. Department of the Interior, U.S. Fish and Wildlife Service.

Gross, M. G. 1982. *Oceanography.* 3rd Edition. Prentice Hall, Englewood Cliffs, New Jersey.

Hutchinson, G. E. 1957. *A Treatise on Limnology.* Vol. 1. John Wiley, New York.

Lippson, A. J., Haire, M. S., Holland, A. F., Jacobs, F., Jensen, J., Moran-Johnson, R. L., Polgar, T. T., and Richkus, W. A. 1979. *Environmental Atlas of the Potomac Estuary.* Environmental Center, Martin Marietta Corp., Baltimore.

Lovelock, J. 1979. *Gaia, A New Look at Life.* Oxford University Press, Oxford.

Lucid, D. 1989. Effects of dissolved inorganic nitrogen concentrations on primary productivity, nitrogen fixation, and community composition of coral reef algal turf: A microcosm study. University of Maryland, M.S. thesis.

Moe, M. 1989. *The Marine Aquarium Reference.* Green Turtle Publications, Plantation, Florida.

Seitzinger, S. 1988. Denitrification in freshwater and coastal marine ecosystems: Ecological and geochemical significance. *Limnol. Oceanogr.* **33**: 702–704.

Spotte, S. 1979. *Sea Water Aquariums, The Captive Environment.* John Wiley, New York.

Sverdrup, H., Johnson, M., and Fleming, R. 1942. *The Oceans, Their Physics, Chemistry and General Biology.* Prentice Hall, Englewood Cliffs, New Jersey.

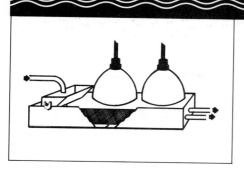

CHAPTER 12

CONTROL OF THE BIOCHEMICAL ENVIRONMENT
Filters, Bacteria, and the Algal Turf Scrubber

In Chapters 9–11, we discussed the enormous role that organisms play in the chemistry of the local water environment as well as the surface of Planet Earth as a whole. In the terrestrial environment and in shallow, well-lighted waters, plants generally dominate over animals in their chemical role and provide a high-oxygen, high-pH, and low-nutrient situation to which most larger complex plants and animals are adapted. There are "natural dumps," marshes, mud flats, forest streams, and the deep, poorly circulated and generally muddy zones of oceans, lakes, and some bays and estuaries. Here organic matter accumulates and microbe activity and the activities of specialized animals that feed on microbes become important. In addition, below the well-lighted zones the microbe activity dominates over plants and the more complex animals, and "inverse chemistry"—low oxygen, relatively low pH, and high nutrients—rules.

Deeper ocean waters form an enormous storehouse of nutrients that are returned to the surface at upwellings. Generally, given the vagaries of the dynamics of the earth's lithosphere and long-term climate, many of the organic "dumps" as well as their rich load of organics pass into semipermanent geological storage. Over geological time photosynthesis and plant production have greatly exceeded animal production and microbe breakdown, producing an oxygen-rich atmosphere and abundant organic storage. Over 1600 times as much carbon exists in the form of coal, oil, limestone, and related deposits within the lithosphere as exists in the biosphere (see Chapter 10).

In the shallow, well-lighted, and well-circulated surface zones of both marine and fresh waters, much patchiness exists. Surfaces and bottom communities provide for greater living activity on the ecological scale, just as surfaces (membranes and organelles) provide the means for higher, more active biochemical activity in the cells of organisms. On a rich kelp bed or high-biomass coral reef, even if plant production exceeds animal use and microbe breakdown, implying export elsewhere or a seasonal buildup cycle, little oxygen is stored. With relatively slow diffusion from the atmosphere available oxygen will be insufficient at night to support respiration of the large biomass. Likewise, a large mussel bed derives its food from the overlying water being moved by the tides and waves. With food brought in, excreted nutrients must be exported with the same tides and waves. Many aquatic communities, particularly rich bottom communities, rely on interchange with a larger, open body of water in which biomass is lower, oxygen levels are high and generally stable, and nutrient levels are low, sometimes even limiting to planktonic plant production. In model ecosystems, creating this interchange with the adjacent ecosystem is a primary requirement and is the basic function of filtration.

In microcosm and mesocosm simulations of wild ecosystems, the aquarist is attempting to re-create all of the physical and biological characteristics of the wild ecosystem involved. In a reasonably accurate simulation, filtration or chemical control should be involved only in providing the effects of the missing, larger body of water. Sometimes, when a mesocosm is located adjacent to such a body of water, this can be done simply by pumping or otherwise exchanging with the existing ocean or lake. Some of the most successful public aquaria, such as the Monterey Bay Aquarium, use this approach. Otherwise "filtration" or other control is necessary. The small aquarium situation is generally similar, but the problem is more acute for a variety of reasons: (1) even smaller size, (2) often inaccurate environmental simulation, and (3) the tendency to exaggerate biomass, particularly large animal biomass for reasons of display.

In this chapter we will discuss the traditional methods of handling this problem, primarily those that rely on microbe management. We will also discuss several recent variations on the traditional approaches as well as higher plant and algal methods of solving the problems.

Sterilization Methods

First, an approach that has been used to keep fish and a relatively few invertebrates is the sterile or "hospital" approach. These methods rely on chemicals, treatments including ozone sterilization, and physical processes such as ultraviolet sterilization. They are equivalent to greenhouse culture in which a few species are maintained and all others are killed using insecticides and herbicides or constant weeding. The sterile approaches deal with a few species, and they are inherently unstable, requiring constant attention to avoid "pests" and environmental collapse. Sterile methods never approach natural ecosystem simulation, and we will not discuss them further. The reader is referred to the many aquarium manuals on the market, including those cited below, for information on these methods. It is the view of the authors that equivalent methods applied to an ever-increasing area of our biosphere for farming and landscape "beautification" combined with full habitat destruction in cities and industrial zones are leading to the destruction of the ability of the biosphere to support higher animals including man.

Bacteriological Filtration

Our discussion of traditional "natural" methods will be brief. Virtually every aquarium manual discusses the bacteriological or biological filter. Recent quality treatments are Hunnan (1981) and Mills (1986). Particularly extensive treatments of biological filtration are those provided by Spotte (1979) and Moe (1989).

Some bacteria, including cyanobacteria or blue-green algae, are photosynthetic. Except for the blue-greens, photosynthetic bacteria are more or less rare, being largely confined to specialized environments. They do not split oxygen, using sunlight to acquire hydrogen ions, and are mostly obligate or facultative anaerobes. The reader is referred to Rheinheimer (1985) for a discussion of the photosynthetic bacteria. Cyanobacteria (blue-green algae), on the other hand, while having bacterial structure and lacking well-defined nuclei, plastids, and most other organelles of more advanced cells, carry out photosynthesis as do the higher plants. Blue-green algae are abundant in virtually all lighted environments. In difficult aquatic and marine situations [very low light, high light, low water quality, exceptionally high water quality (i.e., low nutrient levels), etc.] the blue-greens tend to dominate over other plants that require a more moderate environment. Otherwise blue-green algae are a subsidiary but omnipresent element of most plant communities. In this book, we will treat blue-greens as plants in general, as algae in particular, but recognizing that these critical organisms are probably on the evolutionary line between bacteria and algae.

Heterotrophic microbes, bacteria, and fungi that rely for their nutrition on dead plant and animal tissues are truly ubiquitous in aquatic and marine environments. Microbes are capable of breaking down to carbon dioxide, water, and inorganic salts (nutrients) virtually every organic compound known. Some substances are difficult, breaking down only slowly (e.g., lignin, chitin, hydrocarbons). They are the province of certain specialized bacteria. In normal wild aquatic and marine environments, typically 10–30% of all primary production is eventually cycled through microbes (see Chapter 19). In strongly eutrophic situations, up to 90% may be cycled through microbes (Rheinheimer, 1985). Most heterotrophic microbes function as animals in that the breakdown process includes respiration and requires oxygen. Well supplied with food, microbes are capable of pulling virtually all oxygen out of an ecosystem, and while some heterotrophic bacteria and fungi are capable of anaerobic breakdown, these processes (glycolysis and fermentation) are generally incomplete. All bacteria and fungi digest their food exterior to their bodies. Digestive enzymes are released into the environment, and only the products of digestion are taken up. In large measure the toxicity of some bacteria lies in these released enzymes or toxins rather than directly in the bacteria itself.

All organisms and biological communities produce organic wastes, chemical compounds such as urea, ammonia, organic phosphates, and fats and oils, as well as dead bodies and undigested foods. Bacteria will be present to break down all of these at various rates to their basic state of carbon dioxide, water, and nutrients. Biological and chemical activity is much greater on surfaces than it is free in an aqueous medium. The basic principle of biological (bacteriological) filtration is to provide surfaces within the system for bacteria and fungi to attach to, and over which water can pass. This provides a more efficient breakdown process. The filter can be as simple as a bed of sand or a more complex structure using plastic wools or similar material to increase surface. Simple centrifugal pumps are usually used to move water over the filter bed and back to the aquarium. Air-lift pumps are also capable of moving water in systems with minimum filtering requirements. The air lifts in this case serve a double function, also providing for aeration.

Within the filter bed, surfaces become coated with a bacterial film or slime, the equivalent of the aufwuchs community in wild ecosystems (see Chapter 19). These are shown diagrammatically in Figures 1 and 2. All of the breakdown processes that we discussed in earlier chapters occur on filter beds. These can be most simply demonstrated by relative bacterial numbers (Fig. 3). Some efforts have been made to calculate carrying capacities, the number of organisms and their weight relative to area, volume, and water flow in a filter, for example. Both as a table and an equation, the ability to determine carrying capacity is provided by Spotte (1979). These calculations are suitable for aquacultural type situations where one or perhaps a few species are being maintained in dense culture. Spotte, however, notes that "no method exists presently for calculating the carrying capacity of a low-density seawater aquarium containing an animal population of different species."

Virtually all marine and aquatic ecosystems have permanent planktonic components as well as numerous swimming reproductive phases.

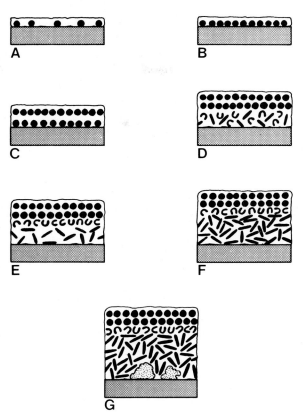

Figure 1 Formation of bacteria slimes on filter bed material; dots are aerobes, cups are aerobes breaking down under anaerobic conditions, rods are anaerobes; A–G is in the order of weeks; G shows beginning of sloughing off of mature film. After Spotte (1979). Reprinted by permission of John Wiley & Sons, Inc.

The animal or zooplankton elements feed upon microscopic plants as well as organic particulates and even bacteria in the water column. All plankters, both plant and animal, as well as the particulate organics, are fed upon by numerous attached or bottom-dwelling filter feeders within the ecosystem (see Chapter 18). Any filtration processes will remove plankton and swimming reproductive states as well as the organic particulates and ammonia and urea that the filter is designed to "catch." The physical adsorption processes, such as foam fractionation and granular activated carbon, that are designed to assist biological filtration in removing organic particulates, suffer from the same difficulty. An additional problem with adsorption processes is that they seriously remove trace ions from solution. According to Spotte (1979), both physical adsorption methods probably remove enough trace ions to interfere with the successful culture of algae. In addition, extensive bacterial cultures in a filtering

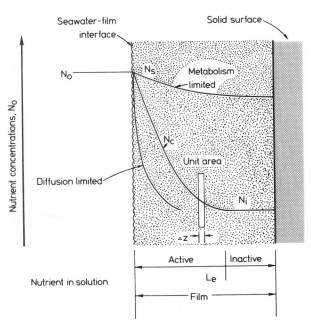

Figure 2 Schematic diagram of nutrient concentrations (e.g., ammonia) in a bacteria film. N_o and N_s, concentration in liquid and at film surface; N_c = balanced gradient between capability of bacteria for uptake and nutrient supply. The upper curve represents the case of a thinnish film and nutrients not fully metabolized. The lower left curve represents an old, thick film with insufficient nutrient supply. After Spotte (1979). Reprinted by permission of John Wiley & Sons, Inc.

situation place bacteria in competition with algae and other plants for organic nutrients, particularly in the form of ammonia and urea.

The most serious difficulty of biological filtration is that the process unbalances most potential ecosystems. In noneutropic wild ecosystems 10–30% of primary production is cycled through bacteria. Even in rich sediments, the wild equivalent of the biological filter, bacterial biomass is typically two orders of magnitude lower than infaunal (clams, worm, and crustacean) biomass (Rheinheimer, 1984). If a biological filter is used, a much larger proportion of production (or feed) is cycled through bacteria. Thus more oxygen is required from the water column and oxygen concentrations are lower—inevitably much below saturation no matter what aeration method is employed, short of bubbling oxygen. In addition, bacteria and fungi release digestive enzymes and toxins that help provide competitive advantage. While these are present in a wild ecosystem, they are lower in concentration and largely limited to "natural" substrates, where microbes are competing with numerous other organisms for a "place in the sun." Filter bacteria, like animals, release carbon dioxide, driving water to levels of lower pH or higher acidity. While this excess

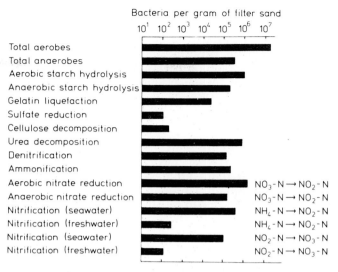

Figure 3 Populations of different functional types of bacteria in aquarium filters. After Spotte (1979). Reprinted by permission of John Wiley & Sons, Inc.

load can be compensated to some extent by using calcium carbonate chips within the flow of the filter system, the environment will remain below saturation for calcium carbonate. Even in the best run system, this inevitably provides a water column with the lower end of the pH range and stresses organisms, particularly calcifying organisms, that are adapted to the carbonate saturated environment.

Finally, biological filter systems inevitably provide a high-nutrient environment. There are many aspects to the eutrophication problem, including unbalance to some opportunist algae and microbes, cancer in humans and undoubtedly many animals, and many effects as yet unknown. Spotte (1979) sums up the situation relative to aquaria: "Overall, eutrophication in seawater aquariums reduces the carrying capacity of the water, interfering with its ability to support life."

The great enhancement of microbe breakdown, using filters designed to maximize the water contact with bacterial communities, is designed to rapidly mineralize (reduce to carbon dioxide and basic nutrients) all organic material once its "function" (culture for food or viewing) is accomplished. It is perhaps an extension of a human view that resources are best when they are easily attained. In the pre-human biosphere, most nutrients were not in the free state. Some were located in living organisms. The vast majority were at least dynamically "locked up," in storage, in the deeper oceans, in soils and sediments, and in the geological state. The long-term internal stability of the biosphere probably derives from this "locked up" state.

Reef Systems

During the 1980s, a variety of improvements to aquarium standard filtration methods were introduced (Moe, 1989). Generally, these methods ("reef systems") increased water flow, aeration, and light, and somewhat reduced dissolved nitrogen concentrations (Fig. 4). "Reef systems" have been partly successful in keeping invertebrates, particularly some corals, that previously could not survive in aquaria, although fish loadings must remain minimal to do so. Since many of the invertebrates typically kept in these systems have zooxanthellal algae in their tissues, they are effectively high plant biomass cultures requiring relatively high light intensities. Although a tremendous advance for keeping many invertebrates, "reef systems" still fall short of many basic requirements for maintaining aquatic ecosystems.

Denitrification

In recent years a new variant to the biological (bacterial) filter systems has been devised that also includes a bacterial denitrification unit. Bacterial denitrification, particularly under anaerobic conditions, is quite capable of removing nitrogen to the atmosphere, and in any ecosystem, microcosm, or aquarium with anaerobic or near anaerobic sediments, this process will be active. However, at the nutrient levels of most natural situations, denitrification rates are at least an order of magnitude below nitrogen removal in photosynthesis (see Seitzinger, 1988; also see Chapter 11). Standard "reef" aquarium methods now in use can reduce dissolved nitrogen to less than 1 ppm (Moe, 1989; Smith and Schuman, 1991). However, this is about 70 μM N, far above that of most natural aquatic ecosystems. Denitrification as a partial process also solves none of the remaining problems of filtration. Since bacterial filtration, even when including denitrification, has many drawbacks and in addition is poor at handling phosphorus overload, except with a version of the foam fractionator, one is led to ask: Why not treat the "problem" as it has been handled in the wild for eons, by plant photosynthesis?

Plant Methods

Plants, including marine and freshwater algae, require quantities of nitrogen to build their tissues. Ammonia, nitrite, nitrate, amino acids, and even urea and related compounds are sources of nitrogen for many algae (DeBoer, 1981). Phosphorus is generally taken up by plants as phosphate, though in some cases organic compounds may also be useful to some

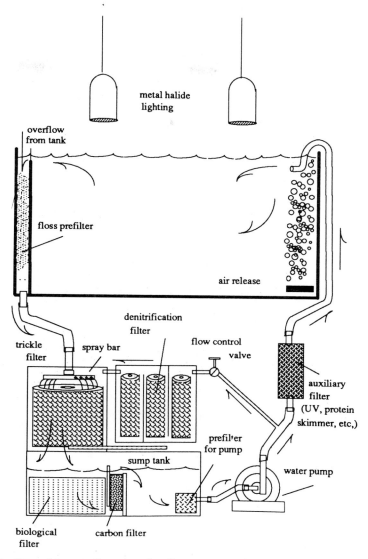

metal halide
lighting

overflow
from tank

floss prefilter

air release

denitrification
filter

trickle
filter

spray bar

flow control
valve

auxiliary
filter
(UV, protein
skimmer, etc,)

prefil'er
for pump

sump tank

water pump

biological
filter

carbon filter

Figure 4 General layout of modern "reef" system. Not shown is typically used trickle bed through a calcium carbonate gravel. After Moe (1989).

algae. Algae absorb these substances from the surrounding water, effectively removing them from the environment by locking them in their tissues. Carbon dioxide also is required for photosynthesis, and is used to build storage sugars and eventually a wide variety of tissues in plants. Through plant metabolism, these animal wastes become converted to forms that once again can be utilized as food by other animals. Algal metabolism also produces oxygen and a variety of essential vitamins that animals and filter bacteria alone generally cannot manufacture.

To simulate the high water quality of a noneutrophic larger body of water for ecosystem modeling in microcosms and mesocosms and to close the gaps in the natural recycling of organic compounds within a closed aquarium system, a variety of techniques that utilize these attributes of the algae as well as those of their higher plant, freshwater, or marine cousins have been developed.

Though the concept of employing algae for aquarium water conditioning is not new (see Spotte, 1979; Honn and Chavin, 1975), it is rarely used to any great extent and certainly has not been developed to its full potential. Encouraging microscopic plant growth within aquaria has become standard practice in recent years. "Live rock" is primarily a fragment of rock or dead coral with its included algal or plant growth. In some cases, to increase the beneficial effects of algal growth, special culture trays containing stands of larger algae have been connected in line with other filtration and purification equipment (Fig. 5). Although some modern writers on the subject of aquaria encourage the concept of algal "filtra-

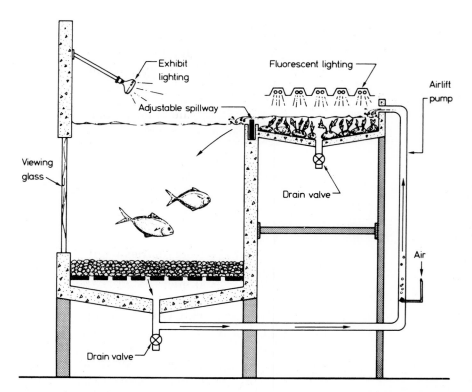

Figure 5 Use of macroalgae for aquarium water conditioning. After Spotte (1979). Reprinted by permission of John Wiley & Sons, Inc.

tion," it is generally regarded as impractical because algal photosynthesis is considered largely uncontrollable and algal production rates are viewed as requiring unreasonably large areas of algae (Spotte, 1979). Another of algae's supposed negative side effects is its "leaky" nature—its tendency to release organic compounds into the water column. The method we will describe is entirely controllable and optimizes algae's potential for water purification, oxygenation, and organic recycling to such a degree that it renders all other methods unnecessary and does not make use of plankton-damaging filtration. In addition, it requires less space and less plumbing than the typical modern "reef" filtration system.

Algal Turfs

The idea of using turf algae instead of a bacteriological filter to purify water for microcosm research was born during research on tropical reefs, home of the earth's most productive (photosynthetic) systems. In reefs, where plants are supplied with abundant sunlight, intense wave action, and strong currents, natural recycling processes occur at a great rate, generating more plant material than in almost any other environment. Compared with the prolific animal life, however, few plants are apparent on most reefs. The high rate of plant production is accomplished mostly by a small, seemingly insignificant group of plants known collectively as algal turfs (Adey and Hackney, 1989). Some photosynthesis also occurs with symbiotic microalgae inside corals, and in large plants such as *Sargassum*, as well as the sea grasses. However, most plant production in the typical shallow water reef is performed by algal turfs.

Algal turfs are short, moss-like mats of algal filaments commonly covering most hard surfaces in shallow reef zones (Fig. 6 and Color Plate 10). They are usually no more than several centimeters tall. Their actual tissue production rate is 5–20 g dry weight per square meter per day—a rate several times that of most terrestrial plants, including carefully tended agricultural crops. Algae, in a well-developed turf, can absorb typically 0.3 to 1.2 g N a day per square meter of screen.

Algal turfs are communities within themselves. A coral reef surface or an attached meter-square plastic screen in a warm, well-lighted, high-water-quality tropical environment would typically contain 30–40 species of plants with red algae, blue-greens, and diatoms dominating (Adey and Goertemiller, 1987). Of those major groups, diatoms tend to be the early colonizers. Later, blue-greens are most important in very high light and reds in more shaded environments. Typically, green and brown algae are lesser but omnipresent components. A similar situation in subarctic-temperature waters would provide the same overall floral composition, though small brown algae tend to be more important and usually about one-half the species number is present. In fresh waters, greens become more important, though with blue-greens generally still dominant. Reds and browns do not appear. In waters of lower quality, the number of species is generally much reduced. Algal turfs grown in raw sewage, for example, have a community that tends to be composed of a few species of blue-green algae (cyanobacteria) and perhaps a green or two.

Algal turfs are the equivalent of grassland communities. They must be grazed (or harvested) or the community will build up to the point where the basal portions die and the turf sloughs off. If larger algae are available, the community will go through a succession to a less productive, higher biomass macroalgal "forest." Algal turfs also have their "locusts" (generally amphipods or chironomid insects), which in the wild are kept moderate in numbers by small fish predators. In the presence of these micrograzers and lacking grazing by large organisms, a macroalgal or higher plant community of moderate to low productivity is inevitable with time. The practical aspects of maintaining algal turfs for managing enclosed ecosystems are discussed below. For a more in-depth discussion of this plant community, its definition, composition, and biology, the reader is referred to Adey (1987) and Adey and Hackney (1989).

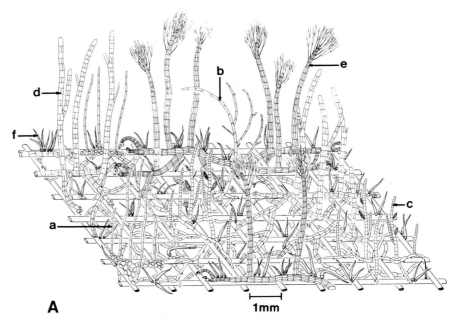

Figure 6 (A) Typical tropical algal turf growing on a 1-mm mesh plastic screen. Genera shown: (a) *Pilinia*, (b) *Cladophora*, (c) *Giffordia*, (d) *Sphacelaria*, (e) *Herposiphonia*, and (f) *Calothrix*. (B) Typical, layered algal community on a fresh water scrubber screen. The lower layer is dominated by the blue-green *Calothrix*, the middle layer by the blue-freen *Oscillatoria* and the green unicell *Gloeocystis*. The upper layer consists largely of long strands of the green alga *Ulothrix*.

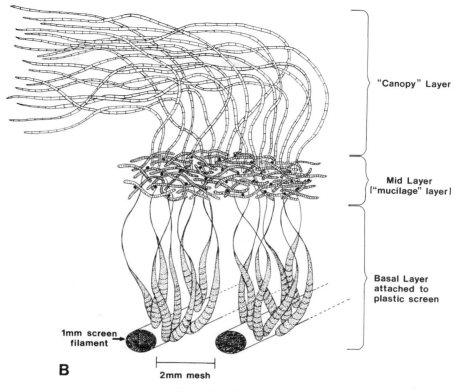

"Canopy" Layer

Mid Layer
("mucilage" layer)

Basal Layer
attached to
plastic screen

1mm screen
filament

2mm mesh

B

Figure 6 (*Continued*)

The wave action and surge-like motion characteristic of reef surfaces are crucial. When turned off, the result is a radical drop in turf photosynthesis because the wave surge boosts the efficiency of the photosynthetic mechanisms by serving as a light "flasher." Algal turfs do not light-saturate at normal levels of solar energy. They are "sun plants" and can use all the sunlight they can get (Chapter 7, Figs. 16 and 17), though there may be a small reduction due to ultraviolet effects at depths less than 20–30 cm under tropical sun (Adey and Goertemiller, 1987).

In addition, wave surge also serves as a strong mixing agent, facilitating the exchange of metabolites needed or excreted by the algae. Another reason for the unusually efficient growth of these algae lies in their structure. The algal filaments are to a large extent simple strands, not differentiated into specialized forms or functions. Most cells in the plant mass are photosynthetic, absorbing light, carbon dioxide, and nitrogenous and phosphorus compounds, and producing oxygen and carbohydrates. In contrast, larger, more complex plants reserve much of their body for other functions such as structural support, reproduction, and protection, leaving a much smaller proportion for photosynthesis.

The very survival of turf algae hinges on their rapid growth rate, for in the wild they are subjected to extremely heavy grazing by herbivores. Aided by their capacity to absorb large quantities of sunlight and metabolites in relation to their body mass, algal turf spores settle, grow, and become reproductive within days or weeks on any available surface. Even when their upright portions are eaten by grazers, their basal filaments remain, having penetrated minute irregularities in the substrate. These filaments rapidly regenerate new growth to replace lost material.

The Algal Turf Scrubber

Overcoming oxygen depletion and waste buildup that occur in the dark in a biological community is a necessity, both in closed systems and in the wild. Shore and bottom communities benefit from the open ocean or open lake, which provides a continuous supply of oxygenated and otherwise high-quality water when photosynthesis stops. But maintaining a large reservoir equivalent to the open ocean is usually not feasible in a captive ecosystem.

Our solution to this problem is the algal turf scrubber, which employs the most efficient photosynthetic components of wild ecosystems and concentrates their activity under optimal conditions. The scrubber, which contains primarily algal turfs, is a separate unit, installed in the microcosm's or aquarium's water-circulation line and lighted in a cycle opposite to that of the main ecosystem (Figs. 7 and 8). While the microcosm is in darkness, the algae in the scrubber are photosynthesizing, insuring a continual supply of oxygenated water, and rapidly removing CO_2 and nitrogenous waste just when it is most abundant. The rate of oxygen production and the uptake of nitrogenous wastes can be easily controlled by changing the length of time the turf algae are lighted.

The scrubber duplicates the balancing process found in the wild, and is effective in full-salt, brackish, and freshwater systems from tropical, temperate, and cold-water environments. As we have discovered through years of research, algal turfs are available in most coastal environments, rocky shores, salt marshes, and estuaries as well as reefs, and need only to be provided with proper conditions to produce a rich growth. Unlike the bacteriological filter, which, in return for the removal of ammonia and some particulates, returns to an aquarium water depleted in oxygen and plankton and high in dissolved nitrates and carbon dioxide, the scrubber keeps oxygen near or above the saturation point and effectively removes all classes of dissolved animal wastes.

Scrubbers also offer the advantage of not destroying plankton or the young stages of bottom-dwelling invertebrates that naturally feed on that

Figure 7 This battery of two meter square algal turf scrubbers is placed on the Great Barrier Reef Marine Park Authority's coral reef mesocosm in Townsville, Australia. At this writing, it has operated as the primary water quality control system on that large model ecosystem for 4 years. Photo by Great Barrier Reef Marine Park Authority.

SCRUBBER

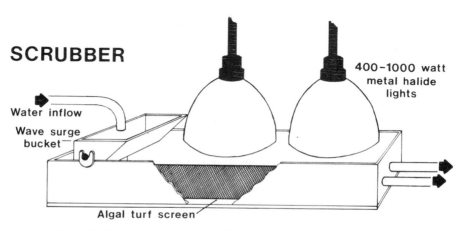

Figure 8 Diagram of standard scrubber showing all essential components.

rich food source. Further, they do not collect small organic particulates, which are better handled by sedimentation as described in Chapter 8, or left to filter feeders within the modeled ecosystem.

The algal turf scrubber is designed to produce the highest rate of plant production in order to remove a large quantity of pollutants. The diagrams shown in Chapter 21, Fig. 2 and Chapter 22, Fig. 1 show how shallow troughs, separate from but connected to the tank being serviced, create optimal conditions for algal turf growth: (1) maximum light; (2) wave surge; (3) controlled grazing; and (4) exclusion of animals that would offset the effects of the algae. This design also provides for coarser and finer tuning of algal functions by varying algal area, wave surge, and light levels and periods. Through harvest, algal turf scrubbing also presents an automatic and controllable source of organic export. Table 1 shows a variety of model ecosystems, microcosms, mesocosms, and aquaria, and the scrubber dimensions and lighting that have been used to successfully manage those closed ecosystems.

Turf algae can be grown on fine-mesh plastic window screen stretched on a frame that fits in the bottom of the trough. A somewhat coarser 2 × 4 mm black polyethylene screen functions as well and is considerably more durable. The screen can be seeded by suspension in an established microcosm or in the natural environment. An alternative, equally satisfactory, approach is simply to stock the ecosystem first with algal turf substrate and, once the scrubber is operating, add the remainder of the community. Because water in the trough is shallow, the turf-bearing screen is situated at or near the surface, and receives maximum light, which otherwise would be filtered through the water column.

For tropical, nonshaded algae, high-intensity metal halide lights or VHO fluorescents are used. For turfs from arctic and temperate regions, one can use high-intensity fluorescent tubes (very high output, VHO; high output, HO) (see Table 1 and Chapter 7). Scrubbers are lighted for 12–18 hours a day. They are best kept in darkness when the light in the community tank is brightest—that is, when the oxygen level in the tank is at its highest and nutrient levels are low.

Pumps deliver water from the main tank to the scrubber trough at a rate depending upon the size of the system, typically ranging from 3 to 10 gallons a minute depending upon the size of the scrubber. At the other end of the trough, water returns to the main tank through gravity drain pipes, or can be used to supply wave generator (see Chapter 25).

Assuming adequate light, algal production is limited only by inadequate exchange of metabolites—oxygen, carbon dioxide, and nutrients—between the water and the cells of the attached algae. We have demonstrated a strong correlation between wave surge and improved metabolic interchange: when we occasionally stop the wave generators in our main reef tanks—while maintaining a constant rate of circulation and level of light—immediately there is 50% reduction in oxygen production. The surge generated by the wave maker produces a back-and-forth motion within the tank, preventing the development of semistagnant boundary layers that occur when a constant flow of liquid passes by a fixed object. A steady current would tend to pin the filaments in an immobile position, and a surface layer of very slow-moving water would develop. Also, while exposed portions of the plants would derive most of the benefits of light and current, interior portions would be shielded, reducing metabolite exchange and photosynthesis. Strong reduction of algal turf production in scrubbers with lack of surge motion has also been directly demonstrated (Adey and Hackney, 1989). It is likely that an algal turf scrubber could be operated without provision for wave surge, but a greater surface area would be required for algal growth to produce an adequate rate of oxygen production and nutrient uptake.

Growing turf communities outside the main body of the model ecosystem protects the algae from ecosystem grazing, which would reduce its effectiveness. However, some small, herbivorous invertebrates can spread to the scrubber trough through the piping and become established. Chief among these are species of amphipods in salt water and chironomid larvae in fresh water. These micrograzers attach themselves to hard surfaces and some species construct small, protective nests. Thus, partially protected from the ecosystem's predators, these invertebrates release larval plankton and young animals, which serve as food for the animals in the main tank, just as they do for larger animals in the wild. However, when these organisms are allowed to proliferate unchecked, they can reduce the efficiency of the scrubber by eating the turf algae and contributing animal wastes to the system. Periodic scraping of all the wetted surfaces of the scrubber unit in addition to the algal screen keeps micrograzers in check, assuring full efficiency of the scrubber.

Periodic harvesting of algal turfs in scrubbers is necessary because these plants are most efficient when young. Those adding new growth use greater quanties of animal wastes than those simply supporting existing tissues. The dense, tangled mass of a mature community shades the turf's interior from light and restricts water circulation, reducing plant production per unit plant biomass. Very heavy algal turf growth also conceals flourishing colonies of amphipods and other invertebrates that consume plant filaments and oxygen and produce more wastes.

Harvesting is accomplished simply by removing the scrubber's screen and scraping away excess growth, with a plastic device like an automobile windshield ice scraper or even a razor blade. This procedure mimics the grazing that the turfs have evolved to counteract. It also effectively removes nutrients from the entire system by collecting nitrogen, phosphorus, and carbon that have become incorporated into the algal biomass. The basal filaments that remain intertwined in the mesh of the screen after scraping quickly send up new growth to replace what has been removed. Optimum harvest rates for all systems that we have worked

TABLE 1

Mesocosms, Microcosms, and Aquaria That Have Been Operated with Algal Turf Scrubbers[a]

	Years in operation (terminated)	Surface area (m²)	Volume (gallons/l)	Scrubber area (m²)	Ratio scrubber area/ system volume (cm²/l)	Scrubber lighting type/no./W	Scrubber power/area (W/m²)	System lighting type/no./W	System power/area (W/m²)	Operational nutrient level[b] (μM ($NO_3 + NO_2$))
Marine										
Tropical reefs										
S.I. Caribbean prototype (μc)	5 (1983)	4.57	2500/10,000	2	2.0	MH/4/1600	800	MH/10/400	875	<1
S.I. Caribbean exhibit (μc)	10 (1990)	11.6	3800/15,280	7.0	4.6	MH/13/5800	828	MH/22/14,800	1276	0.7
Australian I.P. mesocosm	3	646	750,000/ 3,000,000	80	0.26	MH/40/40,000 + natural light	500 plus nat. light	Natural light, Townsville, Aust.	—	<0.5
Home aquarium (reef)	3	0.85	130/520	0.18	3.8	HO/6/300	1670	VHO/6/960	1129	0.5–1.0
Subarctic										
Maine rocky shore and marsh	6	6.98	2500/10,000	1	1.0	MH/2/800	800	MH/14/5600	924	1–10

Estuarine										
Chesapeake mesocosm	4	44.8	15,000/60,000	3.9[c]	0.65	MH/8/3200	820	VHO/64/ + MH/32/42,240	942	0.5–8
Everglades mesocosm	2	130.3	22,000/88,000	3[c]	0.34	VHO/8/1280 MH/12/800 natural light	690 plus natural light	Natural light	—	3–15
Chesapeake home aquarium	1	0.85	120/480	0.082	1.7	HO/4/140	1700	VHO/6/960	816	
Freshwater										
Florida Everglades stream microcosm	2	37.6	4000/16,000	none	—	—	—	—	—	—
Black-water (So. American) stream mesocosm	1.5	7.7	2500/10,000	1.5	7.7	Natural light	—	Natural light	—	0.1–2.3
African pond aquarium	3	0.51	70/280	.085	3.0	HO/2/70	823	VHO/2/230	450	8–16
Black water (So. American) aquarium	1	0.51	70/280	0.072	2.6	HO/2/70	970	VHO/2/230	450	< 1

[a]Dimensions and lighting of both the ecological systems and their scrubbers.
[b]Sometimes operated higher or lower for research.
[c]Does not include "river" input scrubbers.

on to date have ranged from 7 to 20 days. Production rates from harvest to harvest vary widely. A mean standard deviation of a wide variety of microcosm scrubber runs was 26% of the means. *If harvest length is extended to visually fit production, invariably total harvest is reduced. Periodic harvest, preferably at a fixed interval, is necessary for optimum scrubbing.* Also, on screens established in the wild, 6–12 weeks and several harvests are usually required to bring them to full production. While the time required for laboratory or home aquarium screens to seed and come to full production is typically less (3–6 weeks), significant time is still required to allow succession to occur. In the last half of the succession process, normal harvest procedures should be carried out.

In addition to producing oxygen and scrubbing nitrogen and phosphorus, algal scrubbers also remove carbon dioxide and increase pH. In all marine systems (which are strongly buffered) pH increase only improves the match of the model with the wild environment. In fresh waters, unless one desires a highly acid environment, pH increase is normally offset by increasing animal biomass to achieve the desired levels. When a "black-water" acid environment is desired and strong algal scrubbing is employed, it may be necessary to acidify or bubble the environment with carbon dioxide. Addition of tannin-rich top up water can also be effective (see Chapter 24).

The size and scrubbing potential of an algal scrubber must be balanced against the community load to achieve the desired results. To some extent the scrubber algal community and production will adjust against the load, though the dimensions given below for a number of systems will allow a judgement as to approximate size. In underscrubbed systems, nutrients simply rise above desired levels and oxygen will fall below desired levels. In our experience negative feedback or crash situations do not develop. At the other extreme, that of overscrubbing, nitrogen can rarely be driven below about 0.5 μM for N as $NO_2 + NO_3$. Below about 1–5 μM, blue-green algae become more dominant on the scrubber screen and nitrogen fixation increases (Lucid, 1989). Of course phosphorus and other micro-nutrients do not have a natural method of input, except in food or evaporative water input. Once nutrient levels near the lower values found in the wild ecosystem being modeled are achieved, the harvested and dried algae can be returned to the model by a variety of means. We prefer grinding with a mortar and pestle and casting the organic particulates on the water surface. In very large models such return should reflect seasonal cycles—in small systems, daily cycles.

Algal Scrubbers and Synthetic Ecosystems

In microcosms and mesocosms, one is seeking fixed environmental and ecological parameters, namely, those of the wild analog that is to be modeled. A typical goal is a certain nutrient level, and in most of our systems we have used nitrate as a prime indicator. For example, in the 10-year-old Smithsonian Institution 3500-gallon Caribbean coral reef microcosm, a typical plot of nitrate levels for a period of several years is shown in Figure 9. These levels were achieved by five scrubbers having an algal turf area of 1.4 m² each. Nitrate plus nitrate concentrations in the water incoming to a scrubber are typically 0.5–0.7 μM. The water exiting is typically 0.05–0.1 μM, an 80–90% removal rate. Mean production levels are shown in Figure 10. Four of the five scrubbers operated with two 400-W metal halide lamps, the fifth with a single 1000-W lamp. This amount of scrubber area is relatively high per volume of system for research purposes (Table 1). The algae typically present on these scrubbers is shown in Table 2.

In a very different microcosm, the 2500-gallon 5-year-old Smithsonian Institution rocky Maine shore, two scrubbers of 0.5 m² each were

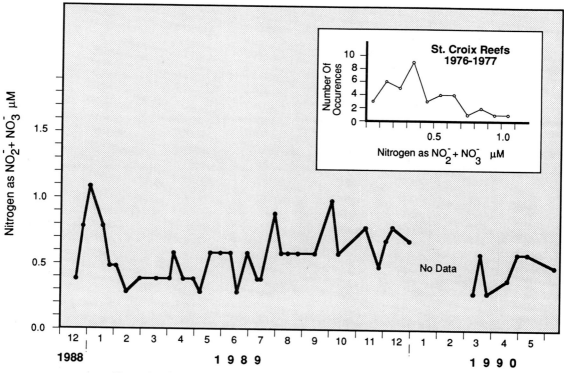

Figure 9 Plot of nutrient levels in Smithsonian coral reef microcosm from December 1988 to June 1990 as compared to dissolved nitrogen over St. Croix reefs in 1976–1977. The range is similar, although mean levels over the wild reef are about 0.2 μM (0.003 ppm) lower than in the microcosm. Note that reactive nitrogen is measured as the nitrogen in nitrite plus nitrate, although nitrite is typically a small fraction of nitrate.

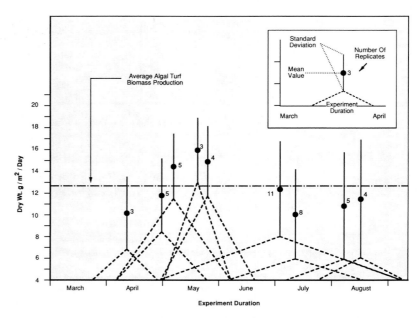

Figure 10 Diagram showing coral reef microcosm scrubber algal turf production for a six-month period.

used. Both of these produce at mean levels of 12.0 g (dry wt)/m²/day (Fig. 11). The scrubber algae dominant are species of *Ectocarpus*, *Enteromorpha*, *Cladophora*, *Polysiphonia*, and *Porphyra*. Blue-greens and diatoms were also present within the dense algal turf. Nutrient levels are typically maintained at 3–10 μM (N − NO$_2^-$ + NO$_3^-$), though on occasion have been driven below 1 μM with these scrubbers.

In small home ecosystem-based aquaria, the situation is rather different. Here, although an analog or type wild system is attempted, the volume is so small that accuracy of simulation is necessarily limited. For example, fish biomass is typically higher than in the wild, feed must be supplied to make up for the lack of forage area, and scrubber-to-volume area must be increased (see for example Table 1). The Australia reef mesocosm, for example, operates at 0.3 cm² of scrubber area per liter of ecosystem water, while our 120-gallon coral reef aquarium operates at 3.3 cm²/l (see Chapter 21). Home-based systems typically will lack precision equipment for measuring nitrate, oxygen, and pH. Nutrient levels for some ecosystem models should be far below what can be measured with simple test kit units. However, nutrient analysis kits can establish an upper level, with a commercial analysis every six months to establish precise values. The aquarist must also rely on the appearance of the organisms to judge whether or not the system is functioning properly. In addition, an understanding of export and its relationship to import is

---

TABLE 2

Common, Persistent Components of Coral Reef Algal Turf
Assemblage in Microcosm Scrubbers

Bacillariophyta
  Licmophora sp.
  Navicula sp.
  Nitzschia sp.
  Thalassiothrix sp.

Cyanophyta
  Anacystis dimidiata (Kutzing) Drouet and Daily
  Calothrix crustacea Schousboe and Thuret
  Entophysalis sp.
  Microcoleus lyngbyaceus (Kutzing) Crouan
  Oscillatoria submenbranacea Ardissone and Strafforella
  Schizothrix sp.

Chlorophyta
  Bryopsis hypnoides Lamouroux
  Cladophora crystallina (Roth) Kutzing
  Cladophora delicatula Montagne
  Derbesia vaucheriaeformis (Harvey) J. Agardh
  Derbesia sp.
  Enteromorpha lingulata J. Agardh
  Enteromorpha prolifera (Muller) J. Agardh
  Smithsoniella earleae Sears and Brawley

Phaeophyta
  Ectocarpus rhodocortonoides Borgesen
  Giffordia rallsiae (Vickers) Taylor
  Pylaiella antillarum (Grunow) De Toni
  Sphacelaria tribuloides Meneghini

Rhodophyta
  Acrochaetium sp.
  Asterocytis ramosa (Thwaites) Gobi
  Bangia fuscopurpurea (Dillwyn) Lyngbye
  Callithamnion sp.
  Centroceras clavulatum (C. Agardh) Montagne
  Ceramium corniculatum Montagne
  Ceramium flaccidum (Kutzing) Ardissone
  Erythrocladia subintegra Rosenvinge

---

crucial. Scrubber algae typically contain 3–6% nitrogen as a fraction of dry weight, the lower percentages characterizing systems operated at very low nutrient concentrations [$<2\mu M(N - NO_2^- + NO_3^-)$]. Most feeds, whether brine shrimp, krill, or flake foods, will contain between 8 and 12% nitrogen of dry weight. Thus, export (of algal turf or removed ecosystem plant biomass) should be two to three times feed input (dry weight). Very low nutrient systems typically are operated with an export–import

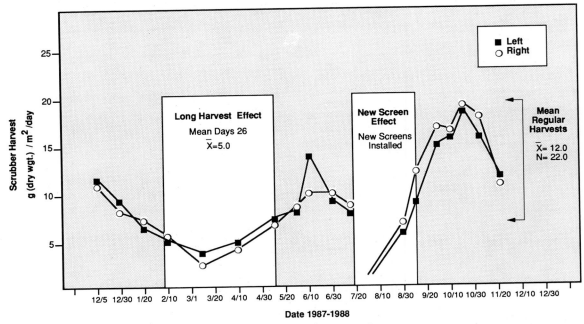

Figure 11   Plot of dissolved nitrogen (NO_2^- + NO_3^-) in Smithsonian rocky Maine Coast microcosm from December 1988 to November 1989.

ratio of 5–7 : 1. This results partly from falling nitrogen percentages in the removed algae and partly from nitrogen fixation. While this process might theoretically over-scrub some micronutrients (the scrubber algae will likely compensate in part as they do with nitrogen and phosphorus), the standard 1–2% water change per month should handle this. In some cases, it may be necessary to add micronutrients. This might seem contradictory to modeling accuracy. However, it is well to bear in mind that many extremely low nutrient situations are nutrient limited in the wild, and this is what blue, high clarity waters mean. Also, high productivity in nutrient deserts (coral reefs) results from large quantities of water with low nutrient concentrations flowing over a fixed point (Adey, 1987). This happens because of the availability of large, total quantities of needed nutrients; in models, this is simulated by adding food and/or nutrients. A variety of small scrubber types for home aquarium systems are shown in Figure 12, and management is discussed in detail in Chapters 21, 23, and 24.

Human culture has gradually developed an extensive use of metals. Some of these, such as iron, are crucial to many organisms in moderate quantitites. Other metals are often required in microquantities. On the other hand, many metals, particularly the heavy metals (lead, mercury,

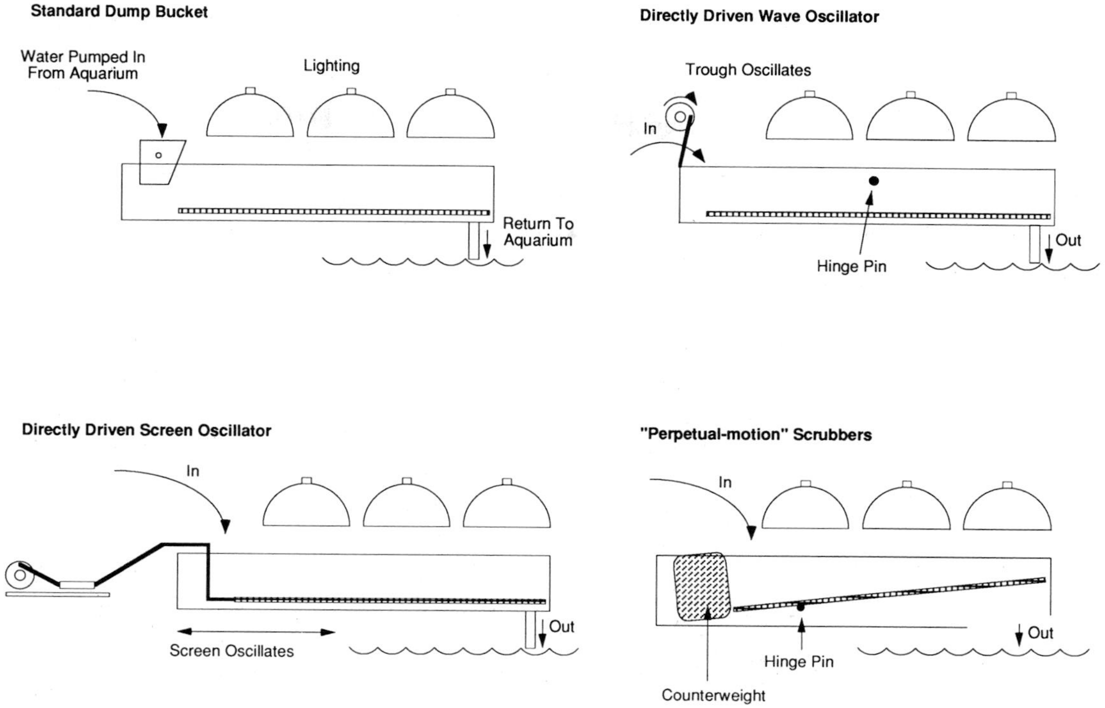

Figure 12 Diagram showing different types of smaller algal turf scrubbers for aquarium operation. The "perpetual-motion" scrubber can also provide wave and surge motion in the aquarium itself.

copper, and zinc), can be highly toxic when they occur in appreciable quantities in the water. Indeed, some human water supplies have serious health problems in the form of dissolved heavy metals, and copper is often used as a poison to reduce algal levels in drinking-water reservoirs.

It has been long known that growing algae have the capability of taking up and concentrating many heavy metals (Green and Bedell, 1989). The algal scrubber process can maintain model ecosystems at acceptably low levels of heavy metals as long as spike additions are not at levels that are toxic to the algae.

Summary

Algal turf scrubbers can manage most aquatic and marine model ecosystems. The data given here for scrubber area and operational procedures will provide the background that will enable the aquarist to adapt the concept to different model ecosystems. Every wild ecosystem varies from its "type," and each microcosm, mesocosm, and aquarium will vary even more from the analog or wild model. These numbers are guidelines developed from over a decade of working experience. When in doubt in a particular case, add 20–30% to scrubber area. It is a simple task to reduce light period if excess scrubbing is occurring.

It must be remembered that algal scrubbing serves two primary purposes: (1) oxygen and pH control, and (2) nutrient control. The first function is primarily a diurnal, or at least a short-term, need. If water volume is very large relative to the area undergoing night-time respiration, and day-time photosynthesis is adequate, scrubbing might not be needed for this function. This applies typically to models larger than several hundred thousand gallons. On the other hand, for smaller mesocosms, microcosms, and aquaria, in most cases any system less than 50,000–100,000 gallons in volume, the gas exchange requirement can be used to calculate required scrubber area. For example, if the aquarist can estimate or can measure and calculate the respiration rate (Chapter 11) of the planned community, then required scrubber area is approximately that needed to supply the oxygen requirements of the model system during the dark period. This method assumes no atmospheric exchange (which for carbon dioxide is likely to be essentially correct) and means that the aquarist must establish a minimum oxygen concentration that is acceptable for the ecosystem in question.

Nutrient scrubbing is a long-term requirement. Depending on system volume relative to feeding, lighting, and other plant or animal export, it is most likely measured in weeks or months for larger model systems. In totally closed systems where no feed is introduced, scrubbers are used to balance seasonal production cycles, or in mesocosm situations where watersheds are included, to return elements to the watershed simulating a geological function. Practically, it is almost impossible to calculate the nutrient scrubbing requirements of a given system. It is best to use the guidelines of Table 1 to determine scrubber area and lighting. These are based on approximately 40 system-years of operation. This can be followed up with monitoring of nutrient concentration as the system matures. If the aquarist does not have access to adequate nutrient monitoring facilities, a safety factor can be added and light period advanced or reduced in accordance with visual observation of sensitive organisms.

References

Adey, W. 1987. Food production in low nutrient seas. *Bioscience* **37**: 340–348.

Adey, W., and Goertemiller, T. 1987. Coral reef algal turfs—Master producers in nutrient poor seas. *Phycologia* **26**: 374–386.

Adey, W., and Hackney, J. 1989. Harvest production of coral reef algal turfs. In *The Biology, Ecology and Mariculture of Mithrax spinosissimus Utilizing Cultured Algal Turfs*. W. Adey (Ed.). Mariculture Institute, Washington, D.C.

DeBoer, J. A. 1981. Nutrients. In *The Biology of Seaweeds*. C. D. Lobban and M. Wynne (Eds.). University of California Press, Berkeley.

Green, B., and Bedell, G. 1989. Algal gels or immobilized algae for metal recovery. In *An Introduction to Applied Phycology*. I. Akatsukai (Ed.). S.F.P.B. Academic Publishers. The Hague.

Honn, K., and Chavin, W. 1975. Prototype design for a closed marine system employing quarternary water processing. *Marine Biol.* **31**: 293–298.

Hunnam, P. 1981. *The Living Aquarium*. Ward Lock Ltd., London.

Lucid, D. 1989. Effects of dissolved inorganic nitrogen concentrations on primary productivity, nitrogen fixation, and community composition of coral reef algal turf: A microcosm study. M.S. thesis. University of Maryland.

Mills, D. 1986. *You and Your Aquarium*. Alfred Knopf, New York.

Moe, M. 1989. *The Marine Aquarium Reference*. Green Turtle Publications, Plantation, Florida.

Rheinheimer, G. 1985. *Aquatic Microbiology*. John Wiley and Sons, New York.

Seitzinger, S. 1988. Denitrification in freshwater and coastal marine ecosystems: ecological and geochemical significance. *Limnol. Oceanogr.* **33**: 702–724.

Smith, T., and Schuman, A. 1991. The denitrator in the reef aquarium. Freshwater and Marine Aquarium **14**: 66–67.

Spotte, S. 1979. *Sea Water Aquariums, the Captive Environment*. John Wiley and Sons, New York.

PART III

Biological Structure

CHAPTER 13

COMMUNITY STRUCTURE
The Framework

This section begins the most critical part of this book. Chapters 2–8 focused on the physical factors that limit and control ecosystem function, whether in the wild or in a living model. The overall message that we tried to present was not to underrate or omit a physical factor because its effects are poorly understood. Also, there are many engineering "tricks" to environmental simulation in synthetic models. Some are old and well understood; others we introduce or at least place in new light in our discussion.

Part II (Chapters 9–12) treated the organic "pollution" both in the wild and in the limiting water volumes of mesocosms and aquaria. Newer methods of aeration, water movement, lighting, and bacterial filtration have made great strides in allowing the keeping of invertebrates that could not be previously kept in closed systems. However, we also introduce controlled use of plant photosynthesis and biomass production, through the use of algal turf scrubbing. This process can manage most "pollution" or control problems and produce a chemical environment that matches virtually any wild ecosystem.

Here we introduce the series of chapters that discuss the biological communities of ecosystems and how they relate to living in model ecosystems. The chapters of Part III treat in detail the manner in which the community members interact with each other through a food chain or trophic structure. Most important, this chapter introduces the great problem not only of ecosystem modeling in the ecological context, but of how we, as a human population, interact with the populations of our natural world. In wild nature, the physical parameters already exist. Our modifications are more or less obvious. Human chemical pollution we can control—whether we do or not is a matter of economics. Space is another matter. There are already too many of us—and each of us wants more space. There is little question but that we are "crowding out" the natural world, and have already forced many hundreds or perhaps many thousands of species into extinction.

The development of a living community in a large or small box brings us face to face with the same problems that we deal with in conservation as a whole. To place this issue in focus, the remaining semiwild country in the seemingly generous allotment of "wild" habitats in South Florida is clearly not enough for the Florida panther to maintain a viable population. It does appear to be quite sufficient for the equally large alligator, at least if hunting is greatly restricted. On the other hand, neither panther nor adult alligator could survive as viable populations in any conceivable mesocosm. Even in a very large microcosm or ecosystem aquarium, a very young alligator might be kept but it would soon outgrow the local environment.

A primary characteristic of natural ecosystems is the relationship between organism size and the numbers of that organism (Fig. 1). Even though there are fewer species and fewer numbers of those organisms large in size, they occupy larger home ranges (Bonner, 1988). Some sea birds and marine mammmals have ranges reaching tens of thousands of miles. Thus, in modeling an ecosystem, it is clear that the larger animals (though perhaps not plants), depending on the size in question and the model system, will have to be accounted for by the intervention of the human manager. This has already become true in some of our larger U.S. national parks, though there is some question that this needs to be the case (Chase, 1986). In a well-designed and operated model ecosystem, protozoa, amphipods, small fish, and crabs will be "at home" just as much as in the wild. On the other hand, the largest animals will probably have to be simulated, and that points up the basic difference between our three model categories. In the largest mesocosms, it is only the adult alligators, sharks, and snapping turtles that have to be omitted. In the microcosm, many predators—for example, pickerel, barracuda, whelks, and octopi—may need to be omitted. In the aquarium ecosystem, on the other hand, great care must be taken with any predator that feeds on organisms larger than a few centimeters in length. These problems can be handled, but some ecological understanding is required. Since that same understanding is also required for our as yet feeble attempts to maintain natural ecosystems and communities in the "real" world, the knowledge acquired in building and maintaining model ecosystems by scientists, hobbyists, and school children alike may well be an important element to our continued well-being on Earth.

The Community

Community structure is the ecologist's term for indicating what organisms are present in a given environment, in what numbers, and how they relate

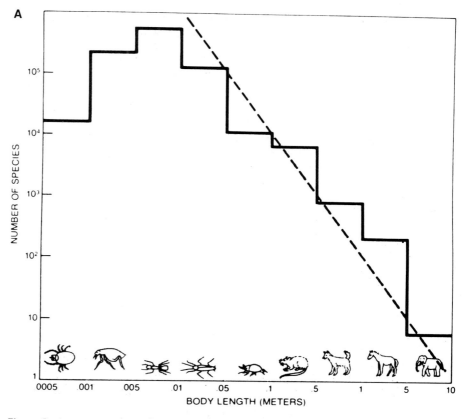

Figure 1 Larger animals are fewer (B) and are represented by fewer species (A). The drop-off in species numbers in the smallest sizes is probably real and is a function of minimum size that can be accomodated by a body plan. Adding protozoa or even bacteria would reduce the position of the peak, but the drop off would remain. After Bonner (1988, from May, 1984, and Peters, 1983).

(Figure continues)

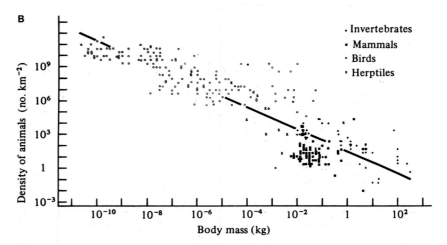

Figure 1 (*Continued*)

to each other. Another way to look at a community is as a collection of niches or slots that organisms can fit into in order to "make a living." In the next chapter on ecosystems and tropic structure, or food chains, we discuss rates of production, of feeding, and of energy and material flow. There is no sharp line between the subject matter of these two chapters. This one refers more to static and behavioral aspects, Chapter 14 to dynamic aspects. Indeed, it is wise to remember that to the ecologist, the term community is rather arbitrary. Its boundaries can be as wide or as narrow as one chooses to make them. For example, a rocky-shore community could refer to a thousand miles of coastline or to a kelp community in a band tens to hundreds of feed wide on that coast, or the kelp epiphyte community could refer to those organisms growing on single kelp plants. Reasonably sharp community boundaries are required for such a designation, but even that criterion is sometimes rather fuzzy. Traditionally, a biological community is named for a dominant element or elements. An *Acropora palmata* reef community is conspicuously dominated by that single coral species as a major element of biomass or structure, controlling many other organisms by its presence. A *Batis–Distichlis* salt marsh community is a salt marsh in which the saltwort *Batis* and the salt grass *Distichlis* more or less equally provide the primary vegetation and therefore cover in the marsh.

The Biome

Biomes are the largest scale terrestrial communities (Fig. 2). They are defined primarily by the effect climate (temperature and precipitation) has on the dominant community structure-creating plants. In this context, climate is a matter of altitude as well as latitude, mountain ranges, and geographical position on a continent. A secondary type of biome relates to the special effects of unusual rock and soil type. Since plants, especially forest trees, but even rushes and grasses, greatly modify the environment and establish a physical structure within which animals live, terrestrial biomes and communities are largely defined and often named by the dominant type of plants.

In this book on the physical modeling of communities, or synthetic ecology, we have concentrated on the marine and aquatic worlds, though often wandering out of the strictly submerged and intertidal habitats and into the transitional wetland communities. The large-scale marine and aquatic parameters that define the underwater equivalent of biomes are rather different from those in the terrestrial world. First, water, instead of

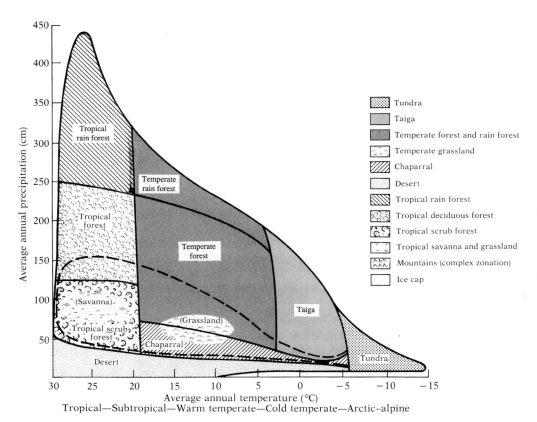

Figure 2 Biomes, the largest scale terrestrial communities, diagrammed in relation to mean annual air temperature and mean annual precipitation. After Erlich and Roughgarden (1987, from Odum, 1971, and Whittaker, 1975).

being one of the two primary limiting parameters, is in great abundance. The only equivalent to the precipitation limitation is salt in the water, since in a sense it renders water relatively unavailable to organisms adapted to fresh water. Temperature is certainly important. However, because its short- and long-term range is more limited than on land, its influence is considerably more muted. Temperature effects on large-scale structural elements certainly are present as exemplified by kelp beds in colder waters as opposed to coral reefs in the tropics. However, in the water world the basic factors controlling the development of a community framework are so different that it is necessary to develop an entirely separate biome classification system.

On land, the vast majority of plants must be rooted in the ground to obtain water. Since light is just as crucial as water, the terrestrial communities we look at today are the result of 300 million years of competition, natural selection, and evolution directed to building more or less massive cellulose and lignin structures rooted in the ground and reaching to the sky. Plants were the first terrestrial builders, hundreds of millions of years before humans evolved. However, in the same sense as human communities, but much more extensively, plants have built natural community frameworks over most of the earth's land. These frameworks have become occupied by a host of animals, not only those feeding on the plants, as we discuss in the next chapter, but animals using the plant structures as a home. And in this context, home is used in all the ways that humans use that word.

Under water, most of this is different. First, a major volume of deep ocean and deep lake water is unavailable to plants, rather like high-altitude rocky ledge or Arctic ice on land. Relative to other biomes the organisms of these areas are determined primarily by physical factors because of the total lack of plants. Also, unlike in air where few organisms can fly and even those only temporarily, many water plants and animals can float and swim more or less indefinitely. Water characteristics and its three-dimensional space thereby provide parameters that control all aspects of animal structure and their interactions. Thus, we can start by designating two deep-water, nonphotic, lightless biomes: nektoplanktonic (swimming and floating) and benthic (bottom). These biomes occupy approximately 50% of the earth's surface at all water depths below about 100–200 m (see Fig. 3 in Chapter 8). However, they are not likely subjects of synthetic ecosystems in the near future. Because it is so extraordinary, a primary production community not based on light, and probably ocean wide in a linear respect, the hydrothermal vent community of mid-ocean ridges is also worthy of the biome designation (Fig. 3).

In the photic nekto-planktonic zone, rarely are plants important that are not microscopic. Thus, while occupying a critical position in plank-tonic food webs the plants themselves do not generally provide a community framework. It is the characteristic of water itself as a volume, the characteristics of light penetration, mixing, and temperature that provides

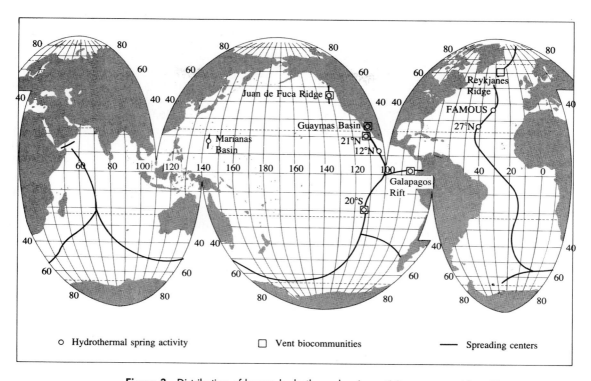

Figure 3 Distribution of known hydrothermal spring activity on ocean ridges. Vent communities probably occur along a narrow zone on most of these ocean ridges. After Thurman and Webber (1984). Copyright © 1984 by Scott, Foresman and Company. Reprinted by permission of Harper-Collins Publishers.

the structuring elements. The *Sargassum* brown algae of the Sargasso Sea is an obvious exception, but being rather limited in extent and probably rather transitory in time and space, it is best left as an interesting community of the photoplanktonic biome. Planktonic mesocosms and enclosures are variants that have been the subject of a considerable amount of scientific research. While we discuss the role of plankton in benthic systems, our focus has been on benthic and wetland communities. For readers interested in the living simulation of planktonic communities we suggest the following references as a guide: Pilson and Nixon (1980) and Grice and Reeve (1982).

Thus, we finally come to the bottom biomes within lighted zones of ocean and fresh water. Even here, while there are some parallels with the terrestrial world, to a large extent the controlling factors remain different. First, on land, wind in the form of tornadoes and hurricanes can cause considerable breaking up of plant structure. However, rarely is the major part of biome or community destroyed. A ship on land would remain until it rotted or rusted away. On the ocean, a ship is only ephemeral.

When power and intelligence are not used to move and orient the ship, it is destroyed. The point is that ocean waves are not to be resisted by any structure of cellulose. Mangroves, the only marine equivalent of terrestrial forests, occur only on quiet, protected shorelines. Kelp, rockweed, sea grass, and submerged or floating aquatics are the equivalent of land forests, but they survive by being flexible, not by developing massive structures that resist. As we will discuss in Chapter 22, kelp forests are certainly deserving of the name in a community sense. However, kelp plants last only a few years at most before they are on the beach, part of the drift, with their biomass designated for another community. Also, under conditions of unusual abundance of urchins, kelp zones can become crustose coralline algae zones. This condition can persist for several years or even decades at a time.

Coral reef and algal ridge communities are true parallels with the forests of the land. In these cases, associations of animals and plants together build calcium carbonate structures that greatly modify the environment and can stand up to the waves (Fig. 4). The parallel with forest building on land is far-reaching, because it is both directly and indirectly that photosynthesis and the subsequent removal of carbon dioxide makes massive construction of calcium carbonate skeletal material possible. Because of the wide variety of new niches created by coral growth, coral reefs and algal ridges together, with as a subset worm and bryozoan reefs, are worthy of biome status.

Wherever there is sufficient wave action and sufficient shore relief without abundant sand, rocky-shore communities occur the world over and have many features in common. Kelp, rockweed, and Irish moss beds are characteristic and have important structure-creating features that suggest biome status (Fig. 5). However, there are other biotic elements that are equally important, namely, barnacles, mussels, and coralline algae. Attached to the hard rock substrate, these organisms also build a low-lying structure that creates habitat and modifies the environment. Thus it seems reasonable that the rocky-shore supratidal and subtidal should be accorded biome status. It is true that these are narrow linear features, a few hundred yards to at most a few miles in width, unlike the thousands of square mile areas accorded some terrestrial biomes. However, it is the nature of light and wave action control, having replaced water as a molding key environmental factor, to occur in narrow coastal bands. The rocky shore biome extends into the tropics. Subtidally, *Sargassum* and a variety of reds replace the kelp beds. Intertidally the intense sun and drying effects along with heavier herbivory prohibit any equivalent of the rockweed communities. However, blue-green algae produce extensive black mats that are fed on by snails. Pleistocene limestones are common in this shoreline zone, having been developed as terraces at the last higher sea-level stand about 100,000 years ago. Between solution and scraping by the snails, these limestones develop an extremely irregular microkarst or iron shore. This community is part of the worldwide rocky-shore biome.

Sandy shores, intertidal as well as subtidal, are also world-wide features in which environmental control, wave action, and an abundant supply of sand from a wide variety of sources create a unique, extremely difficult biome, virtually free of plants. This biome is characterized by a few species of rapidly burrowing filter- and detritus-feeding macroinvertebrates of several phyla, mostly bivalves and worms, and many species of very small (< 1 mm) meiofauna. Many of the meiofauna are from diverse "worm" phyla. Some are tiny arthropods belonging to many crustacean groups as well as other classes. In fresh waters, tiny insects and insect

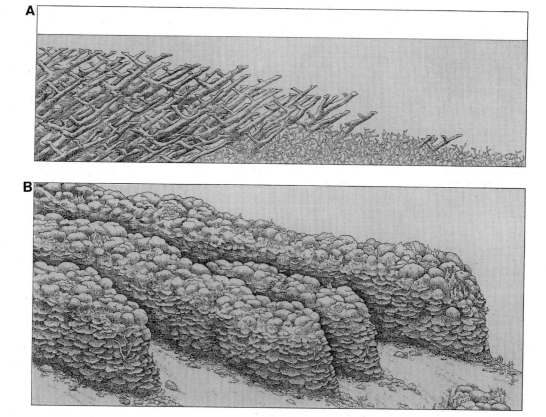

Figure 4 Coral reef frameworks forming community structure. (A) *Acropora palmata* and *Acropora cervicornis* frameworks in shallow water (< 20 m); (B) *Montastrea annularis* buttresses in deeper water (> 20m); (C) a few of the larger organisms occupying the surface of reef framework—many more are subsurface. (A,B) From Coral and Coral Reefs by T. J. Goreau. Copyright © 1979 by Scientific American, Inc. All rights reserved. (C) after Thurman and Webber (1984). Copyright © 1985 by Scott, Foresman and Company. Reprinted by permission of HarperCollins Publishers.

(*Figure continues*)

C   DAY

NIGHT

Figure 4 (*Continued*)

larvae become members of the sand-dwelling meiofauna (see Chapter 8, Figs. 7–10).

Finally, we reach the mud bottom, which may or may not be vegetated. Mangroves, submerged aquatic vegetation, marshes, and swamps (tree'd wetlands) are dominated by their plants. The plants form the structures within which the animal members live and find their food. (Figs. 6–8) Each is worthy of biome status. Finally, even "bare" mud bottoms in the photic zone usually have a diatom film on the surface. However, the basic point is that a muddy bottom frequently supports a very rich fauna that is

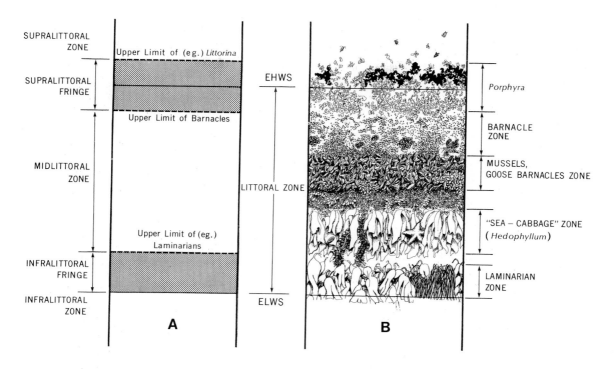

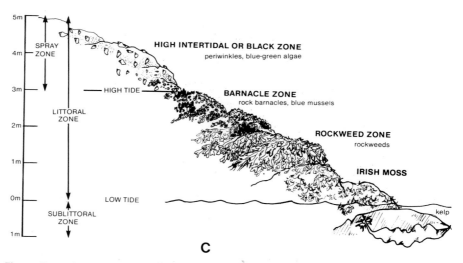

Figure 5 Rocky intertidal zones. (A) Generalized worldwide relationships; (B) west coast of North America; (C) Maine coast. Revised including data from Stephenson and Stephenson (1972), Rickets *et al.* (1985), Fefer and Schettig (1980), and Carefoot (1977). Similar zones extend subtidally with typically a kelp zone, a leafy red algae zone, and finally a crustose coralline zone. Plant structured communities end at 10–50m depending upon turbidity and the sediment supply.

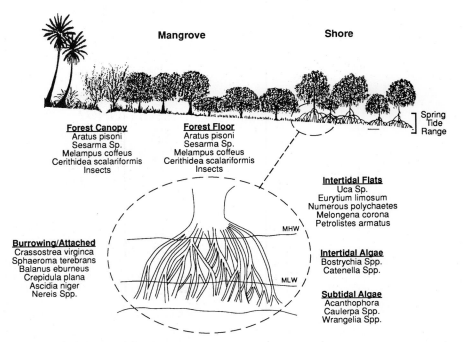

Figure 6 Organisms living within Florida red mangrove communities. After Drew and Schomer (1984) and Britton and Morton (1989).

adapted in many ways to life in this environment. It is the physical–chemical characteristic of the soft bottom rich in organics and low in oxygen and with an overlying water mass usually rich in plankton that establishes community structure (see also Fig. 9).

Table 1 provides a classification for marine and fresh water biomes. It certainly is not perfect anymore than the terrestrial biome classification. However, in accordance with the above discussion, the classification is based on the same principles.

In Chapters 21–25, selected communities, marine, estuarine, aquatic, and wetlands, are treated in detail relative to the placement of those communities in mesocosms, microcosms, and aquaria. In the remainder of this chapter we discuss general features of communities and their establishement in synthetic systems.

Features of Communities

Under a given set of environmental conditions, the development of the structural elements of the community from a "bare" surface (seagrass,

Figure 7 Dominant macro organisms of a North American, Gulf of Mexico, freshwater coastal marshland. (A) Cat-tail, *Typha latifolia*; (B) Bullrush, *Scirpus maritimus*; (C) Dwarf surf clam, *Mulinia lateralis*; (D) Common rangia, *Rangia cuneata*; (E) Tube-building amphipod, *Corophium* sp.; (F) Small-mouthed hydrobiid, *Texadina sphinctostoma*; (G) Hydrobiid, *Texadina barretti*; (H) *Probythinella louisianae*; (I) Freshwater shrimp, *Macrobrachium acanthurus*; (J) Red winged blackbird, *Agelaius phoeniceus*; (K) Snow goose, *Chen caerulescens*; (L) Canada goose, *Branta canadensis*; (M) American widgeon, *Anas americana*; (N) Blue-winged teal, *Anas discors*; (O) Greenwinged teal, *Anas crecca*; (P) Gadwall, *Anas strepera*; (Q) Mottled duck, *Anas fulvigula*; (R) Northern pintail, *Anas acuta*; (S) Lesser scaup, *Aythya affinis*; (T) Shoveler duck, *Anas clypeata*; (U) Florida marsh clam, *Polymesoda maritima*; (V) Marsh clam, *Polymesoda caroliniana*; (W) Theadfin shad, *Dorosoma petenense*; (X) American eel, *Anguilla rostrata*; (Y) Alligator gar, *Lepisosteus spatula*; (Z) Tarpon, *Megalops atlanticus*; (A1) Big mouth sleeper, *Gobiomorus dormitor*. From Brittan and Morton, 1989.

FRESH WATER

BRACKISH/SALT WATER

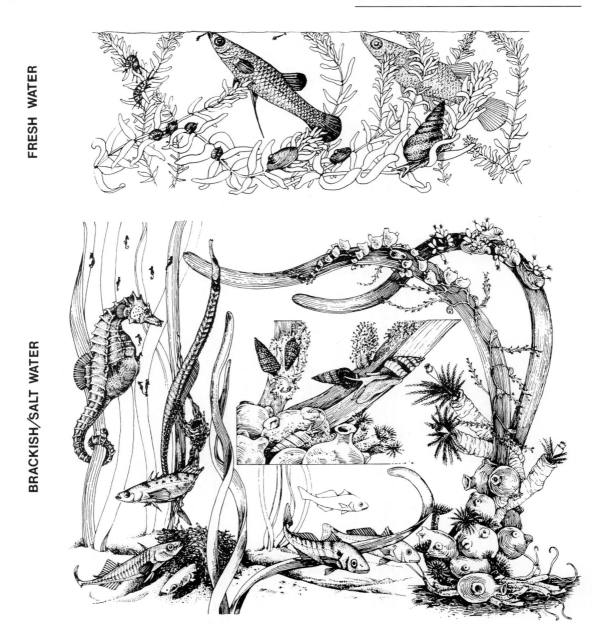

Figure 8 Organisms living within the submerged aquatic vegetation communities of Chesapeake Bay. After Lippson and Lippson (1984).

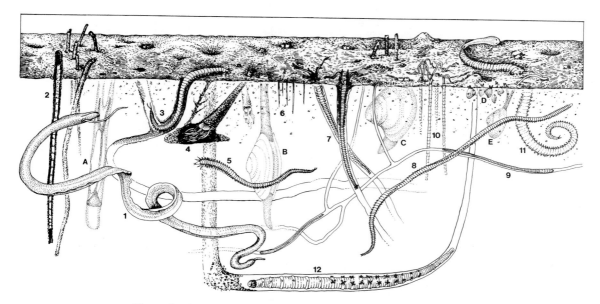

Figure 9 Organisms living within a nonvegetated sandy–mud bottom. After Lippson and Lippson (1984). (A) Stout Razor Clam; (B) Soft-shelled Clam; (C) Hard Clam; (D) Gem Clams; (E) Baltic Macoma. Worms below the intertidal flats: (1) Milky Ribbon Worm, *Cerebratulus lacteus* (to 4'); (2) Common Bamboo Worm, *Clymenella torquata* (to 6"); (3) Clamworm, *Nereis succinea* (to 6"); (4) Trumpet Worm, *Pectinaria gouldii* (to 2"); (5) Freckled Paddle Worm, *Eteone heteropoda* (to 4"); (6) Barred-gilled Mud Worm, *Streblospio benedicti* (to 1/2"); (7) Red-gilled Mud Worm, *Scolecolepides viridia* (to 4"); (8) Opal Worm, *Arabella iricolor* (to 2"); (9) Capitellid Thread Worm, *Heteromastus filiformis* (to 4"); (10) Glassy Tube Worm, *Spiochaetopterus oculatus* (to 2 1/2"); (11) Bloodworm, *Glycera sp.* (to 15"); (12) Lugworm, *Arenicola cristata* (to 12").

kelp, coral reef) changes the local environment. This allows new organisms that require the cover, the substrate, the food organisms, and more light or less light than that provided by the substrate to enter. Thus, the community alters its own environment and slowly drifts to a different structure. This process, called succession, finally reaches a substable state, the climax community, which may take many months or many years to attain. Normally, disturbing factors such as storms, ice, floating logs, or larger predatory animals continually "knock" the community back to the pioneer or intermediate stages of succession. However, such disturbance usually happens in a patchy way and gives rise to heterogeneity in a community. One should remember that a wild community is dynamic in its composition; the aquarist should not expect a model to be greatly different.

A now classic example of these dynamic processes, succession and disturbance, was described for the rocky intertidal shore of northwestern North America by Dayton (1971). In that region and many other temperate/

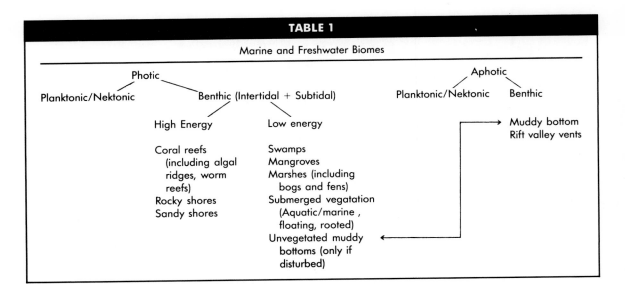

TABLE 1

Marine and Freshwater Biomes

boreal rocky intertidal zones, mussels (e.g., *Mytilis* spp.) can become the primary structure-creating member of a climax community covering enormous areas. However, logs or ice driven by waves, and the feeding of the seastars (e.g., *Pisaster*) and gulls on the mussels, constantly removes large patches of these attached bivalves, allowing several species of barnacles to colonize. As a general principle, great disturbance reduces the number of species. However, a moderate level of disturbance results in the highest species numbers (Fig. 10).

A feature of community and tropic structure not yet discussed is the tendency of more complex communities to be characterized by guilds. A guild is a group of species populations that occupy the same or closely connected niches. This feature has been most studied in birds and insects but certainly characterizes marine communities such as coral reefs. An example of a bird guild and the resultant resource partitioning is shown in Figure 11. This becomes particularly critical in model systems where scaling does not allow the inclusion of all members of the guild and niche overlap allows the utilization of one or a few members to satisfy the need for the niche without overburdening the limited size of the resource in the model.

Within the biomes discussed above, many communities could be delimited. Particularly within terrestrial ecology, considerable dispute has been engendered over the past several decades as to whether or not communities and their boundaries exist. The interested reader is referred to Ehrlich and Roughgarden (1987) and Strong *et al.* (1984) for a review of these issues. Regardless of the situation in the terrestrial environment, we feel that marine, aquatic, and wetland communities do tend to have sharp

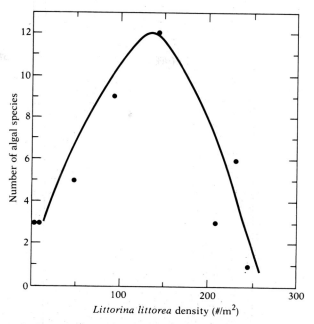

Figure 10 Control of algal species diversity by disturbance created by the grazing of littorinid snails. After Ehrlich and Roughgarden (1987, from Lubchenco, 1978).

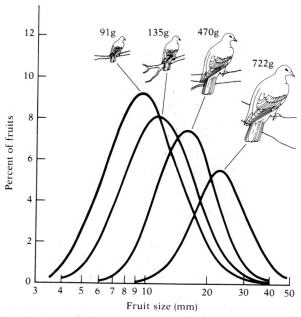

Figure 11 Fruit sizes eaten by different fruit-eating pigeons in the South Pacific. After Ehrlich and Roughgarden (1987, from Diamond, 1975).

boundaries. In part this is because water surface, light, and substrate, which is often wave controlled, are the primary environmental community determiners. These parameters usually have considerably sharper boundaries than the temperature precipitation boundaries of the terrestrial environment, though there would be some parallel where topography and rock type are critical controlling factors. In very extensive sand or mud-bottom biomes, the same difficulty in sharply delimiting communities is encountered.

Community Structure and Ecological Models

Every ecosystem has to be studied and modeled on its own terms. Nevertheless, there are some general principles that can be applied to the development of an ecosystem in a microcosm, mesocosm, or aquarium. When properly carried out, an ecosystem results from the placing of a community or organisms within its appropriate structure, whether physically or biologically derived, and within the framework of a given set of physical parameters—temperature, salinity, waves, current, etc. Although one would necessarily establish all of the physical parameters of a model before adding the living community, the planning of community structure and the engineering techniques needed to achieve the proper physical parameters go hand-in-hand. The following list is a general pattern of approach to establishing community structure within living models:

1. Establish the boundaries of the wild community or communities that you would like to model. Learn as much about them as you can by actual observation and library study. Field guides, especially when provided with an ecological overview, can be extremely helpful, and having a wild analog to compare can be a critical element in success. Also, determine the source of the energy (food) that drives the ecosystem. In many cases this energy supply will result from establishing community structure. In other cases it will have to be supplied (see Chapter 14).

2. Develop species lists and determine what species are the structuring elements. What is it that provides primary modification of physical space and supplies habitat? Is it plants, animals, substrate, or physical factors such as breaking waves?

3. What are the relative numbers of each species? Most important, what larger plants or animals are always present in dominating size or numbers?

4. Begin community development by physical structuring with the dominant plants that are present and/or substrate or animal skeletal units. Provided the proper environment is present, these units will normally survive for some time without significant animal interaction. In the few cases where live animals (rather than skeleta) must provide the physical

structure (e.g., gorgonians), planktonic food may have to be accounted for early on.

5. Carry out structuring in a block transfer process. That is to say, do not try to remove small microorganisms, soil, mud, or any small invertebrates associated with the bases of the plants or other substrate. Bring them along with the structuring elements.

6. Repeat the process in as many small units or injections as possible. Each injection will successfully introduce more microspecies or young plants and animals.

7. As the semipure plant or shell communities establish themselves, begin introduction of the symbionts (e.g., live corals) and the smaller animals and fish, also by a multi-injection process.

8. Finally, tentatively experiment with the larger animals to learn the ecological capacity of the system. When overgrazing or overcropping is apparent, remove the offender.

9. Watch for intense competition for limited resources such as space or food and cull when necessary.

Scaling and Reproduction

We have discussed the difficulties of keeping larger organisms, particularly higher predators in model ecosystems. Size is of obvious importance, but so is behavior. The predator or specialized herbivore that requires a large territory to obtain enough food can only with special manipulation, if at all, be included in a model. Reproduction is a critical form of behavior. For many small plants and invertebrates, reproductive success occurs abundantly in ecosystem models, though in cases with larger, longer-lived planktonic larvae this can depend upon the nature of pumping and filtration. For some fish particularly those that protect the eggs and hatchlings in some manner, such as cichlids, reproductive success in ecosystem models is likewise assured. However, for many fish a fairly large territory, in the form of low egg or larval predator density, sometimes the open ocean, is required. If these species are preyed on, or as they eventually age, they must be occasionally added to the model, from the wild or from special hatch or perhaps refugium environments.

In the wild, species diversity is a function of area and while half the total species of a given biome might exist for centuries in a few tens of acres, the remainder would require hundreds of square miles. Nevertheless, local populations of many species are forever on the verge of extinction due to random or periodic unfavorable events. Very large areas, up to the largest continents and oceans, show more species not just because of the variety of habitats but because extinction is unlikely to occur in many local populations at the same time. We have a much better understanding of these matters in the terrestrial world where forest and mammal preservation has become a matter of great concern. Harris (1984) has developed a model for conservation of the biotic diversity of forests that is based on an archipelago of protected habitats or refugia. In the ecosystem modeling described in this book, plant communities are established in accordance with species dominance and abundance in the wild analog. Some changes in model populations occur with time and generally the plant community is allowed to self adjust. The larger animals, particularly those high on the food chain or with complex life cycles, are manipulated to maintain their role, while the remainder of the animals able to successfully fit the model are allowed to freely develop populations, based on several months of repeated community block or individual injections. While studies of the population ecology of ecosystem models and how they relate to the wild are certainly needed, it is assumed that the basic principles of Harris' model apply. Thus, after economic constraints have determined the maximum size of the model, then refugia are added with the intention of circumventing obvious population constraints of selected elements of the community. Most of our refugia have been directed toward freeing attached fleshy algae, soft bottom invertebrate populations, and plankton from severe predation by fish and larger invertebrates (particularly crabs and lobsters). Several of these refugia are described in Chapters 21–24. In general, the semi-stability that is achieved in community and population structure in the wild over large areas, and at time scales of centuries and millenia, is achieved in microcosms, mesocosms, and aquaria by the manipulation of space (refugia) and when necessary the populations themselves.

Community Establishment Notes

This chapter is specifically concerned with establishing community structure. However, it is assumed that physical parameters (e.g., temperature, salinity, pH, oxygen, nutrient levels) can be kept in bounds during the community-establishing process. While this is not usually difficult, in some cases, particularly where adjacent communities of very different character are joined (a lagoon or mangrove to a reef, for example), the disturbance occasioned to one community might be too intense for the other. In this case the communities might have to be established separately in space or time and joined together later, when disturbance is minimal.

Establishing community structure in a model system requires the bringing together of all the information that we have tried to supply in this book. Communities of organisms and their trophic structure (or feeding patterns—next chapter) go hand-in-hand, and both elements must be considered together for success. The examples given in Chapters 21–24 will help to provide background on the approaches that we have taken over the years. The remainder of this chapter is a series of notes on aspects of the modeling of community structure and its parallel trophic structure, notes that will perhaps help to foster ideas as to the elements and problems involved.

The marsh and mangrove community (or communities) of the Florida Everglades, as established in our mesocosm model (see Chapter 23), can be used as an example of the community development process.

Here, the entire ecosystem was divided into six units, basically salinity-controlled and, to a lesser extent, wave-controlled sections. Except for the muddy tidal channels, wave-washed worm reefs, and sandy–shelly beaches, these communities were dominated by relatively few species of plants densely established in sandy or muddy substrates. In coastal, high-salinity waters, seaward of the worm reef and sandy–shelly beach, the seagrasses dominate, particularly *Thalassia*, *Syringodium*, and *Halodule*. Landward of the beach, on the dune, many flowering plant species are present, but the dominants that we particularly wished to establish were sea oats (*Uniola*), alligator weed (*Alternanthera*), seaside mahoe (*Thespia*), bay cedar (*Suriana*), salt meadow cordgrass (*Spartina patens*), and sea grape (*Coccoloba*). Continuing into brackish waters, a broad band, almost a monoculture, of the red mangrove (*Rhizophora mangle*) dominates the intertidal and supratidal, though with extensive tidal channels. Further landward, where the tidal channels almost disappear, black mangroves (*Avicennia germinans*), sometimes with extensive ground covers of saltwort (*Batis*), glasswort (*Salicornia*), sea purslane (*Sesuvium*), and sea daisy (*Borrichia*), develop a thick, dark forest. Scattered through the extensive stands of black mangrove, especially near the inner margin, are salt marshes with a wide variety of herbaceous species, but particularly dominated by the grasses (*Spartina* and *Distichlis*) as well as *Batis* and *Sesuvium*. Finally, with traces of salt nearly absent (depending upon the season), the terrain becomes a very wet prairie. Cattails (*Typha domingensis*) and sedges (*Eleocharis*) become the community structuring elements.

Our first step in developing the mesocosm was the establishment of the physical tank system that would provide appropriate light, temperature, and of course tidal water of appropriate salinity for each community. Then nutrient-controlling algal scrubbers were placed in operation, with attached settling traps to sink fine sediments. Community development of this mesocosm was started by bringing into the model, in plastic lugs or trays approximately 1 foot by 2 feet in size, the dominant plant species elements that we discussed above. Where several species were involved, these were stocked approximately in proportion to their coverage in the wild. The depth of the mud or soil substrate in this case was about 3 feet. These dominant community elements, with whatever species of generally smaller plants and animals present, were established at whatever tide-related level they occurred in the wild. Subsoils or sediments were placed under the community blocks to achieve the proper height. Once the dominant plants were established, secondary plant species were added, also mostly by block transfer, in accordance with their abundance and location in the wild (margins of tidal channels, tree falls, etc.).

The above-described process of introduction tended to provide very high turbidity levels in the tidal channels. However, these began to settle out after several weeks, and the structuring of the animal skeleta-dominated subtidal communities was initiated. Thus, in the wild, in the inner part of the more brackish red mangrove community, oysters (*Crassostrea virgninica*) form small, coherent, reef-like structures in the tidal channels. Also, at about the spring low tide line on the exposed shores, the skeleta of vermetid, worm-like gastropods form a low, reef-like structure. Both of these animal-structured subtidal communities were also established by repeatedly injecting large-lug-size blocks of oyster and worm reef. Whatever small invertebrates traveled with these minicommunities were also included.

Following the establishment of these "structural" elements by about 2 months, intermediate-sized invertebrates, clams, snails, and crabs (such as littorinids and mangrove crabs), were added individually. Finally, the fish were added, beginning with the "feeder" fish and herbivores (killifish, mollies, silversides, mullet, etc.) and bottom feeders (grunts, eels). Eventually, we began to experiment with a few smaller predators (lizardfish, pipefish, toadfish). At this level, depending upon the size and complexity of the model being constructed, every system is different and must be constantly and carefully adjusted for the effects of the mid- to upper-level predators (see Chapter 17).

In general, plants and their grazers dominate most shallow-water ecosystems. On a large scale, such as along hundreds of miles of coast, a pattern of dynamic stability usually exists between plants and grazers. However, imbalances often occur in a patchy way as one or the other gets ahead in the struggle. Developing and then maintaining this stability in aquaria is an ongoing task but is usually easily accomplished by removing or restricting some key element such as an urchin. Such control is sometimes more difficult at higher levels because of scaling factors or the larger size of the community.

Community structure is relatively simple to develop in a typical cold-water, hard-bottom ecosystem where a very few species of grazers eat a small number of plant species. At the other extreme, community structure in a tropical reef is very complex. However, a good part of the complexity is made of ecologically similar guild species. While "food web" becomes a more appropriate term than "food chain," even in a model system, this can still be built from a simpler to a more complex state.

At one extreme of a reef, on the wave-washed, turbid, and high-nutrient situations of high-island reef flats, several species of the relatively large *Sargassum* dominate, though several layers of understory contain numerous other plants, including several *Gracilaria* species. In this case, the reef-flat community is totally dominated by algae, standing crops of 3–5 kg/m² wet weight, which generally sets conditions for higher elements of trophic structure. These "reefs" are more like wetlands or terrestrial communities in their structure.

On the other hand, on more typical lower-energy and low-nutrient reef flats the basic plant element of the community is provided by a dominant microturf ranging from a few milimeters to a few centimeters in height. This consists of many filamentous algal species including *Herposiphonia, Polysiphonia, Sphacelaria, Bryopsis,* and *Calothrix.* Any space not occupied by a coral, sponge, or other encrusting plant or animal is covered by this turf, typically 30–80% of the surface. Less prevalent are a few smaller macroalgae including *Hypnea, Acanthophora, Laurencia,* and *Dictyota.*

Along the reef crest and in the uppermost fore reef, the calcareous coralline red algae *Porolithon, Lithophyllum,* and *Neogioniolithon* become important. Finally, in deeper water, other corallines including *Paragoniolithon,* a noncalcareous red crust *Peyssonnellia,* and the omnipresent brown leafy alga *Lobophora* provide the plant element at the base of the food chain. Thus, while free-living plants are richly abundant in the coral reef and are the primary energy-providing elements to the community, their community structure role can be somewhat limited. In the typical Caribbean coral reef, stony and soft corals as well as gorgonians and sponges provide the dominating organic structure. However, most of the spatial heterogeneity or habitat structure is provided by the dead skeleta of the corals. Since corals are richly endowed with algal symbionts and would not calcify without them, the analogy with a terrestrial forest is stronger than one might think at first sight.

Reef grazers range from tiny ostrocods and amphipods a few millimeters long to numerous limpets, chitons, and urchins, and small fish and finally up to ½-m-long parrot fish. The total number of species of grazers that are used to develop community structure in a model is very much a function of system size and has to be determined for each case. It is essential not to allow a model system under development to be overgrazed. This would remove the critical energy supply and an important part of the water quality control system at a very sensitive time.

Moving up the reef food chain, small-grazer predators such as grunts, wrasses, squirrel fish, spiny lobsters, and hermit crabs are accommodated and more or less easily controlled even in an aquarium. However, the task becomes more difficult for species that prey on larger grazers and smaller predators. A very large aquarium is required for higher predators such as jacks, snappers, barracuda, and sharks, which, when mature, range widely over a reef in search of food.

Also, when dealing with fish, a critical behavioral element that sometimes needs to be managed is schooling behavior and the question of mating and production of young. Usually the latter is not difficult, if young are wanted and can survive. However, if a fish is comfortable only in a school and only one or a very few of that species is desired in the food chain, it is perhaps best to chose another species from the same guild.

After the enormous food production of a shallow-water, tropical-reef system has passed through several steps of the food chain exchange or web, it has been reduced from some 10–50 g (dry wt)/m²/day to a much more modest 0.01–0.5 g (dry wt)/m²/day. Thus, scaling is more easily achieved in much larger aquaria (in the 100,000–200,000 gallon range). But even in the small aquarium some of this scaling problem can be handled by manipulation, for example, by "external" feeding. Easily raised freshwater fish such as goldfish are ideal in this case because their introduction to salt water produces instant stress and immediate predation by such fish as jacks and barracuda. In these cases where "packing" of the community is desired to include more elements, and external feeding beyond that necessary to simulate input from the adjacent planktonic community is employed, algal scrubbing is the key. As long as removed or exported algal biomass exceeds all added feed so that equivalent nitrogen removal is achieved, then imbalances will likely be avoided.

In the wild tropical reef lagoon, Mother Nature usually plays a major role in controlling grazers. Even here, however, the impact of grazers is obvious. The halos of lagoons, that is, the barren sand areas around reef patches, are a result of the continual presence of grazers that have retreated to the reef during the day, seeking refuge from such higher predators as barracuda and trigger fish.

In a reef or lagoon tank containing a *Thalassia* turtle grass community, parrot fish and urchins help prevent smaller algae from overgrowing the grasses. However, without higher predator checks, some species such as *Sparisoma radians* and *Diadema antillarum* can quickly decimate a grass bed. Thus, the operator's adjustments of a model system becomes the stabilizing element that maintains the small patch of a model community in the average state accomplished naturally, though often in a very patchy way, in a much larger wild community.

On most reefs of moderate to low wave energy, intense grazing is essential to the overall community structure of the whole reef. A coral reef will become an algal reef if grazers are removed or hindered by strong wave action. On Caribbean reefs, *Diadema*, the long-spined urchin, and a host of crabs, parrot fish, and tangs are all dominant and omnipresent grazers. Another common type of grazer, the damselfish, prevents many reefs from becoming overgrazed by its protective territorial activities. It is a farmer to some extent. The primary production and normal function of aquaria simulations of typical reefs can rely on the liberal presence of damselfish to prevent overgrazing.

Considering the above discussion of coral reef community structure, it might seem impossible to establish a stable community in a small aquarium. This is not necessarily the case. In Chapter 21, we describe a coral reef aquarium that has function in a highly stable kind of way for over 3 years. A "rare reef lobster" (*Enoplometopus*) is the higher predator/scavenger that controls community structure of the fish community in this tank. Only when there is enough space in the fish community can new fish be added (simulating additions from the plankton in the wild). Grazing is very heavy in this tank, but many calcareous and otherwise protected algae develop a dense low "forest." Fleshy algae like *Sargassum* and *Hypnea* survive only in refugia.

In cold-water ecosystems, plant families and genera and many controlling factors differ markedly from those on tropical reefs. Nevertheless, the same large and basic differences in the plant elements occur between communities and subcommunities. In the Gulf of Maine, for example, flowering plants such as the *Spartina* and *Juncus* species dominate marshes while *Zostera* (eel grass) determines the character of shallow muddy bottoms. On hard, or more exposed, bottoms, in the same area, the plant elements and standing crops range widely from the moderate-sized but quite dense rockweeds (*Fucus* and *Ascophyllum*) of the upper subtidal, to the large kelps (*Laminaria*) in the upper subtidal, and finally to a small red algal canopy, *Ptilotia*, and the coralline *Lithothamnium* somewhat deeper.

In this case, while the habitat-structuring elements are similar to those occurring on algal dominated reefs, energy flow or trophic structure is quite different. In many communities, including such familiar types as northern land forests and kelp forests, much of the plant production is not eaten by grazers. Instead, it is released from its normal habitat, often assisted by wind and wave, and accumulates as organic litter or detritus. This accumulation is then broken down by bacteria and fungi and devoured by a wide variety of smaller invertebrates, particularly small crustaceans and worms. It is not just semantics that differentiates between live-plant grazers and plant-litter detrivores. The process typically involves a shift in space and time and changes the characteristics of communities in major ways.

Major grazers can be missing altogether in the dense rockweed forests of northern rocky intertidal waters. There, at least three-fourths of the massive plant production is eventually torn off and washed ashore, piling up as large windrows. These windrows eventually break down and return to the water as fine detritus. Despite the fact that only 10–20% of the rockweed is generally removed by grazing, that grazing constitutes the most important element of this community's structure and largely determines its appearance and function

The small amount of rockweed grazing that does occur is accomplished by several species of the plentiful Littorinid snails that live amidst the brown algal canopy. These snails are also responsible for much of the breakage of the rockweed because one of the species has acquired a taste for one particular piece of the main stem.

In the subtidal of the same rocky-shore community, grazing is still limited—with one exception. The green sea urchin *Strongylocentrotus drobachiensis*, generally common in the rocky subtidal of the North Atlantic, is rarely matched by other invertebrates in its intensive grazing ability. The "drobach" can be extremely abundant and can remove virtually all macroalgae in a wide area when unchecked by wave action or higher predators. Establishing a balance between "drobach" and production from a standing crop of kelp is one of the major challenges in arriving at a viable cold-water benthic ecosystem in a tank. Thus, to establish model community structure in this type of an ecosystem is relatively easy, since there are only a few grazers to control. However, the the problem becomes more difficult in a trophic sense. If the equivalent rocky-shore drift beds and then mud-flat storage are not available in a model that contains both communities, then the plants that break off (or are cropped) must be removed from the system. In an aquarium, this exported plant material can be removed along with scrubber algae and used to determine or perhaps even provide input in the form of feed. Likewise, the mud-flat community can be easily structured in a community sense. However, the particulate food supply that in the wild derives largely from another community (rocky shore) must be supplied.

In our community structure notes we have referred to marine or estuarine systems. As we discuss in Chapter 24, the principles are basically the same for fresh water. The primary difference here is the presence of insect populations, many of which fly at some stage and greatly complicate system management in the home or laboratory. Equivalent to our small coral reef mentioned above, we have maintained an "African" tank for many years in a highly stable state. The fish community dominated by highly aggressive cichlids, with protective hatching and raising of fry, achieves reproductive success only when living space becomes available in the 70-gallon tank by the death of a usually older fish.

Most of our discussion of maintaining community structure in this chapter has involved feeding or food chains. However, a wide variety of other animal interactions in a model, including competition for space, is almost as important as predator/prey relations.

By feeding on mussels, for example, the starfish incidentally opens up terrain for barnacles, which are at a competitive disadvantage to the mussels. Also, reef damselfish help to keep total reef productivity higher than it might otherwise be by keeping away herds of those often voracious grazers, the parrot fish. By maintaining cover, damsels also help to make available a different habitat for numerous small invertebrates.

Corals and anemones actively vie for space, and some species are capable of killing or forcing back other species from territory already gained or even built. Since these organisms are generally sedentary, one has to avoid placing them adjacent to each other. Fish, on the other hand, are quite mobile and, depending as always on system size, space-competitive fish pairs, or at least one of the pair, would have to be omitted. Other factors such as environment and growth rate can play a significant part in the equation. For example, the acroporids are among the least aggressive of corals but they are able to dominate many reefs, presumably because of rapid growth rates. Others of these factors are discussed in coming chapters. Many manipulations are learned by keeping a microcosm, and in so doing, the aquarist occupies the position of "chance" factors in the wild environment, learning and manipulating as necessary.

Model Diversity

Finally, with regard to the total number of species within a living model it must be remembered (from studies of the biogeography of islands, which presumably includes patches or separate communities) that species number is a function of area. There is little question that species diversity is a function of latitude. Colder, higher-latitude equivalent ecosystems have fewer species because they are geologically younger, have greater large-scale seasonal disturbance, and have less total productivity or available energy. However, within this framework one can use the formula $S = kzA$ where S is the number of species (total or in a major group), k is a constant depending on type of ecosystem or major group of organisms

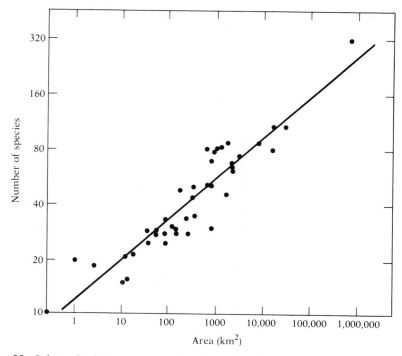

Figure 12 Relationship between area and species number of birds on islands in the vicinity of New Guinea. After Ehrlich and Roughgarden (1987, from Diamond, 1973).

compared, A is area, and z is a constant (typically $z = 0.2$ to -0.35). This is shown for birds on South Pacific islands in Figure 12.

For a modeling example, Adey (1983) in his comparison of the St. Croix (Caribbean) reef and a microcosm of that reef used $z = 0.2$ for stony corals and found an expected microcosm number of 1.3 species. With a half-dozen long-term coral members in this system, it is presumably "packed" with corals. This could be stated a little differently—according to this relationship, in randomly selected areas, the size of the microcosm taken from the wild reef would average only 1.3 species of stony corals. Thus, in practice it would be desirable in scientific modeling, wherever possible, to collect data that would allow this kind of modeling factor to be evaluated. In small systems such as aquaria, which are likely to be packed for many taxa, the excercise is simply instructive and places the effort of developing an ecosystem in a tank in perspective.

References

Adey, W. 1983. The microcosm: A new tool for reef research. *Coral Reefs* **1**: 193–201.

Bonner, J. T. 1988. *The Evolution of Complexity.* Princeton University Press, Princeton, New Jersey.

Britton, J., and Morton, B. 1986. *Shore Ecology of the Gulf of Mexico.* University of Texas Press, Austin.

Carefoot, T. 1977. *Pacific Seashores.* Douglas Ltd., Vancouver.

Chase, A. 1987. *Playing God in Yellowstone National Park.* Harcourt, Brace, Jovanovich, San Diego.

Dayton, P. 1971. Competition, disturbance and community organization: The provision and subsequent utilization of space in a rocky intertidal community. *Ecol. Monogr.* **41**: 351–389.

Drew, R., and Schomer, N. S. 1984. An Ecological Characterization of the Caloosahatchee River/Big Cypress Watershed. Fish and Wildlife Service, FWS/OBS-82/58.2. 225 pp.

Ehrlich, P., and Roughgarden, J. 1987. *The Science of Ecology.* Macmillan, New York.

Fefer, S., and Schettig, P. 1980. An Ecological Characterization of Coastal Maine. Department of the Interior, U.S. Fish and Wildlife Service. FWS/OBS-80/29, 6 vols.

Goreau, T., Goreau, N., and Goreau, T. 1979. Corals and coral reefs. *Sci. Am.* **241**: 124–136.

Grice, G., and Reeve, M. 1982. *Marine Microcosms.* Springer-Verlag, Berlin.

Harris, L. 1984. *The Fragmented Forest. Island Biogeography Theory and the Preservation of Biotic Diversity.* Univ. of Chicago Press, Chicago.

Lippson, A. J., and Lippson, R. L. 1984. *Life in the Chesapeake Bay.* Illustrations by Alice Jane Lippson. Johns Hopkins University Press, Baltimore.

Pilson, M., and Nixon, S. 1980. Marine microcosms in ecological research. In *Microcosms in Ecological Research.* J. Giesy (Ed.). pp. 724–741. Technological Information Center of the U.S. Department of Energy.

Rickets, E., Calvin, J., and Hedgpeth, J. 1985. *Between Pacific Tides.* 5th Ed. Stanford University Press, Stanford, California.

Stephenson, T., and Stephenson, A. 1972. *Life between the Tide Marks on Rocky Shores.* W. H. Freeman, San Francisco.

Strong, D., Simberloff, D., Abele, L., and Thistle, A. 1984. *Ecological Communities, Conceptual Issues and the Evidence.* Princeton University Press, Princeton, New Jersey.

Thurman, H., and Webber, H. 1984. *Marine Biology.* Merrill Publishing, Columbus, Ohio.

CHAPTER 14

TROPHIC STRUCTURE
Ecosystems and the Dynamics of Food Chains

Most direct human relationships with the management of living organisms involve a paternal or protective role that attempts to keep the plants and animals of interest free from interference by other organisms that are not of interest or from negative physical or chemical effects. Usually, unless the creature involved is a pet, used for power, like a horse or oxen, or only partially harvested, as with cows' milk, this protective role comes to an abrupt end at harvest time when it is expected that most energy and resources held by the cultivated plants or animals are to be converted as efficiently as possible to human use.

The modern culture of organisms, whether a pet in the home or a cow or fish in a pen, requires massive feeding, usually of specially crafted feeds that supposedly provide all the vitamins, essential amino acids, elements, and other requirements to avoid deficiency diseases. In the study of ecosystems and organic evolution as a whole, the role of food composition has tended to be neglected. We are beginning to recognize that, as a population adapted over millions of years to a variety of wild foods, we are now "creating" the major modern human diseases such as many cancers and heart disease by building foods that are efficiently produced for energy supply, but equally important directed to some kind of human need for simplicity and order, even in foods (Crawford and

Marsh, 1989). If we cannot, through modern technology and with the support of a major part of our GNP, craft a food supply that is suitable for the human species, how can we hope to artificially supply an ecosystem of many different species with often very different food needs. We should probably put a major social/economic effort into reducing human populations and developing more natural human food supply patterns. In any case, it is clear that ecosystems wild or in model, must be operated through a natural food chain of high organism diversity. Out of economic necessity we are likely to often simulate a portion of a model ecosystem food supply with an artificial feed. However, we should always remember that where this is done we are probably reducing veracity.

The culture of an ecosystem is rather different in basic concept from the culture of individual organisms. To emphasize this basic point, we quote from Wyatt (1976):

> Despite all efforts to survive, the natural destiny of almost all animals and plants in the sea is to be eaten before maturity is attained. It is only the few which live long enough to perpetuate their species. The best that most species can do is remain one step ahead of those which prey on them, and so avoid extinction. So eating and being eaten are central to the whole science of ecology, and it is these two topics which are emphasized here.

While in a captured or cultured ecosystem some manipulation will probably be employed to adjust longevity of certain species, as a whole, an ecosystem will not function without energy flow and the daily pattern of predator eating prey at all levels. Trophic or feeding structure must develop and be maintained or there is no ecosystem. If "eat and be eaten" seems the antithesis of the reason for culture in the first place, before getting into the basic subject of the chapter we would remind our readers of several critical elements discussed earlier.

The primary purpose of this book is to assist scientists, aquarists, and hobbyists in the re-creation of marine, aquatic, and wetland ecosystems in living models. While establishing ecosystems in model may be more or less difficult to accomplish, depending upon size and complexity, this is a crucially important endeavor. On the scientific side, in microcosms some very important ecological principles have been established; and microcosms should receive broader use (Odum, 1984). Although computer and mathematical modeling of ecosystems has been extensively used in the past several decades, some well-known ecologists consider the mathematical model of little value (Andrewartha and Birch, 1984). On the hobbyist side, a broader, heuristic, day-to-day understanding of ecosystem function by a larger public can be a crucial element that alerts mankind, not just environmentalists, to the primary problems of the twenty-first century. Some scientists maintain that "sustained life under present day conditions is the property of an ecological system rather than a single organism or species" (O'Neill et al., 1986). This suggests that while the culture of individual organisms may be important to understanding the intricacies of function of each species, only by working with whole ecosystems will we truly begin to understand the whole as well as the parts. Beyond any consideration of the ethics of total human domination of earth, it is highly unlikely that humans can be totally supported by our own monocultures. Since we are dominating the earth to the point where, increasingly, natural ecosystems do not have the freedom to function on their own, we must quickly learn the techniques of culture of ecosystems. Probably, we will just as quickly learn that our survival will ultimately depend upon reducing human population and returning a fair part of the surface of the earth to a more natural state.

An ecosystem is a structured relationship of living organisms that receives and utilizes a flow of energy. Many chemical elements also flow or are "cycled" through an ecosystem, and system function is usually measured by the flow of carbon, oxygen, nitrogen, or phosphorus. However, energy flow (the creation and degradation of food) is the crucial element.

While only a few species might be present in a very simple ecosystem, "complexity of organization seems to be a key ingredient in ecological persistence" (Conrad, 1976). O'Neill et al. (1984) argued "for a correlation between ecosystem stability and the functional redundancy of the system" and cited a microcosm study of van Voris et al. (1980) as proof of the concept. That species diversity equals ecosystem stability has been debated for years and now tends to be unacceptable to many ecologists. However, in this new light, multiple species at all levels in an ecosystem are what provides the stability. Hundreds of herbivores and few producers or carnivores may constitute relatively high diversity, but provide a recipe for disaster. This concept is crucial for the operation of model ecosystems because total species number is much more limited than it is in the wild. While many species are desired, only if a working trophic structure can be established will the model be successful. This in itself is not necessarily difficult. A number of studies of insect community reestablishment have shown that trophic structure, a food chain, is developed regardless of the species involved (O'Neill et al., 1984). In a model, however, the cultured ecosystem must be open to a wide range of immigration for this to happen. In short, the builder or operator must allow the possibility for the introduction and establishment of a high diversity of species at all levels in the food chain or web.

Energy Capture and Flow

All organisms, and therefore all ecosystems, require a source of energy to operate. Virtually all of the energy of earth's ecosystems derives from plant photosynthesis. Plants in the most efficient of cases, such as wetlands and coral reefs, convert about 5–6% of available sunlight to chemical energy. Generally the efficiency of solar energy "capture" is less, typically 2–3%. In any case, the bodies of plants become the food for all animals, sometimes by simple routes, in other cases by complex routes. For animals, the energy comes in the form of food. Typically, some of the energy in the available food is used for motion, construction of tissues, internal circulation, reproduction, etc. and then discarded as low-level and thus unuseable heat. The remainder, typically anywhere from 20 to 80% of the supply, is simply passed on, altered in character perhaps, but available for another user. This process is repeated over and over again and unless some of the original biomass or "food" is buried in geological storage, such as in a swamp, all of the original energy either is temporarily stored as structural energy in organisms (potentially more food) or is lost as low-level heat to the environment. While this loss might be said to represent a low efficiency of transfer, it must be remembered that most of the loss was used to "drive" the system, to move muscles and carry out the necessary building of required chemistry. Ecologists speak of the transfer efficiency of an organism, which is typically 10–20%. This efficiency is calculated as the new weight or biomass developed, as a percentage of the amount of food eaten. In an adult animal, efficiency could be zero or close to it, but in truth 50% or more may have been used to drive the organism. If one is producing chickens or cows, and weight is money, the transfer efficiency is extremely important. If one is running an ecosystem, it is not of overriding interest.

In ecosystem modeling there should be two measures of efficiency:

1. The percentage of either lost (exported) or stored biomass as a function of that in the wild community being modeled. A model that exported or stored twice as much biomass as the wild analog would have a 50% efficiency.

2. The species diversity of the model as a percentage of that in the wild (in a measure that uses the biomass of the organisms per unit area).

Thus, a completely efficient model ecosystem would be one that uses as much energy as the wild analog and processes that energy in as complex a manner. In a scientific modeling endeavor, these numbers can be estimated; in the aquarium effort they will likely never be calculated. Once thought about, however, these concepts will place the modeling effort in proper perspective. A wild community may grow for a few years, perhaps even a century or two in biomass. This process, called succession, sooner or later must stop (without geological storage or export to another community) and all input energy must be used and converted to low-level heat that escapes. The model ecosystem should be no different from the wild. The time frame for balance is shorter, generally much shorter.

Depending on the arbitrary definition of an ecosystem, all such systems must produce their own food (thus energy) through plants or receive it from producer communities. If one is operating a producer community and a consumer community separately (e.g., a kelp bed and a mud flat), the mechanisms for food transfer and nutrient return must be allowed for. If the producer community is not present, and the aquarist is adding food, then either the food and feces must be all used, and the excess nutrients removed as they would be flushed away by the ocean, for example, or the true equivalent of geological storage is required. If the consumer community is not present when stable biomass maximum is reached, then assuming required food is being added, excess biomass (leaf fall, for example) must be removed, or again the true equivalent of geological storage must be made available.

In Chapters 9–12, we discussed wild and captured ecosystems in terms of the cycling of the major required, and then discarded, elements. In Chapter 13, the ecosystem was treated as a community; we talked about the living organisms themselves, how they behave and react to each other. Here we treat the ecosystem and its modeled counterpart as a functioning machine. We choose to look at this machine from the point of view of how it captures and uses energy or how it produces and then uses food. In a structural sense this is what the ecologist calls trophic structure, a food chain or a food web.

Food Webs

In the mid 1950s H. T. Odum carried out a detailed study of the trophic structure of a freshwater spring and its resulting stream in central Florida. Dividing the organisms in the spring into producers (plants), herbivores, carnivores, top-level carnivores, and decomposers, Odum produced the biomass pyramid shown in Figure 1. In this case the amount of plankton present was minimal and it was ignored.

This now classic and often cited study shows a food chain or trophic structure in a static kind of way. However, it is probably typical for very simple ecosystems and not far from what one would expect to find in well-simulated microcosms and aquaria. In this very simple pyramid one can see one of the problems of all food chains. One can also see part of the reason why they are often called food webs and are typically oversimplified. The stumpknocker (Redear sunfish, *Lepomis microlophus*) eats small invertebrates, especially snails, that are in turn feeding on algae. However, it also eats the algae and the mixed small animals (aufwuchs) that coat the stems and leaves of the aquatic higher plants, the macrophytes. Also, the pyramid concept does not really include the decomposers. The decomposer biomass line in this case is simply set off to the side, presumably to indicate that it receives from all other elements of the pyramid.

In another diagram, developed for a Georgia salt marsh (Fig. 2), a

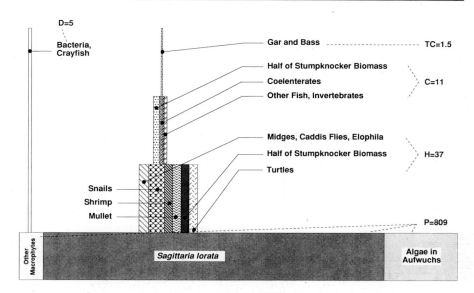

Figure 1   Biomass pyramid for Silver Springs, Florida (gm/m² dry biomass). P, Primary producers; H, herbivores; C, carnivores; TC, top carnivores; D, decomposers. After Odum (1957).

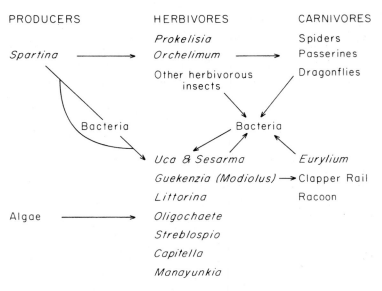

Figure 2   Food chain in a North American East Coast salt marsh. After Levinton (1982), from Teal (1962). Reprinted by permission of Prentice Hall, Englewood Cliffs, New Jersey.

slightly different approach to presenting a simple food web is taken. Here the primary decomposers, the bacteria, are included within the web. However an attempt is made to add the obvious complexity that many "herbivores" are also eating bacteria that are the primary decomposers of organic wastes. This system and its food web could easily be simulated in microcosms. However, the place of the birds and the racoon would be taken by the aquarist. A plankton community, brought in by the tide, could also be added and the aquarist could generally simulate this aspect by artificial feeding.

A more complex food web for a rocky, subtropical shore is given in Figure 3. In this case a large part of the food chain above the producer community is made up by molluscs and finally by a top carnivore, a starfish. Fish and birds undoubtedly play a subsidiary role in this chain. This ecosystem would also not be difficult to achieve in model. The plankton equivalent would have to be added and the aquarist would harvest and control the top predators.

An interesting variant is the food web for the North Sea herring shown in Figure 4. The web is centered on this single, albeit extremely important, fish, but shows how its feeding habits change as it matures.

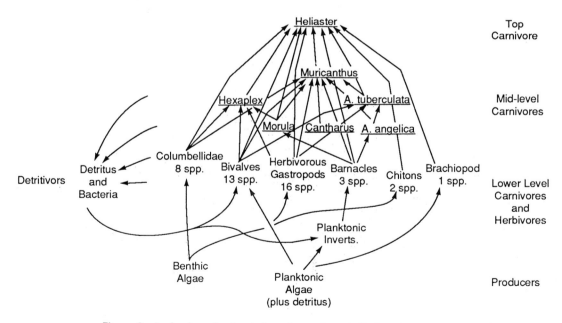

Figure 3 Rocky shore food web from the northern Gulf of California. This web is dominated by the starfish *Heliaster*. The Columbellidae are the dove shells, algae-eating snails; *Heraplex* and *Muricanthus* are murex snails; *Morula* and *Cantharus* are small snails that feed primarily on barnacles; and *A. turbeculata* and *A. angelica* are predatory snails of the genus *Acanthina*. After Levinton (1982, from Paine, 1966). Reprinted by permission of Prentice Hall, Englewood Cliffs, New Jersey.

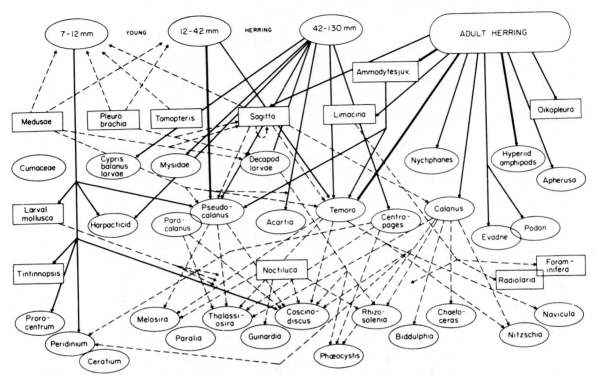

Figure 4 Planktonic food web of the North Sea herring. The food utilized by the herring gradually changes as the fish matures. After Cushing and Walsh (1976, from Hardy, 1924).

This web does not indicate the many predators on the herring itself, nor does it complete the cycle by showing the decomposer part of the local ecosystem. It begins to demonstrate how complicated a realistic trophic structure can become for a wild community and why some scientists would argue that formal trophic structures are only heuristic devices. They can only help to provide a better feeling for the dynamic aspects of ecosystems. This food web would obviously be very difficult to achieve even in a very large mesocosm.

Finally, so as not to neglect fresh waters and to indicate that freshwater food webs can be as complicated as those in the marine world, we provide a web for the upper Nile River in Uganda (Fig. 5). In this case, the crocodile becomes the top predator, feeding on virtually all organisms in the middle and upper levels of the web. With the aquarist playing the role of the crocodile itself as well as the birds and mammals, this web and its community probably could be simulated in a mesocosm.

A full understanding of energy flow in a model ecosystem is likely to be undertaken only in a mesocosm-scale research project. An ecosystem energy flow characterization is a quantification and summary of diagrams

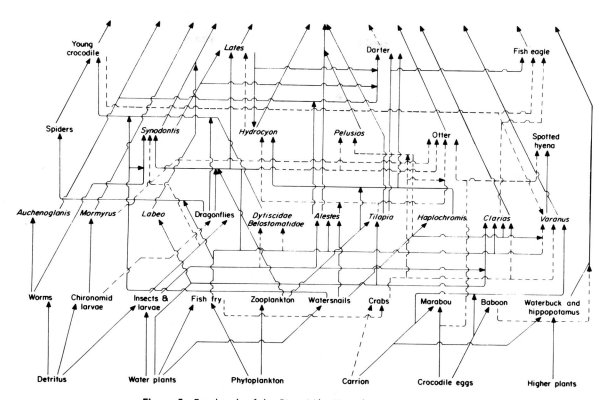

Figure 5 Food web of the River Nile, Uganda. *Varanus* is the monitor lizard. The remaining generic names are fish. After Moss (1980, from Cott, 1961).

like Figures 4 and 5. In Figure 6, for Lake Ontario, the producer community is not shown. This is a benthic community and the total energy uptake (6856 kcal/m) represents organic detritus from plankton and land runoff. This detritus, rich in bacteria, is eaten by a variety of detritivores from which the carnivores and top carnivores derive their food and energy resources. Notice in this case, besides the large amount of unutilized and eventually buried energy in the form of subfossil organic material, a small quantity is also exported in the form of emergents (flying insects).

Food Webs in Model Ecosystems

Whether carried out formally or informally, a food web should be developed for all model ecosystems simply to provide a guide. In Figures 7, 8, and 9, we show such simplified webs for a microcosm coral reef, a Maine rocky shore and bay microcosm, and a subtropical Florida Everglades

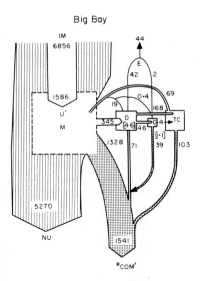

Figure 6 Energy flow in an arm of Lake Ontario, Canada. IM, incoming organic matter; NU, not used (geological storage); R$_{COM}$, respiratory losses; E, flying insect loss. The boxes represent biomass (in kcal/m²) of microorganisms, detritivores, carnivores, and top carnivores respectively. The width of the arrows represents energy flow as cal/m²/d. After Moss (1980, from Johnson and Brinkhurst, 1971).

mesocosm, respectively. It should be particularly noted that in two of these systems there is an energy supplement (feeding) that is intended to simulate primarily the plankton supply from the larger body of water that it is not possible to present to the model. This is quite acceptable as long as the equivalent energy (and nutrients) is exported, or possibly sent to a fossil reservoir within the system. As we discuss in Chapter 12, this is the primary function of the algal turf scrubber, although any full and controllable method of export could perform the same function. One might simply harvest parts of plants from the community itself and remove them, though by that means it would not be possible to smooth out diurnal effects. The primary difficulty with traditional filters is that they do not provide for full export of all elements. While carbon may be released as CO_2 and if eutrophic enough, some nitrogen as N_2, much nitrogen, phosphorus, and other nutrients remain. Also, even if filters were to be frequently and automatically back-washed to partly overcome this objection, they cannot discriminate between desirable and undesirable particles, including living organisms in the water column.

Establishment of Food Webs

Initially, some knowledge both of the feeding capabilities and preferences of organisms and the trophic structure of the community in question is

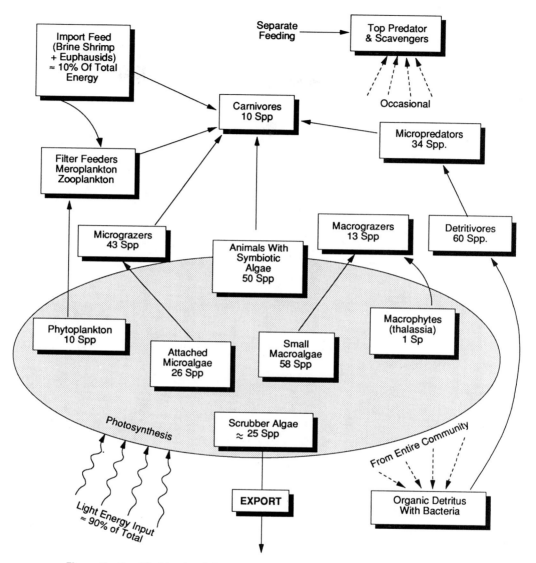

Figure 7 Simplified food web for the Smithsonian coral reef model ecosystem (see also Chapter 21).

necessary to properly manage a microcosm of a natural community. As we discussed in the previous chapter, it is probably possible to model, in closed form, any marine or aquatic ecosystem. However, the systems most interesting to the aquarist, and those with which we now have considerable experience, are shallow, benthic communities in which the base of the food chain—or web—is formed largely by benthic (attached) algae or

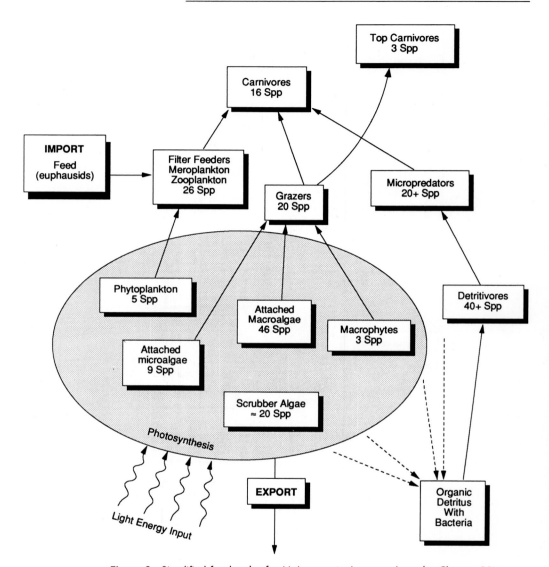

Figure 8   Simplified food web of a Maine coast microcosm (see also Chapter 22).

invertebrates with symbiotic algae. On soft bottoms flowering plants are often included with the algae.

The problems of handling attached plants in a closed microcosm are relatively straightforward. In the next chapter we will discuss the requirements of a variety of marine and freshwater algae and flowering plants. Generally, managing this lowest or producer level in the microcosm in-

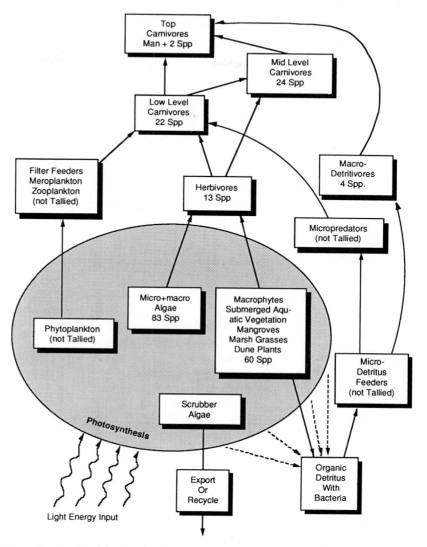

Figure 9 Simplified food web of a Florida Everglades mesocosm (see also Chapter 23).

volves four primary factors related to the natural system being modeled: type of substrate, available light, grazing rate (or rate of removal of plant biomass), and the cover and biomass of the plants themselves. Since these four elements are interrelated, establishing a suitable balance among them is the primary requirement for a successful microcosm, once the overall physical factors are established.

Prior to the introduction of the first plant samples, it is assumed that essential ingredients considered in the first seven chapters are in place, including sufficient light of appropriate quality, proper temperature, and water motion. Some plant species will be immediately successful. Others will require repeated introduction. A sufficient standing crop of plants, algae, or higher plants approximating that in the community being modeled should be allowed to develop before the introduction of a significant colony of grazers.

There are several ways of gauging the correct ratio of the microcosm's standing crop to that of the wild: visual approximation, plant length, dry or wet weight, and primary production measured by oxygen release (see Chapter 10). Larger grazers such as fish, urchins, and crabs can be added according to their approximate abundance or biomass in the community being modeled. Because this involves guesswork and trial and error, it is advisable to work initially with too few rather than too many of the larger herbivores. If plants become too abundant, grazing can be supplemented with harvesting by hand. Additional grazers then can be added one by one until the desired biomass or standing crop of plants is achieved.

Any system being modeled is likely to contain not only larger grazers, typically fish, urchins, snails, and crabs, but also numerous micrograzers such as amphipods and isopods and, in fresh water, insects. Some element of control of these micrograzers is often necessary. Relative micrograzer/macrograzer abundance often determines plant community structure—specifically, the abundance of fine-turf algae in relation to crustose algae, macroalgae, and higher plants.

Micrograzers can be controlled, though often with difficulty, by simply removing them when they become too plentiful. However, carnivores usually are required if one is attempting to simulate either an analog or synthesized ecosystem. When establishing a new microcosm, usually the first carnivores introduced are predators of the micrograzers—wrasses in a reef system; killifish, tom cod, or sculpins in a cold-water marine system; and a wide variety of shiners, bluegills, tetras, barbs, and mollies, to name a few, in fresh water.

Several weeks or months into the stocking process, as plankton begins to build up, especially the swimming reproductive and larval stages of organisms already established, filter and suspension feeders such as sponges, bryozons, worms, and *Chromis* fish in reefs, and hatchetfish in fresh water, can be added. Finally, as the buildup of their prey warrants, such mid-level predators as butterfly fish, angelfish, and grunts in a reef tank, and discus, larger tetras, and barbs in fresh water can be placed in the growing system.

Up to this point, at least for most benthic microcosms, scaling factors have not been a major consideration. In any moderate-size tank, it will usually be possible to add several elements of each of the trophic levels that we have discussed so far. However, when dealing with the higher carnivores such as jacks, barracuda, and sharks in a reef, large cods and haddock in marine cold water, pike and bass in cold fresh water, and a wide variety of cichlids in tropical fresh water, scaling difficulties will invariably arise even in the largest of tanks. These top predators, especially adults, need a fairly large foraging territory to obtain required food. Using small members of a species will help to alleviate the problem. However, the feeding patterns of the young of a top predator may be quite different from those of the adults. In the Marine Systems Laboratory reef tank, we have generally been successful in keeping barracuda by feeding them supplementary goldfish. When introduced to salt water the goldfish immediately emit distress signals and are attacked by the barracuda within seconds. The feeding of barracuda is somewhat analogous to that of snakes and larger cats (short feeding periods alternating with long quiescent spells), and is easily managed. However, smaller fish in the system have been subject to attack in the absence of a reasonable control element. For top carnivore management, it is essential to provide prey with the protection of proper cover, reef structure, or abundant plants.

We have briefly discussed the benthic—or at least bottom-oriented—elements of trophic structure in some of our microcosms. Except for the top carnivores, these elements are all "within scale" for moderate-size tanks. In the beginning of our trophic discussion we bypassed filter feeding of plankton. This is also primarily a scaling problem in that some filter feeders require an amount of plankton that would be derived from large quantities of overflowing water. We have introduced brine shrimp into our reef tanks (see Chapter 17) based on our calculations of plankton input into a St. Croix reef. *Chromis* and numerous coelenterates and worms rely on this input. In our cold-water system, to keep mussels and scallops for long periods—especially in the first year—it was necessary to periodically introduce a phytoplankter such as *Isochrysis*. In the second year, in a 3000-gallon tank, this proved unnecessary. Depending upon the level of ecosystem simulation being sought, over time such refinements tend to be less required. As the ecosystem matures and many animals become fully adapted, the wide variety of reproductive stages present, as well as particulate organics released by many plants and animals, can fulfill most, if not all, requirements of generalized filter feeders. Some specialized filter feeders may not find the size or kind of planker needed and would have to be fed separately or omitted.

Detritus feeders are usually not a problem in aquaria of any size. However, these will have to be introduced to the system. Worms and microelements such as protozoans are usually present if the tanks are "seeded" from equivalent wild environments. Larger detritivores such as sea cucumbers in the sea and catfish in fresh water can be provided according to the amount of detrital load. As unused detritus begins to build up in the system, typically several months into stocking, these animals can be added in numbers based on a visual analysis.

As we discussed in Chapter 12, the primary function of an algal turf scrubber is to maintain water quality on a diurnal basis by simulating the large, relatively low-biomass volume of water in the wild that usually lies adjacent to the ecosystem of interest. However, the scrubber also performs another critical function: removal of excess nutrients. These nutrients might first be stored as excess biomass or particulates on a marsh or mud flat. Generally excess nutrients are the result of adding food to an aquarium or microcosm to simulate either a larger area of plankton production or a larger area of small-invertebrate or insect production (typically freshwater). It can also result from overloading the tank with fish or other higher trophic elements either purely for esthetic reasons or to drive a system faster so as to carry out scale modeling in a smaller space or lesser time. If the algae removed exceeds added food by 2–5 times (see Chapter 12), the system will remain balanced. The algal scrubber is a vital element in overall trophic structure and in system control and must be carefully considered in assembling a model ecosystem.

Among the protists or protozoans there exists a food web in miniature, further complicating the understanding of the trophic structure as a food web. Some of these tiny flagellated and ciliated animals have photosynthetic pigments and organelles and, like many corals, for example, function partially or wholly as plants. Other protozoans are grazers, primarily of algae and bacteria, and many others, including large amoebas familiar to anyone who has taken a biology course, are predators on the smaller protozoans. All of these protists can be and often are eaten by higher invertebrate micrograzers or micropredators or even by fish. This part of the food chain is always present, but while fascinating to microcosm operators, goes unnoticed by all but the most specialized aquarists. Rather than attempt to operate a substable ecosystem with a very limited pool of microbes, it is preferable to stock a tank with "infusions" from sediment and water of the system being simulated—or even from similar systems.

There is a rather informal theory among ecologists that says that community structure can be very different when there are even rather than odd numbers of levels in a food chain. As we have discussed, a complex and stable ecosystem should be more appropriately characterized as a web rather than a chain. Nevertheless, the food webs shown in this chapter for several model systems could be idealized as three-level systems. (The fourth level is normally simulated or strongly managed by the aquarist so that it does not overpower the level below). According to the theory, since the simple carnivore level controls the grazers (and micrograzers), plants are able to maintain biomass and productivity. Indeed, most of the model systems which we have developed have high plant biomass. On the other hand, in our 130-gallon home reef (see Chapter 21), the top predator, the Rare Reef Lobster, has been given a "free hand" for many years. There is heavy grazing pressure in this system. However, the plant community (perhaps equivalent to thorn-scrub) is strongly dominated by calcified and other tough algae. Thus, there is a high plant bio-mass and moderate productivity in spite of the heavy grazing.

Analyzing, modeling, and predicting ecosystem behavior are diffi-cult for ecologists because of the complexity of even the simplest eco-systems. Modeling by measuring energy flows through an ecosystem—a variant of trophic structure analysis—is one of the most useful ecological tools devised in the last 50 years. The process, which goes several steps beyond trophic structure analysis, determines the energy equivalent of food passed from step to step by considering the controls on the amount of energy flow between each step. While perhaps beyond the interest of the average aquarium owner or microcosm or mesocosm operator, such an analysis may interest the advanced hobbyist. Energy flow is particularly adaptable to computer analysis, and a personal computer can be an effec-tive tool in providing management information for the more sophisticated aquarium system.

Trophic Structure in Aquaria

Even in modest-size aquaria, sufficient plant production is not usually difficult to manage if light and water motion are adequate. Grazer abun-dance and standing crop of algae and submerged aquatic vegetation have to be carefully monitored, but manipulation to attain a balance is usually not difficult. Likewise, the plankton component is mostly simulated by dried food, brine shrimp or the equivalent, and top predators, which cannot normally be present, must also be simulated by the aquarist. As long as a sufficient number of hideaways within or separate from the primary ecosystem as refugia are available, the smaller grazer and car-nivorous and detritivorous invertebrates can typically maintain their numbers in the face of predation. The difficulty usually lies with mid-level fish. A few species can reproduce and maintain their numbers in the face of predation pressure in the community tank—some cichlids are quite amazing in this respect and provide exciting observation. However, most fish cannot achieve successful reproduction in a modest-size aquar-ium with numerous other species present, although they frequently pro-vide an important feed component with their eggs and/or young.

Thus, sooner or later as mid-level fish are either preyed upon or die from old age, they must be replaced. This can be done through wild capture, purchase, or by separately culturing the species desired. This process is another case of simulation to achieve the effect of a much larger water volume.

In Chapter 24, we discuss an African pond tank of 70 gallons that has been left for many years to come to its own population level. After enough of the oldest fish eventually die, a reproducing pair of cichlids, after many failures due to predation on the young, will finally succeed in getting a few young to a size where they will not be eaten by the adult fish in the tank.

The Organisms

The major trophic levels for several types of ecosystems are discussed in the next six chapters, along with descriptions of selected organisms. Identification manuals, typical examples being given in the references for each chapter, are available for most plants and animals inhabiting coasts, ponds, and lakes likely to be simulated. These manuals often provide feeding information for the very abundant species. For others, inferences can often be made based on the morphology of a close relative. Often, when information is unavailable elsewhere, it can be gained simply from observing the animals themselves.

Our understanding of how organisms relate to each other structurally and evolutionarily has been increasing by leaps and bounds in the last twenty to thirty years. Beyond simply more anatomical work on less well-known organisms and rapidly improving additions to the repetoire of well-known fossils, the electron microscope (EM) and scanning electron microscope (SEM) have had a very major impact on this process. No longer can organisms be thought of as being either plant or animal. We strongly recommend to any hobbyist or aquarist not familiar with these advances the book called *"Five Kingdoms, an illustrated guide to the phyla of life on earth"* by Lynn Margulis and Karlene Schwartz.

To the average aquarist, much of the information presented in this chapter may seem overly complicated. However, complex ecosystems containing a fair number of species having different feeding strategies are remarkably adaptable to variations in trophic patterns. A mistake that would be intolerable in conventional systems is often no more than an interesting variant in the "dynamic aquarium."

References

Andrewartha, H., and Birch, L. 1984. *The Ecological Web*. University of Chicago Press, Chicago.

Conrad, M. 1976. Patterns of biological control in ecosystems. In *Systems Analysis and Simulation in Ecology*. B. Patten (Ed.). 430–456. Academic Press, New York.

Crawford, M., and D. Marsh. 1989. *THe Driving Force, Food, Evolution and the Future*. Harper and Row. New York.

Cushing, D., and Walsh, J. (Eds.) 1976. *The Ecology of the Seas*. Saunders, Philadelphia.

Levinton, J. 1982. *Marine Ecology*. Prentice Hall, Englewood Cliffs, New Jersey.

Margulis, L. and Schwartz, K. 1988. *Five Kingdoms, an illustrated guide to the phyla of life on earth*. 2nd Ed. 376 pp. Freeman and Co., New York.

Moss, B. 1980. *The Ecology of Fresh Waters*. John Wiley & Sons, New York.

Odum, H. 1957. Trophic structure and productivity of Silver Springs, Florida. *Ecol. Monogr.* **27:** 55–112.

Odum, E. 1984. The mesocosm. *Bioscience* **34:** 58–562

O'Neill, R., DeAngelis, D., Waide, J., and Allen, T. 1984. *A Hierarchical Concept of Ecosystems.* Princeton University Press, Princeton, New Jersey.

van Voris, P., O'Neill, R., Emanuel, W., and Shugart, H. 1980. Functional complexity and ecosystem stability. *Ecology* **61:** 1352–1360.

Wyatt, T. 1976. Plants and animals of the sea. In *The Ecology of the Seas.* D. Cushing and D. Walsh (Eds.). Saunders, Philadelphia.

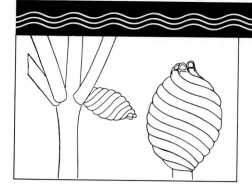

CHAPTER 15

PRIMARY PRODUCERS
Plants That Grow on the Bottom

The functioning of the vast majority of the earth's wild ecosystems is ultimately dependent upon plant photosynthesis. In many ways the really critical impact of human activities on the planet has been on the earth's photosynthetic organisms and their needs and products. In closed-system simulations of ecosystems, whether at the mesocosm, microcosm, or aquarium level, proper function is likewise dependent upon plants. The primary chemical roles played by plants in aquatic ecosystems are discussed in Chapters 8–11. Here, we discuss the plants themselves in their role as providers as habitat and food for animals. This most critical subject has tended to be very much undertreated in older aquarium texts.

The plankton community, including its plants, will be covered separately in Chapter 18. This is partly because of its great difference in form and function and partly because most simulated systems, especially smaller ones, are likely to be benthic or bottom-oriented. Also, although the general subject of this book includes living models of swamp wetlands, and Chapter 24 deals with mangrove and cypress swamps, this is a rather specialized topic that we will leave for other chapters. Here, we will cover three major areas of interest: (1) the benthic algae (of many major groups, divisions, or phyla); (2) submerged aquatic vegetation (submerged vascular plants); and (3) emergent aquatic vegetation (mostly marsh plants growing from waterlogged soils into the atmosphere). A rough indication of the importance of the major groups of benthic plants in the primary environments considered in this book is shown in Figure 1.

As we discussed earlier under community structure, plants are often far more than just the primary food or energy source for the biosphere. In most terrestrial ecosystems and in many marine and aquatic ecosystems,

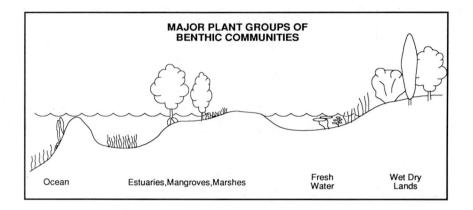

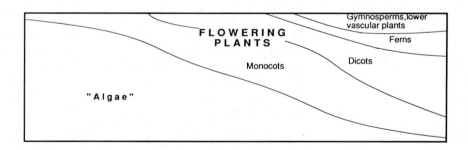

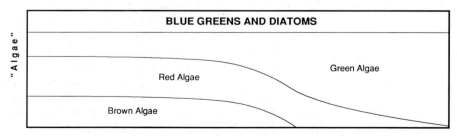

Figure 1 General distribution of the major benthic plant groups from terrestrial environments to the sea.

plants also provide a major element of habitat structure. The plants act to provide cover or protection, sometimes direct attachment, and often are modifiers of environmental characteristics such as oxygen, carbon dioxide, available light and/or drying and ultraviolet effects, pH, waves, wind, temperature, etc.

Throughout this chapter, and indeed throughout the book, we refer to algae as plants. Thus, we are employing the systematic language generally in use prior to the 1980s. However, we are generally in agreement with modern thinking that would separate most algal phyla from the "green" plants. Whether the Kingdom Protista (or Protoctista) is appropriate for all algal groups or, as we would prefer, a separate kingdom for most algal groups except the greens is as arbitrary as the boundaries between ecosystems. While we will adhere to the older terminology as generally being more comprehensible to our readers, we again refer all readers not familiar with the new evolutionary ideas of the 1980s to *Five Kingdoms* by Lynn Margulis and Karlene Schwartz.

Benthic Algae

To the layman familiar with green trees, shrubs, and herbaceous land plants, marine and aquatic algae may seem to be on the small side and at the same time rather simple in morphology. However, the algae of the sea can be very large, some many tens of feet in length, elaborate in construction, and sometimes reproductively very complex. As one might expect, algae generally lack the vascular, water-conducting and food-conducting tissue of the land plants. These structures are not of general use in a watery environment. It is of interest in this context that vascular plants that have become secondarily adapted to the aqueous environment also have very reduced or even no conducting tissue. On the other hand, those higher plants that have evolved back into the water environment often have a very special ability that allows many vascular plants to root into anaerobic muds. This is a secondary gas exchange system made up of intercellular spaces. Also, among the the largest algae, the kelps whose bases of attachment lie heavily shaded in deeper water, an entirely different conducting tissue has evolved to carry sugars from the more productive fronds near the surface to the holdfasts. This is the equivalent of the phloem, or inner bark, conducting tissue of many land plants. The xylem, the often woody water-conducting tissue, is of course absent.

While most algae are anatomically simpler than terrestrial plants, they show great diversity of color, virtually every color of the rainbow. Nevertheless, like terrestrial green plants these underwater relatives (mostly very distant relatives) photosynthesize, using chlorophyll and the energy of sunlight. Though generally using the same photosynthetic process, algae bring water, carbon dioxide, and nutrients together to make their own food and build their own tissues. Many algae also have accessory photosynthetic pigments, which are mostly responsible for the wide range of color characteristics and allow a greater efficiency of light capture in water where light spectra are often quite different than on the land (see Chapter 7). The algae involved are highly diverse, particularly in the sea where there are some 40,000 species belonging to very different groups. While in fresh water there are fewer major groups and species of algae, still there are large numbers, on the order of 10,000 species.

It might seem relatively easy to separate technically the "lowly" algae from the higher plants. However, some algae have conducting tissue, some have structures appearing as roots, stems, and leaves and even the "pièce de resistance" of the dictionary differentiation, a special cell layer surrounding the egg cell, does not apply to the Charophytes (freshwater green algae). Indeed, only with careful semantics does it apply to some of the advanced browns and reds. The major groups of algae are differentiated by their photosynthetic pigments, type of storage food, wall structure, the type of flagella, and to some extent reproduction (Table 1). It seems clear that the higher plants, particularly mosses, ferns, gymnosperms (e.g., conifers) and angiosperms (flowering plants), have been directly derived from the green algal line (Fig. 2). Algae show a wide range of structural organization, most of which are found to varying degrees in all the major groups of benthic algae, as we will show below.

In recent years, blue-green algae (Cyanophyta or Cyanobacteria) have been treated as close cousins of bacteria in the Kingdom Monera. We will continue to refer to them as algae or plants in this book since they photosynthesize as other algae do, even having chlorophyll and accessory pigments similar to the red algae. Nevertheless, blue-greens exhibit many primitive features, including the lack of a membrane-enclosed nucleus, chromatophores, and other cell organelles. Blue-greens can be any color, including purple, red, and black, but are frequently blue-green. They are almost always present in aquaria and can be fairly easily identified with a microscope by the lack of a nucleus or chromatophores, lateral cell walls (between adjacent cells) that are difficult to see, and relatively small cell size. Superficially, they tend to be slimy or silky in appearance. Reproductively, these primitive algae bud off cells or short filaments. Like bacteria the presence of sexual reproduction is not seen but can be demonstrated genetically. Blue-greens are ubiquitous in the wild and some species occur in the most extreme of environments, often those too hot, too nutrient-rich, too-nutrient poor, too variable, etc. for other algae. They are also typically important elements of the algal turfs of scrubbers. A variety of blue-greens and their reproduction is shown in Figures 3 and 4. A current and widely accepted theory holds that the numerous cell organelles of more advanced algae are partly derived through the acquisition of smaller blue-green cells by the larger more "advanced" cells (Margolis and Sagan, 1986).

TABLE 1

Summary of Some Algal Divisions and Their More Significant Characteristics[a]

Division	Common name	Pigments and plastid organization in photosynthetic species	Storage product	Cell wall[b]	Flagellar number and insertion[c]	Habitat[d]
Cyanophyta	Blue-green algae	Chlorophyll a; c-phycocyanin, allophycocyanin, c-phycoerythrin; β-carotene and several xanthophylls	Cyanophycin granules (arginine and aspartic acid); polyglucose (glycogenlike)	α- and ε-Diamino-pimelic acid, glucoseamine, alanine, etc.	Absent	fw, bw, sw, t
Prochlorophyta		Chlorophyll a,b; seven carentoids, of which β-carotene and zeaxanthin are major	Starchlike	Peptidoglycan	Absent	sw
Chlorophyta	Green algae	Chlorophyll a,b; α-, β-, and γ-carotenes and several xanthophylls; 2–5 thylakoids/stack	Starch (amylose and amylopectin) (oil in some)	Cellulose in many (β-1, 4-glucopyranoside), hydroxy-proline glucosides; xylans and mannans; or wall absent; calcified in some[e]	One, 2–8 many equal, apical	fw, bw, sw, t
Charophyta	Stoneworts	Chlorophyll a,b; α-, β-, and γ-carotenes and several xanthophylls; thylakoids variably associated	Starch (amylose and amylopectin)	Cellulose (β-1, 4-glucopyranoside); some calcified	Two, equal, subapical	fw, bw

(continued)

TABLE 1

(Continued)

Division	Common name	Pigments and plastid organization in photosynthetic species	Storage product	Cell wall[b]	Flagellar number and insertion[c]	Habitat[d]
Euglenophyta	Euglenoids	Chlorophyll a,b; β-carotene and several xanthophylls; 2–6 thylakoids/stack, sometimes many	Paramylon (β-1,3-glucopyranoside), oil	Absent	1–3 (–7) apical, subapical	fw, bw, sw, t
Phaeophyta	Brown algae	Chlorophyll a,c; β-carotene and fucoxanthin and several other xanthophylls; 2–6 thylakoids/stack	Laminaran (β-1,3-glucopyranoside, predominantly); mannitol	Cellulose, alginic acid, and sulfated mucopoly-saccharides (fucoidan)	2, unequal[f] lateral	fw (very rare), bw, sw
Chrysophyta	Golden and yellow-green algae (including diatoms)	Chlorophyll a,c (c lacking in some) α-, β-, and ε-carotene and several xanthophylls, including fucoxanthin in Chrysophyceae, Bacillariophyceae, and Prymnesio-phyceae; 2 thylakoids/stack	Chrysolaminaran (β-1,3-gluco-pyranoside, predominantly); oil	Cellulose, silica, calcium carbonate, mucilaginous substances and some chitin; or wall absent	1–2 unequal or equal apical	fw, bw, sw, t
Pyrrhophyta	Dinoflagellates	Chlorophyll a,c; β-carotene and several xanthophylls; 3 thylakoids/stack	Starch, α-1,4-glucan (oil in some)	Cellulose or absent; mucilaginous substances	Two, one trailing, one girdling	fw, bw, sw

Phylum	Common name	Pigments	Storage product	Cell wall	Flagella	Habitat
Cryptophyta	Cryptomonads	Chlorophyll a,c; α-, β-, and ε-carotene; distinctive xanthophylls (alloxanthin, crocoxanthin, monadoxanthin); phycobilins; 2 thylakoids/stack	Starch, α-1,4-glucan	Absent	Two, unequal subapical	fw, bw, sw
Rhodophyta	Red algae	Chlorophyll a (d in some Florideophyceae); R- and C-phycocyanin, allophycocyanin; R- and B-phycoerythrin. α- and β-carotene and several xanthophylls; thylakoids single, not associated	Floridean starch (amylopectin-like)	Cellulose, xylans, several sulfated polysaccharides (galactans) calcification in some; alginate in corallinaceae	Absent	fw (some), bw, sw (most)

[a] Underlined groups are discussed in text. After Bold and Wynne (1985.)

[b] In terms of cell wall chemistry, the vegetative cells have received most attention. Spores, akinetes, dormant zygotes, and other resting stages have not been studied, but it is clear that their walls may contain other substances (e.g., waxes and other nonsaponifiable polymers and phenolic substances).

[c] In motile cells, when these are produced.

[d] fw, freshwater; bw, brackish water; sw, marine; t, terrestrial (soil, rocks, etc.).

[e] Others are wall-less or have xylans, mannans, other glucans, some silica, or protein. Also, nearly all skeletal polysaccharides (cellulose, xylans, mannans) are accompanied by one or more mucilanginous substances (e.g., arabinogalactans and sulfated mucopolysaccharides).

[f] Except the uniflagellate sperms of Dictyotales.

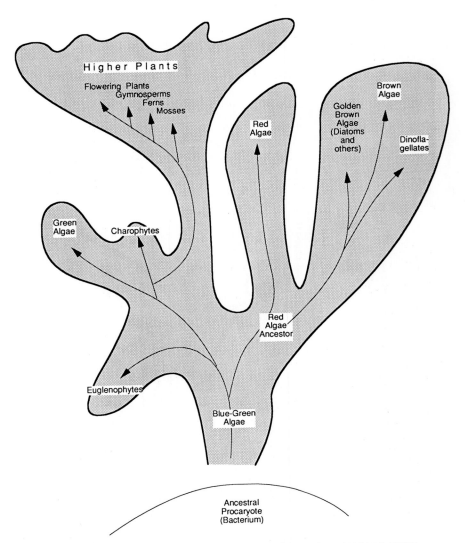

Figure 2 Simplified family tree of plants. Partly after South and Whittick (1987).

Green algae (Chlorophyta) are physiologically close to the land plants. Relative to other multicolored algal types, especially blue-greens, they are sometimes referred to as grass green. Greens occur more or less equally in fresh, estuarine, and marine environments. However, they consist of different species, and often different genera, and sometimes different orders. The green algae are the archetypal algae in the minds of laymen, who are more likely to be familiar with small eutrophic streams or ponds. Not as large as many browns or as generally complex structurally as both reds and browns, greens nevertheless are ubiquitous if not

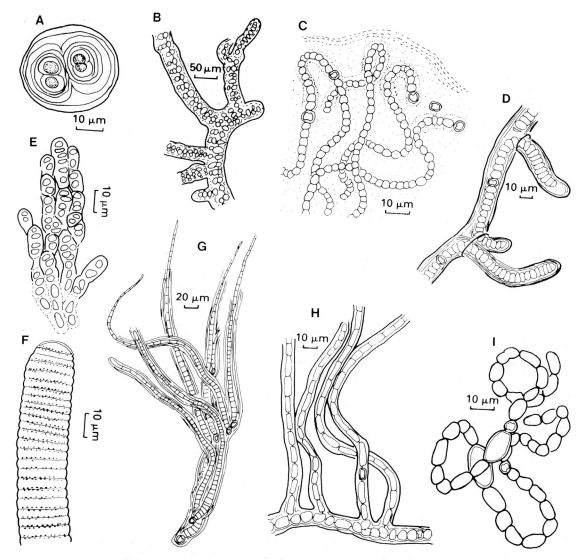

Figure 3 Major types of structural organization (anatomy) in blue-green algae (A) *Gloeocpsa*, (B) *Stigonema*, (C) *Nostoc*, (D) *Scytonema*, (E) *Pleurocapsa*, (F) *Oscillatoria*, (G) *Calothrix*, (H) *Fischerella*, (I)*Anabaena*. After South and Whittick (1987).

dominant plants. They show most simpler forms of reproduction known in the algae, more typically with both male and female flagellated cells set free in the water. However, green algae also range up to full oogamy (a large, nonmobile egg on the female plant and small motile sperm) and in a few cases even direct fusion of cells of adjacent filaments. Figure 5 shows a variety of types of green algae. An offshoot of the green algae, now

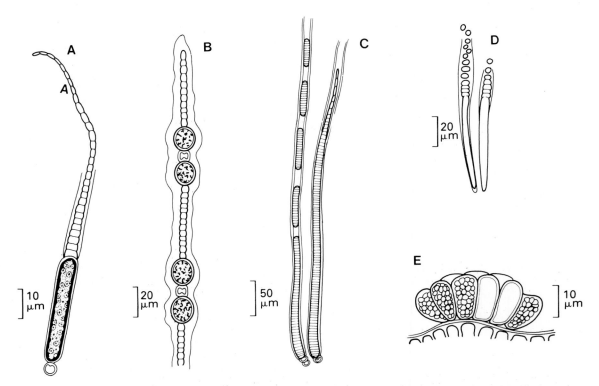

Figure 4   Major types of asexual reproduction in blue-green algae. (A) *Gloeotrichia;* (B) *Wollea;* (C) *Calothrix;* (D) *Chamaesiphon;* (E) *Dermocarpa*. After South and Whittick (1987).

usually placed in a separate division, the Charophyta, is notable because it is generally thought to be on or near the green algal evolutionary line to higher plants. *Chara* and *Nitella,* common plants of high-quality, cal-cereous waters, are partly calcified and develop large eggs enclosed in a sheath of protecting cells (Fig. 6).

An unusual structural type of green algae, well developed in fungi but uncommon in plants, is the siphonaceous or coenocytic body. Based on filaments with rare cell walls and abundant nuclei, some genera of the marine Siphonales become elaborate and often calcified (Fig. 7). Based on deep-core drillings in Pacific atolls several decades ago it was concluded that a strongly calcified member of this group, *Halimeda,* and not the corals, was the dominant carbonate contributor to these enormous, organically built, but now geological structures. Other related genera, such as *Udotea* and *Penicillus,* along with the flowering "sea grasses," form the often dense vegetation of tropical lagoons. These siphonaceous algae along with some surface living algal cells such as diatoms are com-petitors with the flowering plants in the sandy or muddy submerged

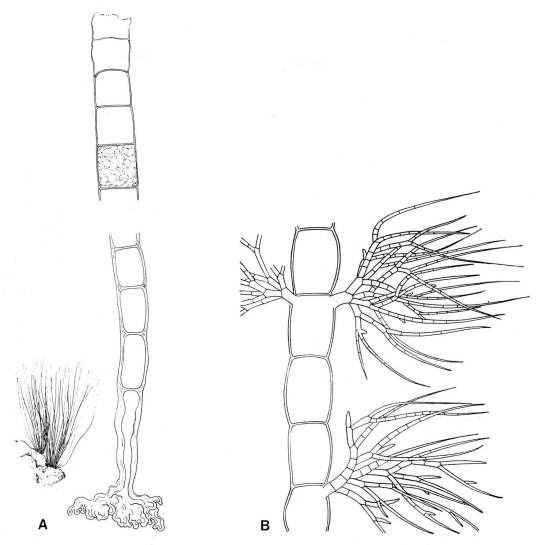

Figure 5 Primary anatomical types of green algae: (A) *Chaetomorpha area* (marine), (B) *Draparnaldia glomerata* (fresh water), (C) *Cladophora* sp. (marine), (D) *Anadyomene stellata* (marine), (E) *Ulva lactuca* (marine), (F) *Enteromorpha* sp., (G) *Oedogonium* (fresh water). [A, C, E, F after Taylor (1957); B, G after Prescott (1951); D from Dawes (1981).] Reprinted by permission of John Wiley & Sons, Inc.

(*Figure continues*)

C

Figure 5 (*Continued*)

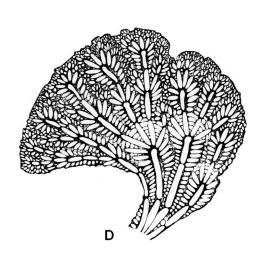

D

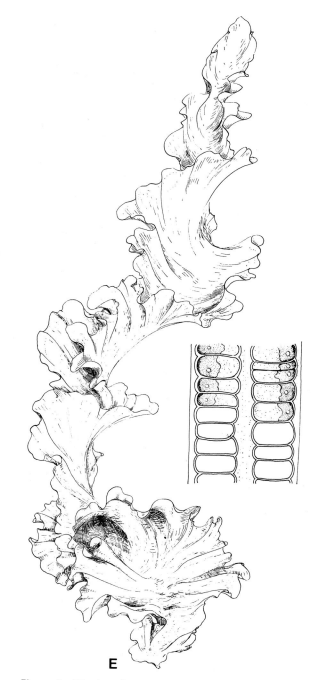

E

Figure 5 (*Continued*)

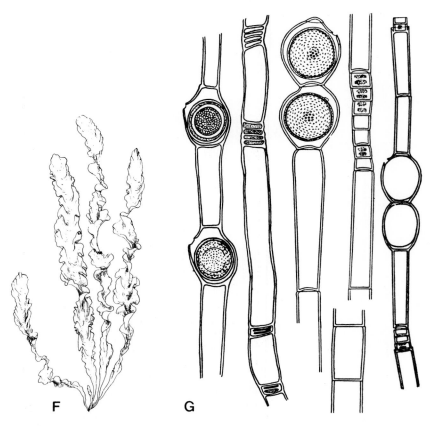

F   G

Figure 5 *(Continued)*

aquatic vegetation biome. Most other algae are limited to the rocky-shore biome and coral reefs.

The strictly marine or saltier estuarine brown algae (Phaeophyta) are the giants of the algal world, some being as long as large trees though never of similarly great mass. Although virtually all structural types are represented in the browns (Fig. 8), the branched filament *Ectocarpus*, occurring in a wide range of coastal environments, arctic to tropics, are especially ubiquitous, and sometimes cover large, shallow, quiet areas with a brownish cottony coating. The large, complex parenchymatous type of body represented by the kelps and rock weeds dominates many rougher water areas. Although usually thought of as the primary plants creating the rocky-shore communities of colder waters in both the northern and southern hemispheres, the large browns are well represented in the tropics by *Sargassum* and *Turbinaria*. Also, the brown genera *Dictyota*, *Padina*, and *Lobophora* are among the commonest mid-size algae occuring in coral reef environments.

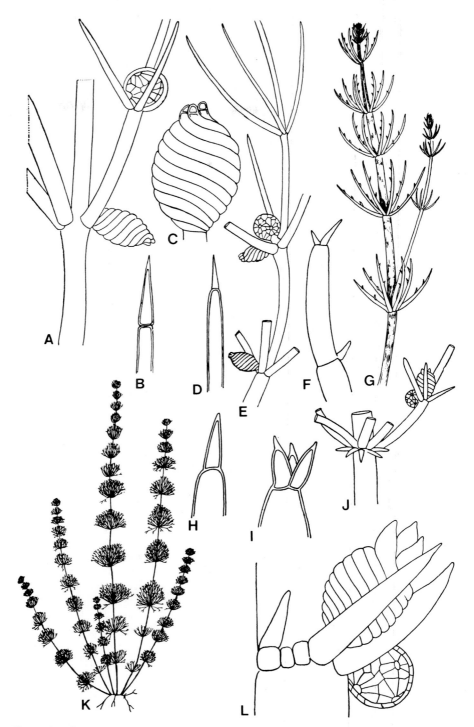

Figure 6 Charophytes, *Chara* sp. (F,G,I,J,L) and *Nitella* sp. (A–E,H,K). A,C,E,J,L, with oogonia. A,E,J,L, with antherida. After Prescott (1951).

Brown algae show virtually all types of sexual reproduction known in the algae. Many of the simpler and smaller types alternate generations between spore-producing, diploid (double chromosome set) plants and haploid, male and female plants bearing sexual cells. However, the larger browns such as the rockweeds show a reproductive pattern that is not

Figure 7   Three members of the marine green algal order Siphonales, a group anatomically built on a coenocytic plan of few cell walls. (A) Uncalcified *Caulerpa verticillata;* note trabecullae (wall supports), W. (B) Weakly calcified *Udotea cyathiformis.* (C) Heavily calcified *Halimeda discoidea;* note flexible connection (G) of calcified segments. All after Dawes (1981). Reprinted by permission of John Wiley & Sons, Inc.

(Figure continues)

C

Figure 7 (*Continued*)

unlike the higher plants and animals. Sperm are produced in conceptacles on diploid male plants, and eggs in similar conceptacles on female plants. The spermatia are released and swim or are carried by currents to the releasing eggs where fertilization takes place (Fig. 9). The fertilized egg after partial division is released to settle and form a new plant.

The reds (Rhodophyta) are primarily small to mid-size marine algae, although a few genera occur in fresh waters, particularly clear streams. They are usually red in color although under some circumstances can be green or black. One family common on coral reefs, the Champiaceae, is characterized by a brilliant iridescence that often includes shades of blue. The red algae are unusual among algae, and all plants, in several ways. Although sometimes appearing parenchymatous, or built of a mass of cells dividing in all directions, almost all reds can be shown from careful observation to be built by the elaboration of filaments (Fig. 10). Many reds are simply branched filaments, though more complex branched fleshy types and leafy forms are not unusual. The red crustose coralline algae, with or without the corals, are capable of building large reefs or algal ridges (Adey, 1986).

Red algae have no motile stages. Even the sperms do not have flagella, though they move rather like amoebas. On the other hand, the egg (or carpogonium) is relatively large with a protruding structure, the trichogyne, to "catch" sperm. In the more complex types, the eggs are borne in special organs called conceptacles, and in most reds the result of fertilization is a complex series of fusions and growths that give rise to large

A   a   b

Figure 8 Primary morphological characteristics of brown algae (Phaeophyta). (A) Filamentous *Ectocarpus* sp., a, habit, b, microscopic view showing gametangia. (B) Pseudoparenchymatous *Acrothrix* sp. developed by cortication of a single axial filament. (C) Fully parenchymatous *Laminaria* (kelp) spp. (D) *Fucus vesiculosus,* a common rockweed. (E) *Sargassum filipendula* showing stems, bladders, and leaves. All after Taylor (1957).

(Figure continues)

numbers of spores (carpospores). These fertile structures are often seen as distinct knots or cysts (cystocarps) on the female plant (Fig. 11). While many types of life cycle are found in the red algae, the most common type is an alteration of similar-appearing generations. The diploid generation bears spores, usually tetraspores, sometimes also in conceptacles (Fig. 12). The haploid or sexual generation generally has both male and female plants producing their respective eggs or sperm.

Finally, a crucial and omnipresent group of the golden-brown algae or Chrysophyta is the Bacillariophyceae or diatoms. [Some writers would

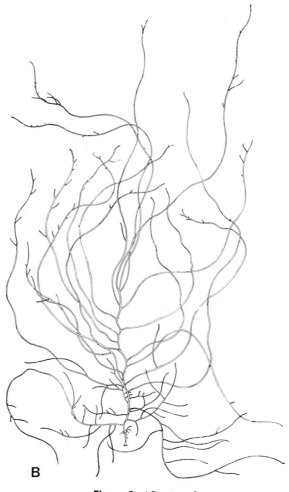

B

Figure 8 (*Continued*)

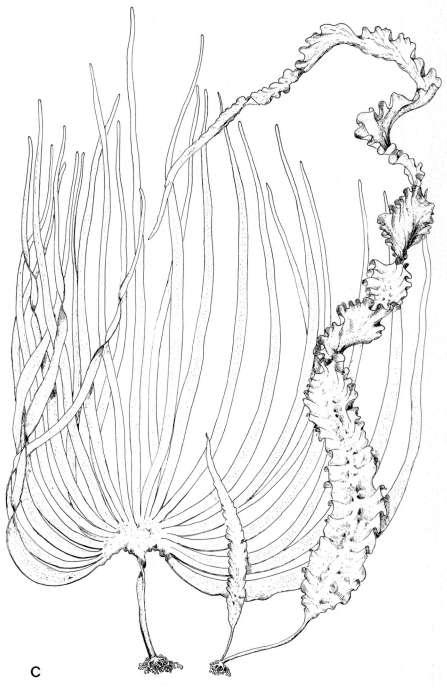

C

Figure 8 (*Continued*)

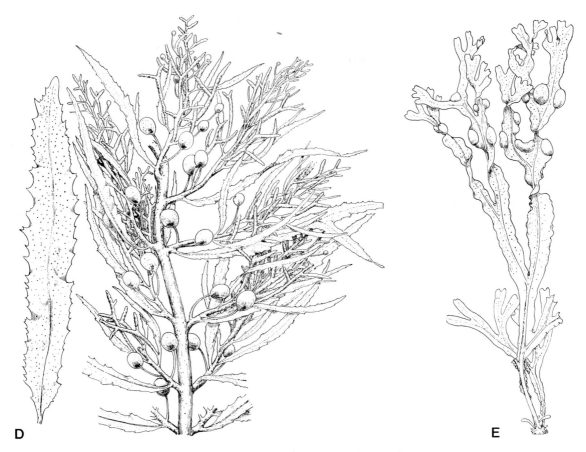

D   E

Figure 8 (*Continued*)

place diatoms in their own phylum: Bacillariophyta (Margulis and Schwartz, 1988.)] Different species of diatoms are at home in both fresh and salt water and are both benthic and planktonic (see Chapter 18). Primarily unicells, some diatoms are also filaments (Fig. 13). Sometimes called the "grass of the sea," primarily for their offshore planktonic abundance, the diatoms are the first algal colonizers and are usually responsible for the coating that snails graze (and the aquarist must clear) on the glass of a well-lit, well-run aquarium.

There are a number of additional algal phyla or divisions that are primarily microscopic. We refer the interested aquarist to several modern texts for a full treatment of the major groups we briefly discussed above as well as these omitted taxa (Bold and Wynne, 1985; Dawes, 1981).

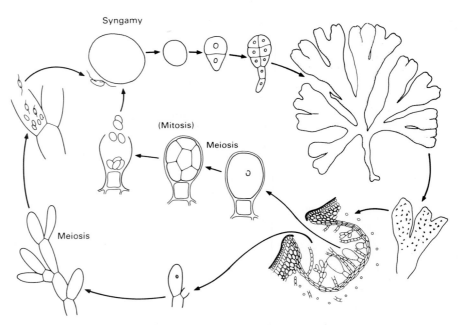

Figure 9 Sexual reproductive structures in northern rocky shore *Fucus* sp. This is a monoecious species (eggs and sperm borne in the same conceptacle). Some *Fucus* species are dioecious and consist of separate male and female plants. After South and Wittick (1987).

Algae in Model Ecosystems

Algae are considered to be the bane of most culture systems. This is primarily because many algae are, reproductively, highly efficient opportunists in wild ecosystems, waiting for the more or less unusual spike of nutrients. This very characteristic, used in a controlled manner, is the basis of the highly efficient algal turf scrubber. A few algae become undesirable as "blooms" in many culture systems partly because nutrients are higher than they should be and, in some cases, partly because water motion and or grazing are insufficient. Algae, as part of a properly run model ecosystem, only contribute to the productivity, food availability, and energy flow, as well as to the overall stability of the system. It is not possible to provide enough background here to fully understand and identify all the algae in any given synthetic system under development. The basic references in this chapter's bibliography plus the field guides for the community modeled can, however, identify and provide some of the natural history information required. Following our description of the basic benthic algal groups, we simply provide some notes with regard to our experience with algae in a variety of mesocosms, microcosms, and aquaria. We also refer the reader to Chapters 21–24.

To the casual observer a colder-water rocky shore often has a dense kelp forest subtidally and a thick carpet of rockweed in the intertidal. These plants are the primary elements structuring many rocky shore communities and around which the remainder of the flora and fauna are oriented. However, forming a substory or understory under, and in bare spots around, the large browns that form the canopy is a wide variety of much smaller red, brown and green algae, as well as attached diatoms and

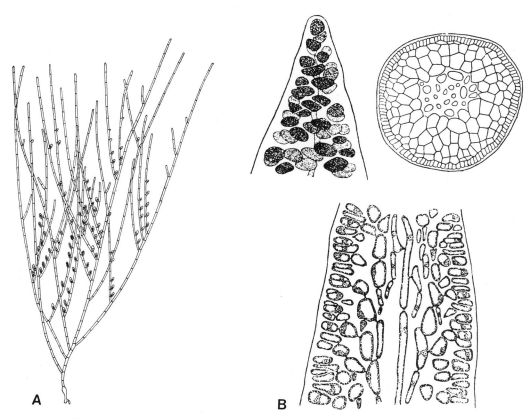

Figure 10 Red algae (Rhodophyta) of increasing structural complexity but all based on a filamentous building plan. (A) *Acrochaetium zosterae,* a branched filament with a filamentous rhizoidal base. (B) *Cystoclonium purpureum,* a single, heavily corticated filament giving rise to a lax, fleshy macrophyte. (C) *Hypoglossum woodwardii,* a leafy macrothallus based on the single axial filament structure. (D) *Ahnfeltia plicata,* a heavily branched thallus. (E) *Mesophyllum* spp., multiaxial, heavily calcified crusts, a member of the Corallinaceae. (A) After Taylor (1957); B–E Fritsch (1952, 1956).

(Figure continues)

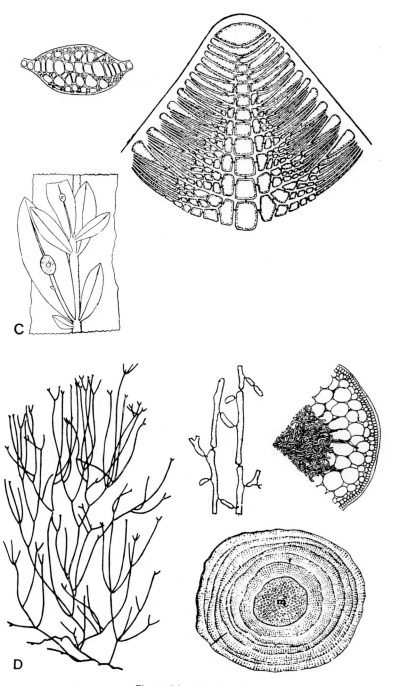

C

D

Figure 10   (*Continued*)

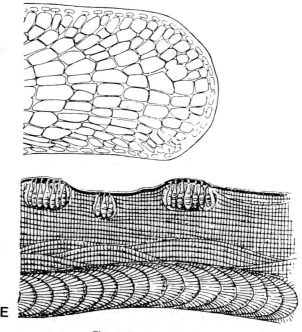

Figure 10 (*Continued*)

generally inconspicuous blue-greens. Under some circumstances, particularly when grazing by sea urchins is heavy, calcified crustose red algae (the corallines) can also be widespread dominants below low tide. A typical algal zonation pattern for a moderately exposed shore in the Gulf of Maine is shown in Chapter 13, Figure 5. Obviously, even to the level of the division, the algae at different elevations on the rocky shore are quite different. Blue-greens (along with some lichens) are most important at the uppermost reaches of the tide. Browns dominate in the intertidal and extend below low water typically to 10–30 feet. Reds really come into their own in the deeper water, mostly below 10–30 feet. Although green algae are present, especially in upper levels and tide pools, they are generally not conspicuous except around small streams or freshwater seeps. The algae being considered for a microcosm of this rocky shore can be quite diverse, depending upon the vertical and wave energy extent of the community being simulated.

Several times in recent years we have modeled subarctic rocky-shore communities (Gulf of Maine) ranging from the highest tide levels to about 50 feet (15 m) of depth (see Chapter 22). Elevation and depth were simulated physically by tide and by variation of light levels in the tank. The larger brown algae (kelp and rockweeds) characteristic of moderately open to protected shores have been generally successful in these microcosms. In the highest intertidal, the blue-greens *Oscillatoria*, *Lyngbya*, and *Spirulina* dominate, with the peculiar green *Prasiola* also being present in abundance. In the mid-intertidal, the rockweed *Ascophyllum nodosum* dominates. Although reproductive and persistent, *Fucus* species are not generally as important as they are in the wild, probably because of the generally lower levels of wave action in these small systems. Beneath the

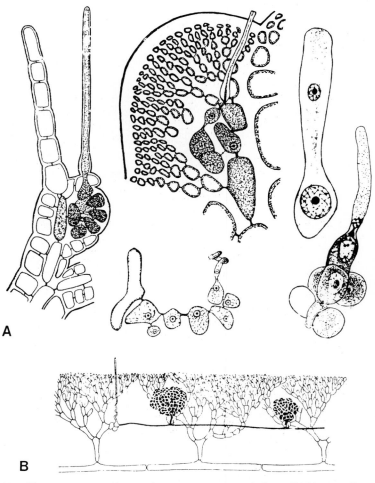

Figure 11 Characteristic sexual reproductive patterns in red algae. (A) Variety of carpogonia (egg cells) with projections (trichogynes). (B) Fertilized carpogonium-producing tube, fusing with auxiliary cells to give rise to cystocarps and carpospores. (C) Crustose corallines with male (c,d) and female (a,f,g,h) cells being borne in separate sexual organs (conceptacles). (e) represents carpospore production after fertilization. After Fritsch (1952).

(*Figure continues*)

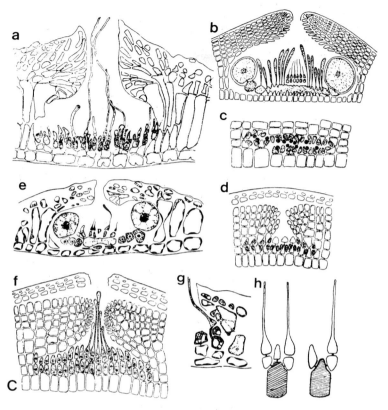

Figure 11 (*Continued*)

Ascophyllum in the mid to lower one-third of the intertidal range, *Chondrus crispus* (Irish moss) forms an extensive low canopy. While this is also true in many places on the Maine coast, in the longest operated microcosm it has become more extensive than when the system was stocked 5 years earlier. Perhaps the generally higher humidities in the closed microcosm environment allow *Chondrus* to broaden its coverage slowly and extend further up the intertidal. The brilliant green filament, *Chaetomorpha melagonium*, with extraordinarily large cells is a persistent minor element in the substory along with *Chondrus*. This is the normal pattern in the wild.

Subtidally the largest rocky-shore microcosm has always contained the kelp *Laminaria longicruris* in abundance, and this is typically the most abundant species in the Gulf of Maine. On the other hand, *Laminaria intermedia*, which is particularly abundant on the most exposed areas of wild coast, does not persist in the microcosm. Perhaps, like *Fucus vesiculosus* in the intertidal and *Alaria esculenta* at the lowest low-tide

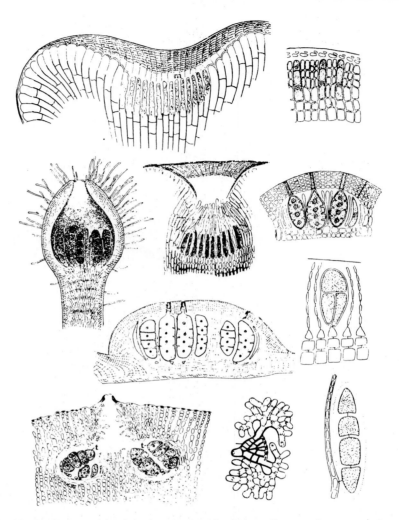

Figure 12 Tetraspores of the diploid generation of red algae. Some are borne at random in the tissue, others in special organs (conceptacles). After Fritsch (1952).

mark, it requires more wave action. The model has urchin grazers in abundance. For these subtidal species, urchin grazing in a quieter environment may well be crucial. In the deepest water in the microcosms the reds *Phycodrys* and *Phyllophora*, as well as the small perforated kelp *Agarum cribosum*, are quite successful in evading urchin grazing. It is likely that these species are moderately distasteful to urchin and snail grazers. A detailed list of the algae characteristic of the rocky Maine coast microcosms is given in Chapter 22.

Coral reefs, subarctic rocky shores, and tropical ponds are all highly productive of plant growth, the reef generally being the most productive because of a higher year-round level of incoming solar energy coupled with the driving or continual mixing effects of the constant trade winds and their seas and currents. There is a tendency in the human mind to equate plant mass (biomass or standing crop) with productivity. This is

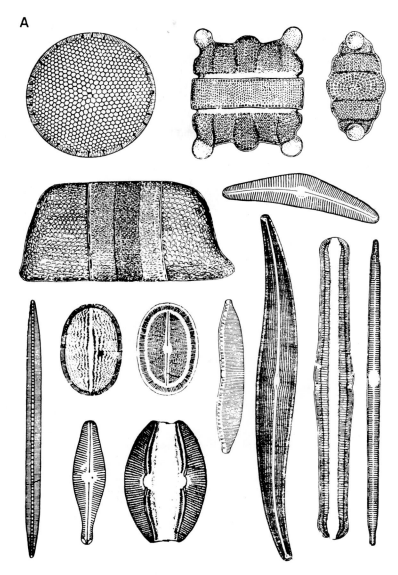

Figure 13 A variety of diatom frustules: (A) unicells, (B) filamentous types. After Fritsch (1956).

(*Figure continues*)

B

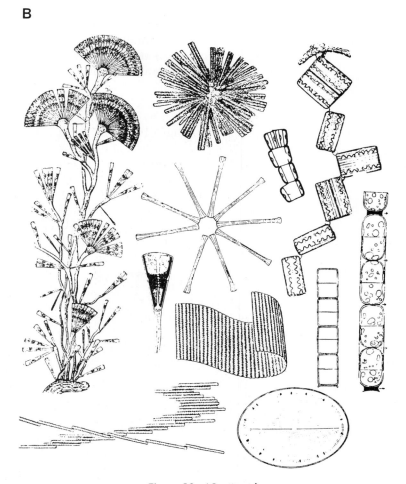

Figure 13 (*Continued*)

not necessarily the case. Algal turf production on a reef is extremely high. However, it is eaten almost as fast as it grows and rarely shows a high biomass. In an opposite situation, a northern conifer forest can have an enormous plant biomass. However, yearly productivity is less than one-tenth of that on a reef—of the plant material produced, little is eaten directly, and breakdown of the plant biomass is very slow. The algal flora of a tropical reef is generally quite different in appearance from that of a northern shore. While *Sargassum*, a close relative of the rockweeds of northern waters, can be abundant under heavy sea conditions, and some larger greens, particularly the calcified *Halimeda*, resists grazing and covers large areas, most algae are relatively inconspicuous. A coral reef algal turf, in spite of its diminutive dimensions, can be extraordinarily diverse and can include an incredible structural and reproductive variety representing virtually all of the major algal groups. Furthermore, these small algae form a plant community as productive as any on the earth, both in the wild and in aquaria and microcosms.

In the coral reef microcosm at the Smithsonian Institution the same basic algal pattern follows as on a reef. In the areas of lesser wave action, larger algae are subject to heavy grazing and do not persist. However, in refugia protected from fish and urchins, many mid-size species survive. They can also be found in small crevices on the reef itself. *Halimeda* is a persistent algal dominant in the microcosm as long as very large parrot fish are not included. The algal turf community, as on a wild reef, is quite rich. Eighty-five species have been tallied, including several diatoms and blue greens. The greens *Cladophora*, the browns *Ectocarpus* and *Dictyota*, and the reds *Hypnea*, *Laurencia*, *Acanthophora*, and *Ceramium*, along with several corallines, are all common members of this model reef community as they are in the wild (see list in Chapter 21).

Before briefly discussing a home reef aquarium with abundant algae from most of the major algal divisions, it may be best to briefly discuss attitudes of the home aquarists with marine systems toward algae. Prior to the 1980s, light levels were extremely low in most home aquaria and nutrients were very high. The only algae that were common were blue-greens, particulary those of reddish or blackish hues. In the 1960s, these were discouraged by most writers (they were not very esthetic—i.e., were slimy in appearance), though in fact similarly slimy blue-green colonies are extremely abundant in many wild reef environments. Later, aquarists were encouraged to keep these algae (any photosynthesis helped). Most recently, writers for the home aquarium audience (e.g., Moe, 1989; Thiel, 1988) have supported the green algae as being particularly beneficial as opposed to the reds. Careful reading generally indicates that by greens are meant *Caulerpa* and related species and reds refer to the reddish blue-green coatings (and not members of the division Rhodophyta).

It is our view that there are not uniquely beneficial or problem algal groups for aquarium systems. As we have indicated, if only blue-green algae are present, either the aquarium in question has not been adequately provided with algae or the environment is marginal. Otherwise, most environments that an aquarist is likely to model should include many different types of algae; the greater the variety, the more viable and stable the tank is likely to be. To many ecologists, abundant green algae indicates eutrophic conditions, although this is by no means an established indicator. We have seen marine ecosystems with unusually low nutrient concentrations with abundant *Ulva* and *Enteromorpha*, the greens most often the subject of this view.

For the past 3 years we have operated a small 120-gallon reef aquarium that has had its algae stocked directly from the reef microcosm described above. Generally the calcified *Halimeda* and corallines such as *Jania* and *Mesophyllum* dominate, though algal turfs are abundant in the crevices and of course on the algal turf scrubber. It is clear, however, that heavy grazing, particularly as it relates to the wave action present, and cover for the noncalcified algae are the chief factors controlling algal growth. Tough algae such as the green *Dictyosphaeria* are abundant, and the fleshy browns and reds are stunted by constant cropping. In this small reef system, diatoms also coat all surfaces, where they are subject to sporadic grazing by snails and limpets. However, the glass must be scraped to remove this limited "aufwuchs." When this is done the grazing fish such as tangs and damsels are waiting to harvest the scrapings. This is discussed again in the next chapter and in Chapter 21, but clearly the primary energy supply for the aquarium (on the order of 95–98%) is supplied by algae. These aquarium reefs are particularly rich in amphipods and mysid shrimp. As in the wild, the algal cover provides protection for these small invertebrates from the predatory activities of angel fish, rock beauties, damsels, and butterfly fish.

Submerged Aquatic Vegetation

Thirty-three higher plant families with submerged aquatic members and 124 genera and over 1000 species can be attributed to submerged aquatic vegetation (SAV) (Table 2). There are two families of sea grasses (see Fig. 14), both of which have many freshwater members. Only 48 species and 12 genera of these vascular aquatics have made it into the sea. "Sea grasses" occur sometimes as enormous beds that probably exceed the rocky-shore biome in area. However, here they are included under the SAV category rather than as a separate marine listing. Their ecological parameters are quite similar to the freshwater equivalents and they form taxonomically only about 5% of the total SAV.

SAVs are lower vascular plants (such as ferns) and flowering plants from several dozen families (less than 10% of the total terrestrial plant families) that have acquired the ability to enter the watery world. In this environment, drying out is rarely a problem but oxygen and light are often limited and competition from the algae can be extreme. In this situation many of these species have undergone a great reduction in vascular (transport) and supporting tissues. Concomitantly many SAVs have developed air spaces and passages as a very simple kind of respiratory system to support rooting tissues buried in oxygen-poor sediments.

Most SAVs in fresh water raise their flowers above the water surface. Many are insect pollinated, and a few are wind pollinated (see Table 2). In a sense, while vegetatively the tissues have evolved considerably to fit the

TABLE 2

Summary of the Major Families and Genera of Submerged Vegetation Showing Pollination Characteristics and Distribution[a]

Family	No. of: Genera	No. of: Species	Genera	Life form[b]	Spore production (Pteridophytes) or pollination (Angiosperms)	Geographical and habitat range
A. Pteridophytes						
1. Isoetaceae	2	c. 60	Isoetes Stylites	E, S	Heterosporous	Cosmopolitan (but mainly at high altitudes in tropics)
2. Ceratopteridaceae (Parkeriaceae)	1	c. 6	Ceratopteris	E, S, Ff	Homosporous	Some spp. terrestrial for at least part of year; all aquatic spp. in fresh water Pan-tropical and subtropical Fresh water
3. Marsileaceae	3	c. 70	Marsilea Pilularia Regnelidium	E, S, Ff	Heterosporous	Cosmopolitan Some spp. semiterrestrial; all aquatic spp. in fresh water
4. Azollaceae	1	c. 6	Azolla	Ff	Heterosporous	Tropical and warm temperate Fresh water
5. Salviniaceae	1	c. 12	Salvinia	Ff	Heterosporous	Tropical and warm temperate Fresh water
B. Dicotyledons[c]						
6. Nymphaeaceae	8	c. 60	Nymphaea Barclaya Brasenia Cabomba Euryale Nelumbo Nuphar Victoria	Fl, E, S	Entomophilous; few autogamous	Cosmopolitan Fresh water
7. Ceratophyllaceae	1	c. 6	Ceratophyllum	S, Ff	Hydrophilous	Cosmopolitan Fresh water
8. Elatinaceae	2	c. 30	Elatine Bergia	E, S	? Entomophilous; many perhaps autogamous	Cosmopolitan (but Elatine mainly temperate, Bergia mainly tropical) Fresh water

(continued)

TABLE 2

(Continued)

Family	No. of:		Genera	Life form[b]	Spore production (Pteridophytes) or pollination (Angiosperms)	Geographical and habitat range
	Genera	Species				
9. Trapaceae (Hydrocaryaceae)	1	4	Trapa	Ff	Entomophilous	Palaeotropical and warm temperate Eurasian Fresh water
10. Haloragaceae	6	c. 100	Haloragis Laurembergia Loudonia Meziella Myriophyllum Proserpinaca	E, S	Anemophilous	Cosmopolitan (but especially south temperate) Some spp. semiterrestrial; all aquatic spp. in fresh water
11. Hippuridaceae	1	1	Hippuris	E, S	Anemophilous	North temperate and cool South American Fresh water
12. Callitrichaceae	1	c. 25	Callitriche	E, Fl, S	Anemophilous; some hydrophilous	Cosmopolitan Some spp. semiterrestrial; all aquatic spp. in fresh water
13. Menyanthaceae	5	c. 35	Menyanthes Fauria Liparophyllum Nymphoides Villarsia	E, Fl	Entomophilous	Cosmopolitan Fresh water
14. Podostemaceae (Podostemonaceae)	c. 25	c. 120	Podostemum Dicraea Griffithella Indotristicha Mniopsis Mourera Tristicha Terniola Willisia, etc.	S	Entomophilous, anemophilous or autogamous.	Tropical (rarely subtropical) Flowing (often torrential) fresh water
15. Hydrostachyaceae	1	c. 10	Hydrostachys	S	?	Tropical & subtropical African Fresh water

Family	No. genera	No. species	Genera		Pollination	Distribution / Habitat
16. Butomaceae	5	c. 10	Butomus, Hydrocleys, Limnocharis, Ostenia, Tenagocharis	E, Fl	Entomophilous; some probably autogamous	Temperate & tropical (except Africa south of equator) Fresh water
17. Hydrocharitaceae	14	c. 90	Hydrocharis, Blyxa, Egeria, Elodea, Enhalus, Halophila, Hydrilla, Lagarosiphon, Limnobium, Nechamandra, Ottelia, Stratiotes, Thalassia, Vallisneria	S, Ff	Entomophilous, anemophilous, hydro-anemophilous or hydrophilous	Cosmopolitan (mainly in warm regions)—some genera Old, others New World Fresh & salt water
18. Alismaceae	12	c. 70	Alisma, Baldellia, Burnatia, Caldesia, Damasonium, Echinodorus, Limnophyton, Luromum, Machaerocarpus, Ranalisma, Sagittaria, Wiesneria	E, Fl, S	Entomophilous; ? few anemophilous or autogamous	Cosmopolitan (but especially north temperate) Fresh water
19. Scheuchzeriaceae	1	1	Scheuchzeria	E	Anemophilous	Cold north temperate Fresh water
20. Juncaginaceae	4	c. 15	Cycnogeton, Maundia, Tetroncium, Triglochin	E	Anemophilous	North & south temperate (few in American tropics) Fresh & brackish water

(continued)

TABLE 2

(Continued)

Family	No. of: Genera	No. of: Species	Genera	Life form[b]	Spore production (Pteridophytes) or pollination (Angiosperms)	Geographical and habitat range
21. Lilaeaceae (Heterostylaceae)	1	1	*Lilaea*	E	Anemophilous	Pacific North & South American Fresh (alkaline) water
22. Posidoniaceae	1	2	*Posidonia*	S	Hydrophilous	Mediterranean, S.W. Asiatic & Australasian coasts Salt water
23. Aponogetonaceae	1	c. 30	*Aponogeton*	Fl, S	Entomophilous or autogamous	Palaeotrpical & southern African Fresh water
24. Zosteraceae	2	c. 12	*Zostera* *Phyllospadix*	S	Hydrophilous	Temperate coasts (except South American & West African) Salt water
25. Potamogetonaceae	2	c. 90	*Potamogeton* *Groenlandia*	Fl, S	Anemophilous; few hydro-anemophilous	Cosmopolitan Fresh (rarely brackish) water
26. Ruppiaceae	1	3	*Ruppia*	S	Hydrophilous	Temperate & subtropical Brackish & salt water
27. Zannichelliaceae	6	c. 25	*Zannichellia* *Althenia* *Amphibolis* *Cymodocea*	S	Hydrophilous	Cosmopolitan (but marine genera mainly tropical) Brackish & salt (rarely fresh) water

	No. of genera	No. of species	Genera	Life form[b]	Pollination	Distribution and habitat
			Halodule *Syringodium*			
28. Najadaceae	1	c. 35	*Najas*	S	Hydrophilous	Cosmopolitan Fresh (rarely brackish) water
29. Mayacaceae	1	c. 10	*Mayaca*	E, S	Entomophilous	Tropical American & West African Fresh water
30. Pontederiaceae	7	c. 30	*Pontederia* *Eichhornia* *Heteranthera* *Hydrothrix* *Monochoria* *Reussia* *Scholleropsis*	E, S, Ff	Entomophilous (few probably autogamous)	Pan-tropical & temperate American Fresh water
31. Lemnaceae	4	c. 28	*Lemna* *Spirodela* *Wolffia* *Wolffiella*	Ff	? Unspecialised	Cosmopolitan Fresh water
32. Sparganiaceae	1	c. 15	*Sparganium*	E, Fl	Anemophilous	North temperate and Australasian Fresh water
33. Typhaceae	1	c. 10	*Typha*	E	Anemophilous	Cosmopolitan Fresh water

[a]After Sculthorpe (1967).
[b]Life forms are indicated as follows: E, emergent; Fl, floating-leaved; Ff, free-floating; S, submerged.
[c]The order in which the families of angiosperms are arranged in the table corresponds with the relative positions they occupy in the phylogenetic system of classification proposed by Hutchinson (1959).

335

very different aquatic environment, the reproductive process has remained mostly the same as that of their terrestrial ancestors. It is interesting to note that while very few freshwater aquatics produce underwater flowers and utilize water transport for their pollen, most marine SAVs have totally committed to the water life and keep their flowers submerged. This partly relates to the higher wave energies in the ocean. Most protected lagoons are rarely less than 6 feet in depth and often have considerable currents and some wave action. (Where they are shallower and less open, they tend to become hypersaline.) This would provide a great structural hurdle for the plant rooted in the bottom and attempting to raise its flowers above the surface.

As we discussed in Chapter 13, there are many reasons to regard the SAV-dominated bottom as a biome of many communities extending world wide. The presence of sea grasses and freshwater SAV communities greatly modifies the environment by trapping organic-rich sediments (and therefore nutrients) and by providing substrate (epiphytic algae and animals), cover (for fish and many invertebrates, both surface and infauna), and food (both trapped organics and the primary production of the plants themselves).

There is a very critical environmental and cultural reason to have a special interest in SAV communities. Algae can be highly productive and indeed respond quickly to the availability of light and nutrients in the water column. Algae also provide a generally high-quality food to many waterfowl, fish, and invertebrates. However, algae mostly lack the tough supportive tissues of terrestrial plants and break down almost as quickly as they produce. Thus, an algal-dominated environment can be a very erratic one, moving quickly with environmental changes. The stable type of algal community occurs in high-energy, low-light, or low organic supply situations. SAVs are slower growing and the plant tissues are slow to break down. SAVs are able to oxygenate the organic-rich substrate (which they have trapped) and to extract the nutrients through their rhizomes and root hairs. Low turbidities, moderate organic sediment supply, and moderate energy levels in bays and protected areas lead to stable species-rich SAV ecosystems. Even when storm or winter conditions produce windrows of sea grasses on the beach, breakdown is slow and nutrient return to the water column is stretched out over a long period of time.

Man's activities in semiprotected coastal bays and estuaries, particularly in increasing both nutrients and turbidities, have resulted in sometimes disastrous losses of SAV communities. This provides unstable ecosystems with more limited productivites and a reduction in effective use of the available energy supply.

Although SAV biomass levels may be high as herbaceous land plants go, they do not match up to the biomass levels of land plants or even kelp or rockweed beds. Nevertheless, SAVs can be among the most productive of plant communities (Table 3). It is particularly interesting that sea grasses in a generally poorer nutrient environment are more productive on an areal basis than freshwater aquatics. Although it has been debated whether or not the availability of carbon dioxide (or bicarbonate) is relatively limiting in fresh waters (see Chapter 10), it seems more likely that a generally higher physical energy level (Chapters 5 and 6) and greater light availability (because of lower turbidity) are the primary factors. As we have discussed, higher freshwater nutrients may actually favor attached algae (as aufwuchs) and limit aquatic plant production. Much of the SAV nutrient supply is probably derived from the sediments. Sea grasses generally have a higher root to shoot ratio (> 1) as compared to their freshwater counterparts. This probably relates to the lower nutrient levels and higher wave and current energies in the ocean.

Sea-grass beds can be heavily grazed, although exactly how much is hotly debated (Stevenson, 1988). It seems clear that, at least in the temperate environment, much of the SAV production goes to detritus rather than grazers. However, much of algae production in the temperate and colder environment also goes to detritus rather than grazers. Certainly in tropical lakes and rivers there are many fish grazers, and this has great importance to the aquarist. These matters we discuss in the next chapter.

The microcosm or aquarium is often operated in a way similar to that of the very troubled and unstable farm pond, bay, or estuary: excess higher-food-chain animals supported by artificial feeding and high nutrient levels, low light intensities, and limited invertebrate fauna, especially filter feeders. In closed ecosystems, whether fresh or salt water, it is crucial to maintain a moderate to low nutrient environment rich in stabilizing SAVs where appropriate. In Chapter 12, we discussed methods for maintaining normal nutrient levels. Even without sophisticated nutrient measurements, one can easily determine when nutrient levels are normal by noting the balance between epiphytic algae and the SAVs. Most SAVs receive a major portion of their nutrients and carbon dioxide (as HCO_2^-) from their root systems. Thus, if the water column has low to moderate levels of nutrients, the algae are unable to grow so rapidly that the young growing tips and mid-aged leaves of SAV become heavily coated with an aufwuchs of algae, sediment, and small invertebrates. As we will discuss in the next chapter, a moderate level of grazing (both algae and SAV) is important in this equation. However, low nutrients provide the critical starting point.

Unlike our treatment of the algae, we will not provide descriptions of the major groups of SAVs. To some extent these are available in the better-quality aquarium literature. We cite four references for those wishing to gain a more in-depth understanding of these plants (Muhlberg, 1982; Phillips and Menez, 1988; Sculthorpe, 1967; Stodola, 1967). Many local field guides that describe these plants to species are also available in book stores specializing in natural history. In the remainder of this section, we will provide a general description of SAVs and then describe some specific

TABLE 3

Comparison of Biomass and Productivity of Selected Freshwater, Brackish, and Marine Macrophytes

Species/location	Peak biomass (g dw m^{-2})			Rates of productivity		NPP (g m^{-2} d^{-1})
	Above	Below	Total	(mg O$_2$ h^{-1} g^{-1} lf)	(g C m^{-2} d^{-1})	
Freshwater and brackish						
Vallisneria spiralis						
L. Merrimajeal, Aust.	463	66	529			2.1
Vallisneria americana						
New York State	50	13	63			
Wisconsin	344	64	408			1.5
Mississippi R.	180	38	218	3.8–10.6		3.2
Maryland	53	38	91			
Myriophyllum spicatum						
Wisconsin	407	39	446			
Wisconsin	220				3.35	
Wisconsin				2–17	2–4	
New York State	387				1.2–1.8	
New York State	878				4.44	
Maryland				8.0		
Hydrilla verticillata						
Florida	161.4					
Florida	280					
Maryland	322	4	326	2.8–5.4		
Ruppia maritima						
Chesapeake Bay, Va.	80–150					0.92
Redfish Bay, Texas	160	50	210		0.98	
Borax Lake, Calif.	64					
Caimanero Lagoon, Mex.			1,000	0.9–9.7		

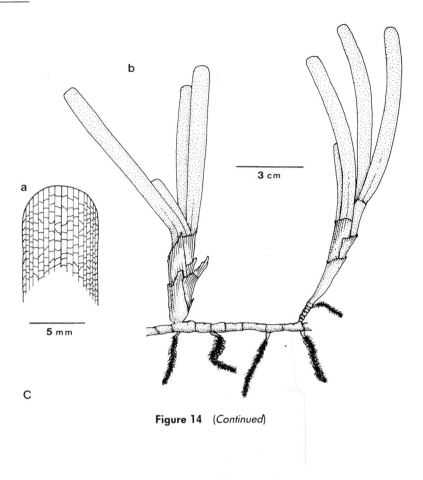

Figure 14   *(Continued)*

sediment must be brought with them, at least 4 or more inches thick, and that the rhizomes should not be allowed to wash out during transportation.

More recently we have attempted to establish *Zostera marina* and *Ruppia maritima* in a Chesapeake Bay mesocosm (Chapter 23) and *Zostera* in a Maine coast microcosm (Chapter 22). Although we have apparently been successful in accomplishing stocking in the Chesapeake system, the constant "rooting activities" of crabs and fish in the Maine tank has thwarted our efforts. This may well be a scaling problem: too many root-grubbing species perhaps without higher predator control in a small system. In a more recent larger mesocosm system (Florida Everglades, Chapter 24), with a richer invertebrate and fish fauna and better wave action (sandy mud rather than mud), we are experimenting with establishment of all the Florida sea grasses. At this writing, several years

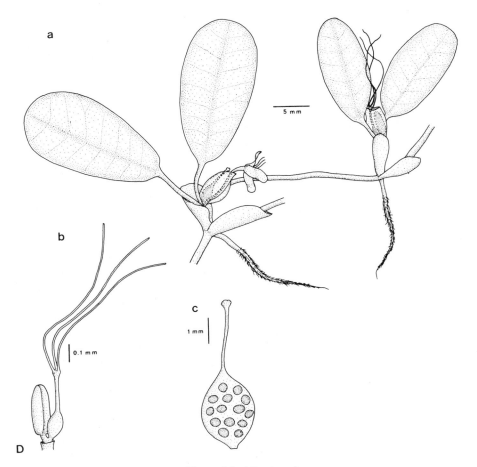

Figure 14 (*Continued*)

after stocking, *Thalassia, Halodule,* and *Syringodium* have been success-fully introduced in spite of the considerable number of rooting crabs and fish.

Freshwater SAVs and Model Ecosystems

It is clear that plants, having established a foothold on the land and finally, some 100 million years ago, having evolved flowers as an effective means of reproduction, underwent an enormous adaptive radiation, eventually producing tens of thousands of species in hundreds of major lines or families. In addition to the ferns and fern allies, who were never far removed from the water environment in any case, members of many flowering plant families have returned to the water environment. It is striking that only a limited number of morphologies work in the aquatic situation, and these can be arranged approximately in order of reduction or adaptation to the generally less demanding conditions (Fig. 15). The end points of this reduction are (1) the tiny duckweeds, more like the generally simpler algae except for the miniature tell-tale flowers, and (2) the totally submerged Naiads with their water-borne pollen. This morphological array with selected examples is shown in Figures 16–23. Though his examples were regarded as mnenomic devices, Stodola, in 1967, developed a similar scheme and listed the morphological characteristics of many of the common aquarium and backyard pond species of SAV. Those are included in Figures 16–23.

On a typical lake with moderate erosion and slope, SAV vegetation occurs lakeward of a more-or-less narrow band of emergent vegetation or marsh (Fig. 24). Depending on the depth and turbidity, the morphological forms occur in an irregular zonate pattern: partly emerged, floating, and totally submerged, followed by algae at the greatest depth that plants are found.

In recent years we have developed several freshwater mesocosms and aquaria that have included SAVs. These are described more fully in later chapters. Here, we simply make some general points. Never have we used undergravel filters and rarely have we used a substrate other than that from which the plants were collected. In common with all of our benthic establishment techniques, the requisite bottom was collected with its plants and invertebrates as intact as could be managed in an often difficult collecting situation. Our attempt was always to stock initially with every member of the analog community and to let our particular ecosystem variant "select" the appropriate species. Often multiple insertion of key SAV species was carried out before success was achieved. In some cases in the first year of operation, a bloom of one or two species developed. This can be controlled by selective weeding or disturbance.

We have tried several aquarium simulations using a variety of SAVs. As long as the water column itself was scrubbed to moderate nutrient levels, even with heavy fish and artificial feed loadings, no serious problems with aufwuchs coatings or plant loss have been encountered. A few algae eaters, such as snails and stick-cats (*Farlowiella*), insures the minimizing of aufwuchs. In one case, the base of a 70-gallon aquarium was stocked with several inches of barnyard soil and cow manure overlain by several inches of gravel. SAV growth was particularly successful in this case over several years of operation.

The primary problem with freshwater SAV maintenance is that of

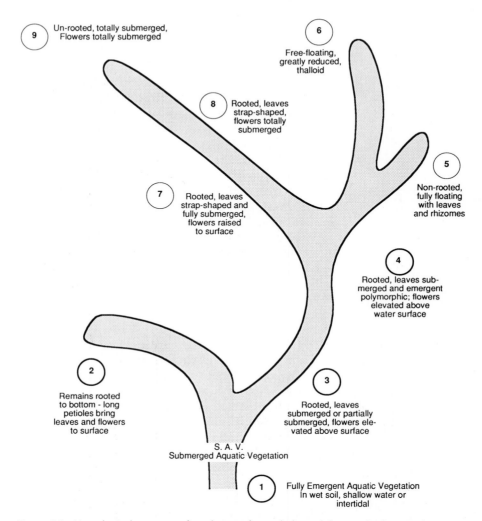

9 Un-rooted, totally submerged, Flowers totally submerged

6 Free-floating, greatly reduced, thalloid

8 Rooted, leaves strap-shaped, flowers totally submerged

5 Non-rooted, fully floating with leaves and rhizomes

7 Rooted, leaves strap-shaped and fully submerged, flowers raised to surface

4 Rooted, leaves submerged and emergent polymorphic; flowers elevated above water surface

2 Remains rooted to bottom - long petioles bring leaves and flowers to surface

3 Rooted, leaves submerged or partially submerged, flowers elevated above surface

S. A. V.
Submerged Aquatic Vegetation

1 Fully Emergent Aquatic Vegetation In wet soil, shallow water or intertidal

Figure 15 Hypothetical patterns of evolution of morphological forms of submerged aquatic vegetation from emergent aquatic ancestors. Type forms shown in Figures 16–23.

SAV grazing fish. To some extent this can be thwarted by using particularly tough species of SAV and periodically "feeding" a preferred softer species. However, even a small headstander or a silver dollar, for example, can raise havoc with the plant standing crop. These matters are discussed in the next chapter.

Gentianaceae
  Nymphaeoides spp.
Primulaceae
  Samolus floribundus
Daucaceae
  Hydrocotyle vulgaris
Nymphaeaceae
  Barclaya longifolia
  Brasenia schreberi
  Nuphar spp.
  Nymphaea spp.
Butomaceae
  Hydrocleis nymphacoides
Aponogetonaceae
  A. distachyus
Araceae
  Orontium aquaticum

Figure 16 Type 2 SAV. Plants rooted in the bottom but having long petioles that bring floating leaves to the surface and stalks that float or elevate flowers. Drawing (*Nymphaea elegans*) after Godfrey and Wooten (1981); plant list after Stodola (1967).

Alismataceae
  Sagittaria spp.
  Alisma spp.
  Echinodorus spp.
  Elisma natans
Aponogetonaceae
  Aponogeton spp.
Araceae
  Anubias spp.
  Lagenandra spp.
  Cryptocoryne spp.

Figure 17 Type 3 SAV. Plants roots in the bottom, a rosette of submerged leaves, sometimes with aerial leaves little modified, flowers aerial, usually insect pollinated. Drawing (*Sagittaria graminea*) after Godfrey and Wooten (Copyright 1979 by the University of Georgia Press. Used by permission.); plant list after Stodola (1967).

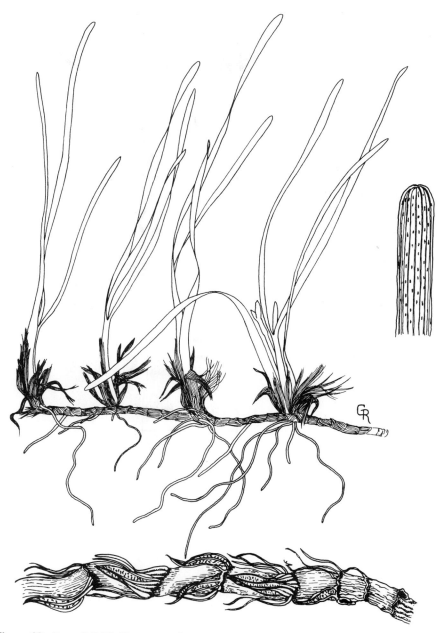

Figure 22 Type 8 SAV. Plants rooted, generally rhizomatous, leaves mostly elongate, strap-shaped, flowers submerged, water-pollinated. Drawing (*Thalassia testudinum*) after Godfrey and Wooten. Copyright 1979 by the University of Georgia Press. Used by permission.

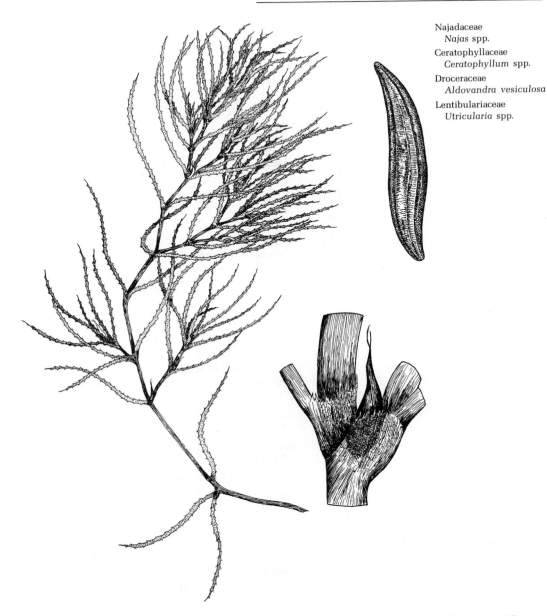

Najadaceae
    Najas spp.
Ceratophyllaceae
    Ceratophyllum spp.
Droceraceae
    Aldovandra vesiculosa
Lentibulariaceae
    Utricularia spp.

Figure 23 Type 9 SAV. Unrooted, totally submerged with water-pollinated submerged flowers. Drawing (*Najas minor*) after Godfrey and Wooten (Copyright 1979 by the University of Georgia Press. Used by permission.); plant list after Stodola (1967).

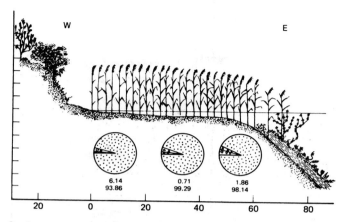

Figure 24 Section across exposed eastern shore of Lake Kisajno, Poland. An emergent vegetation on the beach terrace of *Phragmites australis* is succeeded on the dropoff by *Ranunculus circinatus, Elodea canadensis,* and Charophytes. After Hutchinson (1975). Reprinted by permission of John Wiley & Sons, Inc.

Emergent Aquatic Vegetation (EAV)

In the transition from land to open water, a band often narrow but sometimes extraordinarily wide, there exists a zone of soil or substrate that is generally waterlogged. This band is flooded periodically either daily by tides or seasonally by variations in rainfall. If wave energy is high the band is occupied by rock or moving sand and the appropriate algal vegetation or no vegetation at all. On the other hand, if wave energies are low, fine, relatively stable sediment collects. Often this sediment becomes vegetated, dominantly by flowering plants but usually with an understory or coating (aufwuchs) of algae. The community that occupies this zone is called a marsh if low and mostly herbaceous, or a swamp if dominantly occupied by woody plants. Although it might seem that this community should be heavily occupied by relatives of the same plants that have adapted to the submerged environment, in fact this is only rarely the case. The dominant herbaceous plants in the marsh are grasses, sedges, rushes, and cattails, families with very few submerged relatives. While the woody plants of swamps come from a wide variety of flowering plant families, and cypress, an extremely important swamp component in North America, is a conifer, virtually none of the woody members or their relatives have made it under water. We will not discuss woody swamp plants in any depth in this book. Mangroves and cypress, along with a few other groups, however, are treated in Chapter 24 on the Everglades mesocosm.

In Chapter 13, we placed submerged aquatic vegetation including sea grasses in a biome separate from emergent marsh vegetation. Although sometimes with very sharp boundaries, these biomes often grade into each other over a linear zone, sometimes tens of feet wide. While this gradation might seem excessive to allow biome differentiation, especially considering the linear character, it is helpful to remember that terrestrial biomes intergrade over tens, sometimes hundreds, of miles. Also, while marshes, even if of great length, typically range from tens of feet to several miles wide. This narrowness results from a physical limitation and the narrowness of proper conditions in the transition from land to open water. In Florida, for example, where the terrain is of very low relief, marsh with its characteristic plants covers thousands of square miles. Likewise, in the shallow offshore, sea grasses occupy even greater areas. Within the largely marshy fresh and salt Everglades complex, patches of open water, sometimes richly endowed with submerged aquatic vegetation, do exist. However, this mosaic of SAV biome within a much larger emergent vegetation biome reflects the fact that water abundance, light, and substrate, as the primary determiners of biome characterization in the water world, have a very different character of distribution as compared to temperature and precipitation on land.

Marshes are among the most productive communities on earth (Fig. 25). Normally endowed with an overabundance of water, light, nutrients, and carbon dioxide, it is only the very basic physiological and biochemical limitations of plant structure that put bounds on photosynthesis. Many are sun plants in that they do not saturate for photosynthesis but rather show higher and higher rates to the highest light intensities. While grazers of the higher plants of marshes are numerous in kind (deer, muskrat, birds, insects, and snails, to name a few), rarely is a major portion of the primary production removed by grazing. In freshwater marshes, except in patchy muskrat "eat-outs," typically no more than 10 or 20% of the net primary production is consumed by grazers. In marine and estuarine marshes, the figure is even less. Most of the production eventually either degenerates in place or is removed by currents and tides to basins or beaches where gradual breakdown produces a bacteria-rich detritus. This marsh plant production becomes a critical, largely aseasonal food source for the adjacent lake or estuary (see Chapter 18).

The dominating marsh plants, fresh- and saltwater, whether grasses (Fig. 26), sedges (Fig. 27), rushes (Fig. 28), or cattails (Fig. 29), have a number of characteristics in common that point to the requirements for success in this very rich but in some ways difficult environment. All are monocots with long, thin, or even rounded leaves. Basically the leaves are mounted on rhizomes, which, once established, provide an enormous colonizing potential. Also, all are wind pollinated, an important feature for environments adjacent to large lakes or the ocean, areas that are often windswept. Similar to SAVs, emergent aquatics often have lacunae or air spaces that provide for the movement of oxygen from leaves to rhizomes and roots. Some have been shown to oxygenate the generally oxygen-deficient soil in which they grow.

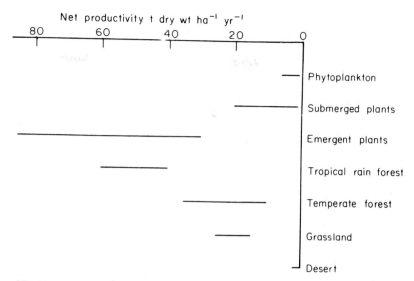

Figure 25 Net primary productivity (biomass developed) of submergent and emergent aquatic vegetation as compared to phytoplankton and major terrestrial communities. After Moss (1980).

The dense growth of emergent aquatics, the resultant tendency to accumulate often organic-rich silt, and a general lack of grazing with accumulation of detrital biomass all lead to a rather extreme environment in the marsh (Fig. 30). Extreme conditions are exacerbated in a salt marsh, where landward and away from channels, salinities can become very high to the point of causing species zonation and sharply reduced growth (Fig. 31). Some salt-marsh plants have salt-excreting glands and, lacking rain or wind, the excreted salt can be seen on the blades (Phleger, 1977). In dry climates, the zone of flooding at the highest tides becomes a salt flat or barren, devoid of plant life.

In lakes, the location of the emergent aquatic vegetation, or reed beds as they are sometimes called, lies in the transition zone between land and water. However, the extent of development ranges widely depending on wave exposure and steepness of the shore (Fig. 32). Under some conditions of extremely rapid and matted growth, the "reed bed" can develop into a floating platform.

Although the aquatic vegetation, emergent and submergent, has been treated as largely the province of higher plants that have returned to the watery environment, algae play a critical roll. As the often structural element of the "aufwuchs" community that coats all submerged objects (Fig. 33), algae become an important primary producer, often the major direct plant food supply to fish and invertebrates (see Chapter 14, Fig. 1).

Figure 26 Emergent marine and aquatic grasses. Saltmarsh: (A) *Puccinellia* spp. from northern Europe. B, Freshwater *Glyceria borealis*, C, *Leersia oryzoides*, D, *Zizania aquatica*, from New England. A, after Fitter et al. (1984); B–D After Hutchinson (1975). Reprinted by permission of John Wiley & Sons, Inc.

(Figure continues)

Figure 26 (*Continued*)

Figure 27 Emergent sedges of the genus *Scirpus* from British streams and ponds. Note triangular format and "cones." After Fitter et al. (1984).

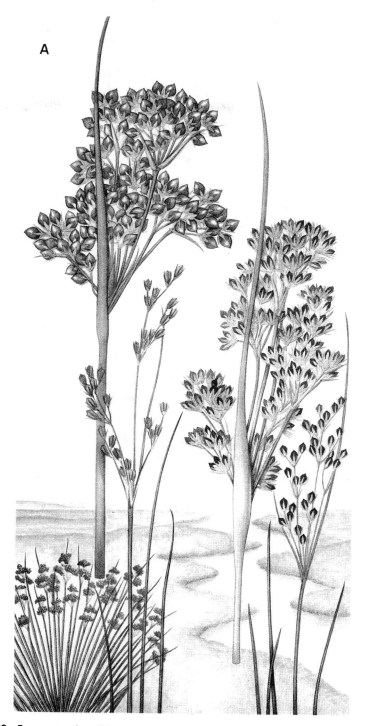

Figure 28 Emergent rushes of the genus *Juncus* (A) from British salt marshes and (B) New York lake margins. A, after Fitter et al. (1984) B, after Hutchinson (1975). Reprinted by permission of John Wiley & Sons, Inc.

(*Figure continues*)

B

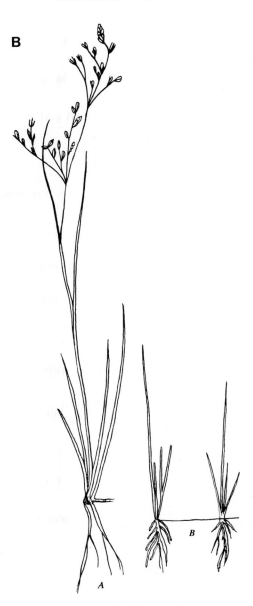

Figure 28 (*Continued*)

Figure 29 Freshwater emergent aquatic vegetation of the genera: (A,B) *Typha* (cattails) and (C) *Sparganium* (bur-reed). After Hutchinson (1975). Reprinted by permission of John Wiley & Sons, Inc.

Color Plate 4 Archimedes screw pump. Installation on the Smithsonian Florida Everglades mesocosm. Screw pump (green shaft in center) raises sea water from the Gulf of Mexico (lower right) to distribution tower (black). From the tower, water flows to scrubbers (upper left), wave generator (lower left), and the tidal estuary (far right).

A

B

Color Plate 5 Photographs of a "dump scrubber" providing a combination of wave surge and algal scrubbing to minimize space in mini ecosystems. This unit on a 130-gallon coral reef aquarium, similar to that in Color Plate 3, is shown (A) in the up or fill position—note flow or "ripple-back" of the water and (B) in the dump position. Photos by Nick Caloyianis.

Color Plate 6 Salt marsh on inner Gouldsboro Bay, Maine. In addition to the mud flat fringing the main tidal channel and numerous salt pans (salt pools), tall *Spartina alterniflora* (smooth cordgrass) flanks the tidal channels (dark green bands). At progressively higher levels, short *S. alterniflora* and then *Spartina patens* (salt hay-yellowish tint) dominate. Along the forest edge, flooded at only the highest spring tides, a narrow band of *Juncus gerardi* (black needle rush) occurs. Photo by R. Craig Shipp.

Color Plate 7 Tidal controller in Everglades mesocosm (plastic box). The controller (see Chapter 23, Fig. 9) determines the level of water in the estuary proper (behind wall) by raising and lowering the outflow hose (visible at the base of the wall). Sea water from the Gulf of Mexico tank (foreground) is constantly pumped to the estuary. It returns through the tidal hose. Vegetated dune to right, red mangrove in background.

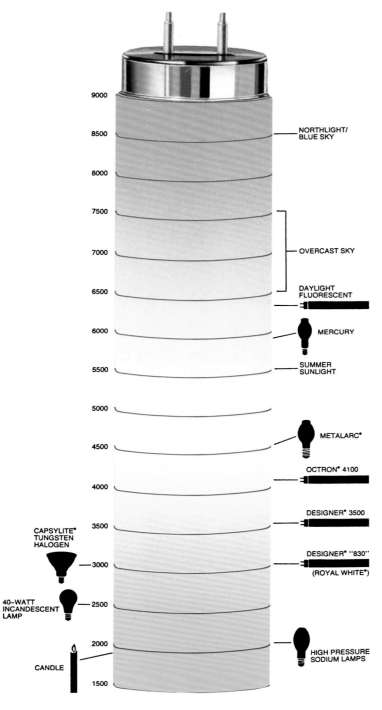

9000

8500 — NORTHLIGHT/
BLUE SKY

8000

7500

7000 — OVERCAST SKY

6500 — DAYLIGHT
FLUORESCENT

6000 — MERCURY

5500 — SUMMER
SUNLIGHT

5000

4500 — METALARC®

OCTRON® 4100

4000

DESIGNER® 3500

3500

CAPSYLITE®
TUNGSTEN
HALOGEN

DESIGNER® "830"

3000 (ROYAL WHITE®)

40–WATT
INCANDESCENT
LAMP

2500

2000 HIGH PRESSURE
SODIUM LAMPS

CANDLE

1500

Color Plate 8 Spectral characteristics of selected "color types" of fluorescent lamps, including HO (high output) and VHO (very high output). After GTE/Sylvania (1987).

HID Lamp Types	CRI	Color Temp.
Mercury Lamps		
Clear	22	5900K
Brite White Deluxe	45	4000K
Warmtone	52	3300K
High Pressure Sodium Lamps		
Unalux®	20	2100K
Lumalux®	22	2100K
Metal Halide Lamps		
Metalarc® Clear	65	4000K
Metalarc® Coated	70	3700-3900K
Metalarc® 3K Coated	70	3200K
Super Metalarc® Clear	65	3700-4200K
Super Metalarc® Coated	70	3100-3700K
Super Metalarc® 3K Coated	70	3200K

Mercury Lamps

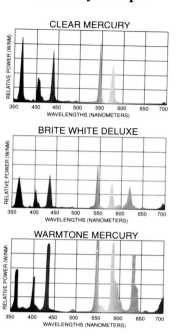

High Pressure Sodium Lamps

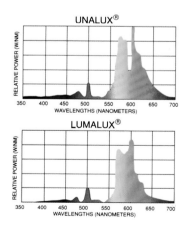

Color Plate 9 Spectral characteristics of selected mercury vapor, solium, and metal halide lamps. After GTE/Sylvania (1987).

Metal Halide Lamps

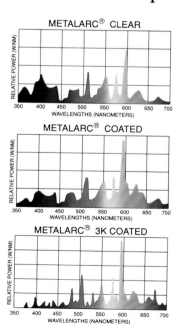

Color Plate 10 Dense algal turf growing on a high-wave-energy reef pavement. This pavement developed behind an algal ridge in the eastern Caribbean (field of view approximately 1 foot across). In this environment, where grazers are greatly restricted by wave energy, more typically macro genera such as *Sargassum*, *Turbinaria*, *Halimeda*, and *Avrainvillea* are beginning to compete with turf genera such as *Coelothrix*, *Gelidiella*, *Centroceras*, *Polysiphonia*, *Jania*, *Laurencia*, *Dictyota*, *Giffordia*, and *Cladophora*. This turf is 10–15 cm thick. See also Color Plate 11.

Color Plate 11 Algal turf areas on the (A) forereef and (B) backreef zones of the Smithsonian coral reef microcosm. The Stoplight Parrotfish *Sparisoma viride* grazes both zones. However, on the forereef, grazing time is very much limited by wave action, and the *Coelothrix–Gelidiella* algal turf community is dense and up to several cm long. On the backreef, on the other hand, the same community, intensively and continuously grazed, becomes a stubble, a few mm long. Photos by Karen Loveland.

Color Plate 12 Terminal male of the striped parrotfish *Scarus iserti* grazing short algal turfs on carbonate rubble at the lagoon base of the Smithsonian 3000-gallon Caribbean coral reef. A school of these fish (note the female partly hidden by reef rubble) were the dominant grazing (rather than browsing) fish on this model reef for many years, just as they are on many wild Caribbean reefs. These fish were easily accommodated in this size microcosm. Larger parrotfish (such as the Stoplight Parrot), tried later, were difficult to manage, not so much because of grazing pressure but because of the enormous rain of fine carbonate feces they spread over the reef surface.

Color Plate 13 The Western Atlantic Stoplight Parrotfish, *Sparisoma viride*. Note this is a terminal male (a female was shown in Color Plate 11). These animals are able to perform considerable scraping of the carbonate reef surface, enabling them to feed on the rhizomic and embedded bases of coral reef turf algae. Photo by R. Stuart Cummings.

Color Plate 14 Yellow tang, *Zebrasoma flavescens*, browsing algal filaments hidden among *Halimeda* in a 130-gallon coral reef aquarium. A large, well-lit tank with abundant algae can support several of these fish. However, as adults, one fish can become dominant and prevent its competitors from feeding to the extent of starvation. Photo by Nick Caloyianis.

Color Plate 15 Blue devil, *Abudefduf* sp., lying over the red algal crust *Peysonnellia rubra* in a 130-gallon coral reef aquarium. These small damsels will occupy territories in any model system, and some initial experimentation is usually required to determine the size of the population that a given system will take. If a predator of small fish is present, the community will usually self-adjust. Photo by Nick Caloyianis.

Color Plate 16 High-back headstander, *Abramites hypselonotus*. This freshwater-plant browser prefers young shoots of flowering plants and will often clip mature leaf stalks as well. Adults, even if provided a preferred vegetation supplement, such as lettuce, can destroy the productivity of small aquaria. Photo by Nick Caloyianis.

Variegated platy *(Xiphophorus variatus)*

Platy *(Xiphophorus maculatus)*

Atlantic herring *(Clupea harengus)* 1 ft (30 cm)

Black-barred livebearer *(Quintana atrizona)*

Sailfin molly *(Poecilia latipinna)*

Pacific sardine *(Sardinops sagax)* 7 in (18 cm)

Shortfin molly *(Poecilia mexicana)*

Merry widow *(Phallichthys amates)*

Atlantic menhaden *(Brevoortia tyrannus)* 10 to 12 in (25 to 30 cm)

Color Plate 18 Filter-feeding species of the Clupeidae (herrings). After Migdalski and Fichter (1976).

Mosquitofish *(Gambusia affinis)* 2 in (5 cm)

Color Plate 17 Several genera of the family Poecilidae. Note the characteristic up-turned mouth. After Migdalski and Fichter (1976).

Color Plate 19 Body form and dentition of typical barracuda, *Sphyraena barracuda*, from the tropical Atlantic. This common reef fish can reach 2 m. Photo by R. Stuart Cummings.

Color Plate 20 The Lyretail wrasse, *Thalassoma lunare*, in a 130-gallon coral reef aquarium. With sufficient algal production and a thick standing crop to provide abundant amphipods, isopods, and other crustacea, a square meter or so of reef surface will keep one wrasse well fed. Constantly foraging, and in rapid motion, these fish will often ignore artificial "plankton feed" if the food web is working properly. The finger coral at the lower left is *Eusmilia fastigiata* and the head coral to the right is *Montastrea cavernosa*. Photo by Nick Caloyianis.

Color Plate 21 Queen Angelfish, *Holacanthus ciliaris*, in the Smithsonian coral reef microcosm. This fish entered the system within the first year or two of its operation and remained there until its closure in early 1991. It feeds almost entirely on benthic sponges, anemones, and small invertebrates within the system, extensively searching in crevices and under overhangs to find its prey. Photo by Karen Loveland.

Color Plate 22 Spiny lobster, *Panulirus argus,* on a shallow water Caribbean reef. These animals sense and "uproot" small invertebrates in sediments or algal turfs or mats with hairs on the tips of their walking legs. When small, they are suitable for even quite small ecological models, though larger animals can destroy the surface of a reef or lagoon system when only several square meters of territory is available. Photo by R. Stuart Cummings.

Color Plate 23 Green crab, *Carcinus maenus,* from a barnacle-encrusted Maine intertidal with a heavy cover of the brown rockweed, *Ascophyllum nodosum.* Although these animals work well as mid-level carnivores in larger microcosms with strong algal biomass and a heavily creviced surface providing small invertebrate protection, we have had trouble keeping eelgrass, *Zostera marina,* on muddy bottoms where these animals are present. Photo by Karen Loveland.

Color Plate 24 Common sea star, *Asterias vulgaris,* on a coralline-covered pebble-shell bottom. These echinoderms feed almost exclusively on mussels. If sufficient cover and refugia are available and the system is richly endowed with organic particulates for the mussels, then there is little chance for population destruction in moderate-size mesocosms. Because of the ease of capture of the sea stars, food web manipulation by removal or separate feeding is easily accomplished. Photo by Nick Caloyianis.

Color Plate 25 The deep angelfish, *Pterophyllum altum,* feeding on the underside of a dense, floating cover of *Salvinia* and *Nymphaea.* This is a 70-gallon, blackwater tank, where dried flake food addition simulates aquatic and terrestrial insect input to the system. Constant algal scrubbing and export of the rapidly reproducing *Salvinia* balances the flake food input. Photo by Nick Caloyianis.

Color Plate 26 Dense Turtle Grass (*Thalassia testudinum*) bed in the attached lagoon of the Smithsonian coral reef microcosm.

Color Plate 27 Close up photograph of lagoon sand shown in Color Plate 26. Note extensive burrowing by a rich infauna, primarily polychaete worms and amphipods.

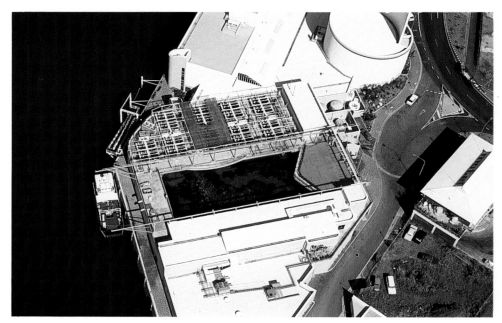

Color Plate 28 Aerial photograph of the Australian Barrier Reef Aquarium. The main reef unit is the large, uncovered tank in the center of the photograph. The predator tank (covered with blue plastic sheet) lies to the right. The scrubber batteries and refugia ("mesh" area) in the upper center manage both reef and predator tanks.

Color Plate 29 Photograph of a 130-gallon home reef aquarium using an algal turf scrubber. The *Tridacna* (giant clam) (center right) has doubled its diameter since being placed in the tank. Note the "growth rings" on the new shell, probably resulting from greater variations of temperature and salinity in the tank as compared to the wild.

2-2-88   3-26-88   5-20-88   7-12-88   9-12-88

11-12-88   1-19-89   3-24-89   6-6-89   8-20-89

10-29-89   1-1-90   4-21-90   7-21-90

Color Plate 30   Two and one-half years of molts of the reef lobster (*Enoplometopus*) from a 130-gallon home reef aquarium. Photo by Nick Caloyianis.

Color Plate 31 Maine coast microcosm mud flat and salt marsh 6 years after establishment. The photo was taken in mid-winter at low tide neaps. In this very warm (for air temperature) winter regime, some species "brown out" as they do in the wild; others are evergreen. All of the original plant species are present and all flower. However, in this extremely small area, the established zonation (Chapter 6, Fig. 7) has been lost and the original boundaries of the zones blurred. Photo by Nick Caloyianis.

Color Plate 32 Maine coast microcosm rocky intertidal at low water neaps. *Ascophyllum nodosum* (foreground) and *Fucus vesiculosus* at higher levels (background). The black coating on the vertical rock (center background) is the lichen *Verrucaria*. It has been slowly recolonizing vacant surfaces in the supratidal range, above high water neaps. Photo by Karen Loveland.

Color Plate 33 Upper Irish moss/coralline zone in the Maine microcosm. The fleshy red is *Chondrus crispus* (Irish Moss), the red coralline crust on the mussel (*Mytilis edulis*) is *Lithothamnium glaciale*. The coarse green filament is *Chaetomorpha melagonium*. Photo by Nick Caloyianis.

Color Plate 34 Tuft of *Gymnogongrus norvegicus* (about 4 inches in diameter) in the lower *Chondrus crispus* zone of the Maine coast microcosm. This occasional red alga in the wild does particularly well in the microcosm, probably possessing compounds defensive against grazers. Photo by Nick Caloyianis.

Color Plate 36 Red alga zone at the base of the Maine coast microcosm. Encrusting red algae, particularly *Lithothamnium glaciale* and the fleshy reds *Phyllophora membranifolia*, *Phycodrys rubens*, and *Ptilotia serrata*, abundantly encrust pebbles and the horse mussel *Modiolus modiolus* in this zone. Loose fragment of *Laminaria longicruris* to left. Photo by Karen Loveland.

Color Plate 35 The kelp *Laminaria longicruris* in the Maine coast microcosm. This kelp species has performed extremely well in this system, repeatedly reproducing and growing to maturity especially immediately under and attached to the wave generator. This is the dominant kelp in most of the kelp band in eastern Maine. Photo by Karen Loveland.

Color Plate 37 The calcified annelid worm *Spirorbis borealis* encrusting the side of a cobble in the lower *Chondrus crispus* zone of the Maine microcosm. The green filament in the lower left is *Chaetomorpha melagonium* at about 1 mm in diameter. Photo by Nick Caloyianis.

Color Plate 38 Photograph of the high-salinity end of the Chesapeake Bay mesocosm at the Smithsonian Institution. Note the metal halide lamps over the marsh and the VHO lighting units over the deeper portions of the tank (background). The shallow water salinity gates between each salinity segment extend down the wall to the left. Photo by Karen Loveland.

Color Plate 39 Photograph of Smithsonian Chesapeake mesocosm showing fans, temperature control vents, wave buckets, and tidal control reservoir (overhead). The large white pvc pipe hanging from the tidal tank holds the bellows that actually controls tide levels.

Color Plate 40 Photograph of high-salinity *Spartina* marsh in the Chesapeake mesocosm in the late autumn. Also pictured are *Salicornia* (glasswort) (red), *Aster dumosus* (foreground), and *Juncus roemerianus* (black needle rush) (right). Photo by Karen Loveland.

Color Plate 41 Photograph of the mid-salinity *Juncus–Distichlis* marsh of the Chesapeake mesocosm. Note the dense green tufts of *Juncus roemerianus* on the side foreground of the tank and the salt bush, *Iva frutescens*, embedded in a field of *Distichlis* (background). Photo by Karen Loveland.

Color Plate 42 Tidal fresh section of the Chesapeake system during the late autumn. Although much of the vegetation is reduced at this time of the year, the more moderate temperatures in the model environment allow some species to be evergreen all winter (*Pontederia*—Pickerelweed, *Scirpus*—three square, and *Polygonum*—smartweed). Photo by Karen Loveland.

Color Plate 43 *Callinectes sapidus*, the blue crab, feeding on a ribbed mussel, *Geukensia demissus*, in the Chesapeake mesocosm. Photo by Matt Finn.

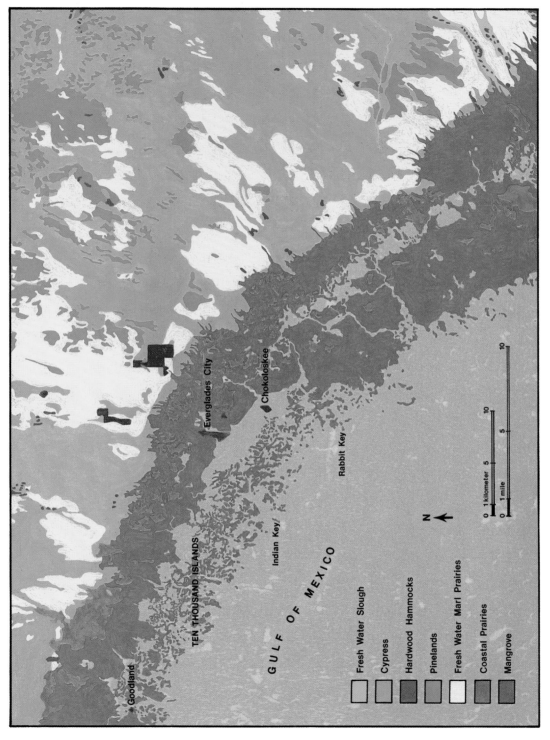

Color Plate 44 Map of the Ten Thousand Islands region of the Florida Everglades showing the principal ecological communities. The section modeled and discussed in Chapter 23 is in the Ten Thousand Islands area in the upper left and the black mangrove, salt marsh, and prairies and hammocks lying to the northeast.

Color Plate 45 Photograph of one of three fans used to achieve wind in the Everglades mesocosm. The plastic box in the lower right contains the tidal control system. Photo by Karen Loveland.

Color Plate 46 Photograph of the "worm reef" and vegetated dune in the Everglades mesocosm with the red mangrove zone behind. The base of the dark, algal-covered "worm reef" in the foreground marks the level of lower low tide springs. Photo by Karen Loveland.

Color Plate 47 Photograph of the salt marsh of the Smithsonian Everglades mesocosm with black mangroves behind. The tree with the lighter green leaves in the upper middle is the white mangrove, *Laguncularia racemosa*. Photo by Karen Loveland.

Color Plate 48 Photograph of the oligohaline marsh area (mid right) of the Everglades mesocosm in winter. The single tree in the left foreground is a buttonwood. The tall blades (behind) are those of the cattail, *Typha domingensis*. The dense, brown grasslike growth in the closer part of the tank (lower right) is that of the spike rush, *Eleocharis robbinsii*. Photo by Karen Loveland.

Color Plate 49 Photograph of a 130-gallon, 20–25 ppt aquarium of lower Chesapeake Bay in mid-winter. The triangular chamber at the lower left is the tidal reservoir. The greenish-brown area at the upper left is a *Distichlis*-dominated salt marsh. The deeper part of the tank is an oyster bed with the red beard sponge, *Microciona prolifera*.

Color Plate 50 Photograph of the pond community lying at the base of the freshwater Everglades mesocosm at the end of the wet season. Lacking large grazers in this model, dense stands of *Pistia*, the water soldier, (foreground) and *Eichhornia*, water hyacinth, (center rear) along with the smaller, floating *Salvinia*, *Azolla*, and *Lemna* cover the stream and pond. These plants are periodically "grazed back" by hand cropping and forming a compost heap in the corner of the pond.

Color Plate 51 Photograph taken from the drier end of the Florida Everglades prairie, hammock, and stream mesocosm. The foreground communities are *Andropogon* (broomsedge) prairie on the left and pine–palm (*Pinus elliottii–Sabal palmetto, Roystonea elata*) on the right. Wetter prairie (saw grass, *Cladium jamaicensis*) and hammock fern (*Blechnum*) communities are in the background.

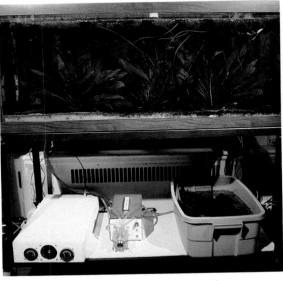

Color Plate 52 Photograph of a 70-gallon black water stream community using a small algal turf scrubber for environmental control. Beneath the tank are the timer/ballast unit (left) and the automatic evaporative replacement unit (center and right). The reservoir contains black swamp water as described in the text.

spring range of about 12 feet. The microcosm has a 38-cm spring tide range, with a neap to springs high tide difference of 9 cm. *Spartina alterniflora* was established in the upper half of the neap range and *S. patens* in the spring tide zone above neap highs. Both species established easily and have been functioning since May 1985.

The second set of marshes including full-salt, brackish, and a tidal fresh marsh was established in an artificially lighted Chesapeake Bay mesocosm in 1987. Transfer was also entirely by block methods, and the system has a mature diversity of about 50 species of emerged aquatics. Most of the dominants, including *S. alterniflora, S. patens, Distichlis spicata, Juncus roemerianus, Iva frutescens, Limonium* sp, *Scirpus* spp., *Carex* spp., *and Pontederia cordata*, are present in this model estuary (see Chapter 23). Although *Typha angustifolia* has remained present in this system, following its original stocking, it has never been dominant, probably because of the small space and interference by other species. This mesocosm has a maximum tide range of 6 inches, as compared to 1–2 feet throughout most of Chesapeake Bay. Establishment of block surfaces was directly proportional to tide levels. *Spartina alterniflora* did not survive the first and second winters in the polyhaline zone but did survive in the

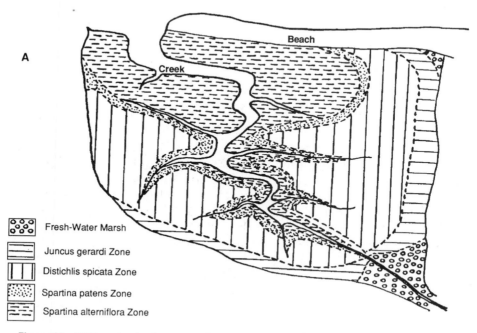

Figure 31 (A) Zonation by dominating species in a New England salt marsh. (B) Cross section of a mid-Atlantic coast salt marsh showing species zonation, biomass, and soil salinity. A, After Chapman (1960); B, after Haines and Dunn (1985).

(Figure continues)

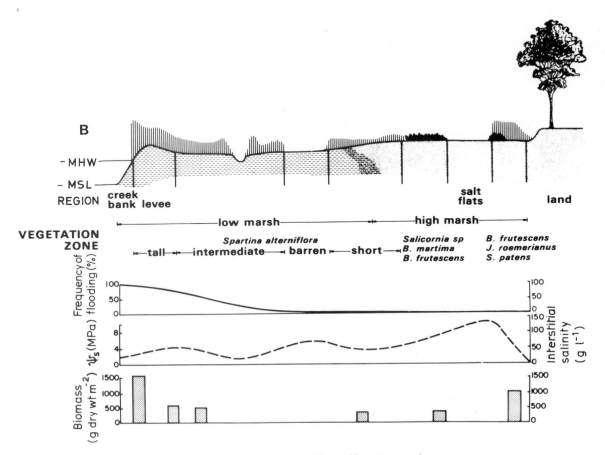

Figure 31 (*Continued*)

marine zone. It was successfully reestablished in the spring. Perhaps insect predation is involved, a problem we had not found in earlier mesocosms. Otherwise, no significant difficulties were encountered. A miniature, 70-gallon version of a Chesapeake Bay salt marsh at 26 ppt was later established separately. In this aquarium, also with a 6-inch semidiurnal tide, the two *Spartina* species, *Distichlis spicata*, *Limonium* sp., and *Salicornia* sp were similarly established with ease. Here, the difficulty lay with a small *Juncus roemerianus* colony that has provided virtually no growth in about a year of operation. Since *J. roemerianus* grew extensively in the large system, flowering frequently, this may be a negative effect of one species on another in such a small system. Peak light levels are about 800 $\mu E/m^2/s$ in the aquarium as compared to about 1600 $\mu E/m^2/s$ in the mesocosm. This also could be a factor.

Finally, in 1988, an Everglades mesocosm with both salt and fresh

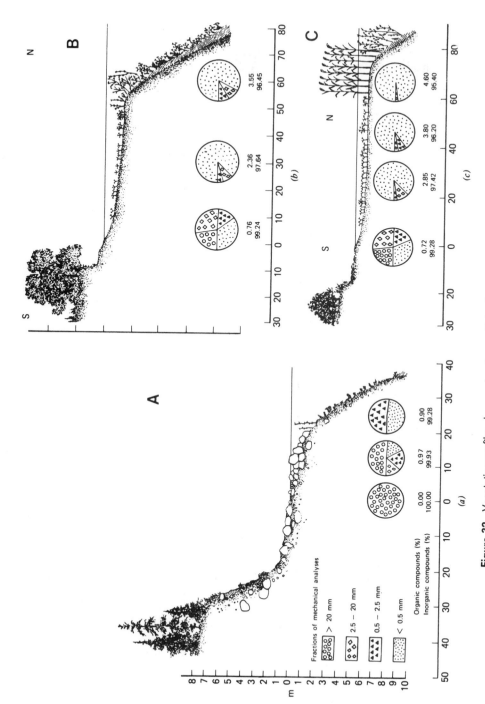

Figure 32 Vegetation profiles along various types of lake shores. (A) High-energy, eroding steep shore, emergent vegetation lacking submergent vegetation minimum. (B) Moderate-energy, steep shore with emergent vegetation lacking, but strongly developed submergent vegetation. (C) Moderate-energy, low-slope shore with a narrow, partly protective emergent vegetation zone and submergent aquatics on both protected shelf and on the lake slope. (D) Protected, low-sloping lake shore with a rich development of emergent and submergent vegetation. (E) Very-low-energy, low-sloping shore developing a floating shelf of emergent vegetation and with a floating leaf bed of submerged aquatics. After Hutchinson (1975). Reprinted by permission of John Wiley & Sons, Inc.

(Figure continues)

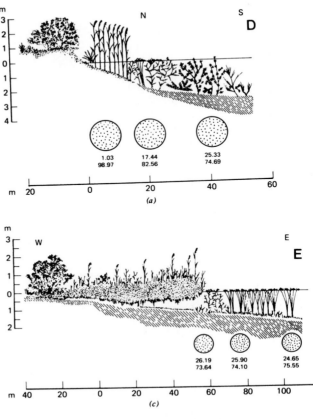

Figure 32 *(Continued)*

marshes was established in a greenhouse at Washington, D.C. This system with its salt marsh and mangrove communities is described in Chapter 24. The three primary mangroves in the Western Atlantic (red, *Rhizophora mangle*; black, *Aricennia germinans*; white, *Laguncularia racemosa*) have all flowered and set seed in this mesocosm after 3 years of operation. Here brief mention is also made of both oligohaline (seasonal salt to 6 ppt) and fresh "prairie" marshes. The oligohaline marsh is dominated by *Typha domingensis* and *Eleocharis robbinsii*, both of which have successfully flowered and vegetatively expanded. Thirty-four other flowering plants including *Conocarpus erectus* (buttonwood mangrove) and *Myrica cerifera* (wax myrtle) are also included. The "prairie" marsh, ranging from a flooded period of 2 to 10 months, includes approximately 100 species. The wetter portions are strongly dominated by *Cladium jamaicensis* (saw grass), as the Everglades prairie is in the wild. This plant has vegetatively expanded and flowered for two seasons, indicating a successful modeling.

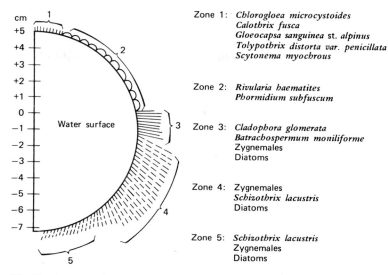

Figure 33 Floating pipe from a European lake showing the algal communities that typically develop the "aufwuchs" that coat emergent and submergent aquatic vegetation. After Hutchinson (1975). Reprinted by permission of John Wiley & Sons, Inc.

References

Adey, W. 1986. Coralline algae as indicators of sea level. In *Sea Level Research: A Model for the Collection and Evaluation of Data*. O. van de Plassche (Ed.). pp. 229–280. Geo Books, Norwich.

Bold, H., and Wynne, M. 1985. *Introduction to the Algae*. 2nd Ed. Prentice Hall, Englewood Cliffs, New Jersey.

Chabot, B. and H. Mooney, 1985. Physiological ecology of North American plant communities. Chapman and Hall. New York. 351 pp.

Chapman, V. J. 1960. *Salt Marshes and Salt Deserts of the World*. Interscience, New York.

Dawes, C. 1981. *Marine Botany*. Wiley-Interscience, New York.

Fitter, R., Fitter, A., and Farrar, A. 1984. *Collins Guide to the Grasses, Sedges, Rushes and Ferns of Britain and Northern Europe*. Collins, London.

Fritsch, F. E. 1952. *The Structure and Reproduction of the Algae*. Vol. II. Cambridge University Press, Cambridge.

Fritsch, F. E. 1956. *The Structure and Reproduction of the Algae*. Vol. I. Cambridge University Press, Cambridge.

Godfrey, R., and Wooten, J. 1979. *Aquatic and Wetland Plants of the Southeastern U.S. Monocotyledons*. University of Georgia Press, Athens, Georgia.

Godfrey, R., and Wooten, J. 1981. *Aquatic and Wetland Plants of the Southeastern U.S. Dicotyledons*. University of Georgia Press, Athens, Georgia.

Haines, B., and Dunn, E. 1985. Coastal marshes. In *Physiological Ecology of North American Plant Communities*. B. Chabot and H. Mooney (Eds.). Chapman and Hall, New York.

Hutchinson, G. E. 1975. *A Treatise on Limnology*. Vol. III *Limnological Botany*. John Wiley and Sons, New York.

Margulis, L., and Sagan, D. 1986. *Microcosmos*. Summit Books, New York.

Margulis, L., and Schwartz, K. 1988. Five Kingdoms, an illustrated guide to the phyla of life on earth. 2nd Ed. 376 pp. Freeman and Co., New York.

Moe, M. 1989. *Marine Aquarium Reference*. Green Turtle Publications, Plantation, Florida.

Moss, B. 1980. *Ecology of Fresh Waters*. Halsted Press, New York.

Muhlberg, H. 1982. *The Complete Guide to Water Plants*. EP Publishing Ltd., German democratic Republic.

Phillips, R. C., and Menez, E. 1988. Sea grasses. *Smithsonian Contribs. Marine Sci.* **34**: 1–104.

Phleger, F. B. 1977. Soils of marine marshes. In *Ecosystems of the World—Wet Coastal Ecosystems*. V. J. Chapman (Ed.). Elsevier, Amsterdam.

Prescott, G. 1951. *Algae of the Western Great Lakes Area*. Cranbrook Institute of Science Reprint 1982 Koeltz.

Sculthorpe, C. 1967. *The Biology of Aquatic Vascular Plants*. Arnold. Reprint 1985 Koeltz.

Stevenson, J. C. 1988. Comparative ecology of submerged grass beds in freshwater, estuarine and marine environments. *Limnol. Oceanogr.* **33**: 867–893.

Stodola, J. 1967. *Encyclopedia of Water Plants*. T. F. H. Publications, Neptune City, New Jersey.

South, R., and Whittick, A. 1987. *Introduction to Phycology*. Blackwell Science Publishers, Oxford.

Taylor, W. R. 1957. *Marine Algae of the Northeastern Coast of North America*. University of Michigan Press, Ann Arbor, Michigan.

Thiel, A. J. 1988. Keeping and growing marine macroalgae. Freshwater and Marine Aquarium **11**:98–118.

Williams, S., and Adey, W. 1983. *Thalassia testudinum* Banks ex Konig. Seedling success in a coral reef microcosm. *Aquatic Botany* **16**: 181–188.

CHAPTER 16

HERBIVORES
Predators of Plants

Virtually all organic primary production on the earth and in its waters is accomplished by plants. It is the food value of plant bodies that provides the energy to operate the remainder of the biosphere. Depending upon the ecosystem, some of this plant production becomes detritus after the demise or partial demise of the plant. In some cases, the percentage of the plant biomass that becomes detritus is very large (see Chapter 19). However, in some ecosystems large amounts of plant production (10–90%) are eaten directly by herbivores such as grazers, browsers, suckers, and seed or fruit eaters.

Most people are generally aware of the phenomenon of overgrazing. When there are too many plant-eating animals, whether in a pasture or on a range or countryside, without natural or human limitation, the plants are eaten faster than they can grow. The net result is a relatively barren landscape, or one minimally occupied by inedible plants. Total plant production is then far below potential and the grazers cannot find enough food. In the dry terrestrial environment, deserts may result, and often a feedback loop is present. The presence of vegetation encourages rain. Once it is gone, in the marginal situation, rainfall and plant productivity are greatly reduced. In another situation, a reduction of higher predators, such as grizzly bears and mountain lions, can lead to overpopulation by deer or elk. This also results in overgrazing and a panoply of negative ecological effects (see Chase, 1985). In a synthetic ecosystem modeling a well-lighted aquatic ecosystem, plants will likely be present, as will herbivory. While the desertification process is not a consideration, reduced plant production in the marine aquatic environment often results in a degradation of environmental quality. With few plants, oxygen levels are lower, pH values are more acid, nutrient levels are higher, cover and substrate are reduced, sedimentation is slower (thereby reducing light), wave and current action are increased, and of course potential food is reduced. Thus, the balance of plant growth and herbivory is crucial to model ecosystems and microcosm and aquarium management alike.

Grazing fish can significantly reduce algal vegetation. For example, Power *et al.* (1985) describe how a grazing minnow (*Compostoma anomalum*) is capable of causing considerable reduction of algal biomass as well as changing algal community structure in a freshwater stream. Significantly, those same authors were also able to show that where the algal turf was reduced, the addition of predator bass caused the minnows to decline and the algal biomass to increase. Any aquarist who has kept headstanders, parrot fish, or tangs with plants has experienced the same basic phenomenon, though not usually with a fish predator present to control the grazer.

At first sight, it might seem that plants can only sit quietly and be eaten, and the limitation to total plant slaughter is predators of the grazers. On the contrary, some plants are actually helped by grazing, for example, prairie grasses and reef algal turfs (D. Owen, 1980; Adey and Hackney, 1989). These plants are adapted to being eaten—they grow fast from their base, reproduce quickly, and are resistant to environmental factors (wind and waves) that may damage other plants (Color Plate 11). Most important, their competitors, larger plants, do not do well when constantly snipped off at small size. Also, many plants attract specialized predators of pollen, fruit, or seeds so as to gain fertilization or distribution of seeds. This becomes more properly a symbiotic relationship. However, even beyond these predator-adapted plants, except where human effects have unbalanced ecosystems, or where seasonal effects of cold or dry conditions or other weather anomalies have considerably affected plants, only 10–50% of net plant production is normally eaten (J. Owen, 1980). In some cases, such as salt marshes, the amount directly eaten by grazers can be extremely small (Parsons and de la Cruz, 1980).

Plants have developed many defenses. Woody plant tissue as well as much coarse greenery is generally inedible by animals. As most people know, in the woods one can get very hungry with abundant potential plant food all around. While herbivores abound, plants protect themselves in many ways. Particularly in the marine environment some algae develop tough, cellulosic or even calcium carbonate walls making grazers pay a high price for their meals. An overgrazed situation in warm, calcium carbonate-rich waters is likely to show reduced photosynthesis. However, between algal turfs and those algae with tough walls, the reduction is likely to be small. Meanwhile the herbivores have to constantly guard against their own predators. Thus, as a whole "the world is green."

Under water, in shallow zones, plant growth exceeds animal growth, as it must, except where food input from another community is important. However, in the aquatic, estuarine, and marine coastal environments, the human factor enters into the plant–herbivore equation. Man's building and farming activities on the land release enormous quantities of silt from eroded soils into natural waters. These sediments produce elevated turbidities, which reduce light and therefore aquatic plant photosynthesis. Along with lower light levels, water quality degeneration resulting from herbicides and numerous other chemicals as well as excess nutrients also hinders balanced plant growth. Under these conditions, herbivores, particularly fish and waterfowl, often overgraze and then become strongly limited themselves.

Types of Herbivores

Table 1 lists the major orders of animals with herbivorous members and indicates what form the herbivory takes. Many phyla have at least a few herbivores, and a large number of orders have all or most of their members at that trophic level. In orders primarily characterized by aquatic and marine members, an even higher percentage is characterized by all or most of their members being herbivorous.

In the terrestrial environment, because of the special adaptions of plants that are required to stand up into a dry atmosphere and carry out sexual reproduction, the range of types of herbivory is very broad. These include grazing, browsing, the rasping of bark or stems, sucking of plant juices, predation on flowers (nectar or pollen), and feeding on fruit and seeds as well as plant roots. In wetlands, marshes, and swamps, virtually all of the same types of herbivory occur. However, in the truly submerged environment, while all these same types could be identified as a few special cases, herbivory is largely restricted to grazing and browsing. Because submerged plants generally lack the strengthening requirements of land plants (lignin and massive cellulose), the adaptations for grazing and browsing are not as extensive as they are in land plants. On the other hand, a major form of plant defense under water is calcification, and some animals (chitons, limpets, and parrot fish, for example) are adapted, sometimes extensively, to overcome these defenses to varying degrees.

Plant Defenses

As mentioned above, some plants (all algae) have developed "rock-hard" skeleta of calcium carbonate, in large measure to reduce grazing. These

TABLE 1

Major Orders of Marine and Aquatic Animals with Herbivorous Members[a]

Phylum	Class or Order	Frequency of herbivores within group	Name	Example	Example of tissue eaten	Mode of feeding
Invertebrate herbivores						
Protozoa	Several	Many	Amoeba	*Amoeba dubia*	Diatoms	Cytoplasmic engulfing
Nematoda	Several	Many	Nematodes	*Dorylaimida*	Algae	Cell contents, suck
Echinodermata	Echinoidea	Many	Sea-urchins	*Echinus esculentus*	Seaweeds	Rasping
Mollusca	Gasteropoda	Most	Slugs, snails	*Deroceras reticulum*	Leaves, fruits	Rasping
	Amphineura	Virtually all	Chitons	*Ischnochiton ruber*	Algal turfs, corallines	Rasping
Tardigrada		All	Bear animalcules	*Macrobiotus macronyx*	Sphagnum moss	Sucking cell contents
Arthropoda	Subclass					
Class Crustacea	Branchiopoda	Many	Water fleas	*Daphnia pulex*	Diatoms	Selective filtering
	Copepoda	Most	Copepods	*Calanus*	Phytoplankton	Selective filtering
	Malacostraca Order					
	Isopoda	Some	Slaters	*Ligia oceanica*	Seaweed	Chewing
	Amphipoda	Some	Gammarids	*Gammarus neglectus*	Microalgae	Browsing
	Euphausiacea	Most	Krill	*Euphausia superba*	Phytoplankton	Selective filtering
	Decapoda	Few	Crabs, lobsters	*Birgus latro*	Fruits, etc.	Chewing

	Order / Subgroup	Abundance	Common name	Example	Food	Feeding
Insecta						
Plecoptera		Few	Stoneflies	Isoperla	Algae	Larval stages
Ephemeroptera		Most	Mayflies	Ephemerella	Turf grazers	by larval stages. Browsing or filter feeding
Trichoptera		Few	Caddis Flies	Neophylax	Algae	by larval stages. Browsing or filter feeding
Coleoptera		Few	Crawling water beetles	Haliplus	Algae	by larval stages. Browsing or filter feeding
Diptera		Many	Midges	Microtendipes	Algae	by larval stages. Browsing or filter feeding
Vertebrate herbivores						
Actinopterygiia		Several	Bony fishes	Carp	Water weed	Browsing
Sarcopterygia	Dipnoi	Few	Lung fishes	Protopterus	Stoneworts	Browsing
Amphibia		Few	Frogs, toads	Rana tadpoles	Water weed	Browsing
Reptilia	Chelonia	Some	Sea turtles	Chelonia myolas	Sea grasses	Browsing
	Squamata	Few	Snakes, lizards	Giant iguana	Seaweeds	
Aves	Anseriformes	Many	Ducks, geese	Anas	Water weeds	
	Passeriformes	Few	Many marsh birds	Agelaius	Marsh and wetland seed eaters	Seed Eaters
Mammalia	Lagomorpha	Few	Marsh rabbits	Silvilagus	Marsh plants	
	Rodentia	Some	Muscrat, marsh rats	Ondatia	Marsh plants	Browsing
	Sirenia	All	Sea cows	Trichechus	Sea grasses	Browsing

ᵃModified after Crawley (1983).

plants are few in freshwater environments where calcification is generally more difficult and where long-term stable evolutionary pressure to develop calcification is not usually present. The Charophytes (*Chara, Nitella*) are prime examples, though even these plants are more characteristic of hard waters. Sinter or calcium deposits, primarily by blue-green alage in mineral and hot springs, probably are the direct result of pH effects rather than an adaptation for defense against grazing.

Even in the marine environment where calcification can happen simply as a result of photosynthesis, relatively few major groups have developed calcification as a defense. However, where this has happened, primarily the red algal family Corallinaceae and related groups and the green order Siphonales (e.g., *Halimeda, Udotea*) the genera and species involved are quite widespread and ecologically very important. Skeleta of *Halimeda* are generally recognized to be the major contributor to tropical limestone structures such as atolls, and corallines are builders of large reefs (algal ridges). In general, even where they do not dominate, coralline algae are important contributors to coral reef structures and often extensively encrust shallow bottoms well into the Arctic and Antarctic. Particularly in ecological situations where urchins or fish grazers are highly abundant, corallines can develop the equivalent of thorn scrub in sheep country.

The most obvious chemicals providing protection of terrestrial plants are the lignins and tannins present in many vascular plants. So important are these in the terrestrial environment that at least one author (Swain, 1979) has stated, "it seems doubtful whether plants, and hence life as we know it, could have developed on land without the acquisition of the ability to synthesize these two classes of phenolic compounds." Lignins and tannins are also heavily involved in the protection of many marsh and wetlands plants and lead to the great resistance to breakdown of marsh plants. Both are usually thought of as providing protection against fungal degradation. However, they also provide an amazingly effective general protection against animal herbivores. Tannins are so resistant to breakdown that they "wash out" to color bogs, swamps, some streams (blackwaters), and even coastal waters. In addition to providing a measure of direct protection from herbivores, lignin greatly increases the strength of woody plants, allowing them to "stand erect," particularly in the nonaqueous environment.

There is, however, another whole class of defense by plants against herbivores and even against parasites and other plants. These are direct chemical defenses. The chemicals themselves are called secondary compounds because generally no day-to-day metabolic use other than as poisons is known. A wide variety of chemicals are involved. They include cyanide derivatives, alkaloids, and terpenoids. Most higher plants and many algae develop such protective toxic chemicals to varying degrees. Curiously, however, virtually every defensive toxin has led to detoxification mechanism in some herbivores. Since all living organisms share the same basic chemistry, it is unlikely that a plant could develop an ultimate chemical defense against a herbivore. While a few secondary compounds are deadly, most operate on a more subtle basis, reducing an animal's efficiency. It is more a matter of protection to a certain degree. The energetic and therefore growth or reproduction costs of significant amounts of secondary compounds are considerable and in the end are weighed, by natural selection, against percent reduction in grazing. Likewise, for a herbivore to develop a way around a toxin is costly. The evolutionary selection process sometimes leads to coevolution of a plant and a specific grazer.

Many secondary compounds have been identified in algae as well as in the flowering plants. It has also been demonstrated that the effects of these compounds, mostly terpenoids, on grazers can be lethal (Norris and Fenical, 1982). Plants rich in secondary compounds are avoided by grazers. On the other hand, some grazers have come to tolerate the secondary compounds of algae. A few even concentrate these compounds for their own defense. Norris and Fenical (1985) discuss methods for isolation and identification of these mostly lipid-soluble chemicals. However, as we discuss below, in practice, in model ecosystem development and operation, each situation requires the adjustment of herbivore/plant relationships by the human operators.

Theoretically, in the confines of small model ecosystems, secondary toxic compounds could provide serious problems to the general community. However, in our experience this has not been the case. The epiphytic and benthic dinoflagellate *Gambierdiscus toxicus* has been widely regarded as the source of a compound that, when concentrated in higher predators, gives rise to ciguatera. This is a form of severe fish food poisoning that affects the central nervous system and is common in the tropics. *Gambierdiscus toxicus* has been identified in at least one coral reef microcosm where only under certain conditions of heavy blue-green algal growth has the substrate for the dinoflagellate been present and fish distress occurred. In another case of a reef microcosm, damage to the toxic sea cucumber *Stichopus nigricians* caused some fish distress and death (see Chapter 21). However, of the many invertebrates known to be present, none was affected. Thus, the toxin was apparently specific to vertebrates. In general, given an active ecosystem and no unusual concentrations of toxic organisms, the effects of secondary compounds at the ecosystem level are not likely to be a problem in model systems.

Modifications of Marine and Aquatic Herbivores

Herbivores become better adapted to their mode of feeding through a variety of morphological and physiological modifications. These adaptations allow them to forage for food more effectively in the face of plant defenses and the ever-present danger of predators. The modifications for herbivory are almost as numerous as the species involved if one were to examine the matter in great detail. Nevertheless, there are types or guilds of herbivores that can make understanding the process easier. We will examine several fairly well defined types.

Cellulose in a variety of forms is the major constituent of the cell walls of most plants (see Chapters 9 and 15). Cellulose itself, however, can be broken down or digested by animals only with difficulty and generally only with the help of microbes or protists. In most marine and aquatic plants, the very thick and lignified walls of woody plants are absent. Nevertheless, the walls of emergent marsh plants do possess lignin and break down very slowly. Even submerged aquatics and algae are difficult to digest as compared to animal cells. Some herbivores make do by simply crushing the cells and digesting the cell contents. Others, like many land herbivores, have specially modified digestive tracts to allow for the activity of microbes in breaking down cellulose. In the marine environment, calcified surfaces laid down by algae or in which algae grow or bore provide the need for rasping or scraping mouth parts. Calcium carbonate itself provides no organic energy and is not chemically broken down further. However, when dissolved by digestive tract acidity, calcified structures provide calcium in solution, which can be useful for shell construction by the herbivore itself. In the following pages, we will briefly describe eight types of marine and aquatic herbivores to provide a general understanding of the difficulties faced by marine and freshwater herbivores and their adaptations to overcome these problems.

Parrot Fish (Scarids)

Parrot fish (Figure 1 and Color Plates 12, 13) are full grazers in every sense of the word as it is applied to terrestrial mammals. They develop a strongly calcified beak of fused teeth that is used to scrape calcified algae and calcified surfaces that have plants growing into them. Depending on the species and the location, some parrotfish will also scrape coral. While to some degree this makes them carnivores, it must be remembered that coral is richly endowed with an algal symbiont (Chapter 20). Most parrot fish will also browse to some extent, and some species spend much of their time in this mode of feeding. However, when this is the case, the plant involved is typically a tough sea grass or calcified alga. As with most marine and aquatic grazers, presented with the opportunity to capture small invertebrates, parrot fish will also take animal food.

In addition to the obvious rasping/scraping apparatus, parrot fish also possess a pharyngeal mill consisting of internal molar-like teeth that are used to grind up the mixture of carbonate and algae or coral tissue that

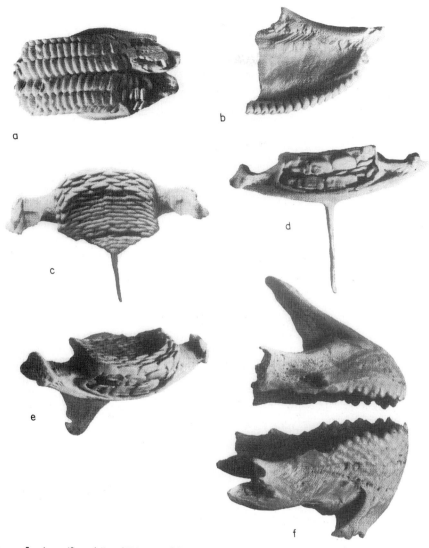

Figure 1 Jaws (f) and "teeth" (a–e) of the pharyngeal mill, grinding complex of a terminal male Western Atlantic Stoplight Parrotfish, *Sparisoma viride*. After Gygi (1975).

is delivered to it from the beak. The effectiveness of the beak/mill combination is such that parrot fish are one of the major degraders of reef hard structure as well as being important suppliers of the resultant fine carbonate silt. In a model system, without sufficient wave and current and settling traps that are the equivalent of lagoons, the supply of fine sediment defecated by parrot fish can be a serious problem for corals. These sedentary animals must expend considerable energy to constantly slough off the sediment.

Many parrot fish also develop behavioral patterns that adapt them to a grazing mode of life in reef environments. Most species operate in schools and graze in roaming "herds." The herds consist mostly of females and different-colored secondary males having a hierarchical social structure that includes a dominant female and a single brightly colored terminal or "super-male." Social grazing tends to confuse and disperse territorial reef species like damselfish, allowing the parrots access to plants they would otherwise be denied. Likewise, as with a terrestrial herd, the schooling behavior makes predation by larger fish more difficult.

Tangs (Surgeon Fishes)

The Acanthurids, tangs or surgeonfish, are extremely abundant tropical and subtropical marine fish of a few genera and species (Color Plate 14). Along with parrot fish and damselfish, these animals are mostly herbivores, and together they form the largest part of the fish biomass of most reefs (Thresher, 1980). Tangs are browsers, with lips and dentation for snipping off the tips and branches of algae. They also have long thin-walled digestive tracts and some species have a sand-filled, muscular gizzard-like foreintestine (Fig. 2). These are all adaptations for foraging on relatively soft algal filaments and blades, though the thorough processing of cellulose seems unlikely. Like parrot fish, tangs are generally strongly schooling, at least as adults. Often they are part of mixed "herds" of parrot fish and tangs. Tangs have unique tail spines for defense, hence the name surgeonfish. These remind one of the horns of some hoofed grazing animals, the spines of porcupines, and the tail spikes of some famous dinosaurs.

Tangs, particularly when young, have a stringent requirement to feed almost continuously, undoubtedly because of a relatively poor utilization of their algal food (Thresher, 1980). While the long-thin-walled intestine is probably well adapted to absorbing the cell contents of crushed cells, it is probably poorly suited to handling cellulose. This has considerable bearing on their survival in model ecosystems, since they must have an algal lawn and brush that allows almost continuous foraging. Also, given a small area previously occupied by other herbivore fishes, tangs may be kept in a corner to the point of starvation.

Damselfish (Pomacentrids)

The pomacentrids fill a wide variety of environments in tropical waters. Some are planktivores (*Chromis*); others are small invertebrate eaters

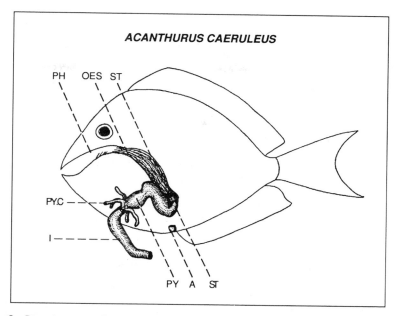

Figure 2 Digestive tract of a tang,*Acanthurus caeruleus*. PH, pharynx; OES, oesophagus; ST, stomach (muscular gizzard); PY, pylorus; PYC, pyloric caecae; I, intestine; A, anus. After Breder and Clark (1947).

strongly associated with anenomies (clownfish). The majority, however, are benthic algal browsers (Color Plate 15) with the usual occasional small invertebrate added to the diet. What is especially interesting about damselfish is their very strong territoriality and an apparent "farming" behavior. Many damsels vigorously defend their territories against larger organisms including scuba divers, but particularly tend to vent their ire on other grazers. They also clear or weed their "farms" of undesirable algae and other objects and have been described as killing off coral within their farms to create surface for algal growth.

Damselfish are ideal animals for model ecosystems since they occupy small territories. However, their very strong territoriality can also make developing community structure difficult in a small tank. If one adds the damsels of the community last, and does not crowd the system, usually there is not a problem.

Headstanders (Abramites, Anastomus)

The headstanders (Color Plate 16) are voracious freshwater feeders of softer submerged aquatic plants such as *Vallisneria*, *Potamogeton*, and *Egeria*. An adult fish will quickly strip the latter to a bare stalk, leaving numerous floating fragments. Tougher plants, perhaps with some secondary compounds such as *Echinodorus* and *Sagittaria,* are usually left untouched. There are apparently no detailed anatomical data for the headstanders. However, one suspects that their primary adaptation to browsing lies in their mode of operation in the water, which is headstanding. Judging by their voraciousness and restriction to soft plants, it seems likely that a long gut, perhaps with numerous pyloric caecae to assist in digestion, is present.

Properly handled, headstanders are excellent higher plant grazers in freshwater models. In a small tank one may have to add *Elodea* or *Egeria* to provide sufficient forage or to reduce predation on desired plant species.

Herbivorous Crab (Mithrax)

Many crabs are herbivorous. The tropical western Atlantic, Caribbean king crab, *Mithrax spinosissimus* (Fig. 3), is strongly adapted to herbivory,

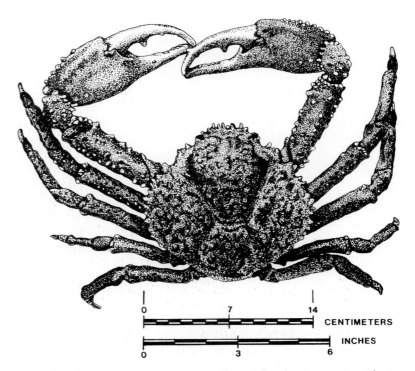

Figure 3 *Mithrax spinosissimus,* a grazing/browsing crab from tropical western Atlantic waters. This large nocturnal crab (Caribbean king crab) occurs abundantly in deeper reef waters when large grazing surfaces and small to midsize caverns are in close proxmimity.

somewhere between grazing and browsing (Adey, 1989). This species, as well as the other members of the genus, provides an excellent example of a marine plant eater among the arthropods, having many special features that fit the animals for a grazing/browsing niche (Fig. 4).

The scooped and serrated tips of the chelae or claws of *Mithrax* are particularly well adapted to the clipping off of turf and small macroalgae as well as the pulling out of larger macroalgae. Numerous mouth parts handle the algal food and present it to a set of cutting or clipping mandibles. Just inside the mouth is a large buccal cavity or forestomach that serves to store the roughage. At the back of the forestomach is a muscular gastric mill with well-developed molar-like teeth that serve to finely crush and grind the plant material, now thoroughly mixed with digestive enzymes. The forestomach acts as the equivalent of the rumen in a cow.

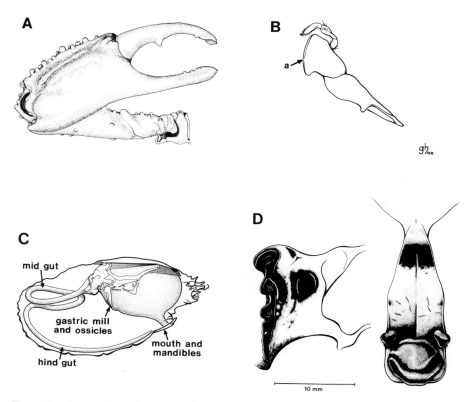

Figure 4 Characteristic adaptations of *Mithrax spinosissimus* to the herbivorous mode of life. (A) Spoon-shaped, serrated chelae for removing and clipping algae, especially algal turfs. (B) Mandibles (single of a pair) for clipping algae into short fragments. a, Cutting edge. (C) Centerline section of carapace showing digestive tract: large gastric mill, midgut (numerous ceca not shown), and hindgut. (D) Gastric mill ossicles (teeth)—center ossicle and one side shown. Drawings by Charlotte Johnson and Gustavo Hormiger. After Adey et al. (1989).

Finally, although the intestine is relatively short and more or less straight, numerous digestive caecae exist at the beginning of the tract to provide a very large digestive volume.

Mithrax species, which are numerous and generally smaller than the large *M. spinosissimus*, are excellent grazer candidates for the model ecosystem. Largely nocturnal, they tend to forage from a protective cave or crevice base. They also tend to have a harem-type structure and a minimum territory as adults. On the other hand, juveniles can be quite territorial, and one animal will become dismembered and sometimes eaten by another in a chance encounter. Adults will occasionally take animal food, disemboweling an urchin, for example, to get it. However, the majority of the food eaten is plant material.

Chitons (Polyplacophorans, Amphineura)

Chitons are the quintessential grazers of algal turfs and crustose algae. Except in shell morphology, they have probably not evolved far from the ancestral molluscs (Yonge and Thompson, 1976). These "coat-of-mail" shells (Fig. 5) with their eight calcareous plates and leathery girdle, often embedded with calcareous spines and small plates, are adapted to remaining on the grazing grounds. Their multiplate flexibility allows extraordinary adhesion to a rough substrate without losing the protection provided by their shells against predators. Like most snails, the chitons feed with a scraping radula consisting of many rows of 17 teeth tipped with the iron mineral magnetite (Fig. 5). The large esophagus has well-developed salivary glands and special starch-digesting "sugar glands." The intestine is very long, filling much of the body cavity, as an adaptation to the requirements of plant digestion. Between the two intestines lies a special sphincter that forms the feces into pellets. Since the chiton must excrete into the semiclosed space created by adhesion of the animal to its substrate, the same space with gills and mouth, this probably provides an isolating disposal mechanism.

Although the more advanced types of molluscs, such as many of the gastropods, can be voracious predators, many of the more primitive gastropods are grazers. While having an adult shell and body plan very different from the chitons, many snails, having undergone a twisting process called torsion when very young, possess the same basic grazing adaptations (Fig. 6).

Herbivores and Model Ecosystems

In a properly designed model ecosystem where direct physical parameters are operating correctly and sufficient species diversity over a broad

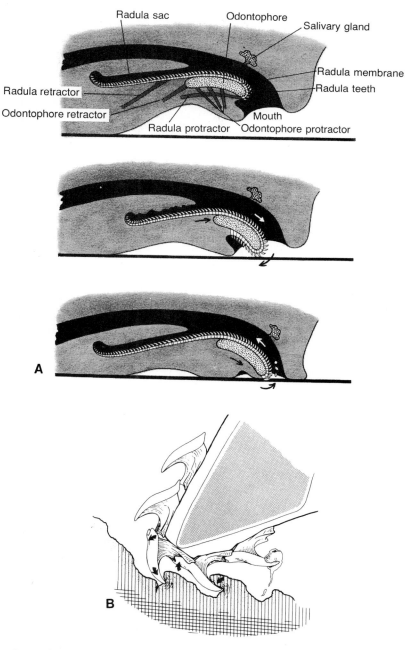

Radula sac   Odontophore   Salivary gland

Radula retractor

Odontophore retractor

Radula membrane

Radula teeth

Mouth

Radula protractor   Odontophore protractor

Figure 5 Feeding and digestive systems in algal turf and crust-eating chitons. (A, B) Diagramatic views of radula action. (C) Longitudinal section through feeding area and foregut of a chiton. (D) Bottom view of chiton. (E) Dorsal view of entire digestive tract of *Lepidochiton cinera*. All after Barnes (1980).

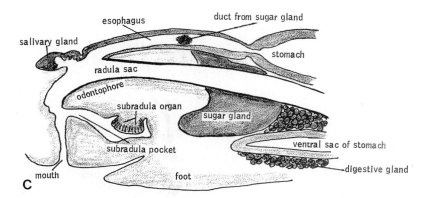

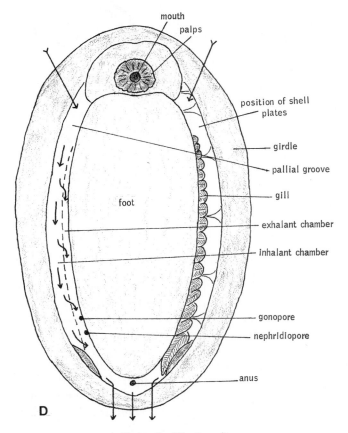

Figure 5 (*Continued*)

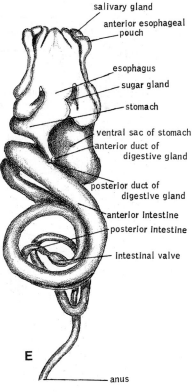

Figure 5 (*Continued*)

trophic range is present, a balance of herbivores and plants becomes the key element to long-term stability. This is particularly true as model size gets smaller and becomes critical at aquarium dimensions. Unless one is working on a specific analog model that is poor in plants and primary production, it is essential to maximize plant productivity in the model. This is the case for most microcosms and aquaria. First one determines that there is sufficient light (Chapter 7) and water motion (Chapter 4). Then the objective becomes the maximizing of plant surface and the reduction of grazing to the level where grazable plant production roughly equals that used. Since there is much room for unbalance, it is important to have a reasonable amount of plant biomass that is defended by a tough skeleton or chemicals. The productivity of the grazable plant material that remains is then what one will balance against utilization. Algal turfs that support heavy grazing can be very important in this respect. However, in

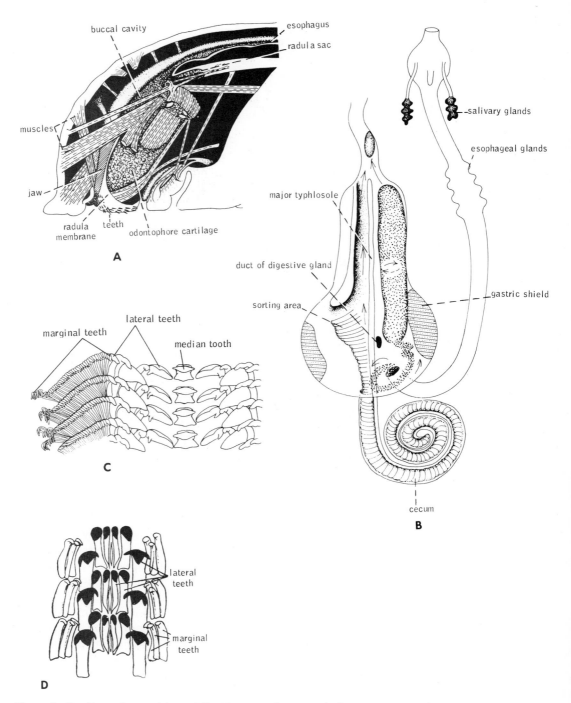

Figure 6 Feeding regions, radula, and digestive tracts of primitive, herbivorous gastropods. Note in (B) the large caecum in which much digestion takes place as well as the process of pellet formation at the end of the tract. After Barnes (1980).

small artificial systems, particularly those in which large, abundant, or voracious herbivores have been placed, it may be necessary to populate the tank largely with inedible plants and artificially feed the browsers with added plant material at the desired level. This added plant material must then be balanced by scrubbed algae (Chapter 12) or some other form of export.

Equally important in this process is the distribution of grazing types so that all types of vegetation can be utilized, assuming enough surface is present to fully support the grazer biomass. Usually this is not a particular problem in mesocosms where one is gradually adding grazers after the vegetation is established. It can become considerably more difficult in systems of smaller dimensions. One particular problem that needs careful observation is that of stocking with young grazers. Young animals generally withstand transport and reestablish better than adults. Thus, they are often used in the establishment of new systems. However, if the grazer and browser biomass increases considerably with time, a well-balanced young ecosystem can become heavily overgrazed at maturity. We have faced this problem several times with our coral reef microcosm at the Smithsonian Institution (Chapter 21).

In some moderate- to small-size model ecosystems, grazing is easily balanced against plant productivity. However, the system is not large enough to support higher predators that will keep grazer reproduction in check. This situation, often encountered with snails in both fresh and salt water, requires the operator to assume the role of the higher predator in periodically cropping the grazers.

Another situation often encountered in the refugia of both fresh- and saltwater models is where micrograzers (mostly amphipods or chironomids) totally consume the smaller filamentous algae, the most productive elements of the ecosystems. This rarely happens in the primary tank because of the presence of many fish predators of small invertebrates. The insertion of a small wrasse will usually solve this problem, though to some extent this defeats the purpose of the refugium. Ideally every ecosystem model would have a series of refugia, each treated a little differently to provide a refuge for a different component, and thereby simulating a larger area.

Although normally spatially separated from the model ecosystem itself, algal turf scrubbers, both fresh- and saltwater, can also suffer from an explosion of micrograzers (see Chapter 12). In fresh water, these grazers are usually the larvae of mayflies and Midges; in salt water the problem is usually herbivorous-amphipods. This kind of overgrazing is generally easy to solve by simple cleaning of the scrubber walls when the screens are scraped. If the problem extends to auxiliary tanks or settling traps, the same wrasse (or bluegill in fresh water) will quickly reduce the problem.

References

Adey, W. 1989. *The Biology, Ecology and Mariculture of Mithrax spinosissimus Using Cultured Algal Turfs.* Mariculture Institute, Washington, D.C.

Adey, W., and Hackney. J. 1989. Harvest production of coral reef algal turfs. In *The Biology, Ecology, and Mariculture of Mithrax spinosissimus Using Cultured Algal Turfs.* W. Adey (Ed.). Mariculture Institute, Washington, D.C.

Barnes, R. 1980. *Invertebrate Zoology.* Saunders College, Philadelphia.

Breder, C., and Clark, E. 1947. A contribution to the visceral anatomy, development and relationships of the Plectognathi. *Bull. Am. Mus. Nat. Hist.* **88:** 291–319.

Chase, A. 1987. *Playing God in Yellowstone National Park,* Harcourt, Brace, Jovanovich, San Diego.

Crawley, M. 1983. *Herbivory: The Dynamics of Animal–Plant Interactions.* University of California Press, Berkeley, California.

Gygi, R. 1975. *Sparisoma viride* (Bonnaterre), the Stoplight Parrotfish, a major sediment producer on the coral reefs of Bermuda. *Eclogue Geol. Helvet.* **68:** 327–359.

Norris, J., and Fenical, W. 1982. Chemical defense in tropical marine algae. *Smithsonian Contribs. Marine Sci.* **12:** 417–431.

Norris, J., and Fenical, W. 1985. Natural products chemistry: Uses in ecology and systematics. In *Handbook of Phycological Methods.* Vol. 4. *Ecological Field Methods.* M. Littler and D. Littler (Eds.). Cambridge University Press, Cambridge.

Owen, D. 1980. How plants may benefit from the animals that eat them. *Oikos* **35:** 230–235.

Owen, J. 1980. *Feeding Strategy.* Oxford University Press, Oxford.

Parsons, K., and de la Cruz, A. 1980. Energy flow and grazing behavior of canocephaline grasshoppers in a *Juncus roemerianus* marsh. *Ecology* **61:** 1045–1050.

Power, M., Matthews, W., and Stewart, A. 1985. Grazing minnows, piscivorous bass, and stream algae: Dynamics of a strong interaction. *Ecology* **66:** 1448–1456.

Swain, T. 1979. Tannins and lignins. In *Herbivores, Their Interaction with Secondary Plant Metabolites.* G. Rosenthal and D. Janzen (Eds.). Academic Press, New York.

Thresher, R. 1980. *Reef Fish. Behavior and Ecology on the Reef and in the Aquarium.* Palmetto Publishing, St. Petersburg, Florida.

Yonge, C., and Thompson, T. 1976. *Living Marine Molluscs.* Collins, London.

CHAPTER 17

CARNIVORES
Predators of Animals

Any organic material is a potential food or energy source for a living organism. Animal flesh in general is usually easily digestible, provides a balance of needed compounds and elements, and is a rich source of energy. It seems likely that 3–4 billion years ago, as soon as the earliest bacteria developed the ability to utilize a nonorganic energy source, a line of evolving offspring became predatory on these primary producers. Likewise, it seems clear that there are very strong selection pressures that lead virtually all evolving lines of organisms to produce some members that are predatory on other living organisms.

To most people, the word predator brings to mind a voracious, fast, and large-clawed and fanged lion or tiger on land and a sharp-toothed shark or piranha in the water. However, the range of predators is enormous and includes tiny protozoans that feed on bacteria and other protozoans as well as the sponge gently filtering plankton from the water column. There are even a few plants, such as the Venus fly trap, that are predators of animals, and many plants are predators (i.e., parasites) of other plants.

In this book, because they mean such different things when handled in model ecosystems, plankton predators, parasites, and predators that are also partly symbiotic with algae are treated in separate chapters that follow. Even so, the range of predator structure and ancestry as well as how they are handled in the model ecosystem is enormous. The first thought of the aquarist when asked about predators is that they are difficult to handle in the model ecosystem because of its relatively small size and need for a large foraging territory. Indeed, a barracuda would normally patrol several thousand square meters or more of coral reef, making inclusion in a model system without supplementary feeding virtually impossible. On the other hand, many voracious predators exist among the protozoa, and these little protists can be happily present in the smallest of ecosystem models. Also, the majority of fish, fresh and salt, that are handled in the aquarium trade are mid-level predators normally feeding on insects in fresh water and small crustacea, worms, and other invertebrates in salt water. In the "balanced aquarium" or model ecosystem, the average predator might be a little more difficult to accomodate to the small ecosystem dimensions than the average herbivore, but in practice animal size is the most crucial factor.

In this chapter, we will first briefly discuss some general characteristics of predators that will perhaps help with the concept of handling these organisms in models. We will then discuss a range of the types of predators that occur in the marine and fresh water environments. Finally we will try to impart some of our experience in dealing with predators in actual model systems.

The Carnivore Predator

Herbivores generally greatly outnumber carnivores in terms of individuals and biomass (J. Owen, 1980). On the other hand, carnivores occur in a wider variety of forms (species). This is partly because each herbivore is subject to more than one form of predation. However, "everybody is eaten by somebody," and a tremendous complexity is generated in any food web by predators feeding on other predators.

There are general patterns of operation that characterize most predation. These can be reduced to finding, catching, and eating. Predators either inherit the recognition or quickly learn to visit the right kind of environment in which their prey are likely to be. Once in the vicinity of appropriate living food, a wide variety of sensory apparatus, some highly specialized, is used to pinpoint the prey for capture. Smell or other chemical sensing, visual sharpness, and the utilization of sound or electrical signals are the typical tools of the hunter. Each species hones these down to the particular combination that works.

Once located, the prey must be caught and secured. This step provides a tremendous range of adaptations for evolutionary success that include ambush, stealth, attraction, traps, special weapons (chemical, electrical, or physical), and speed and strength. The killing and/or holding step has its own set of weapons, whether teeth, claws, venom, or pure muscular ability. In general, the last step, eating, is the least difficult for the predator. Nevertheless, even here some organisms have had to develop special methods for avoiding toxic parts or wholes and even for the more mundane separation of inedible bones, scales, and feathers.

In the competitive evolutionary race in which not only other organisms but also sometimes difficult environmental conditions are present, each predatory individual becomes "a computer with a program" that is analyzing the availability, palatability, accessibility, and profitability of its prey. Energy and time cannot be continually expended in hunting without a food return that not only allows compensation of the energy expended in the hunting itself but also allows enough excess food for reproduction, migration, and the ability to ward off other predators and parasites. The problems and choices faced by organisms can be analyzed mathematically. For example, an interesting analysis of the feeding/predation options of a bluegill are discussed by Stephens and Krebs (1986). Because of the necessity for a predator to optimize its predatory actions, it is often the sick, the old, the injured, and the young of its chosen species that are preyed on first and most successfully.

Predators can be highly specialized to feed on a single species, in some cases even to a single phase in the life cycle of a single species. While such specialization usually increases the rate of success of the individual predator, it is often an evolutionary dead end. If the environment changes and the prey is greatly reduced in numbers, extinction of the predator may be the net result. Prey populations fluctuate due primarily to weather and disease and to their source of food. To a large extent the highly specialized predator is doomed to follow with its own population fluctuations.

The Prey

Animal prey are rarely greatly diminished by a predator. Great reduction could mean disaster for predator and prey alike. Such prey usually have an enormous range of defences, including defensive behavior, defensive organs and the chemicals they produce, social support, and camouflage or mimicry. The relationship between predators and prey has been called a "gigantic biological chess game" (Whitfield, 1978).

As described by Edmunds (1974), the elements of prey defense consist of avoidance (primary defense) followed by a variety of direct active and inactive measures (secondary defense) when simply staying hidden fails.

Many animals stay out of sight within a substrate, reef rock, or a muddy bottom, for example. A whole host of worms, crustaceans, mollusks, and insects in fresh water utilize this mode. An examination of the hundreds or even thousands of individuals of these animals that can be seived from a square-meter grab of muddy bottom shows how effective this strategy is. Nevertheless, organisms as different as fish (such as grunts) and certain diving ducks simply scoop up mouthfuls of this bottom, seiving out the animals hiding therein. Also, there are predators within their midst. A characteristic feature of the fauna of muddy coastal bottoms is types of annelid worms (e.g., Nereids) that have large pincers adapted for harvesting their mud-dwelling mates.

Other organisms remain above the bottom but rely on a wide variety of camouflage. Many flatfish, for example, not only look like the bottom they are lying on, but change colors from place to place as required. Other fish and invertebrates use the military type of camouflage to break up the outline that would identify the organism to a predator. Finally, decorator crabs take the most direct and effective route of "gluing" algae to their carapace to provide a living, traveling plant cover.

The most obvious defenses are sharp spines, sometimes poisonous, such as the long-spined sea urchin (*Diadema*) or the easily recognizable and potent sting of *Millepora* (the fire coral). Finally, as more indirect defensive mechanisms, are those that mimic a dangerous organism such as the "big eye" on the butterfly fish.

Once identified, the prey usually still has the more active option of defense. While there is a myriad of forms of protection from predation, we note just a few here. Many fish and even clams survive by lightning-fast movement. Razor clams are famous in this respect. A more obvious and seemingly inpenetrable method is to build a rock-hard shell of calcium carbonate or chitin. Many mollusks and snails have taken this route. However, as with most forms of defense, in spite of the hard shell, many animals (e.g., boring snails and straddling starfish with extensible stomachs) have circumvented this means of protection. Sea gulls take advantage of their flight capabilities and lift clams and snails to a height from which they will break open when dropped on a hard surface. Indeed, the birds soon develop a sense of the minimum height needed for breakage so as to avoid excess energy expenditure.

A method common among fish and similar to many grazing land animals is schooling behavior. Large numbers of fish swimming in formation can confuse a larger predator, and while some members of the school are lost to the predators, as a whole the method is quite effective. However, man's netting counterbehavior is often devestating to schooling and a principal factor in overfishing. Associations of animals, anenomes growing on hermit crab shells, for example, provide an even more complex method of defense that can provide considerable protection.

Crabs demonstrate how many modes of defense can be utilized. These animals have several defensive tactics against fish. The chitinous shell and often numerous spines are more obvious in their protective effects. However, many crabs take a very aggressive "attack" stance and with their sharp or crushing chelae can severely damage a striking fish. In addition, crabs can autotomize appendages, leaving a leg to the predator while the animal escapes. The leg can usually be grown back after several molts.

Finally, a form of defense particularly characteristic of plants and colonial animals, but occurring virtually everywhere in the animal world, is that of toxic compounds. Sometimes the transfer of the toxins can be accomplished as a sting, the injection of some cone shells being fatal to fish, potentially even fatal to man. Other groups, gorgonians and some sponges, for example, possess toxic or distasteful chemicals throughout the animal body. Thus, the toxin is transferred only if and when the animal is eaten. Even in this case, angelfish for example typically eat toxic sponges, and gorgonions are subject to predation by specialized snails (*Cyphoma*). Thus, as mentioned earlier, predation evolutionarily often develops into an "arms race." As prey develop defenses by natural selection, the predators slowly evolve strategies to circumvent the defenses.

In the next section of this chapter we will briefly discuss a selection of predatory animals. There are thousands of modifications that allow organisms to be successful predators. We will discuss only a few. Examples will be used that are likely to be present or that could be used in a model ecosystem.

Micropredators

Protozoa, or protists as they are now generally called, are single-cell adaptations to all of the complex activities that are embodied in multicellular plants and animals. Some biologists would call them acellular since all functions are accomplished by a single unit rather than by aggregations of specialized units as in the cellular condition. The protist has numerous organelles rather than organs to accomplish all the complex needs of life including motion, food capture, digestion, and reproduction. Virtually every known type of plant and animal activity is also carried out by protists at the microscale. There are plants, grazers, decomposers, planktivores, and of course predators. What is particularly interesting in the context of this book is that these predators, some quite voracious, are tens of micrometers to at the most a few millimeters in longest dimension. Rarely would they be limited by the size of the model ecosystem and indeed they might be quite at home in a liter-sized microcosm.

Figure 1 shows two of these micropredators. Since the *Peranema* is eating a *Euglena*, which some biologists would call a plant, perhaps it should be termed a herbivore. However, *Euglena* has a flagellum and is motile. Thus, it seems more appropriate to regard this situation as

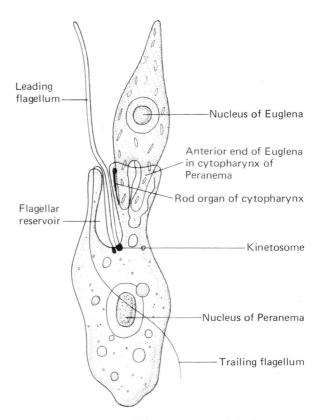

Leading flagellum

Nucleus of Euglena

Anterior end of Euglena in cytopharynx of Peranema

Rod organ of cytopharynx

Flagellar reservoir

Kinetosome

Nucleus of Peranema

Trailing flagellum

Figure 1 The protozoan *Peranema* "swallowing" a captured *Euglena,* a photosynthetic flagellate. After Barnes (1980).

carnivory. The so-called rod organ at the anterior end of the *Peranema* functions as both a proboscis and a hook. Once the *Peranema* has oriented for the attack, the rod organ snaps into the prey and is used to either "suck out" the cell contents of the prey or to pull the prey into a developing inpocketing which eventually becomes a large food vacuole.

Suctorians are rather specialized protozoa that are attached to sedentary, being motile only in reproductive or sporulation states. Figure 2 shows the suctorian mode of predation, which is similar to that seen in many coelenterates on a larger scale. The spines or tentacles have organelles (haptocysts), which apparently bear a potent toxin. When prey protozoa contact the tentacle, the haptocysts inject the poison into the prey cell, immediately causing immobilization. Then the tentacle becomes a tube or proboscis, which is capable of sucking up the cell contents of the captured organism.

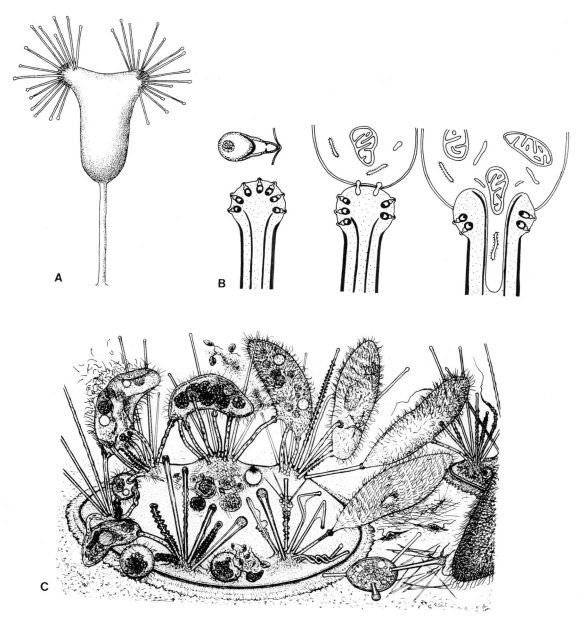

Figure 2 Suctorian weapons of immobilization and capture and mode of feeding: (A) *Acineta;* (B) suctorian tentacle showing haptocyst armament and means of sucking up the cell contents of its prey; (C) disk-like *Heliophyra* capturing and ingesting *Paramecium.* After Barnes (1980).

A Worm Predator—The Bloodworm

Glycera, the bloodworm, is a common intertidal or subtidal marine polychaete worm. Often sought out as fish bait by "worm diggers," it provides an important bait industry in some areas. *Glycera* possess four sharp jaws in an extendable proboscis (Fig. 3). The jaws in themselves are potent armament, but they also inject a poison as strong as a bee sting. The animal forms a series of burrows from which it lies in wait for the numerous worms and small crustaceans that are primarily detritivorous in its rich mud habitat. *Glycera*, along with its worm and crustacean prey, is also sought out by numerous mud-browsing fish and crabs. While it is

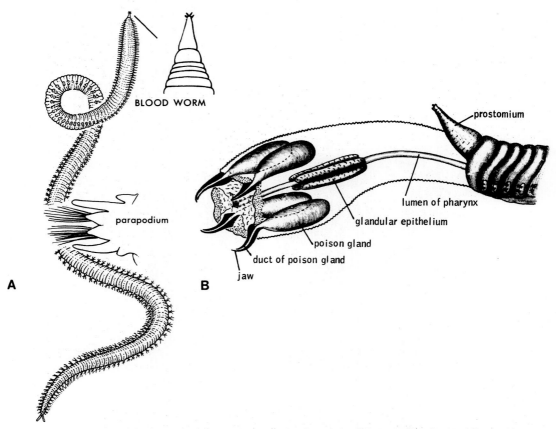

Figure 3 (A) *Glycera americana,* (B) head region with retracted proboscis; (C) extended proboscis showing four jaws with their poison glands; (D) head region; (E) showing enlarged pharynx and beginning of intestine. A after Gosner, 1978; remainder after Barnes (1980).

(Figure continues)

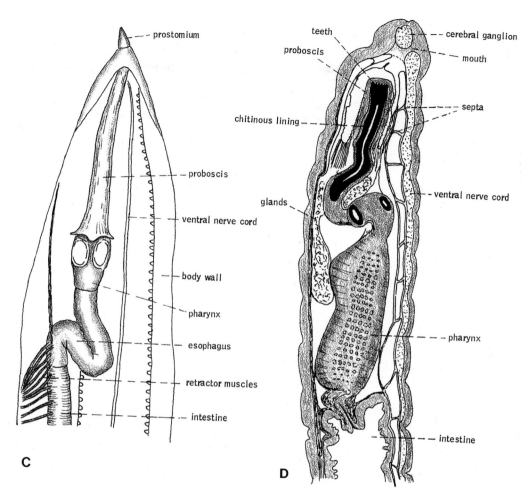

Figure 3 (*Continued*)

thus a potent mid-scale predator, it is at the same time prey for many larger predators. Since in some localities digging for bloodworms has become a major industry, obviously the hiding defenses of the bloodworm were quite adequate. On the other hand, man's ever more voracious appetite for the worms as bait and man's digging and hunting capabilities could lead to overfishing. As long as the process is not extended subtidally with more efficient mechanized apparatus driven by fossil fuels, or broad areas of estuarine and coastal habitat become polluted, bloodworm populations are likely to be maintained from the subtidal.

A Poisonous Predator

Of the two major groups of molluscs, one tends to think of the clams as being filter feeders and the snails, with their rasping radulae, as grazers of algae and other plants. While the clams are not far removed from the stereotype, with some being deposit feeders and virtually no predators (except for a few parasites), the snails or gastropods are far more varied. Indeed, many snails, such as *Buccinum* (Fig. 4A) and *Busycon*, are rather voracious predators on their relatives, the clams. In most of these cases, the snail radula is modified as an organ for rasping a hole in the clam shell. However, there is one family of snails that can be regarded as very high-level predators. Theoretically, some species of the family might be predators of man, even at times killers of men. Such snails are the often beautiful cone shells (Fig. 5) (Yonge and Thompson, 1976).

The family Conidae has about 500 species, most of which prey on polychaete worms and other snails. A few, however, in spite of their still quite characteristic "snail's pace," are predators of fish. The catching of fish is accomplished with a rapidly extensible proboscis armed with poison barbs. While the barbs are modifications of the radula teeth of the other snails, members of this family also possess poison glands and ducts as well as a sac in which to bathe the poison darts. After paralyzing and killing small fish, the snail's proboscis, remaining extended, can engulf and partially digest the fish. Thus armed, the "lowly" snail becomes in turn a predator on some of the more advanced predators of the sea.

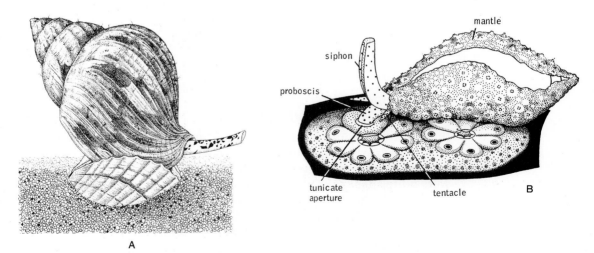

Figure 4   Methods of predation of some gastropods. (A) *Buccinum* prying open a cockle shell; (B) cowie feeding on tunicates. After Barnes (1980).

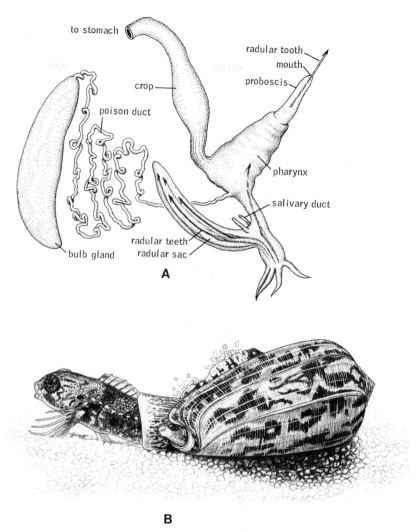

Figure 5 Feeding structures of cones: (A) pharynx; mouth and poison appratus of *Conus striatus;* (B) cone swallowing a paralyzed fish. After Barnes (1980).

Top Invertebrate Predators

Molluscs as a whole are very slow if not virtually sedentary. As we have discussed, even the highly predaceous and sometimes dangerous cone shell is adapted to catching motile prey by its evolution of poison barbs. The cephalopods, particularly the squids, however, show many of the features that characterize aquatic and marine vertebrate evolution. The cuttlefish and squids, with their streamline bodies, a combination of lateral fins and a jet-propulsion funnel, and a highly developed nervous system known for its giant axon nerve cells, are particularly adapted for rapid mid-water propulsion. Also, their eight tentacles and double-suckered arm and, most of all, their highly developed image-forming eyes render them formidable predators (Fig. 6). Indeed, the giant squids become nearly the equals of the topmost predators in the sea, the sperm whales. Once prey is captured by the squids and cuttlefish, as if their expertise at the stalk, pursuit, and capture were not enough, they have a powerful and razor-sharp beak finally capped by a paralyzing toxin to reduce energy-draining prey struggle. Recognizing the "eat and be eaten" status of even highly adapted predators, cuttlefish and squids also have a uniquely variable and complex ability to change skin colors and pattern for disruptive displays and mimicry. Finally, unique in the entire animal kingdom, along with their closer cephalopod cousins, these superinvertebrates possess camouflaging smoke screens in the form of their "ink." In the world of predators, we must truly reach the vertebrates to see even marginal improvements in the archetypic characteristics of the cephalopod predator.

It is among the reptiles and mammals that the topmost carnivores of watery environments are to be found. However, since realistically these are not the common top predators in synthetic ecosystems, at least not in microcosms and aquaria, we will finish our all too brief discussion with two fish.

A Lower-Level Fish Predator

The evolution in the more primitive chordates and early vertebrates of the muscular-ensheathed notochord and eventually vertebrae led to a bilaterally symetrical, head–tail body plan that is ideal for rapid movement through water. Finally, with a brain and spinal cord forming a well-developed central nervous system, image-forming eyes, and a wide variety of sensory apparatus including the lateral line "sonar system," the basic construction of fish is hard to improve upon for a watery predaceous life (Bond, 1979). It has obviously been elaborated upon, evolutionarily, in a tremendous variety of ways (Migdalski and Fichter, 1976).

Although a wide variety of possibilities exist, for a very brief discussion with synthetic ecosystems and aquaria in mind, we choose the family Poecilidae (Color Plate 17). These fish include the mollies, swordtails, guppies, and mosquito fish. Poecilids have relatively recently been evolved in fresh waters of the tropical to temperate Americas, and because

Figure 6 Primary hunting characteristics of squids and cuttlefish. (A) The cuttlefish *Sepia* capturing a shrimp. (B) Basic body plan of the squid *Loligo*. (C) Diagrammatic view of the "jet propulsion" and ventilating system of a cuttlefish. After Barnes (1980).

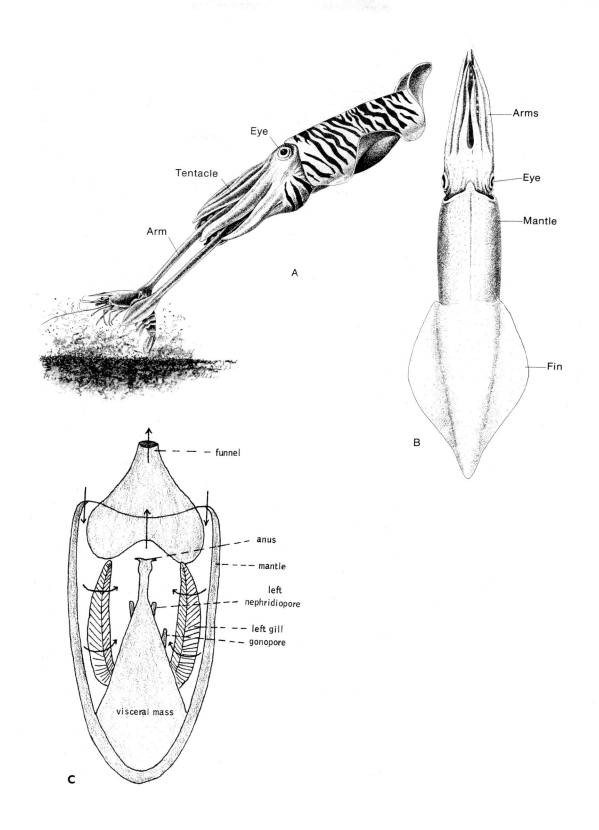

Eye

Tentacle

Arm

A

Arms

Eye

Mantle

Fin

B

funnel

anus

mantle

left
nephridiopore

left gill

gonopore

visceral mass

C

of their use in the aquarium trade and their abilities as eaters of mosquitoes, some have become virtually worldwide in distribution. These small fish are capable of withstanding a wide range of environmental conditions, even high salinities, making them highly adaptable to wet–dry season regimes in which water bodies can dry down to virtual puddles. Many can also move into estuaries and take advantage of the hunting territory offered by enormous marsh–mangrove areas that may be flooded only on spring tides. The poecilids are live-bearing, thus circumventing the free-egg mass and planktonic larval stages, which would likely be less successful in semiephemeral, small water bodies.

A glance at Color Plate 17 shows the primary trophic adaptation of these fish. Their mouths have evolved to make them particularly effective at surface feeding. While their highly generalized characteristics also suggest animals capable of feeding on almost anything organic including algae, they are especially adapted to feeding surface-floating or alighting insects. An enormous food source is potentially available in the form of filter-feeding mosquito larvae. Indeed, it is to these fishes and their cousins the killifishes that we owe a major part of mosquito control. In some areas of the world marshes were drained to reduce mosquitoes. Since that method was not particularly successful, a more modern approach is to ditch not for draining but to allow fish such as poecilids and killifish access. It is usually only in stagnant waters inaccessible to these fish that mosquitoes do exceptionally well. As we will discuss later, we have had great difficulty in achieving even a low level of mosquito production in our Everglades mesocosm because it was designed such that the mollies and killifish have access to every part of the system, at least at high spring tides and seasonal high waters.

A Top Predator

Finally we come to the archetype top-level aquatic predators. This is, without doubt, the most difficult problem for synthetic systems, and we have chosen to illustrate the barracuda (Color Plate 19). For a variety of reasons these streamlined fish work to a degree in a relatively small system. The barracuda in the tropical coral reef environment is generally similar in design to the pickerel or muskellunge in fresh waters. Adapted for quiet cruising or lying in wait for most of the time, its elongate highly muscular body is capable of lightning strikes at other fish. Most important, its dentation is unmistakable. Long, razor-sharp teeth slightly inclined inward prevent the escape of larger fish once they are caught. Smaller or elongate fish are often cut in two on the first pass, the pieces then retrieved at leisure. Like many top predators, marine or terrestrial, these fish particularly hone in on weak or struggling prey. Since it would be an unusually large mesocosm that could truly support such fish beyond the juvenile stage, this characteristic can be used for artificial support, as we shall discuss later.

Predators and Synthetic Ecosystems

We have placed all animals from the upper levels of food webs under the general heading of "carnivores." However, it is likely that the higher they rank the more likely it is that they will pose an increasingly broad range of difficulties for managers of aquaria or microcosm ecosystems. Also, while we lump all free-living, nonplanktonic or detritivorous animal-eating vertebrates and invertebrates in this category, it is necessary to remember that size is at least as important as mode of nutrition. A carnivorous amoeba, worm, or copepod, or even a tetra fish, is unlikely to create serious problems in a model ecosystem, but a large grazing fish, as we mentioned in the last chapter, requires considerable foraging territory. In many ways it is more important to prevent overgrazing than overpredation. Excess carnivory is not only more apparent and more easily guarded against, at least it does not directly damage the energy supply of the system. As we have stressed repeatedly, these are primarily scaling problems due basically to limited microcosm size.

Top carnivores are few and they usually forage over a relatively large territory. It is possible to reduce this territorial difficulty somewhat by using younger members of a species whose feeding habits are reduced in size from those of the adults. On the other hand, young animals may require a larger total mass of food. Also, a constant pattern of removal and reintroduction must be employed. Another approach is to introduce special feeding stocks to stimulate a larger foraging territory for the particular species in question. Selected carnivores of different sizes and positions in the food webs of our coral reef and Maine Coast microcosms are shown in Figures 7 and 8. The food webs were shown in Chapter 14, Figures 7 and 8.

Among the variety of mid-level carnivores in our reef tanks are fish, such as grunts, wrasses (Color Plate 20), butterflies, squirrels, and angels (Color Plate 21), and, among the invertebrates, various worms, lobsters, and predatory snails (Color Plate 22) Although some of these organisms feed on the reproductive stages of a variety of small animals, most feed on small- to moderate-sized grazers, filter feeders, or detritivores. As long as the tank is not overloaded with low-level carnivores and sufficient primary production, particularly algal growth, is present, little difficulty can be expected. However, two special cases are created by the typically nocturnal migratory habits of grunts and lobsters; both frequently migrate into associated reef lagoons to feed and a fair amount of lagoon or equivalent territory is needed for foraging. In some reef tanks we have added a

Figure 7 Selected carnivores in the food web of the Smithsonian coral reef microcosm (see Chapter 21). Drawing by Alice Jane Lippson.

Figure 8   Selected carnivores in the food web of the Smithsonian Maine coast microcosm (see Chapter 22). Drawing by Alice Jane Lippson.

moderately large lagoon area. Another approach, that of providing artificial nighttime feeding (see Chapter 21) can overload the system and thus requires sufficient scrubbers or other equivalent methods of export.

It is possible to estimate animal biomass at different trophic levels in natural reefs and to use this data to adjust the number of lower carnivores in a reef microcosm. In practice, however, both the natural reef estimate and the microcosm measurement are difficult to make. Generally it is possible to determine whether or not sufficient food is present at this level simply by looking at the external physical appearance of fish abdomens. Slack guts suggest the need for increased artificial feeding or enlarging the foraging area. Exceptionally full guts bright colors, and normal behavior call for reducing or stopping artificial feeding.

In some cases involving daytime feeding of lower predators such as butterflies and angels, food availability on a reef is primarily determined by filter feeders such as sponges, worms, and corals. Rarely have we experienced problems with these groups, perhaps because reproduction in a large number of invertebrates and algae is continuously supplying small planktonic stages. In some cases we have artificially maintained plankton levels in our reef microcosms. This will be discussed further in Chapter 18.

Higher-level carnivores in our reef systems have included barracuda and snappers. These high-speed pouncers rely entirely on smaller fish and, except for the smallest sizes, require a larger foraging area than is likely to be available in the small to moderate-size microcosm. Left to their own devices, these fish will soon deplete a tank of all but the largest fish. A continuous supply of small reef-fish added to a reef microcosm can make up for losses to these higher-level carnivores. This is likely to be overly expensive and inconvenient. We have found that goldfish, which are cheap or easily raised, not only provide adequate feeding for barracuda but also get eaten first because of their immediate distress upon entering sea water. Thus, although reef fish occasionally are lost to a barracuda, as in the natural environment, the problem is not serious. Snappers, on the other hand, tend to be voracious and to tackle any other residents from their own size downward. They make poor higher predators for a relatively small system.

When supplementary feeding is used for higher carnivores, a problem, in terms of system maintainance, is the net import of feed and its contained nutrients. However, if an equivalent export is achieved, for example as with algal turf scrubbers, the overall result in water quality is not too different from that in a wild reef. On most wild, shallow-water reefs, flow from the open ocean constantly supplies juvenile fish ready to drop from the plankton. Most of these juveniles are eaten immediately. Also, scrubber action in the model is a simulation of the inefficient grazing of algal turfs in the wild, which results in a constant loss of algal fragments from the reef. These kinds of exchanges at moderate levels probably make little difference, either in the appearance of a model ecosystem or in its biological function. In the end, while the principles can be "stretched" using algal scrubbers if ecosystem simulation is desired, community scaling is the primary problem. Either microcosm size must be relatively great, or mid-level carnivore number and higher-carnivore size must be kept to a minimum, in keeping with the natural situation in the wild.

The cold-water, rocky benthic ecosystem of the Maine Coast is considerably less complex than a reef ecosystem. It is younger, and therefore has had less time to develop diversity and trophic complexity, and, in the case of the Atlantic subarctic, is rather depauperate due to its pattern of biogeographic development with time.

We have kept a number of medium-size carnivores in our cold-water systems, including *Cancer irroratus*, the rock crab, *Carcinus maenus*, the green crab (Color Plate 23), *Asterias vulgaris*, the common sea star (Color Plate 24), and *Buccinum undatum*, the waved welk. *Asterias* is one of the major mid-level animal predators both on the Maine coast and in our tank. Its primary prey is the edible mussel *Mytilis edulis*, although a wide variety of bivalves, gastropods, and barnacles are also potential food sources. As we discuss below, the success of *Mytilis* as a filter feeder in a closed microcosm is determined by how much plankton and organic particulates is available. Because this calls for a relatively large volume of water, it may be necessary to aid artificial plankton, particularly early in tank development. Large quantities of *Mytilis* are readily available from the wild or commercially. Thus overpredation by starfish in a microcosm system is seldom a serious problem if artificial feeding is acceptable. In our 5-year-old Maine coast system *Asterias* is able to feed on most exposed *Mytilis*, but it cannot negotiate beneath rocks and in crevices that harbor a large population of mostly smaller animals. Again this points up the important role adequate cover plays in predator/prey relationships even in aquaria and microcosms.

Along the Maine subtidal and in our cold-water microcosm, crabs (particularly rock crabs on rocky shores and green crabs on the mud flats) are the primary predators of herbivores and detritivores. These animals prey especially heavily on the urchin *Strongylocentrotus*, as well as on a wide variety of molluscs and small invertebrates—particularly worms. Because rock crabs normally forage over many square meters of bottom, scaling problems in terms of prey reduction are introduced into the small cold-water microcosm just as into the reef aquarium. The solution to the problem is basically the same: limit crabs to moderate size, and/or boost grazer populations by adding grazers either from a natural community or from supplementary "nursery" tanks. The lobsters *Homarus americanus* in the Maine microcosm and *Panulirus argus* in the reef microcosm are generalized scavengers and mid- to upper-level predators and do particularly well in our systems. However, typically within a year, if added as juveniles, they grow too large for the ecosystems, causing excessive physical damage due to lack of sufficient foraging territory. Since these are exhibits and the public enjoys seeing the lobsters, about once a year we remove the larger animals replacing them with juveniles of the same species.

In our cold-water systems we have generally included the predatory snail *Thais lapillus,* a higher-level carnivore that preys extensively on barnacles. Once again, we have a case of an organism that contributes to the primary difficulty in microcosm maintenance, which is simulation of a large natural food supply area or volume in a relatively small tank. In this case, for purposes of exhibition or maintaining species for biological work, management problems have been few, and *Thais* populations are adjusted by the highest predator, man, at desirable levels. In this case, the barnacles have great difficulty in maintaining their populations, although some manage because of sufficient cover. The actions of the managers as thinking, top predators are crucial.

Larger carnivores in the Maine coast tank also include several medium-size pollock, cunners, and sculpins. In turn, some of the food for these animals is derived within the system from photosynthesis carried up the food chain through worms and small crabs. However, because of the area involved, some of the food provided for them is added dried krill. This import is of course balanced through export in the algal scrubbers.

Higher predators most associated with fresh-water ecosystems include pike and bass in cold water and piranha in tropical waters. All of these fish are in the same category as the reef and rocky-shore higher predators described above. They all can be kept in well-maintained community tanks operated as ecosystems, but unless the tanks are quite large it is likely that feeding will have to be supplemented and that the tanks will require careful balancing with scrubbers.

On the other hand, mid-level carnivores that feed on insects, worms, and small crustacea in the wild are highly diverse in fresh waters and form the backbone of the aquarium trade. They include many of the tetras, barbs, toothcarps, come cichlids and angelfish (Color Plate 25). At a somewhat higher and a little more difficult level are many of the cichlids and labyrinth fishes. It is interesting to note that while a wide variety of invertebrates are mid-level carnivores in both cold and tropical salt waters, the vast majority of mid- and upper-level carnivores in fresh water are fish, reptiles, birds, and mammals.

While some of the more specialized of these fish can certainly cause problems because of their food requirements, the same basic principles generally apply: the larger the predator and the further its food source from the plant/producer level, the larger the tank should be and the more difficult its management is likely to become. Foraging territory can be simulated, but when live food must constantly be added, ecological balance becomes more difficult to achieve. Through algal scrubbing or truly equivalent technologies, we now have the means to achieve this balance, so that we can concentrate on the real problems, community structure and behavior.

Because of the small size of aquaria, carnivores are generally more difficult to maintain in aquarium-sized models than in microcosms and mesocosms. On the other hand, perhaps the most successful continuous long-term management of a higher predator in any of our systems has been that of *Enoplometopus occidentalis*, the "rare reef lobster," in a 130-gallon coral reef (see Chapter 21, especially Figs. 12 and 13. This animal grew to maturity in the aquarium, and, at this writing, has served for over 3 years, in a very stable manner, as the top predator of the system. Old fish, as they become slower and less active, sooner or later fall prey to this very aggressive lobster, though always at night. Likewise, newly added fish, off balance in a new system and frequently harrassed by other territorial fish, also fall prey to the lobster. If a new fish survives the first night or two, almost invariably it becomes a long-term addition to the system.

References

Barnes, R. 1980. *Invertebrate Zoology*. Saunders College, Philadelphia.

Bond, C. E. 1979. *Biology of Fishes*. Saunders College Publishing, Philadelphia.

Edmunds, M. 1974. *Defense in Animals*. Longman, New York.

Gosner, K. 1978. A Field Guide to the Atlantic Seashore. Peterson Field Guide Series. Houghton Mifflin, Boston.

Grant, E. 1982. *Guide to Fishes*. Queensland Government, Brisbane, Australia.

McClane, A. 1974. *Field Guide to the Saltwater Fishes of North America*. Holt, Rinehart and Winston, New York.

Migdalski, E., and Fichter, G. 1976. *The Fresh and Salt Water Fishes of the World*. Bay Books, London.

Owen, J. 1980. *Feeding Strategy*. Oxford University Press, Oxford.

Stephens, D., and Krebs, J. 1986. *Foraging Theory*. Princeton University Press, Princeton, New Jersey.

Whitfield, P. 1978. *The Hunters*. Simon and Schuster, New York.

Yonge, C., and Thompson, T. 1976. *Living Marine Molluscs*. Collins, London.

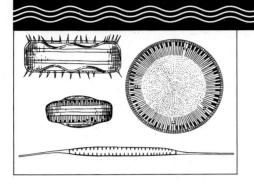

CHAPTER 18

PLANKTON AND PLANKTIVORES
The Floating Plants and Animals and Their Predators

Plankton are floating or weakly swimming plants and animals that are more-or-less passively carried by currents in the natural environment. There is no sharp line dividing the plankton from the nekton, the mid-water swimmers. However, jellyfish and flagellated or ciliated organisms that have only minimum control over their position in the water are usually regarded as plankton. Plankters occur in virtually all natural bodies of water. In the open ocean and in large lakes, collectively the plankton totally dominates in plant primary productivity and biomass. However, even in the smallest bodies of water including those that are man-made and ephemeral, some plankters are normally present.

It is perhaps useful to point out that the larvae of most mosquito species feed primarily on floating plants and animals, even when the quantity of water in their "pool" is extremely small (Gillett, 1971). It is also critical to note that in the open ocean and within large lakes, the phytoplankton replace trees, herbaceous vegetation, and grasses on land in providing for virtually all photosynthesis. Offshore, minute plants become the base of the entire food chain. This environment usually does not appear green, or if it does, it is normally only weakly so because turnover or grazing is high. However, productivity can be moderate even with a low standing crop.

Many types of plankton, particularly phytoplankton, have been cultured, and some microcosms and mesocosms have been dedicated primarily to the open water column and its floating organisms (see especially Nixon *et al.*, 1984; Oviatt *et al.*, 1981). On the other hand, in benthic mesocosms and in aquaria, the primary subject of this book, plankton have largely been ignored. In traditional aquarium technology, physical and/or bacterial filtration is employed. This would effectively trap all of the mid to large sizes of floating plants and animals, including planktonic larvae. Depending upon the filtration system employed, only the smallest plankters escape in this very artificial world.

In the earlier synthetic ecosystems described in this book, plankton have been poorly treated or not examined at all. In most cases, it was clear that their composition was depauperate and limited in biomass. It seems likely that a failure to design for the presence of a viable and diverse population of plankton is the primary limitation of the synthetic ecosystems described. This chapter on the plankton is included primarily to indicate the steps that we have been taking to overcome their limitation in the more recent model systems. Most important, this chapter is meant to encourage careful consideration of all the factors that can lead to establishment of a more normal floating component of organisms in all mesocosms, microcosms, and aquaria.

Plankton Size and Composition

Whether it is the scientist with his or her nets or filters trying to retrieve a sample of a planktonic plant or its predator, or a floating predator attempting to feed, the size of a plankter is crucial. The floating organism can be caught only if the apparatus is the right size. While terminology has varied widely, the terms and size ranges shown in Table 1 are typical of those now in use.

The plankton, whether freshwater or marine, is made up of two general types of organisms. The holoplankton is those plants and animals that normally spend most or all of their life cycle solely within the water column. Meroplankton, on the other hand, consists of the reproductive stages of plants, invertebrates, and fish that are normally either bottom dwelling or included in the fast-moving or neritic community of midwater. The holoplankton or true plankton also consists of abundant bacteria and protozoa, which are often attached to organic particulates, as well as free-living phytoplankton and zooplankton.

TABLE 1	
Size Classification of Plankton[a]	
Size class	Mean diameter
Megaplankton	> 5 mm
Macroplankton	2–5 mm
Mesoplankton	0.2–2.0 mm
Microplankton	20–200 μm
Nanoplankton	2–20 μm
Picoplankton	0.2–2 μm

[a]Note that most bacteria fit in the picoplankton range. However, a large percentage of the suspended bacterial flora is attached to larger organic particles. After Cushing and Walsh (1976) and Stockner (1988).

The Bacteria

Virtually all organic wastes, plant and animal, occurring suspended in fresh or salt waters or delivered to those waters from the terrestrial environment are subject to bacterial breakdown (see Chapters 10–12). It seems likely that the same species of bacteria or at least strains of the same or similar species can carry out the same processes in both fresh and marine waters. However, many marine bacteria when placed in fresher waters appear to be limited by a sodium requirement (Hobbie, 1988). In the open ocean, refractory or difficult-to-break-down organic particulates are few, whereas in fresh or coastal waters, compounds such as lignins and tannins derived from higher plants are more abundant and last for longer periods in the water column. Nevertheless, in all waters, bacteria are present on suspended organic particulates or are truly free-living, using dissolved organics as food source. As we discuss below, a wide variety of organisms from marine and fresh waters feed on these bacteria. This is the end of the detrital food chain in mid-water and is also the base of a secondary food web. In a sense, the bacteria plankton keep bouncing the organic energy supply back up the food web until it is finally fully degraded (Fig. 1).

Because of their lack of a membrane-bound nucleus and other similarly delimited cell organelles, cyanobacteria (blue-green algae) are now generally accepted, evolutionarily, as bacteria. However, functionally and to a large extent morphologically the blue-greens act more like algae, and we will treat them that way in this chapter.

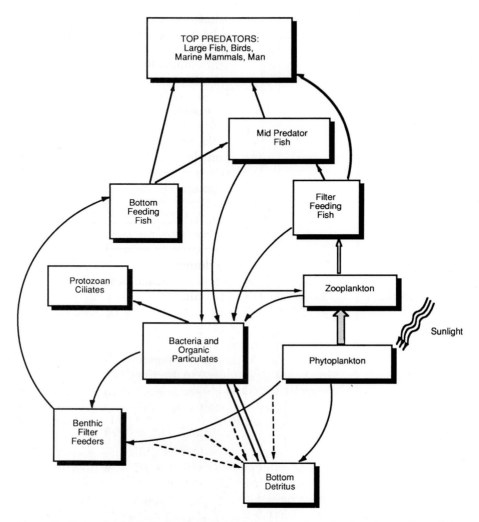

Figure 1 Generalized planktonic food web in shallow waters showing the position of the bacterioplankton.

Phytoplankton

In Chapters 7, 9, 10, and 15, we discussed marine and aquatic plants, the process of photosynthesis, and the role plants play in ecosystems and mesocosms. Here, we wish to emphasize the strictly planktonic plants and their positions in food webs, especially those that should be included in model system design.

The floating plants of open waters, fresh, brackish, and salt, are primarily those of a dozen or so alga phyla or divisions, depending upon one's classification scheme. For an introductory treatment of these groups, the reader is referred to Bold and Wynne (1985); Reynolds (1984) provides an ecological treatment. Some algal groups are almost entirely planktonic, although many have both planktonic and benthic members.

As a whole, in both fresh and salt waters, the diatoms (class Bacillariophyceae in the division Chrysophyta, the golden brown algae) form the dominant plant components of the plankton (Figs. 2 and 3). Diatoms have cell walls of silica and are often highly ornamented. Occurring in two main groups, centric (radially organized) and pennate (bilaterally organized), diatom shells or frustules consist of two interlocking halves or valves. While many genera are unicellular, others are filamentous and a few are complexly branched. Some of the pennate diatoms have an elongate groove or rhaphe, which is used in ways not fully understood to provide a certain gliding-type mobility to the cell. Diatoms can sexually reproduce by forming motile or nonmotile gametes. While some are fully planktonic, many are dominantly benthic and enter the plankton only in reproduction or when disturbance of the bottom temporarily introduces them to the water column.

Any well-lighted and naturally operated model ecosystem that has

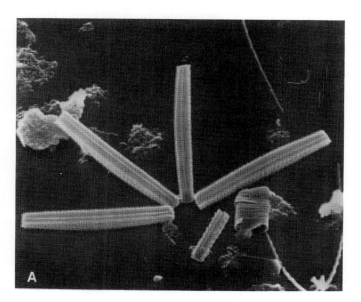

Figure 2   Marine diatoms: (A) pennate diatom *Thalassionema nitzschiodes,* ×500; (B) diagram of frustule construction in a pennate diatom (r, raphe; pn, cn, polar and central nodules); (C) valve of centric diatom, *Actinoptychul senarius* (×1300); (D) colonial centric diatom *Bacteriastrum furcatum* (×700). After Dawes (1981). Reprinted by permission of John Wiley & Sons, Inc.

(Figure continues)

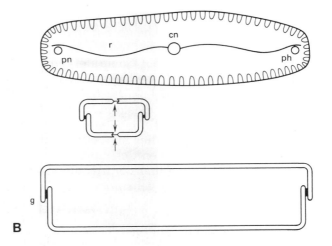

Figure 2   (*Continued*)

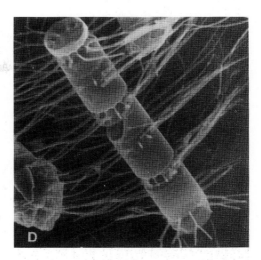

Figure 2 (*Continued*)

had nonsterilized introductions from the wild will have a diatom community. However, if filtered, skimmed, or richly bubbled, the mid-water diatom population is likely to be minimal or virtually absent. In general, diatoms prefer cooler and less nutrient-rich waters. Higher temperatures, higher or extremely low nutrients, or other extremes will usually lead to replacement of the diatoms by dinoflagellates or blue-greens.

Dinoflagellates of the Division Pyrrophyta are abundant in the oceans, particularly in the tropics. A number of genera and species also occur in fresh and brackish waters, although they tend to be less important than in the oceans. Dinoflagellates are relatively primitive algae with a tendency toward a single distinctive structure. Although some benthic unicells and filaments occur in this group and most algae symbiotic in animals (see Chapter 20) are derived from this group, most genera are planktonic. Whether marine or freshwater, dinoflagellates, as the name suggests, are characterized by a plant-like and often sculptured cellulosic cell wall or "armor" of two hemispheres (Figs. 4 and 5). Lying in the groove or girdle between the hemispheres are two flagellae, one extending around the groove and the other projecting downward along another groove, the sulcus, occuring in the lower hemisphere. This provides the great motility characteristic of the group. Many coastal and freshwater genera also have resistant resting spores that serve to carry them through unfavorable environmental conditions, usually resting on the bottom.

Dinoflagellates are characteristic of warmer, often more nutrient-rich conditions. Some genera excrete toxic compounds. Sometimes, under warmer conditions, perhaps additionally supplemented by organics derived from land runoff, these species form "red tides" that can kill fish and/or invertebrates. Shellfish poisoning and ciguatera (tropical fish poisoning) are derived directly by ingesting the dinoflagellates. Food-chain concentrating of the toxins can also render them fatal even to man. Nevertheless, these important phytoplankters are generally desirable members of model ecosystems. In our experience, only a single case of mecososm/microcosm operation has exhibited red tide problems. This occurred in the starting phase of a coral reef system when several fish were

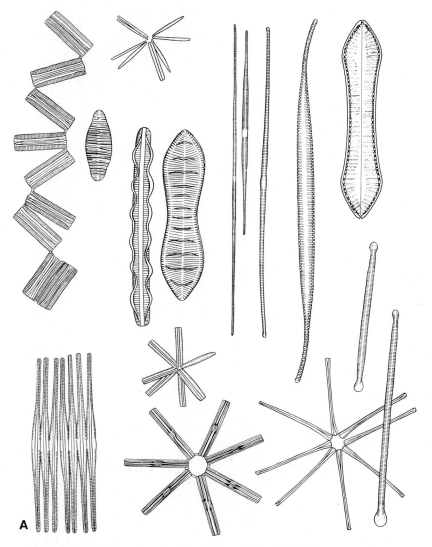

Figure 3 Fresh water diatoms: (A) pennate types, ×137 to ×930; (B) centric types, ×375 to ×750. After Hutchinson (1967). Reprinted by permission of John Wiley & Sons, Inc.

(*Figure continues*)

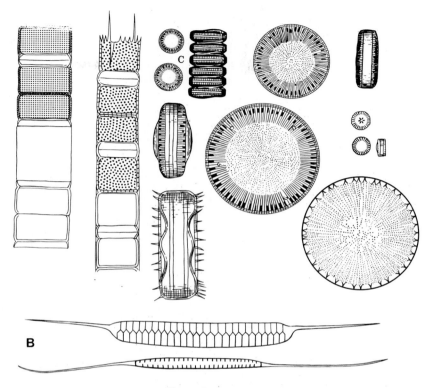

Figure 3 (*Continued*)

lost and a number exhibited neural distress, tending to lose control over swimming and orientation. The large dinoflagellate *Gambierdiscus*, known as a source of ciguatera, was shown to be responsible for the situation. However, gradually over several months, the symptoms disappeared from the mesocosm although the presence of *Gambierdiscus* could still be demonstrated for several years thereafter.

Blue-green algae, generally in more-or-less extreme conditions (Fig. 6), can be important elements of both freshwater and marine phytoplankton communities. Usually thought of as being indicators of rather advanced eutrophic or nutrient-rich conditions in lakes and streams, blue-greens can also be dominant in the extremely low-nutrient conditions of tropical, oceanic waters. Although rarely mentioned in the literature, blue-green-dominated communities can also characterize the waters of extremely pure mountain lakes (Hutchinson, 1967). Certainly the ability of many blue-greens to fix nitrogen from the gaseous state, readily available dissolved in all waters, is partly responsible for their success in situations poor in dissolved nitrogen. At the other extreme, these "algae" appear to be demonstrating their bacterial nature and general ability to withstand extreme environments competitively. For all of the above

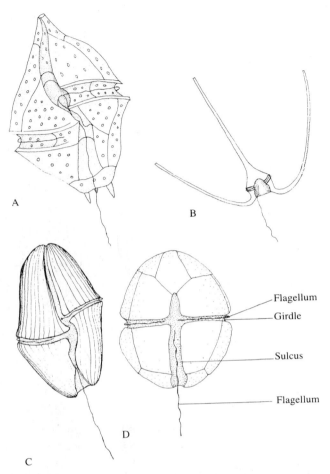

Figure 4 Marine dinoflagellates: (A) *Gonyaulax,* (B) *Ceratium,* (C) *Gymnodinium,* (D) *Peridinium.*

reasons, blue-greens tend to fare well in the open waters of mesocosms if they are not heavily filtered. For the traditional aquarium that is excessively rich in nutrients and with relatively low light levels, this makes sense. For ecologically more "normal" conditions, blue-green dominance does not necessarily carry negative conotations, although a greater diversity of algal groups generally suggests a more even and biologically richer state.

For the remaining algal groups, marine and freshwater plankton communities begin to greatly diverge in their composition. Another major group of the Chrysophytes (golden-brown algae) that often dominates in tropical and temperate oceanic situations is the coccolithophores. These phytoplankters, geologically well known because of their armor-plated, calcareous discs (coccoliths) attached to the outer cell walls (Fig. 7), are

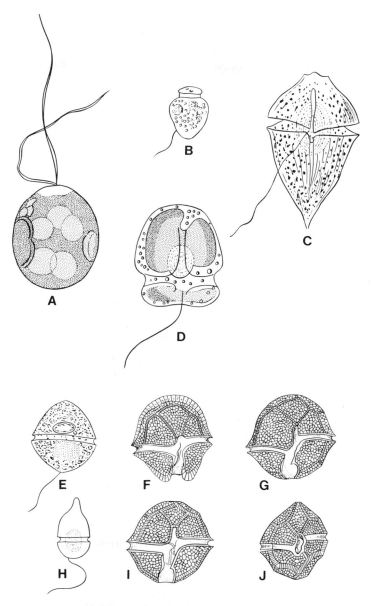

Figure 5 A variety of freshwater dinoflagellates ranging from primitive (A) to more advanced (E–J). Most forms shown are fully photosynthetic; (C) and (F) feed on other small animals or phytoplankton. ×375 to ×1500. After Hutchinson (1967). Reprinted by permission of John Wiley & Sons, Inc.

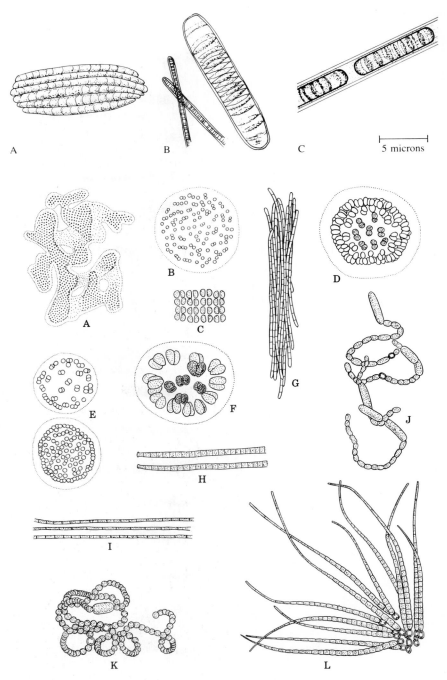

Figure 6 Common planktonic blue-green algae (Cyanobacteria) from saltwater (*top*) and freshwater (*bottom*) environments. (*Top*) (A) *Trichodesmium,* (B) *Oscillatoria,* and (C) *Lyngbya.* (*Bottom*) (A,B) *Anacystis,* (C—F) *Agmenellum* and *Gomphosphaeria* spp., (G) *Aphanizomenon,* (H) *Oscillatoria,* (I) *Lyngbya,* (J,K) *Anabaena,* (L) *Gloeotrichia.* (*Top*) After Thurman and Webber (1984); Copyright © 1984 by Scott, Foresman and Company. Reprinted by permission of HarperCollins Publishers. (*Bottom*) after Hutchinson (1967). Reprinted by permission of John Wiley & Sons, Inc.

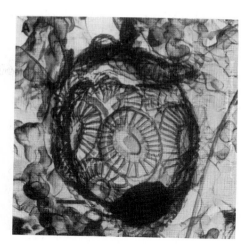

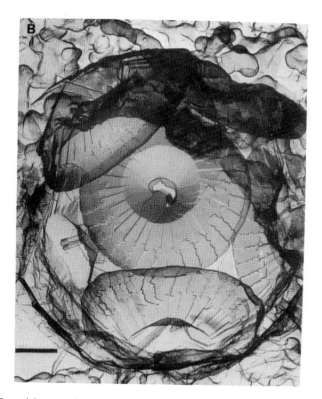

Figure 7 Coccolithes emplaced on the surface of coccolithophores: (A) *Emilisnis huxleyi;* (B) *Umbilicosphaera sibogae,* both bars 1 μm; (C) *Calcidiscus,* ×2800. After Bold and Wynne (1978).

(*Figure continues*)

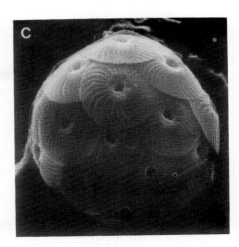

Figure 7   (*Continued*)

motile, typically with a pair of flagella. While abundant enough in the ocean to create geological deposits under certain conditions, only a few coccolithophore genera occur in fresh water, and these are relatively rare. The freshwater equivalents of these coccolith-bearing unicells are a variety of flagellated and nonmotile freshwater green algae (Chlorophyta) (Figs. 8–10). The desmids, many species of which can be benthic, are particularly interesting members of the freshwater plankton. Generally characterizing more moderately acid waters as compared to the high-pH waters of the coccolithophores, the desmids almost invariably have strongly structured cellulosic walls consisting of two symmetrical semicells (Fig. 10).

Most of the photosynthesis occurring in the photic or well-lighted zone of open waters is accomplished by the algal groups described above. These algae also form the base of the mid-water food webs and also support the underlying filter-feeding benthic fauna. For many decades it has been thought that most phytoplankton production was grazed and passed up through a relatively simple but long food chain. Now it is recognized that, similar to many shore and terrestrial communities, much of the production of offshore phytoplankton is not directly eaten by grazers but rather dies and is degraded by bacteria. Some of this detritus "loop" is then passed up a food web by filter feeders and their predators.

The Planktonic Food Web

Filter or suspension feeding is not as simple as copepods eating phytoplankton and then being fed upon by larger invertebrates or small fish.

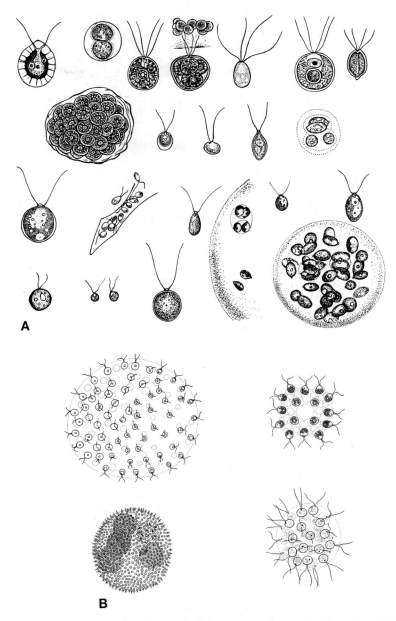

Figure 8 (*A*) Freshwater planktonic unicellular green alage of the order Volvocales. *Chlamydomonas,* ×500. (*B*) Colonial motile members of the same order. *Volvox,* ×177. After Hutchinson (1967). Reprinted by permission of John Wiley & Sons, Inc.

Figure 9   Freshwater planktonic green algae: (*Top*) colonial chlorococcales, (A) *Coelastrum*, (B—F) *Pediastrum*, ×167. (*Bottom*) Miscellaneous genera, (A) *Dictyosphaerium*, (B—H) *Oocystis*, (I) *Spherocystis*, (J) *Gloeocystis*, (K) *Kirchnoriella*, (L) *Tetraedron*, (M) *Elaktothrix*, (N) *Scenedesmus*, ×500. After Hutchinson (1967). Reprinted by permission of John Wiley & Sons, Inc.

(Figure continues)

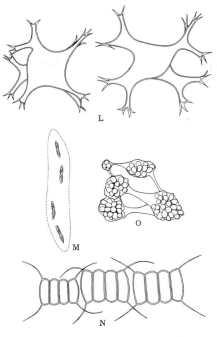

Figure 9 (*Continued*)

Especially if one includes free bacteria and organic particulates with their bacterial flora, a total spectrum of size from fractions of a micrometer to millimeters and of greatly varying food quality typically exists in a plankton community. Often omitted from the more generalized open-water food web are the protozoans that provide a link to the smallest of plankton or particulates. Also, one generally thinks of the planktonic food web, especially in its lower segments, as being one of filtering or filter feeding. However, in addition to the abundant and highly varied filtering processes, at all levels raptorial or "grasping" feeding is also important. Since technically we covered raptorial feeding in the last chapter, for ease of discussion we will concentrate on a more narrowly defined process of filter feeding in this chapter.

Mechanisms of Filter Feeding

Generally, three basic processes of food capture are utilized by all filter feeders. These are netting or more straightforward filtering devices, entangling or flowing mucous threads or trails, and cilia trains. None of these are mutually exclusive. In many cases, various combinations or even all three are used together. The number of types of filtering mechanisms are almost as numerous as the number of filtering genera. Also, we would

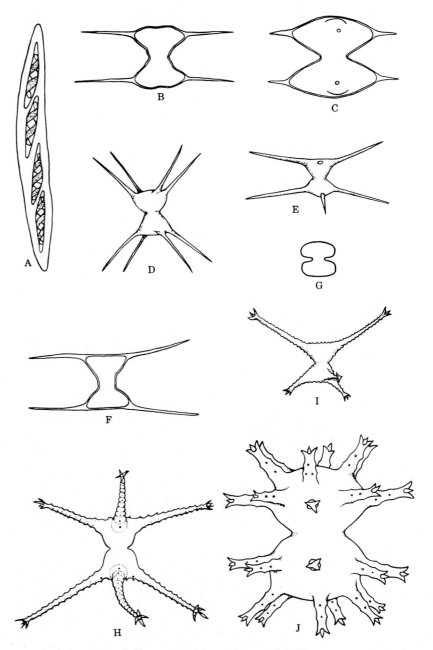

Figure 10   Planktonic green algae belonging to the desmids: (A) *Spirtaenia*, (B–F) *Staurodesmus*, (G) *Cosmarium*, (H–J) Staurastrum, all × 300–400. After Hutchinson (1967).

remind the reader that even in the plankton prey are often not passive. Motility, the presence of spines and unmanageable shapes and even toxins, some extremely potent, are utilized to avoid capture. As in the benthic world, organisms can be arranged along a so-called r–k spectrum. This consists of organisms at one end not wasting energy on protection and putting everything into rapid reproduction (accepting great losses), and at the other extreme, organisms that place considerable energy into construction of skeleta, spines, toxins, etc. and a long life that minimizes reproduction. With the intention of hopefully developing an increasing interest in the supporting of planktonic communities in complex ecological models, whether microcosms or aquaria, we will discuss a spectrum of filter-feeding examples.

Some Protozoans

Two widespread and geologically important protozoan groups are abundant herbivores and carnivores at the lower levels of the planktonic food web. The radiolaria and foraminifera (Fig. 11) are dominantly marine

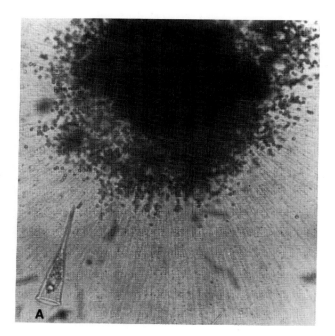

Figure 11 Marine radiolarians: (A) living animal with numerous, spine-like pseudopodia and a captured protozoan (×150); (B,C,D) siliceous tests or shells of a variety of genera, (B) is 300 μm in length. After Thurman and Webber (1984). Copyright © 1984 by Scott, Foresman and Company. Reprinted by permission of HarperCollins Publishers.

(*Figure continues*)

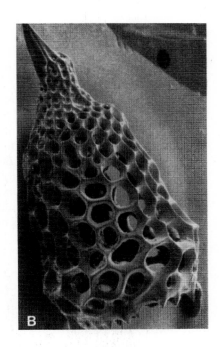

Figure 11 (*Continued*)

Figure 11 (*Continued*)

though with close freshwater relatives. They sometimes secrete incredibly complex shells of silica and calcium carbonate, respectively. In common, these animals somewhat like their distant relatives, the amoeba, extend numerous sticky tube-like pseudopodia. The pseudopodia trap small plankters and organic particulates, ingesting them into food vacuoles.

Filter-Feeding Arthropod Plankters

Many mid-water crustaceans are filter feeders. Prominant for example are most cladocerans such as the well-known *Daphnia*, (Fig. 12) primarily occuring in fresh waters, as well as many of the abundant marine and freshwater copepods (Fig. 13). In these animals, water currents for feeding are established by movement of the numerous appendages, and the filtering is accomplished by "fans" of setae or small spines located on the bases

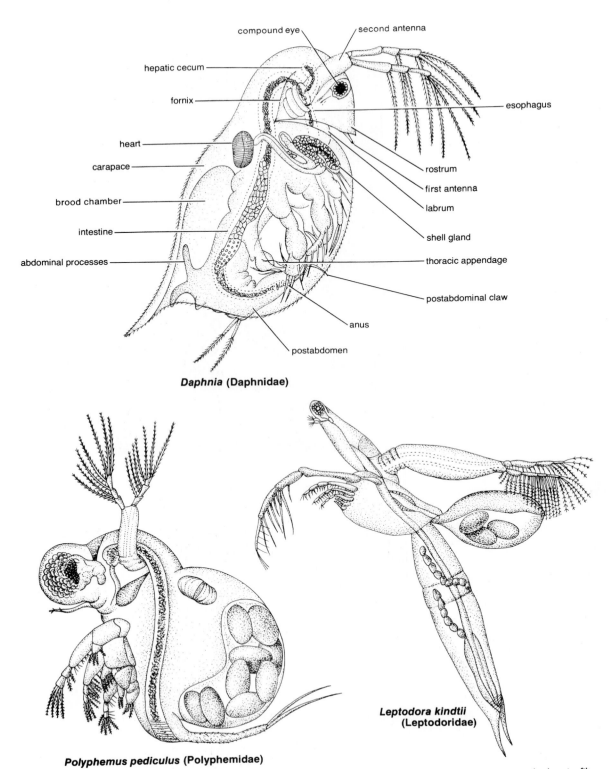

Daphnia (Daphnidae)

Polyphemus pediculus (Polyphemidae)

Leptodora kindtii (Leptodoridae)

Figure 12 Several genera of Cladocerans including *Daphnia,* the common planktonic filter feeder from lakes and rivers. From Parker (1982).

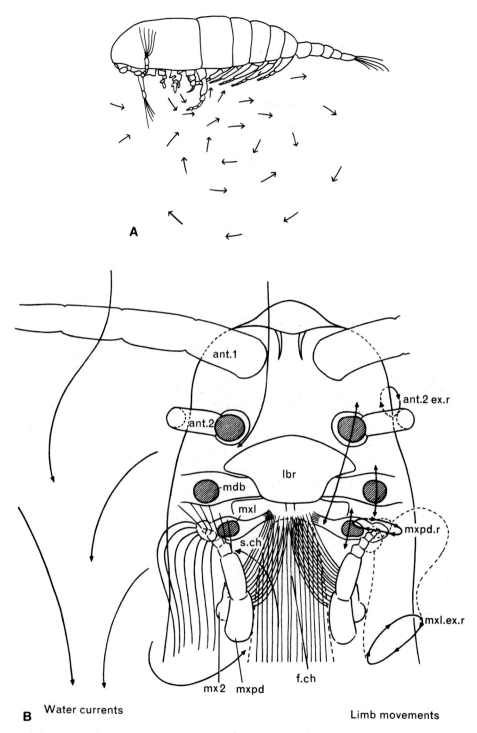

A

B   Water currents   Limb movements

Figure 13 The marine copepod *Calanus* showing (A) vortex of water movement and (B) feeding currents through filtering setae on the ventral side. After Barrington (1979).

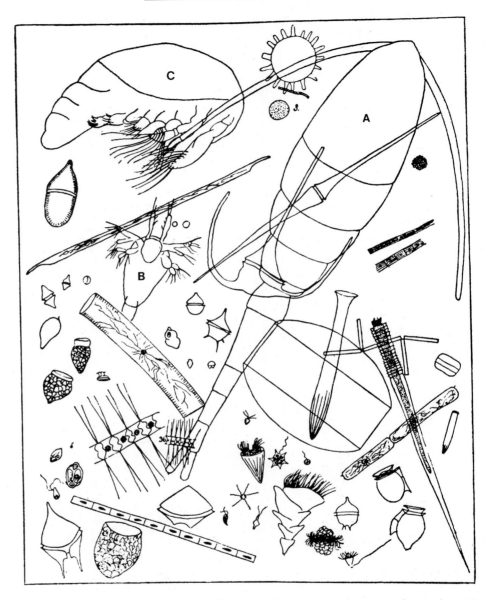

Figure 14 Planktonic copepods and the variety and sizes of phytoplankton on which they would feed: (A) *Oithona similis* (adult); (B) *O. similis* (nauplius); (C) *Temora longicornis* (nauplius). After Jorgenson (1966).

of the appendages. Water is forced through the setal filters, in some cases such as *Daphnia* using the "bivalved" shell cavity. The collected planktonic cells are wiped off the setae and delivered to the mouth parts by specialized appendages. Mucous secretions from glands located at the bases of some appendages can assist in forming the collected particulates or algal cells into more easily handled "food balls." Other appendages and the mouthparts are capable of rejecting particles that are too large, unpalatable or perhaps even toxic. Figures 14 and 15 graphically illustrate the difficulties of discrimination faced by these planktonic herbivores based on size and shape alone.

Many of these animals and related species are also capable of raptorial feeding if the prey is too large for the filtering system. However, the efficiency of capture is usually reduced if the prey must be individually handled. In spite of the difficulties of capture and ingestion, populations of cladocerans and copepods are capable of overgrazing phytoplankton blooms (Porter, 1977). It is typical of northern waters, marine and fresh, that the spring phytoplankton bloom, often of a few particularly well-adapted rapidly growing species, is rapidly depleted by zooplankton as the populations of the herbivores are quickly developed in response to the

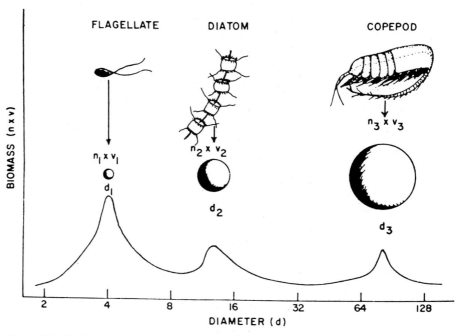

Figure 15 Planktonic copepod and the relative size and abundance of its principal algal prey. After Cushing and Walsh (1976).

presence of abundant plant cells. A group of very small filter feeders that are almost exclusively freshwater is the rotifers. These animals are equally important in the benthic environment and are discussed in the next chapter.

Sponges

Many benthic organisms are planktivorous. While much of the food for these animals is stirred off the bottom by wave action, currents, or the activities of animals, a good proportion is truly planktonic. Sponges are the archetype filter feeders (Figs. 16 and 17). Here, both pumping and

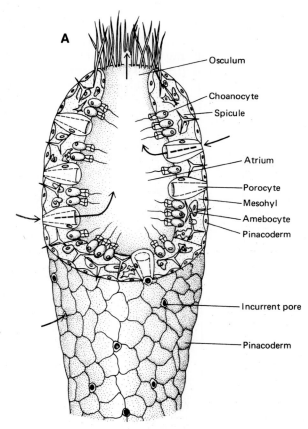

Figure 16 (A) Simple and (B–D) more complex type of sponges showing the "pumping" cells (choanocytes) and the increasing complexity of organization of pumping and flow chambers. After Barnes (1980).

(Figure continues)

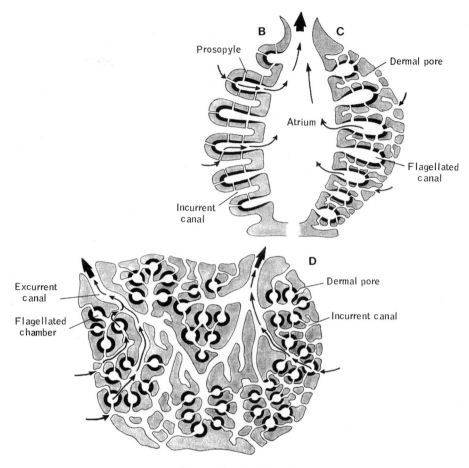

Figure 16 *(Continued)*

filtering are accomplished by individual specialized cells called choanocytes. The remainder of the sponge body provides the walls of the pumping channels, structural strength, protection, and reproductive capabilities. Virtually all feeding activity is concentrated in individual cells,
which is unusual in metazoans, and demonstrates the primitive state of
the sponges.

Chaetopteris, A Mucus-Bag Filter Feeder

Many aquatic and marine invertebrates secrete net bags of mucus to filter
out plankton from the water column. These are the equivalent of spider

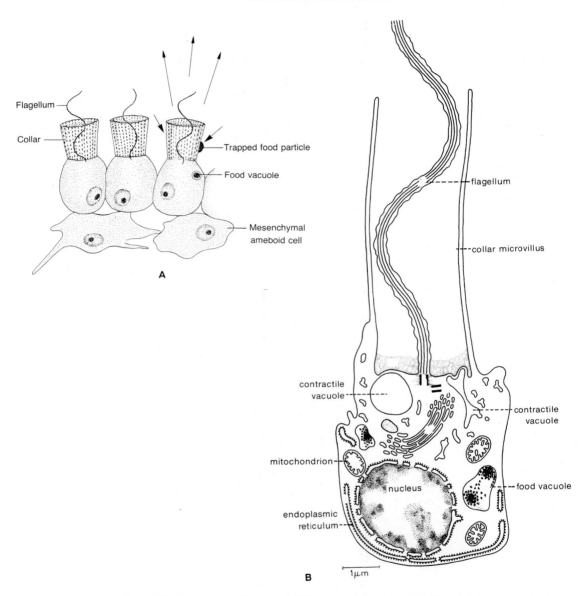

Figure 17   Pumping and filtering cells (choanocytes) of sponges. (A) Group of choanocytes in chamber; (B) ultrastructure of a choanocyte in longitudinal section. After Barnes (1980).

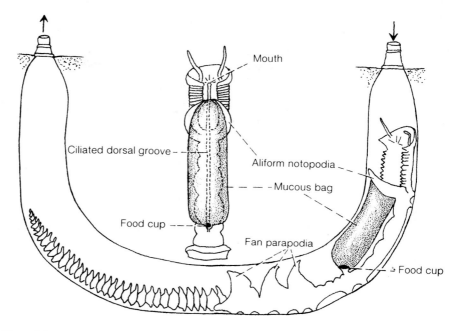

Figure 18 Longitudinal section of the parchment worm, *Chaetopterus*, in its burrow showing the "fanning" parapodia and filtering mucus bag. After Barnes (1980).

webs in the terrestrial world, though in many filter feeders using this method, the mucus bag is often eventually ingested along with its trapped particles.

Chaetopteris, the parchment worm (Fig. 18), is cosmopolitan on muddy bottoms in tropical to temperate seas. Using modified legs to "fan" water through its U-shaped burrow, it continuously secretes a mucus bag, which is collected with its entrained food particles by cilia in a central groove. There it is formed into food/mucus balls, which are eventually passed upward to the mouth in a ciliated groove.

A more web-like example is the feeding system used by vermetid gastropods, calcareous, worm-like, attached snails that are colonial and build small reefs or step-like projections on rocky shores. These animals, which are particularly adaptable to ecological aquaria, secrete a spiderlike web that is eventually retrieved and eaten along with whatever particles have become entangled.

Bivalves, The Master Filter Feeders

Primitive bivalves have feather-like ctenidia or gills that function primarily as devices for gaseous exchange. In more advanced species the gills are developed into elaborate folded lamellae that serve a dual purpose, food gathering as well as respiration (Fig. 19 and 20). The advanced bivalve gill is richly endowed with cilia as well as groups of cilia that function together. Mucus sheets are also secreted by cells of the gill. Together, the mucus and the sometimes highly coordinated patterns of beating cilia move water in intricate patterns over the surface of the gills. This motion collects or rejects particles and eventually transports them to the mouth area. So sophisticated is the typical bivalve gill feeding system that in many cases, organic particles or living organisms larger than 4 μm are routinely removed at very high efficiency. In some cases, particles as small as 1 μm can be removed, while the bivalve retains a highly discriminatory apparatus for extracting undesirable particles.

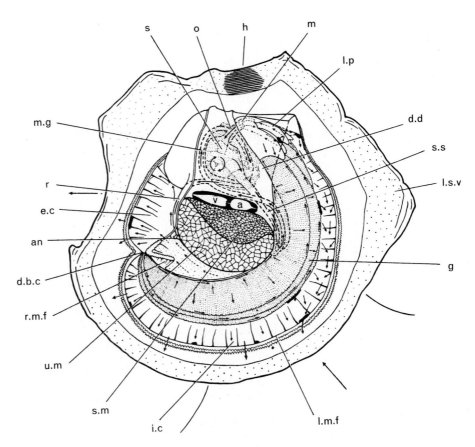

Figure 19 Internal anatomy of *Ostrea edulis* (oyster) with right shell removed. Arrows show current created by cilia on the gills (g). After Barrington (1979).

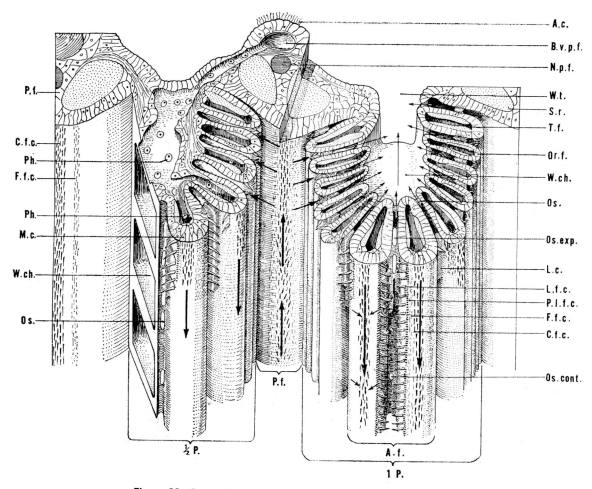

Figure 20 Perspective view of section of gill lamellae showing cilia tracts and current flow. *Crassostrea virginia,* oyster. After Jorgenson (1966).

Fish Planktivores

Many fish snatch small individual plankters (or nekton) from the water column. Both fresh and marine waters abound in *Chromis*, killifish, and tetra-like fish that feed in this more directly predatory fashion. On the other hand, some fish, for example those in the herring family (Color Plate 18), are much more elaborately endowed for direct water-column filtering

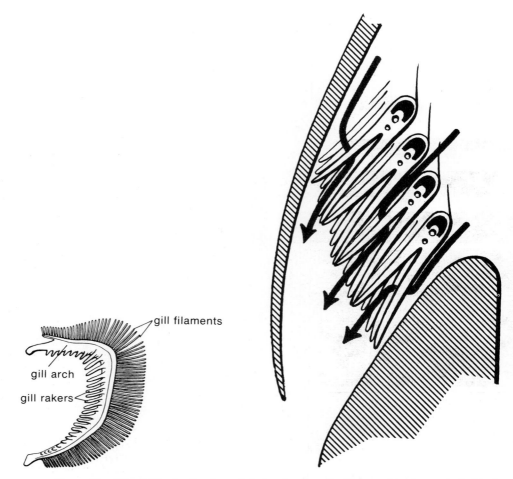

Figure 21 Gills of filter-feeding bony fish showing the gill rakers used to sieve out planktonic food. After Bond (1979).

gill filaments

gill arch

gill rakers

of plankton. Most herrings are marine or estuarine but many are anadromous, swimming up rivers and streams from the ocean when reproductively mature to breed in fresh water. A few herring are landlocked or occur exclusively in fresh water. In all fish, water is taken in through the mouth and passed laterally over fine, blood-rich gills for gas exchange. Inside the often blood-red gaseous exchange part of the gills lie a series of seives or gill rakers (Fig. 21). In the herring the gill rakers are specialized to seive out large or smaller plankton depending upon the species. While zooplankton are undoubtedly the dominant food acquired in this fashion, in some cases the rakers also are fine enough to take the larger phytoplankton.

Plankton and Model Ecosystems

In most synthetic ecosystems, except for the few that have been directed toward simulation of plankton-dominated systems, the plankton have been given minimum attention. In the aquarium fields, whether home hobbyists or professional aquarists are involved, plankton have not only been ignored but the methods employed for water quality control are such that these organisms are largely if not totally destroyed. It has been a major goal of our efforts at ecosystem simulation to avoid any methods that would artificially reduce or destroy floating or suspended plants, animals, protists, and microbes.

In a typical synthesized marine or aquatic ecosystem, the plankton populations are minimal because of the relatively small volume of water as compared to that typically present in the wild. Thus, just as the chemical and physical effects of the larger body of water, ocean or pond, etc., must be simulated with scrubbers and other devices, it is also necessary to simulate the presence of the plankton that would normally be present in that larger body of water. In this chapter, we will deal primarily with the planktonic elements required by benthic communities in microcosms.

Just as tropical lagoon tanks have been attached to our coral reefs and a mud-flat and salt-marsh tank is attached to the cold-water rocky shore, open-ocean microcosms could be developed and attached to either a reef or rocky shore. However, these would require a very large volume of water, and unless an open, high-quality natural body of water is available adjacent to the model it would probably generally be impractical. Our approach has been to limit destruction of plankton as reproductive stages of benthic organisms and to import true planktonic biomass in quantity and, to a more limited extent, in quality, into benthic microcosms. Of course, an equivalent nutrient export is required before this approach can be used. The algal scrubber (Chapter 12) fulfills this requirement along with managing gas exchange, although other plant systems could be used in the same way.

In the model systems described in Chapters 21–24, filtration is not used, the functions of filtration being obtained by algal scrubbers and settling traps. These systems have moderate planktonic floras and faunas. The Everglades mesocosm, which uses an Archimedes screw for saltwater circulation, is particularly rich in plankton including reproductive states.

Plankton is made up both of organisms that spend all or most of their life cycle in mid-water (holoplankton) and the young stages of other organisms that otherwise spend most of their lives either living on the bottom (meroplankton) or swimming. It also includes both plant and animal elements. In a normally functioning microcosm or mesocosm system, reproduction is widespread among the plants and animals of the benthic community. These reproductive and larval stages provide a relatively large meroplankton population in any closed-system microcosm in which they are not constantly filtered out or destroyed by pumps. This is a major disadvantage of bacterial filtration in that, by most methods of application, meroplankton are treated just as unwanted organic particulate matter. In a typical coral reef community, a major part of the water mass that overlies the reef has been driven there recently, usually from the open ocean, by waves or tides. Studies have shown that while the planktonic biomass that arrives over the reef from the ocean is similar in weight to what is lost from the reef to the lagoon, the composition of the plankton arriving is quite different from that departing the reef. In the water mass coming into the reef, the plankton is dominated by fully planktonic elements, those animals and plants that make their whole living floating in open tropical waters. Leaving the reef, the planktonic biomass is dominated by fragments of benthic algae lost from the reef's surface.

In the Caribbean windward reef models that we have operated at the Smithsonian, a constant flow of open ocean water is driven across the reef by the pumping arrangement. The simulated loss of plant material from the reef is easily managed within the framework of the normal algal turf scrubbing harvest. However, while the scrubber system can simulate the effects of "high-quality" incoming ocean water chemically and physically, it cannot supply the holoplankton. Brine shrimps, *Artemia salina*, are easily hatched from cysts purchased in pet and aquarium stores. As 48-h (cysts-in-water) hatchlings, they provide an excellent "mid-sized" plankton supply. One can grow or purchase brine shrimps larger in size, thereby providing a spectrum of mid to larger plankton input. However, we have found this approach time-consuming, and therefore used dried krill to simulate the larger-sized plankton input. It is desirable for more accurate system function to provide a wide range of both species and size input. A holoplankton import of about 2 g/m²/day (dry) was found in a well-developed St. Croix reef, and for lack of suitable additional information, that is the rate that we have used for our reef system. For 4 m² of reef surface our rate of input is about 2 g (dry)/day. It is important to note that this is a relatively small part of the equivalent of approximately 60–70 g (dry)/day input through photosyntheses on the model reef surface as determined by oxygen measurement.

In the Smithsonian Maine coast model system, we have also used planktonic input. Because they are so easily grown, we have used the flagellate *Isochrysis*, although *Dunaliella* or a wide variety of diatoms would do equally well. These algal species are considerably smaller than a brine shrimp or krill.

A wide variety of mostly dried artificial foods is now available on the aquarium market. Many of these have been carefully structured as a balanced food source for aquarium fish. While dried foods are more limiting than live foods since some planktivores may not sense and capture dead material, they nevertheless provide a very convenient simulated planktonic input. When dried foods are used, we recommend widely varying the type and size to maximize availability to filter feeders. In the many home-size aquaria that we have constructed using the methods described in this book, we have almost exclusively followed this technique.

In an open ocean reef, arriving plankton is not restricted by the time of day. The composition often varies markedly from day to night primarily due to vertical migrations, but the standing crop or biomass remains more or less constant throughout the 24-hour day. Fish, in particular, easily modify their behavior patterns with the adoption of fixed "feeding" schedules. It is advisable to approach as closely as possible a pattern of continuous introduction to more closely simulate the natural environment and avoid heavy fish cropping. Insufficient information is available to determine rocky-shore community consumption of open-water plankton. Scrubbers on the cold-water tanks have been used simply to adjust nutrients or water clarity to the desired level rather than to achieve a specific algal export. As we will describe in Chapter 22, large algal fragments are also used to simulate shore drift in the Maine microcosm.

We have developed several devices to accomplish continuous "planktonic" input. The most important of those that we have used are the simple devices shown in Figure 22. A typical rate used with the continuous unit is a 10-s introduction every 5 minutes. When dried foods are employed, a number of electrical or clock-type devices are commercially available to time feeding.

Planktivores of a wide variety of animal phyla occur in virtually all benthic systems whether hard or soft bottom. It would be impossible to construct a benthic community with any degree of realism without this element. A large percentage of reef organisms in particular use planktonic food sources, either true plankton or reproductive stages of benthic species. Sponge, filter-feeding worms, and vermetid gastropods as well as bivalves fall in this category. Most coelenterates, including stony corals, gorgonians, and anenomes, also are included, though strictly speaking, the capture method is more actively predatory. Reef fish of moderate to small size will capture a moderate-sized plankter if the opportunity presents itself. However, a few species such as *Chromis* and *Apogon* are largely plankton-eating. Considering the list of organisms present in such

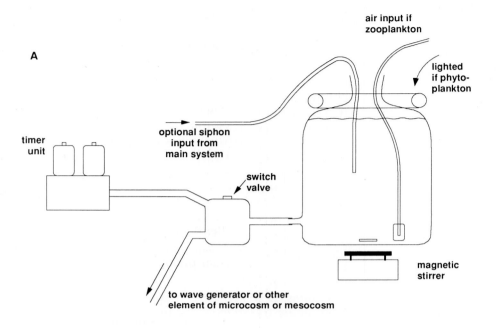

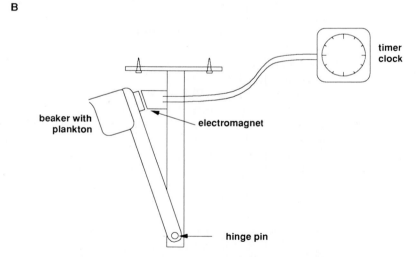

Figure 22 Devices used to provide a continuous input of planktonic food: (A) "continuous" input from a larger reservoir; (B) fixed number of "spiked" inputs.

reef models (see Chapter 21), fully one-quarter can be classified as at least partially planktivorous. This percentage is probably low, as the worm phyla have not been carefully studied and many of those species are also plankton eaters.

Cold-water coastal systems such as the Maine coast are richly provided with nutrients from runoff and from the constant exhuming and mixing of organic materials from the bottom by tides. The mid-water communities are thus proportionally very rich in plankton. The Maine mesocosm, although having a reduced holoplankton, nevertheless successfully supports a plethora of filter-feeding colonial animals such as sponges (*Leucolenia*), hydrozoans (*Hydractinia*), anemones (*Metridium senile*), molluscs (*Modiolus modiolus, Mytilis edulis, Placopectin magellanicus, Aequipectin irradiana, Mya arenaria, Yoldia limatula, Macoma balthica*), barnacles (*Balanus balanoides*), and worms (*Spirorbis* species and *Polydora ligni*) (see Chapter 22). Some of these organisms undoubtedly feed on suspended particulate organics. Most, however, feed on the plant and animal plankton, much of it undoubtedly in reproductive stages in the water.

At the time that this book is being written, much of our efforts are being directed at increasing the planktonic diversity and biomass in our estuarine models. Some of this information appears in Chapter 23; most of the work is "in progress."

Plankton is important to normal ecosystems. While plankton can "explode" even in the wild, in most cases this is because of human interference or an unusual meteorological event. Given good water quality and careful nutrient management, plankton can be highly beneficial elements to simulating benthic communities in aquaria. The stability provided by an appropriate level of algal scrubbing as described in Chapter 12 is such that while phytoplankton blooms can occur when import exceeds export for some period, the change is slow, over weeks or months. Also, it is relatively easy to stabilize an aquarium at moderate plankton densities. The only question the home aquarist need ask is what turbidity level is desired. Using algal scrubbing for control, no apparent negative effects have resulted from this practice even in cases where intense green or yellow blooms have been allowed to persist for several days to several weeks.

Generally, a healthy system with good water quality relative to oxygen and pH should carry a small to moderate load of phytoplankton and zooplankton. Filter feeders such as clams can actually be used to maintain moderate levels. If, when experimenting, plankton levels become too high for desired visibility, diatom filtration can be used to quickly reduce the number of plankton cells, although biological control is more desirable. When diatom-filter "plankton scrubbing" is used, the filter should be monitored closely and removed immediately when desirable levels are achieved. Depending on the capacity of the diatom filter relative to tank volume, this could be as little as 5–10 min or as much as an hour. In any case, such filtration should not be allowed to operate for long periods as not only will excessive quantities of plankton be removed, but also the filtered plankton will break down and the diatom filter will become a bacterial filter, returning nutrients to the tank and defeating the purpose of the operation.

We feel strongly that the primary area of synthetic ecosystem development needing further improvement is that of pumping. As we discuss in Chapter 5, centrifugal pumps are damaging to plankton and defeat efforts to achieve much of what we have discussed in this chapter. While centrifugal pumps are relatively inexpensive, highly reliable, and available in a very wide variety of types commercially, bellows pumps, Archimedes screws, and vacuum pumps are rare to nonexistent in small sizes. It is difficult today to achieve adequate pumping. Nevertheless, it is extremely important that every effort be made to achieve pumping that does not damage plankton. Other environmental limitations are not likely to be significantly improved until we remove the great limitations placed on model ecosystems by the destruction of plankton with pumps.

References

Barnes, R. 1980. *Invertebrate Zoology*. Saunders College, Philadelphia.

Barrington, E. 1979. *Invertebrate Structure and Function*. John Wiley, New York.

Bold, H., and Wynne, M. 1985. *Introduction to the algae*. 2nd Ed. Prentice Hall, Englewood Cliffs, New Jersey.

Bond, C. 1979. *The Biology of Fishes*. Saunders College, Philadelphia.

Cushing, D., and Walsh, J. 1976. *The Ecology of the Seas*. Saunders College, Philadelphia.

Dawes, C. 1981. *Marine Botany*. John Wiley, New York.

Gillett, J. 1971. *Mosquitoes*. The World Naturalist/Weidenfeld and Nicolson, London.

Hobbie, J. 1988. A comparison of the ecology of plankton bacteria in fresh and salt water. *Limnol. Oceanogr.* **33:** 750–764.

Hutchinson, G. E. 1967. *A Treatise on Limnology*. Vol. 2. Wiley, New York.

Jørgensen, C. B., 1966. *Biology of Suspension Feeding*. Pergamon Press, Oxford.

Migdalski, E., and Fichter, G. 1976. *The Fresh and Salt Water Fishes of the World*. Bay Books, Sydney.

Nixon, S., Pilson, M., Oviatt, C., Donaghy, P., Sullivan, B., Seitzinger, S., Rudnick, D., and Frithsen, J. 1984. Eutrophication of a coastal marine ecosystem—An experimental study using the MERL mircocosms. In Fasham, M. *Flows of Energy and Materials in Marine Ecosystems*. M. Fasham (Ed.). pp. 105–135. Plenum Press, New York.

Oviatt, C., Buckley, B., and Nixon, S. 1981. Annual phytoplankton metabolism in Narragansett Bay calculated from survey field measurements and microcosm observations. *Estuaries* **4:** 167–175.

Parker, S. (Ed.) 1982. *Synopsis and Classificaiton of Living Organisms*. McGraw-Hill. New York.

Porter, K. G. 1977. The plant-animal interface in freshwater ecosystems. *Am. Scientist* **65:** 159–170.

Reynolds, C. 1984. *The Ecology of Freshwater Phytoplankton.* Cambridge University Press. Cambridge.

Stockner, J. 1988. Phototrophic picoplankton: An overview from marine and freshwater ecosystems. *Limnol. Oceanogr.* **33**: 765–775.

Thurman, H., and Webber, H. 1984. *Marine Biology.* Merrill Publishing, Columbus, Ohio.

CHAPTER 19

DETRITUS AND DETRITIVORES
The Dynamics of Muddy Bottoms

In the idealized food chain, herbivores eat plants, and with the materials and energy derived from this food they power their movements and build their body tissues. These grazers or browsers are in turn eaten by carnivores. While the herbivores may require bacteria or protozoans in their digestive tracts to accomplish digestion of some of the plant materials, this is internal, a kind of symbiosis, and conceptually the simple food chain remains. For perhaps half of the primary plant production in the biosphere as a whole, the photosynthetic energy accumulated by plants is passed to the remainder of their communities in this more-or-less idealized food chain or web. Particularly in marine and open freshwater ecosystems, most planktonic and benthonic algae eaten by fish and invertebrates are in large part internally digested by those grazers. It is true that the inefficencies of handling the plant food ultimately lead to a major part of the food passing through the gut and therefore this energy going into suspended detritus (Fig. 1). However, as we discuss in Chapter 18, much of this detritus is ecologically treated as additional plankton and, with its bacterial and fungal flora, is grazed or filtered accordingly.

On the other hand, in many coastal and freshwater ecosystems, larger, tougher benthic algae much as kelp and *Sargassum*, as well as higher plants such as submerged aquatic vegetation, sea grasses, marsh

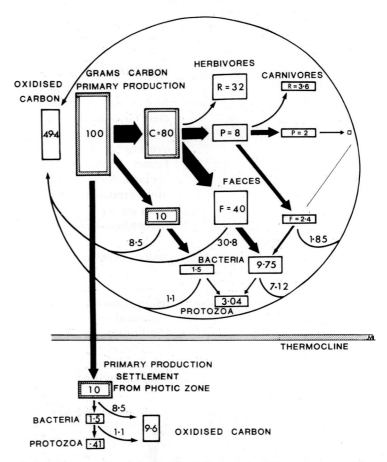

Figure 1 Energy flow (as carbon) in an idealized planktonic food chain. Note that while a majority of phytoplankton production is eaten by herbivores only about 50% is utilized either in respiration or biomass building. The remaining 50% is transferred to the mid water detrital cycle. C, Herbivore consumption; R, respiration; P, growth production; F, feces production; oxidized carbon is that which has passed through detritus. After Newell (1984).

grasses, mangroves, and virtually all land plants washed into the water as leaves, branches, and fruits, are the major source of primary production. The tough stalks, leaves, and woody parts of these plants are not readily digestible by most grazers. Fragmented by grazers or physical degradation, the cellular contents can be exuded or leached as dissolved organic material or particulates to be used by bacteria, fungi, and protozoa or even directly by other plants. However, the cellulosic and lignic cell wall components become larger debris. Then gradually, as physical and organic processes work them over further, they become a finer and finer detritus. These waste plant materials are attacked over and over again, primarily by bacteria, fungi, and protozoans, until they are fully broken down and their energy and minerals redistributed through detritivores to the community. Figure 2 shows a kelp bed, as a simplified example of a detrital-dominated ecosystem in which the particles occupied by bacteria are fed on by filter feeders.

As we discussed in Chapter 8, in typical lakes and coastal eco-systems with protected bays as well as in deeper water on open coasts, fine sediment delivered by streams or eroded from shorelines is deposited in zones of low wave energy. Organic particulates derived from rocky shores or marshes of great plant production are often deposited along with the muds, sands, and clays. A major portion of lake and coastal marine in-shore production is delivered as plant detritus and reworked animal feces to such organic-rich soft bottoms. While organic content can be as high as 10%, typically the percentage of organic materials in these sediments is low, less than 5%. Yet the supply is large and tends to be smoothed, without great seasonal or other periodicities that leave gaps in the food supply. An example of such a detrital-dominated system, where much of the organic particulates are delivered by waves and tides to sandy–muddy flats and bay bottoms, is shown in Figure 3. The epifauna and infauna of these soft-bottom communities can be taxonomically rich and diverse in species. For the example cited, probably a relatively impoverished case, over 60 species of macroinfauna larger than 2 mm and consisting of worms, bivalves, and crustaceans are present (Table 1). The uncounted smaller meiofauna would provide many additional species. These bottoms

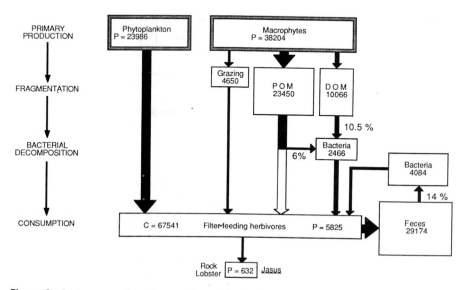

Figure 2 Basic energy flow diagram for a kelp bed in South Africa. POM, particulate organic matter; DOM, dissolved organic matter. Numbers in kjoules/m²/yr; P, produced; C, consumed. After Mann (1988).

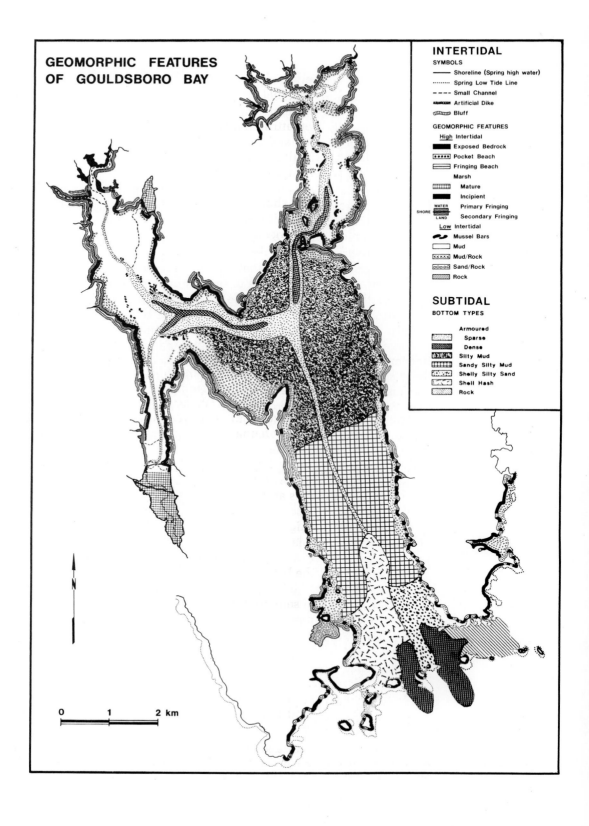

GEOMORPHIC FEATURES
OF GOULDSBORO BAY

INTERTIDAL

SYMBOLS
——— Shoreline (Spring high water)
.... Spring Low Tide Line
– – – Small Channel
Artificial Dike
Bluff

GEOMORPHIC FEATURES

High Intertidal
Exposed Bedrock
Pocket Beach
Fringing Beach

Marsh
Mature
Incipient

SHORE | WATER / LAND | Primary Fringing
Secondary Fringing

Low Intertidal
Mussel Bars
Mud
Mud/Rock
Sand/Rock
Rock

SUBTIDAL

BOTTOM TYPES
Armoured
Sparse
Dense
Silty Mud
Sandy Silty Mud
Shelly Silty Sand
Shell Hash
Rock

N

0   1   2 km

can also be relatively high in animal biomass (Chapter 5, Fig. 1). It is particularly interesting that the marine and estuarine soft bottoms, on the average, have an order of magnitude higher faunal biomass than fresh waters. This has been attributed primarily to the driving or mixing effect of tides in the marine and estuarine environments (Nixon, 1988). Larger waves in general are almost certainly an additional factor.

This chapter examines the abundant organisms of fine sediments where the included organic detritus has collected from the overlying waters and from adjacent shore and land areas. When these fine sediments also lie in shallow, well-lit waters and can support a diatom or blue-green growth, plant-derived organic materials can even originate directly from the sediment surface. It is in these sediments, if they are not so rapidly buried as to become geological deposits, that the last traces of organic energy are wrung from organic particles, further supporting the activities of animals. The size ranges typically in use by biologists and ecologists to define this benthic flora and fauna are given in Table 2. We will briefly discuss a selection of these organisms based on the size classification. To a large extent, microbenthos includes larger protozoans and a very wide diversity of tiny invertebrates from all phyla. Although most meiofauna groups have macrofaunal representatives, a few, such as nematodes, kinorhyncks, and gastrotrichs, are primarily members only of the meiobenthos or of soils. The marine macrobenthos is the domain of a large percentage of known invertebrates. In freshwater sediments, numerous insect larvae join some infaunal phyla that occur in both salt and fresh water. In general, microfauna feed on small organic particulates, meiofauna feed on larger organic particulates, and their host of attached microfauna and macrofauna feed on meiofauna. However, as always in biology and ecology, organisms and their activities are not easily categorized. The food chain is more accurately described as a web, and some macrofauna are capable of feeding on small organic particulates. Also, there is no sharp line between planktonic (suspension) feeders lying within the sediment and feeding on water-borne organisms, those organisms feeding off detritus stirred off the bottom, filter feeders on interstitial water between the sediment grains, and true deposit feeders eating organic particulates that form part of the sediment. Our separation between Chapters 18 and 19 is as much a convenience as a sharp line of demarcation.

Bacteria

Bacteria are abundant in all sediments, freshwater and marine, although generally the finer and more organic-rich muds and clays have a much

Figure 3 Distribution of sedimentary bottoms on a portion of the open rocky coast in eastern Maine. Note that while this is a rocky coast with very rich rockweed and kelp beds, areally soft-bottom communities dominate. After Shipp *et al.* (1987).

TABLE 1

Common Invertebrate Macrobenthos (> 2mm) in the Muddy Bottoms of Gouldsboro Bay, Maine and Its Offlying Waters (no./m²)ᵃ

Range	Genus species	Phylum	Offshore	Nearshore	Outer Bay	Upper Bay	Intertidal
Wide range	Nucula proxima	Mollusc/bivalve	590/100%	14348/100%	8,045/100%	137/22%	0
	Ninoe nigripes	Annelid/polch.	250/80%	645/100%	75/62%	2/22%	0
	Unidentified species	Nemertean	39/60%	89/83%	15/37%	17/11%	0
	Nephtys incisa	Annelid/polych.	57/80%	25/50%	19/75%	2/11%	0
Intermediate range	Scolopus acutus	Annelid/polych.	0	267/62%	167/67%	235/66%	0
	Ophelina acuminata	Annelid/polych.	0	114/50%	206/75%	33/56%	0
	Prionospio steenstrupi	Annelid/polch.	0	44/33%	242/87%	6/11%	0
	Unidentified clam species	Mollusc/bivalve	163/100%	58/83%	12/37%	0	0
	Nucula delphinodonta	Mollusc/bivalve	0	11/33%	148/62%	65/44%	0
	Edotea montosa	Arth./isopod	23/100%	62/83%	4/25%	0	0
	Unidentified worm species	Annelid/polych	23/100%	58/50%	0	0	0
	Ampelisca abdita	Arth./amphipod	0	0	50/62%	167/89%	0
	Tharyx acutus	Annelid/polych.	0	0	15/66%	95/66%	0
	Nephtys ciliata	Annelid/polych.	23/20%	0	10/25%	48/78%	0
	Aglaophamus neotenus	Annelid/polych.	0	0	15/37%	31/66%	0
	Phoxocephalus holbolli	Arth./Amphipod	0	0	31/75%	6/11%	8/17%
Narrow range	Dentalium entale	Moll./scaphopod	30/100%	0	0	0	0
	Diastylis cornufer	Arth./Cumacean	14/80%	0	0	0	0
	Sternapsis scutata	Annelid/polych	23/80%	0	0	0	0

Species	Group					
Yoldia sapotilla	Mollusc/bivalve	74/100%	0	10/25%	0	0
Harpinia propingua	Arth./Amphipod	0	6/33%	77/87%	0	0
Diastylis sculpta	Arth./Cumacean	0	0	667/62%	0	0
Eudorella pusilla	Arth./Cumacean	0	0	471/100%	7/33%	0
Cyclocardia borealis	Mollusc/bivalve	0	0	73/66%	2/11%	0
Orchomonella minuta	Arth./Amphipod	0	0	54/75%	0	0
Anaitides mucosa	Annelid/polych	0	0	23/62%	2/11%	0
Ampharete acutifrons	Annelid/polych	0	0	33/75%	6/11%	0
Lumbrinereis tenuis	Annelid/polych	0	0	25/62%	4/11%	0
Saccoglossus kowaleski	Hemichordate	0	0	0	0	86/83%
Scoloplos fragilis	Annelid/polych	0	0	0	0	150/83%
Corophium volutator	Arth./Amphipod	0	0	0	25/11%	278/66%
Unidentified worm species	Annelids/Oligoch.	0	0	0	6/12%	580/66%
Number of species		68	64	74	32	16
Mean incl/m		2916	17,564	10,440	749	1447
Number of stations		5	6	8	9	6

*a*Numbers given (250/80%) are mean numbers of individual animals per station and the percent of the stations of region at which the species was found. Only species occuring at greater than 60% of the stations within any one region are included (about one third of total). The strongly zonate character of the bottom fauna results from a variety of interacting factors including depth and sediment coarseness, to water climate (the inner bay is strongly subarctic in character with overlying ice and temperatures below 0° C in winter, and summer temperatures of 15—20° C; offshore winter temperatures are in the 0—2° C range while summer temperatures are rarely over 12° C).

TABLE 2	
Size Ranges of Benthic Fauna[a]	
Macrobenthos	> 0.5 mm
Meiobenthos	0.1–0.5 mm
Microbenthos	> 0.1 mm

[a]In practice this usually refers to what will pass through a screen of that mesh size. Thus, the dimensions are generally minimum dimensions: a worm 0.2 mm in diameter, 2 mm long will be treated as meiobenthos. After Levinton (1982).

larger numbers of cells. Typically, half of the bacterial flora is attached to sediment particles and half occurs in the interstitial water. As we discussed in earlier chapters, virtually all organic compounds can be broken down by some bacteria species. Bacterial numbers, especially those of the more generalized decomposers, can be enormous (on the order of 10^6 to 10^9 cells per cm^3) in fine, organic-rich sediments. However, as compared to macro- and meiobenthos, the resulting biomass (and therefore effect on organic processes) is usually relatively low. Rheinheimer (1985) discusses a case of a fine, organic-rich marine sediment in which the bacteria constitute less than 1% of the biomass although numbering 355×10^9 cells in the top 5 mm of the sediment (Table 3). Numbers are generally highest near the surface, or around the tubes and burrows of macrofauna, where oxygen is abundant. Within a number of centimeters from the surface, where oxygen becomes depleted, bacterial numbers are lower and sulphur and methane species become dominant. Here, breakdown of organic materials is incomplete, although constant bioturbation or mixing by the fauna repeatedly brings buried material to the oxygen-rich surface and oxygen to greater depths.

Fungi

Fungi, as well as being important parasites, are ubiquitous decomposers of organic materials, particularly plant materials, in virtually all of the earth's environments. Although generally requiring a rich supply of oxygen, a wide variety of fungi are characteristic of freshwater and marine sediments and the organic materials that are in the process of being incorporated into those sediments. Except for some yeasts, the primary life form of fungi is a mycelium of branching, ramifying, colorless hyphae, 10–30 μm in diameter (Fig. 4). It is the branch or rhizoidal tips of these hyphae reaching into the interstices of cellular or multicellular organic

TABLE 3
Mean Numbers and Wet Biomass of Organisms in a Silt Sample of 1 m² in the Top 5 mm[a]

Organism group	Numbers	Biomass (g)
Large macrobenthos	2.8	3.75
Small macrobenthos	230	3.30
Meiobenthos	146×10^3	1.15
Protozoa	283×10^6	0.02
Diatoms	590×10^6	0.06
Bacteria	355×10^9	0.07

[a]From Rheinheimer (1985, after Zobell, 1963).

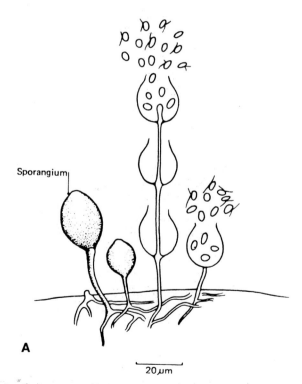

Sporangium

20 μm

Figure 4 Aquatic phycomycetes showing nonseptate mycelium and fruiting bodies releasing flagellated zoospores: (A) *Phythophora,* (B) *Pythium,* (C) *Blastocladia.* From *Freshwater Biology* by L. G. Willoughby (1976), reproduced by kind permission of Unwin Hyman Ltd.

(Figure continues)

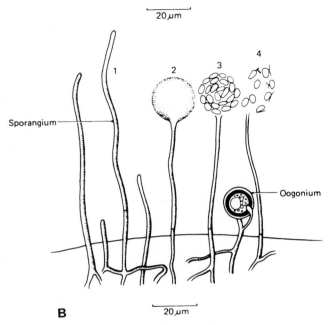

20 μm

1

2

3

4

Sporangium

Oogonium

B   20 μm

Figure 4   (*Continued*)

material and releasing digestive enzymes that results in the breakdown of those materials.

In the terrestrial environment the relatively large fruiting bodies of the basidiomycetes and ascomycetes form the familiar mushrooms and bracket fungi. These fruiting bodies are based on decomposition of mostly dead plant materials by characteristic, septate hyphae. Some ascomycetes and basidiomycetes occur in the marine and freshwater environments (Fig. 5), though here they tend to lack the familiar macroscopic fruiting bodies. However, they still possess the characteristic spore-bearing asci and basidia that allows them to be identified. On the other hand, the unicellular yeasts, which are thought to be very reduced ascomycetes, are abundant in both marine and fresh waters.

The dominant fungi of the aquatic environment are phycomycetes (water molds) and fungi imperfecti. The former possess nonseptate hyphae and generally have flagellate zoospores or gametes at some point in their life cycle (Fig. 4). The latter are higher fungi with septate hyphae but no known sexual reproduction. Many phycomycetes are parasites, but the majority either are decomposers or are capable of following either a parasitic or a decomposer mode of life.

Figure 6 shows a greatly idealized decomposer food chain based on the fungal breakdown of leaves on the surface of a sediment bottom. It is

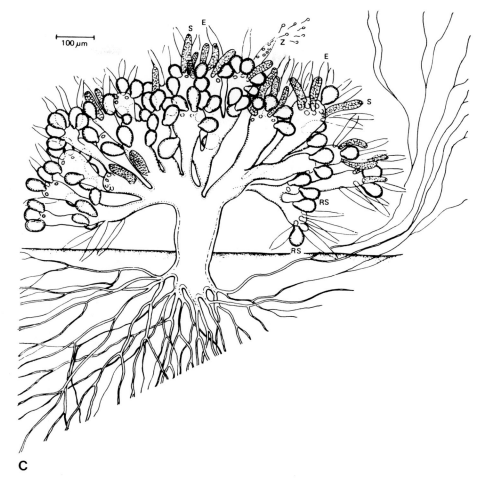

100 µm

C

Figure 4 (*Continued*)

still functioning when the leaf has been degraded to fine detritus. The feeding path of natural ecosystems is typically much more complex.

Protozoa

Some protozoans, members of the Kingdom Protista because of the recognition that the group has a diversity of phyla, are large enough to be members of the meiobenthos. A few, such as the giant ameba, many foraminifera, and even some ciliates, can technically qualify as macrobenthos. However, many protozoans are less than 100 µm in largest

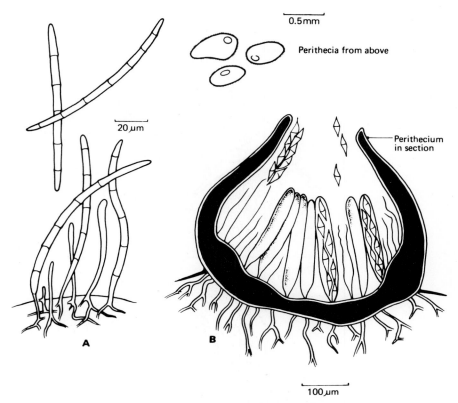

Figure 5 Aquatic ascomycete (*Anguillospora*) showing nonsexual spore production (A) and the "reduced" fruiting body (B) resulting from sexual reproduction. From *Freshwater Biology* by L. G. Willoughby (1976), reproduced by kind permission of Unwin Hyman Ltd.

dimension and are to be regarded as microbenthos. Some protists function as plants. Others are tiny but voracious predators, as we discussed in Chapter 17. However, large numbers of protistan microbenthos are detritus and bacteria feeders within aquatic sediments and wet soils. A typical feeding and digestion process for particulates or bacteria is shown in Figure 7. Since these are a food source for many in the meio- and macrobenthos and an important link in the detrital food chain, we will briefly discuss a few characteristic types.

Mastigophorans are protozoans with flagella, typically one or two but sometimes with many flagella. Some animals placed here are photosynthetic and often included in algal groups, and some are parasitic on a wide variety of animals including man and his domestic animals. However, many flagellates are feeders on detritus either within or on the surface of sediments. A few of these genera characteristic of the Smithsonian Chesapeake mesocosm, described in Chapter 23, are shown in Figure 8.

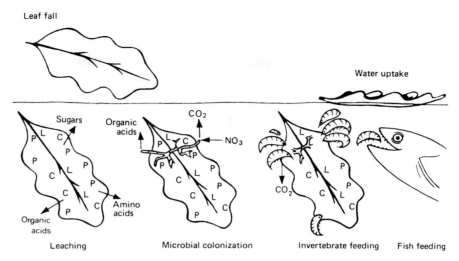

Figure 6 Basic process of breakdown of plant material in aquatic environment. C, Cellulose; L, lignin; P, protein. From *Freshwater Biology* by L. G. Willoughby (1976), reproduced by kind permission of Unwin Hyman Ltd.

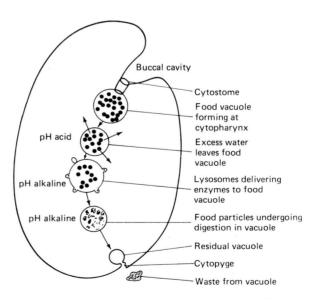

Figure 7 Basic feeding process in a single-celled protozoan. After Barnes (1987).

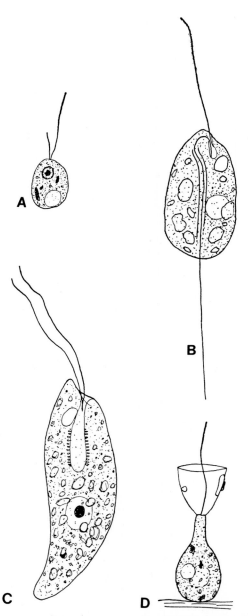

Figure 8 Several flagellated (mastigophoran) protozoa of the Smithsonian Chesapeake meso-cosm: (A) *Monas,* 5–16 μm; (B) *Chilomonas,* 15 μm; (C) *Anisonema,* 15–60 μm; (D) *Monosiga,* 5–15 μm. After Pennak (1953). Reprinted by permission of John Wiley & Sons, Inc.

All of the major protozoan groups have small bacteria and particulate-feeding members that are technically members of the microbenthos. However, the remaining nonflagellate types are mostly larger and strictly speaking are members of the meiofauna though they are often deposit feeders on particulates.

Meiobenthos—Protozoans

Sarcodines are closely related to the flagellates, even though generally motion and feeding are accomplished by pseudopodia and ameboid actions. The familiar *Amoeba* is a sarcodine. Many have flagellated reproductive states, and a few are both ameboid and flagellate. Some sarcodines with well-developed calcareous and siliceous shells, some foraminifera and radiolarians primarily in marine environments and heliozoans in fresh water, are primarily planktonic. Others, particularly among the forams, have many benthic members. Along with some naked amoebas, amoebas that build shells, and the forams, these typically meio- to macrobenthic-sized animals can be extremely important feeders on bacterial-coated particulates as well as diatoms and flagellates within and on the surface of sediments (Fig. 9).

The ciliophorans, the majority of which lie in the meiofaunal range, are the most advanced of the protozoa (Fig. 10). This group includes the *Paramecium*, which nearly every one has seen either in high school biology or in films. Here, the unicellular condition, through the development of a wide variety of cell organelles, becomes as complex in function as many multicellular invertebrates. The cell surface of a ciliophoran is typically fixed in form, being characterized by a complex pellicular structure. In primitive genera, the cilia, which are part of a subpellicular infraciliary system, occur in simple rows. In more advanced forms the cilia can grouped to perform complex feeding or locomotion tasks. In some cases they are grouped in tufts as cirri; in others they can work together to form a well-defined, undulating membrane.

Different species of ciliophorans can engage in a wide variety of feeding modes. Many are bacterial, detrital, or microalgal feeders in either the deposit or filtering mode. These tend to have complex mouth and buccal cavity ciliary relationships. Others are raptorial in feeding style (Chapter 17), preying on small multicellular invertebrates as well as other protozoans. Many of the raptors have specialized stinging organelles or toxicists that function like the stinging cells of jellyfish.

It is difficult to overrate the importance of the protozoan community in a normally functioning wild ecosystem, mesocosm, or aquarium. In a brief survey of the 15,000-gallon Smithsonian Chesapeake mesocosm (Chapter 23), 11 genera of flagellates, 14 genera of sarcodines, and 41

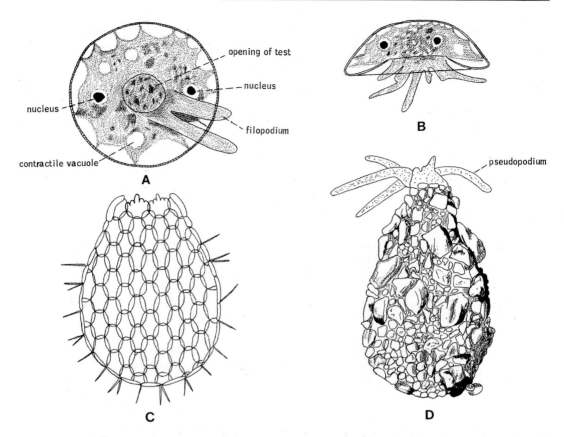

Figure 9 Protozoans in the meiofauna size range. Shelled amoebas (A–D) and foraminifera "shells" (E–J) (Sarcodines). (A,B) *Arcella;* (C) *Euglypha;* (D) *Difflugia;* (E) *Bulimina;* (F,G) *Elphidium;* (H,I,J) *Ammonia.* Scale 200 μm. A–D after Barnes (1987); E–J after Pennak (1953). Reprinted by permission of John Wiley & Sons, Inc.

(Figure continues)

genera of ciliates were counted. Most of these were characteristic of the dominating benthic soft bottoms in this system.

Meiofauna—The Multicellular Invertebrates

Many invertebrate phyla (22 out of 39) have at least a few meiofaunal representatives as adults. Many groups are temporarily present as larvae. However, the major groups occuring only in this fauna of tiny organisms are five peculiar phyla that are almost totally meiofaunal: the Loricifera, Kinorhyncha, Tardigrada, Gastrotricha, and Gnathostomulidae (Fig. 11). Some members of the major invertebrate phyla, Arthropoda and Turbellaria, are also represented here (Fig. 12). Because they are so important

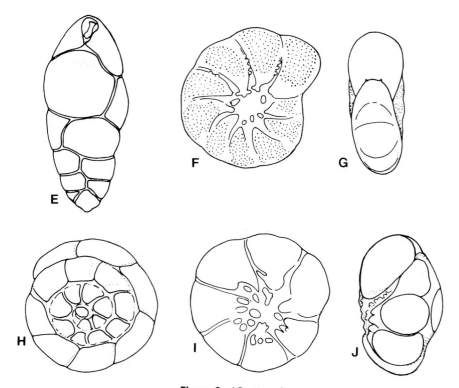

Figure 9 (*Continued*)

and are not discussed elsewhere in this book, the nematodes and rotifers that are ubiquitous meiofaunal denizens deserve special mention.

Many nematode genera are parasites (the roundworms) on animals and plants and include some of man's most serious parasitic infections. Other genera are herbivorous or carnivorous in a wide variety of environments. Both parasites and herbivores can be moderately large, many on the order of several centimeters. Most nematodes, however, are non-parasitic denizens of sediments and soils. In these environments they are generally quite small, less than 2 mm and often less than 0.5 mm. The nematodes of both marine and terrestrial soils and sediments can be extremely abundant, often millions per square meter of bottom.

Nematodes living in sediments are typically round and smooth with a faint indication of annulations on an elastic and rather tough cuticle. The anatomical layout is simple but still fairly complex for such small animals (Fig. 13). Nematodes move through the interstitial spaces of sediments by undulatory motions. Many species feed on diatoms, fungi, bacteria, and organic particles coated with bacteria and diatoms. Others are carnivorous on other nematodes and protozoans and can have complex

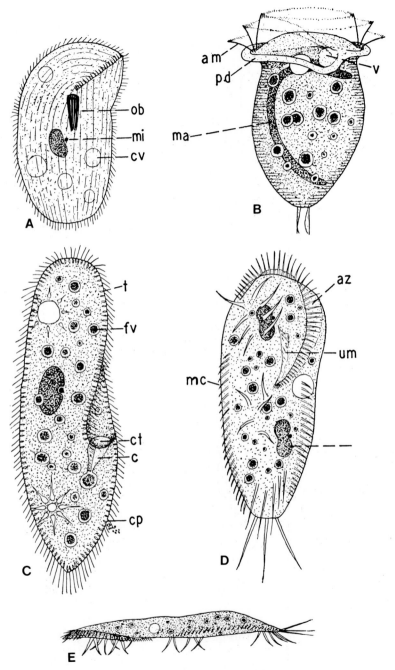

Figure 10 Fresh water ciliates characteristic of the Chesapeake mesocosm. (A) *Chilodonella,* 50–300 μm; (B) *Vorticella,*35–160 μm; (C) *Paramecium* 60–300 μm; (D,E) *Stylonychia,* 50–300 μm. (E is a lateral view.) After Pennak (1953). Reprinted by permission of John Wiley & Sons, Inc.

dentition (Fig. 14). Some carnivores have mouthparts with spears or stylets and achieve poisonous injections of their prey. Nematodes are extremely important low to mid members of detrital food webs. While relatively difficult to identify for the nonspecialist, their presence should be encouraged in synthetic ecosystems as they are strong contributors to detrital recycling and to the stability of marine and freshwater ecosystems.

Finally, since they are ubiquitous elements of freshwater meiobenthos, we will briefly discuss the rotifers or rotatoria. These distinctive little animals, with many species occurring in pond and lake zooplankton, are familiar to anyone who has looked at drops of pond water through the microscope (Fig. 15). The typical paired "whirling" corona of cilia at the head bring food to the mouth. The food is generally organic particulates, bacteria, or algal cells, but some species are predatory on other small invertebrates. The mastax, just interior of the mouth, is a food processing center, often muscular and with tooth-like structures for the grinding or tearing of food particles. In some cases locomotion is derived from the coronal cilia; however, in sediment dwellers, bending motions, contractions, and extension of the body and "foot" are equally likely to provide movement.

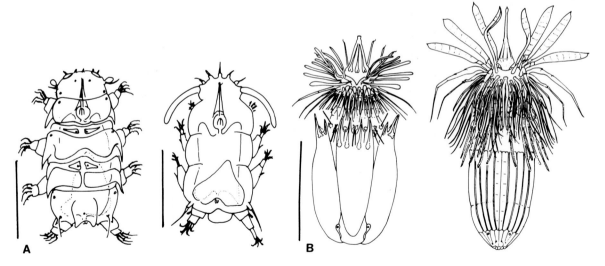

Figure 11 Invertebrate phyla occuring primarily in the meiofauna. (A) Tardigrada, marine and fresh (water bears), bars 50 μm; (B) Loricifera, marine with extrusible mouth and a chitinous skeleton, bar 100 μm; (C) Kinorhyncha, marine/estuarine, extrusible head, strongly segmented, bar 100 μm; (D) Gastrotricha, marine and fresh, move by cilia, bar 100 μm; (E) Gnathostomulida, marine of low oxygen sediments, ciliary motion, and distinctive jaws, longer bar 100 μm, shorter bar 10 μm. All after Higgins and Theil (1988).

(Figure continues)

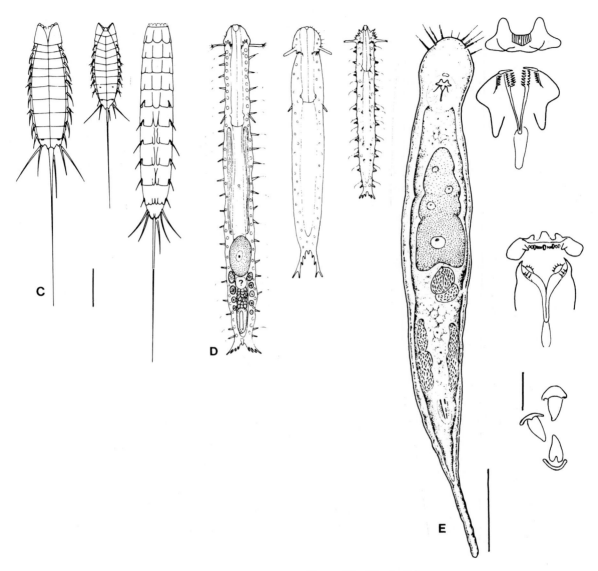

Figure 11   (*Continued*)

Macrobenthos

Virtually all invertebrate phyla have members that occur on or within aquatic sediments. Soft bottoms are often extremely rich in species and are critical feeding grounds for many fish and some diving and wading birds. In addition to predator species, soft bottoms have a broad spectrum of invertebrate types ranging from deposit feeders to suspension feeders.

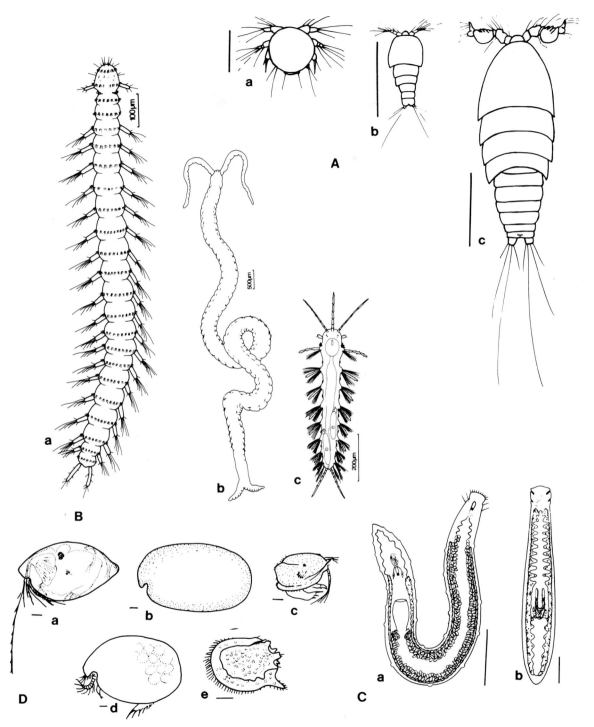

Figure 12 (A) Benthic meiofaunal elements of some major invertebrate phyla. (A) typical harpacticoid copepods, (a) nauplius larva, (b) copepodid larva, (c) adult, bar 300 μm; (B) polychaetes, (a) *Pusillotrocha,* (b) *Saccocirrus,* (c) *Nerilla;* (C) turbellarians, bar (a) = 1 mm, (b) = 200 μm; (D) ostracods, bar 200 μm. After Higgins and Theil (1988).

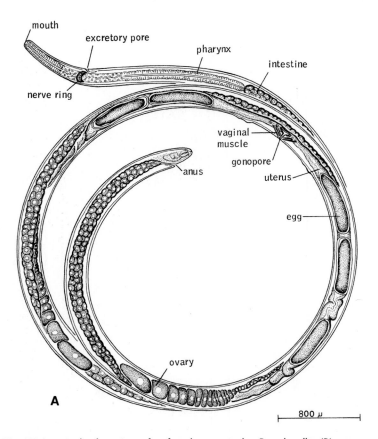

Figure 13 (A) Longitudinal section of a female nematode, *Pseudocella;* (B) cross section of generalized nematode through pharynx. After Barnes (1987).

(*Figure continues*)

As we mentioned in the last chapter, there is no sharp dividing line between the soft bottom and the immediately overlying water column. Many species of fish that live primarily in the water column feed on the bottom. Many clams and worms that live within soft bottoms feed on plankton from the overlying water. Since we touched on filter or suspension feeders in the last chapter, here we will concentrate on a few examples of invertebrate deposit feeders.

Deposit Feeding in Soft Bottoms

As an illustration of the richness and complexity of soft-bottom communities, we will use the relatively well studied, estuarine Delaware Bay

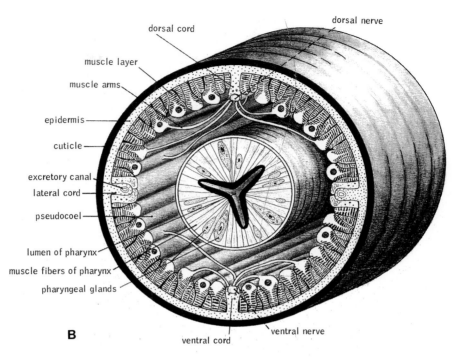

dorsal nerve

dorsal cord

muscle layer

muscle arms

epidermis

cuticle

excretory canal

lateral cord

pseudocoel

lumen of pharynx

muscle fibers of pharynx

pharyngeal glands

ventral nerve

ventral cord

B

Figure 13 (*Continued*)

(Table 4). Of 174 species tallied during a bottom survey, 84 were deposit feeders, 37 suspension feeders, 15 omnivores, and 38 carnivores. In terms of species numbers, the great majority of deposit feeders were about equally divided between polychaete worms and amphipods. However, of the three bivalve mollusk species, two were highly abundant. Thus, in a biomass sense, the fauna is about equally divided between polychaete worms, amphipod crustaceans, and bivalve mollusks.

Annelids are strongly segmented, highly developed worms. Polychaetes in particular have two lateral "feet" or parapodia on most segments, are often rich with spines or setae, and have a well-developed head region with a wide variety of sensory and feeding structures. The role of polychaete worms in deposit feeding has been well studied and two general feeding types have been identified: the burrower, and the surface or selective feeder (Whitlach, 1980). These two types, from the dominant forms of Delaware Bay, are shown in Figure 16. Generally, the burrowers are adapted in a variety of ways for simply taking up mouthfuls of raw sediment, digesting the organic material, bacteria, and any contained meiofauna, and ejecting large quantities of the remaining fecal material. The surface feeders, on the other hand, use ciliated tentacles or palps and

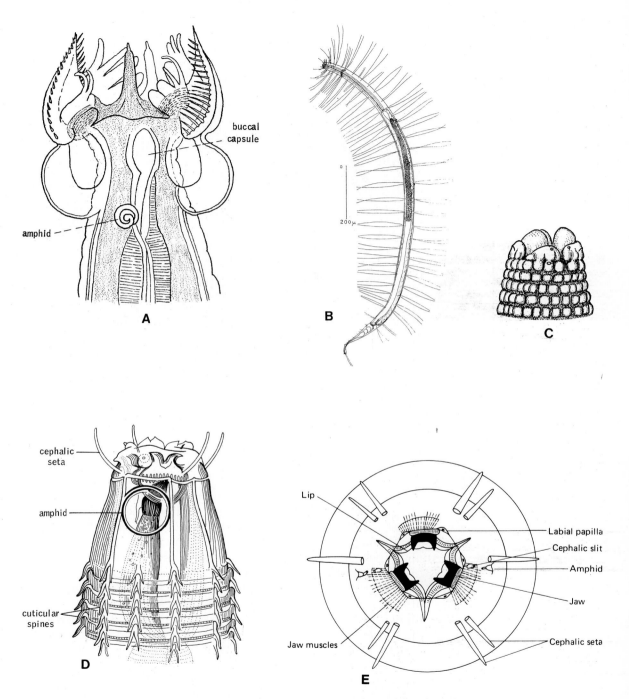

Figure 14 Structural variety in nematodes. (A) *Wilsonema* with elaborate mouthparts; (B) abundantly setate *Trichotheristus;* (C) *Placodira* with armored caticle; (D) *Monopisthia* with spiny cutide; (E) cross-section through head of *Enoplus* showing jaws. After Barnes (1987).

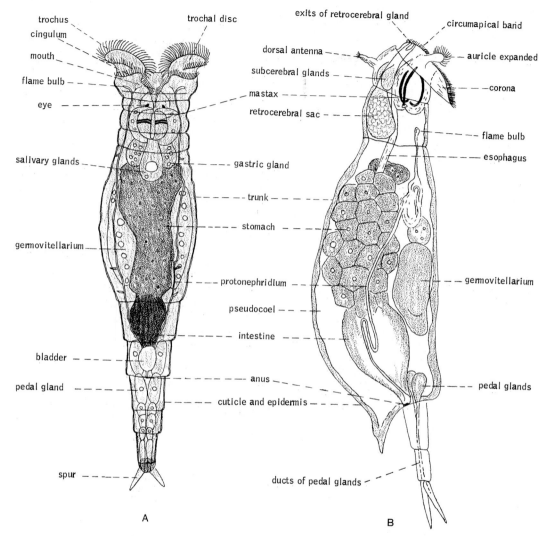

Figure 15 Structure of rotatoria: (A) *Philodina*, (B) *Notommata*. After Barnes (1987).

mucus to select organic particles from the surface of the sediment and transport them to the mouth. in some cases, the tentacles can be numerous and quite long.

Some of the most primitive but abundant bivalve mollusks, including the nut shells, Nuculidae, and the Nuculanidae, are adapted for deposit feeding (Fig. 17). The mechanism is similar to that used in the worm groups using selective deposit feeding with ciliated and mucus-rich tentacles or palps. Unlike the selective-feeding polychaetes, the

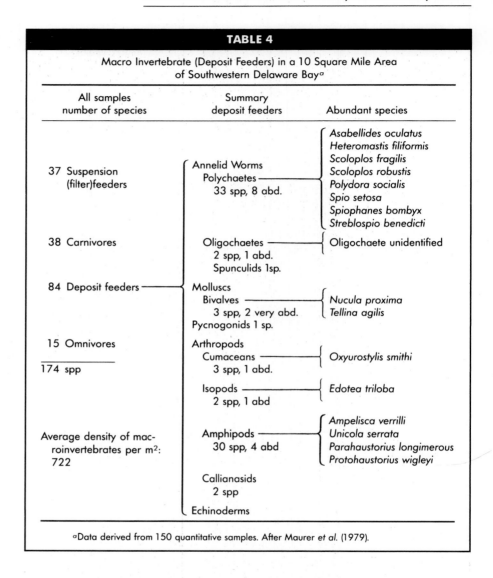

TABLE 4

Macro Invertebrate (Deposit Feeders) in a 10 Square Mile Area
of Southwestern Delaware Bay[a]

All samples number of species	Summary deposit feeders	Abundant species
37 Suspension (filter)feeders	Annelid Worms Polychaetes 33 spp, 8 abd.	*Asabellides oculatus* *Heteromastis filiformis* *Scoloplos fragilis* *Scoloplos robustis* *Polydora socialis* *Spio setosa* *Spiophanes bombyx* *Streblospio benedicti*
38 Carnivores	Oligochaetes 2 spp, 1 abd. Spunculids 1sp.	Oligochaete unidentified
84 Deposit feeders	Molluscs Bivalves 3 spp, 2 very abd. Pycnogonids 1 sp.	*Nucula proxima* *Tellina agilis*
15 Omnivores	Arthropods Cumaceans 3 spp, 1 abd.	*Oxyurostylis smithi*
174 spp	Isopods 2 spp, 1 abd	*Edotea triloba*
Average density of macroinvertebrates per m²: 722	Amphipods 30 spp, 4 abd	*Ampelisca verrilli* *Unicola serrata* *Parahaustorius longimerous* *Protohaustorius wigleyi*
	Callianasids 2 spp	
	Echinoderms	

[a]Data derived from 150 quantitative samples. After Maurer et al. (1979).

deposit-feeding bivalves feed within the sediment rather than on the surface. However, the European *Scrobicularia* clam has a long siphon that sweeps in a radial fashion around its burrow, thus functioning much like the long-tentacled polychaetes.

Of the major groups of detritivors in Delaware Bay, the isopods and amphipods (Fig. 18) need little detailed discussion. These little pill bug-like and shrimp-like crustaceans are richly endowed with feeding apparatus in terms of legs and mouth parts and are well equipped to handle organic particulates. As with the worms, some amphipods are tube builders, some are normally shaped burrowers without constructing tubes, and still others are elongate and quite narrow, allowing movement interstitially between sand grains.

Although not conspicuous in the sediments of Delaware Bay, the Echinoderms have several major groups that are richly endowed with detritivorous abilities. The brittle stars, sand dollars, heart urchins, and

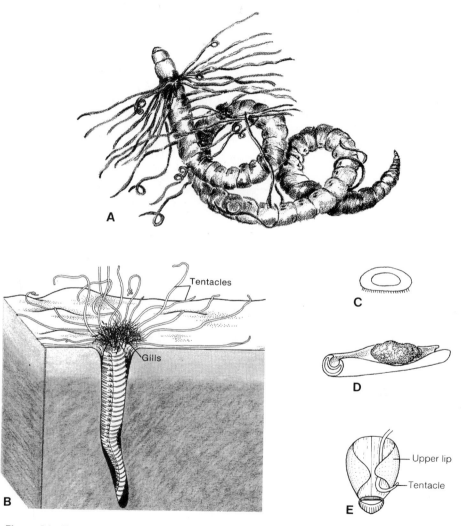

Figure 16 Two main types of marine polycheate worm deposit feeders. Selective surface feeders: (A) *Cirratulus;* (B) *Amphitrite,* in burrow; (C) ciliated tentacle; (D) with food transported, (E) tentacle bringing food to mouth, of *Terebella;* (F) *Spio* with two tentacles; (G,H) *Arenicola,* burrower with extensible pharynx for nonselective feeding. All after Barnes (1987).

(Figure continues)

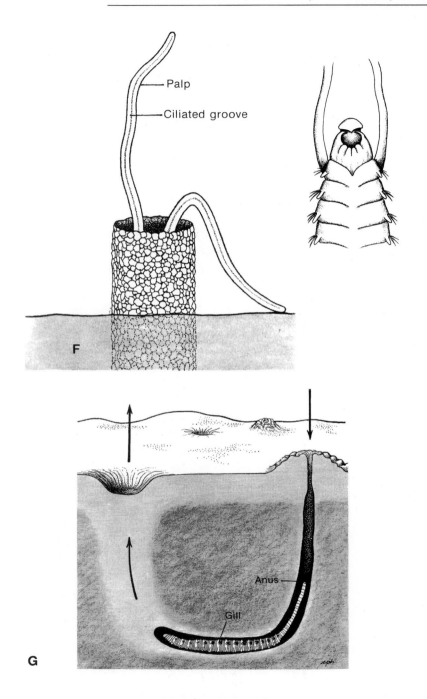

Figure 16   (*Continued*)

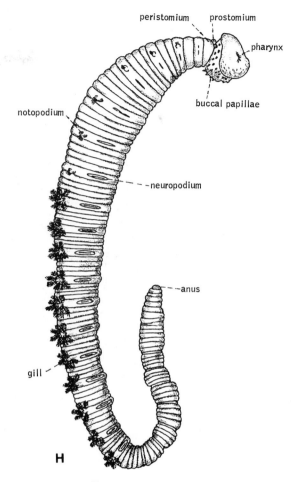

peristomium   prostomium

pharynx

buccal papillae

notopodium

— neuropodium

— anus

gill

H

Figure 16 (*Continued*)

sea cucumbers are often conspicuous on soft bottoms and have basic feeding patterns that are similar to those described for the polychaetes (Fig. 19).

Freshwater Soft Bottoms

The sandy and muddy sediments of fresh water have either the same or functionally similar groups of organisms that feed and perform as do their counterparts in the sea and in brackish waters (Lopez, 1988). However, among the detritivores themselves, there are two major taxonomic differences.

The polychaete worms, largely restricted to the sea, are replaced by oligochaete worms in fresh water. Oligochaetes, which do occur in the sea but to a lesser degree, are segmented and richly setate. However, they lack the lateral "feet" or parapodia of the polychaetes and generally have simpler head and feeding structures (Fig. 20). Earthworms and the tubifex

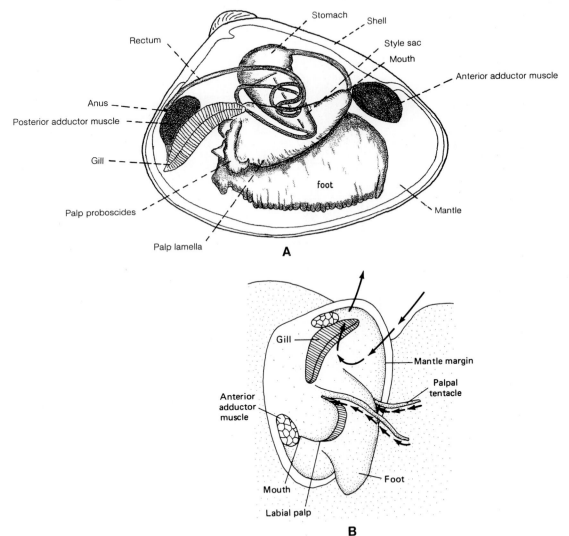

Figure 17 Deposit-feeding bivalves. Here respiration with water flow over gills and feeding are separate. Ciliated palps or tentacles carry out the feeding function. (A) *Nucula;* (B) generalized feeding, respiration process; (C) *Yoldia;* (D) *Nuculana;* ciliary action on palp. All after Barnes (1987).

(*Figure continues*)

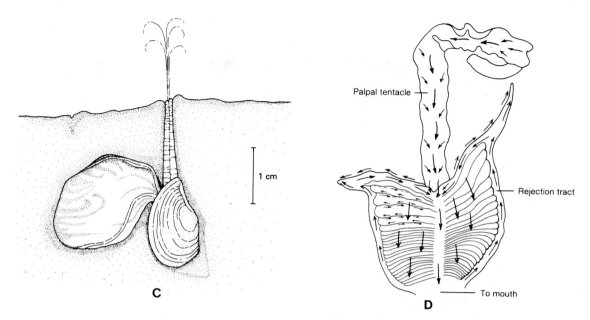

Palpal tentacle

Rejection tract

To mouth

1 cm

C

D

Figure 17 (*Continued*)

worms of the aquarium trade are well-known oligochaetes. Typical tubifex worms are burrowers feeding on bacteria and organic particles in the sediment. However, their posterior extremities, which are red due to an excess of respiratory pigment in the blood, extend up into the water from the often oxygen-poor bottom sediments.

Insects, virtually absent from fully marine environments, become major elements in most fresh waters. Very conspicuous and abundant organisms of freshwater muddy bottoms replacing many polychaete and bivalve members of marine soft bottoms are chironomids, the midges or Tendipedidae (Chapter 8, Fig. 13). Flying adult midges form large mating clouds over ponds and lakes and do not feed. The eggs are laid in water and the larvae (Fig. 21) can be directly herbivorous or filter feeding, depending on the species. However, many chironomid larvae are deposit feeders. Some form tubes like many polychaetes in the marine environment, and others are naked. These typically highly abundant and diverse forms are an important food source for other aquatic insects and for fish. A freshwater microcosm or aquarium that is open even occasionally to the outdoors will certainly develop a midge fauna. In our Chesapeake Bay and Everglades mesocosms, the chironomids are permanent and important members of the fauna. Like amphipods in salt and brackish waters, they also are nuisance micrograzers on algal scrubbers.

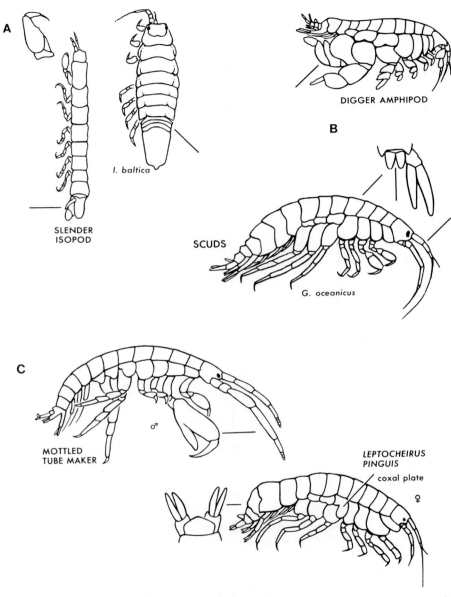

Figure 18 Deposit-feeding crustacea: (A) isopods, *Cyathura* and *Idotea;* (B) burrowing amphipods, *Haustorius* and *Gammarus;* (C) tube-building amphipods, *Leptocheiris* and *Jassa.* From *A Field Guide to the Atlantic Seashore* by Kenneth L. Gosner. Copyright © 1978 by Kenneth L. Gosner. Reprinted by permission of Houghton Mifflin Co.

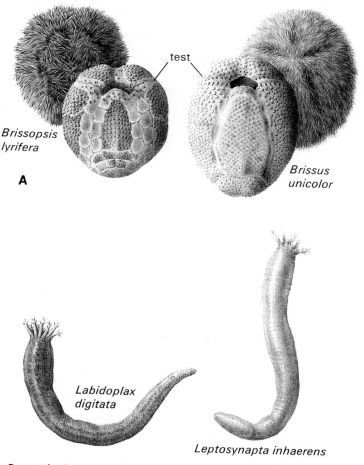

Figure 19 Deposit-feeding echinoderms: (A) heart urchins; (B) sea cucumbers. After Campbell (1976).

Carnivores and the Detritivore Community

Like any complete community, the detritivore-based soft bottom has its predators. Of course, animals basically external to the community, such as bottom-feeding fish, birds, and large roving crustacea such as crabs and lobsters, easily come to mind. However, there are also many carnivores at all levels from protozoa to arthropods that are integral elements of the soft bottom. Note that almost one-quarter of the Delaware Bay soft bottom community discussed above (Table 4) consisted of carnivores. These generally smaller predators also become prey to the top carnivores that are

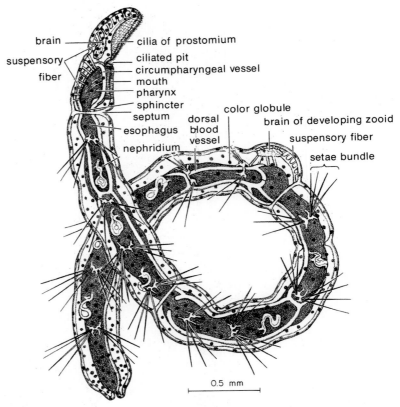

Figure 20 Fresh water eligochaete annelid worm, *Aeolosoma*. Note relatively simple head structure, lack of parapodia but abundant setae. After Barnes (1987).

external to the community. Carnivores in general were treated in Chapter 17.

Detritus and its Role in Model Ecosystems

In all of our microcosms, mesocosms, and aquaria we have included moderate-sized soft-bottom communities. Some have been dominated by soft bottoms. These greatly increase the diversity and the stability of the ecosystem being modeled. While a variety of these systems is described in detail in Chapters 1–24, we briefly make reference here to the detrital aspects of several of these model communities. Finally, we discuss in depth the more traditional approaches to the handling of fine organic sediment in aquaria, and we relate these to the methods that we describe below.

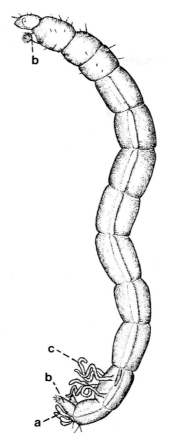

Figure 21 Larval stage of *Chironomus* (*Tedipes*): a, proleg; b, abdominal gills; c, anal gills. After Pennak (1953). Reprinted by permission of John Wiley & Sons, Inc.

Detritus and Coral Reef Microcosms

A large part of the waste organic material of reef communities is repeatedly captured and recycled in mid-water by fish and on the reef surface by many invertebrates including corals. Much of what is not recycled on the short scale, along with the fine carbonate sediment generated by the boring and scraping of organisms, is carried either into deeper water on the forereef or into the lagoon. Lagoons in particular as detrital ecosystems are rich food areas for some fish, such as grunts. In this case, the fish typically ingests a mouthful of sediment and works the sediments over its gills as a seive to extract worms and other included macrofauna, then "spits out" the remaining sediment through its mouth or gill slits. Often, to avoid predators such as jacks and barracuda, grunts in large schools "file off" a reef at night to work lagoon sediments. Some scientists feel that by this process, grunts can be critical in returning nutrients to the reef ecosystem. Since the normal processes of distant removal (down the dropoff or into a broad lagoon) are not as effective in a microcosm or aquarium situation, a rich macrofauna of worms, constantly reworked by fish like grunts, is an essential component of the living reef aquarium. In addition to grunts, most reef fish that are mid- to low-level invertebrate predators will search the surface of available sediment for the occasional worm or crustacean that exposes itself at the wrong time. It is difficult with a very small lagoon in proportion to reef to avoid overpredation of the detrital community. Isolated and partially isolated lagoons as refugia have provided partial solutions to this problem (see Chapter 21).

As can be seen in Chapter 21, Figure 1, and Color Plate 26, a dense *Thalassia* bed has developed in the Smithsonian's coral reef lagoon. This "grass" bed effectively traps further fine sediment, the surface of which is occupied by dozens of detritivore species, primarily crustaceans and polychaetes (Color Plate 27). One particularly conspicuous component is the clouds of mysid shrimp that remain on or within a centimeter or two from the sediment surface. The fish and a spiny lobster constantly re-work the detrital fauna. The support provided by the detrital community is such that the lobster shows extensive growth and has to be replaced with a younger animal every year or two to prevent excessive physical destruction of the lagoon.

In reef microcosm and aquarium environments, where extremely low nutrient levels are generally needed, the newly stocked lagoon community may "out-well" at a higher rate than an established lagoon eco-system. Thus, higher algal scrubbing capability may be required during stocking and early operation to avoid excess nutrients.

Generally, we have had little difficulty in operating coral reef com-munities together with their soft-bottom lagoon counterparts. However, large parrot fish or tangs and some sea urchins, in the absence of high predator controls, need to be avoided if a successful sea-grass community is to be maintained. In the wild situation, lagoons are generally quite large in size in proportion to their reefs and are heavily patrolled by barracuda, jacks, and large trigger fish. Thus, heavy grazing is typically avoided. Just as for predators of detritivores, a semi-isolated lagoon will solve this scaling problem to some degree.

Detritus and Cold-Water Microcosms

In the wild Maine coast ecosystem, a relatively small part of the rich kelp and rockweed production is grazed or browsed in place (on the order of 10–20%). In many areas, it builds up to a large biomass and then is torn off by winter storms and deposited as berm-like drifts along the upper shore. This algal food becomes a continuing store of energy for the entire ecosystem. It is gradually broken down by a variety of crustaceans (such as the amphipods *Orchestia* and *Talorchestia*), by protozoans, and, of course, by bacteria and fungi. Eventually, as fine organic particulates, this organic drift, now a soup, finds its way back into the inner coastal waters, from which it either settles out in deeper water or is delivered by the incoming tide to the mud flats and salt marshes at the heads of bays. These muddy areas, rich in organic sediment, provide for a large diversity of deposit- and filter-feeding animals, which in turn become the feed for a wide variety of fish and other invertebrates. It is particularly interesting in Gouldsboro Bay, Maine, that even though organic particulates in the sediment gradually decrease in abundance seaward, away from the source, species diversity of the infauna increases. These inverse gradients are likely related to the extreme range of weather conditions. Landward the waters are relatively fresh in spring, quite warm in summer, rather saline in the fall, and intensely cold in winter. In any case, large numbers (hundreds to thousands per square meter) of individuals of these fewer species are available to constantly rework the rich energy source.

We have built and operated a number of rocky Maine shore microcosms that included muddy bottoms and small mud flats. The largest of these is described in Chapter 22. This system has provision for the equivalent of algal drift accumulation and breakdown. However, the fine particulates eventually collect at the base of the rocky shore, in the tidal reservoir, and in the semiseparated mud-flat tank. Although not as rich as the mud bottom of a Maine bay, these sediments contain numerous mollusk species (*Hydrobia minuta*, two *Nucula* species, *Yoldia* and *Macoma*), several worms (*Scoloplos fragilis*, *Ninoe nigripes*, *Cirratulus* sp., *Nereis* sp., *Lepidonotus* sp., and *Thelepus* sp.), and a few amphipods (*Corophium volutator* and *Ampelisca abdita*). A rich fauna of protozoa and meiofaunal worms is also present (Chapter 22, Table 6) to fill in the lower protions of the detrital food chain. At the other end of the food chain the green crab and killifish constantly rework the sediment for the larger detritivors.

It is particularly of note that although the muddy bottom in this model was stocked from the wild to a depth of 40–60 cm, the relatively small scrubber is capable of driving nutrients to levels below 1 μM. Nutrient levels can be raised for experimental purposes by disturbing the sediment, but otherwise, even with the extensive surface reworking by killifish, green crabs, and polychaete worms, nutrient export remains at an easily controllable level.

Detritus and Estuarine Mesocosms

In the Chesapeake mesocosm (Chapter 23), just as in the Chesapeake Bay itself, large areas of fresh to salt marshes are present. Productivity in the marshes and other shallow-water areas of the system is very high and produces a dense growth of vascular plants and a surface coating of algae. A reduced seasonal cycle of light and temperature is operated in this tank and at least now, in maturity, the yearly plant production is mostly allowed to break down within the mesocosm (10–20% is exported). Virtually all of this plant biomass is eventually deposited on the five sediment bottoms that dominate the unit. The deeper areas of the system, as well as in the tidal reservoirs, have detrital bottoms. The resulting sediments are richly organic, with a diversity of surface and infauna (not including bacteria) that exceeds 150 species including 66 protozoa, 7 rotifers, 22 annelids, 32 crustaceans, 24 molluscs, and 15 species from miscellaneous invertebrate phyla (see Chapter 23, Table 3).

Anaerobic conditions can be found with the probe of a finger into the sediment of this model, and the constant release of nutrients into the tank water would seem assured. Yet the waters of this mesocosm are constantly saturated or near saturated with oxygen, and the moderately high levels of nutrients entering the model with fresh tap water (simulating river input) are progressively reduced as this water works its way through the estuary. Typically in mid-nutrient-level operation, fresh water enters the systems at between 20 and 30 μM for dissolved nitrogen. By the time salinity has climbed to 15%, dissolved nitrogen is on the order of 5–10 μM. It remains throughout the estuary at this level except for the highest salinities, where algal scrubbing reduces concentrations (simulating ocean flushing) to less than 1 μM. (See Chapter 23, Fig. 5 and accompanying text.)

The Everglades mesocosm is similar in many ways to the Chesapeake model. Perhaps the most interesting additional characteristic of this subtropical system is the presence of mangrove communities. These estuarine swamp builders efficiently trap sediment and have their distinctive invertebrate elements, particularly the fiddler crabs (*Uca* species) and the coffee snail (*Melampus coffeus*). The mangroves themselves provide a constant rain of leaves, which along with the ubiquitous mat of algae provide the primary energy supply.

Most important for all of the systems described above, fine particulates, organic and inorganic, are not filtered out. They are allowed to collect in the appropriate low-energy parts of the models, where they support a rich microbe, protist, and invertebrate fauna that utilizes the energy in organic particulates to drive reasonably complex detrital communities. In all of these cases ambient nutrient concentrations in the water column remain at or below levels characteristic of the wild analogs.

Detritus and the Aquarium

In traditional aquarium technology, fish and a few larger accompanying invertebrates are living in a hotel. The only significant food available is provided by the aquarist. It is hoped that detritus produced by the fish will be minimal, but what does become available is captured by the bacterial filter and, along with ammonia, urea, and unused food, is broken down in the filter. The filter is thus the sewage or septic system for the aquarium. Unfortunately, in this case the water from the "septic system" is constantly recycled to the aquarium. This is an extreme simplication of natural ecosystem processes and as such is subject to radical fluctuations. When such systems are driven to the limits of the bacterial filter, a rapid degeneration of environmental quality often follows. Overloading of a nonpolluted wild community with detritus followed by oxygen crashes and fish death is unusual. Its occurence usually results from an extreme physical situation such as a flood or hurricane. In a typical wild eco-system, a whole spectrum of organisms is available to process detritus without unusual consequences. Even where excess biomass is available, as in a mud flat, and anaerobic conditions dominate below the top cen-timeter or two, a suite of animals is present to work over this available food under these conditions and in turn become prey to fish and inverte-brates working the sediment from the water column.

The ecologically normal situation for a soft-bottom community is low in energy and relatively high in organic particulates. Using the under-gravel filter, which has become standard for much aquarium work, the normal place for soft-bottom communities is kept low in organics and organisms just like a coarse sandy beach or a gravel bed in a fast-running stream. This arrangement greatly reduces the potential for development of a normal detrital community, as organics and organisms alike tend to be delivered to the main bacterial filter for processing (Fig. 22). Even where undergravel filters are not used, the potential for developing a normal detrital community is still reduced, in part by starvation for organics and by removal of plankton and reproductive stages.

In typical modern aquarium processes that include increased oxy-genation and pH control by a variety of trickle filters, increased light for photosynthesis, and possibly even a denitrification filter, organic particu-lates are still delivered to a wide variety of trapping (filtering) devices where primarily bacteria and to a lesser extent fungi and protozoans are expected to process all organics to carbon dioxide, water, and nutrients. Thus, the filter is equivalent to the detrital communities of wild eco-systems. However, as we discussed above, bacteria, while extremely abun-dant in numbers, typically are a minor biomass element on most active and fully viable soft bottoms. The filter process prevents the development of a full detrital-based food web with its host of invertebrates. It also prevents the "kicking upstairs" of a considerable part of the energy supply to the fish and larger invertebrates that would normally feed on the com-munity and continually rework the organics to full degradation. Also, as we discuss in depth in Chapter 12, complete bacterial processing of de-trital organics in the aquatic environment without full return to plant biomass (the algal turf scrubber or similar system) reduces water quality and does not allow the export of nutrients to match food input.

For many communities to be simulated in microsms and aquaria, it

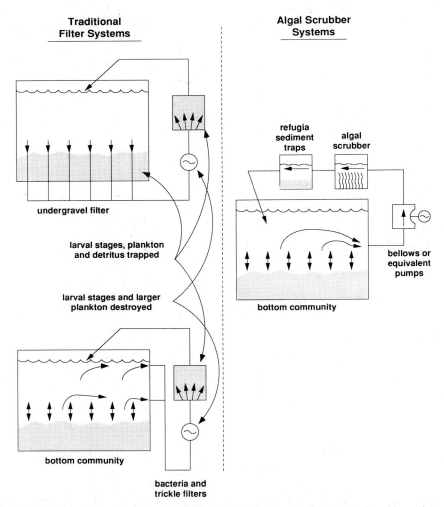

Figure 22 Schematic comparison of traditional filter systems with the algal turf scrubber relative to benthic infauna.

can be inconvenient to maintain a detritus-based soft-bottom community of proper proportion to the bottom feeders in the primary system of interest. When this is the case, the sediment is overworked and the detritus-based community of invertebrates can be kept too low to properly process the organics. At the same time the sediment tends to be excessively thrown up into the water column without the extensive mucal binding that is characteristic of such communites. The problem can be solved to some extent by including sediment traps or detrital community refugia within the piping system. This procedure allows the full development of an un-predated detrital community, which can then provide larval stages for the main system. While it would be preferable to provide a proper balance of communities, this approach avoids an all-or-nothing situation and simulates a distant detrital bottom.

References

Barnes, R. 1987. *Invertebrate Zoology*. 5th Ed. Saunders, Philadelphia.

Campbell, A. 1976. *The Seashore and Shallow Seas of Britain and Europe*. Country Life Books. Feltham, England.

Gosner, K. 1978. *A Field Guide to the Atlantic Seashore*. Peterson Field Guide. Houghton Mifflin, Boston.

Higgins, R. P., and Thiel, H. 1988. *Introduction to the Study of Meiofauna*. Smithsonian Press, Washington, D.C.

Levinton, J. 1982. *Marine Ecology*. Prentice-Hall. Englewood Cliffs, N.J.

Lopez, G. 1988. Comparative ecology of the macrofauna of freshwater and marine muds. *Limnol. Oceanogr.* **33**: 946–962.

Mann, K. 1988. Production and use of detritus in various freshwater, estuarine and coastal marine ecosystems. *Limnol. Oceanogr.* **33**: 910–930.

Maurer, D., Watling, L., Leathan, W., and Kinner, P. 1979. Seasonal changes in feeding types of estuarine benthic invertebrates from Delaware Bay. *J. Exp. Marine Biol. Ecol.* **36**: 125–155.

Newell, R. 1984. The biological role of detritus in the marine environment. In *Flows of Energy and Materials in Marine Ecosystems: Theory and practice*. NATO Conf. Ser. 4 Marine Science 13. Plenum Press, New York.

Nixon, S. 1989. Physical energy inputs and the comparative ecology of lake and marine ecosystems. *Limnol. Oceanogr.* **33**: 1005–1025.

Pennak, R. 1953. *Fresh Water Invertebrates of the United States*. Ronald press, New York.

Rheinheimer, G. 1985. *Aquatic Microbiology*. 3rd Ed. Wiley and Sons, New York.

Shipp, C., Staples, S., and Adey, W. 1985. Geomorphic Trends in a Glaciated Coastal Bay: A Model for the Maine Coast. *Smithsonian Contributions to the Marine Sciences* **25**: 1–76.

Whitlach, R. 1980. Patterns of resource utilization and coexistence in marine intertidal deposit-feeding communities. *J. Marine Res.* **38**: 743–765.

Willoughby, L. 1976. *Freshwater Biology*. Hutchinson, London.

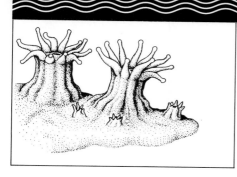

CHAPTER 20

SYMBIONTS AND OTHER FEEDERS

Innumerable intimate relations have developed between members of different species of plants and animals. This can be as occasional but important as the grapsis crab that frequently finds protection from marauding fish while hiding beneath the long spined sea urchin *Diadema*. It can be a whole way of independent life such as the cleaner wrasse that removes parasites from many different species of willing fish, or it can be as fixed and complex as the parasite that must have several specific hosts to complete its life cycle. Symbiosis is a term meant to include all of these, as well as mutualism, where both parties benefit, commensalism, where one species benefits and to the other species the relationship is of no consequence, and parasitism, where one species benefits and the other is harmed. Parasitism is arbitrarily and with some difficulty separated from predation. It is generally accepted that a parasite is smaller and more numerous than its host. Also, the parasite consumes only a small part of the host. However, if that part is critical, or if other predators or parasites are introduced as a result, the death of the host can result.

Symbiosis has generally been regarded as a biological curiosity (mutualism) or a serious and usually correctable problem (parasitism), but not a centerline factor of organic evolution. In recent years, it has come to be recognized that symbiosis, or the joining together of separate organic lines, is a major factor in evolution (see e.g., Smith and Douglas, 1987).

Of the thousands of interspecies relationships that can be called symbiotic, many do not involve feeding directly (Fig. 1). Since the focus of

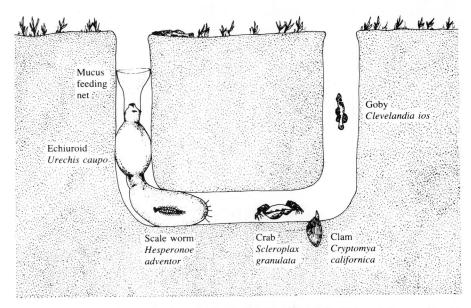

Figure 1 Commensal organisms living within the burrow of the Echiurid (annelid) worm *Urechis caupo.* After Thurman and Webber (1984). Copyright © 1984 by Scott, Foresman and Company. Reprinted by permission of HarperCollins Publishers.

our biological discussion for ecosystem modeling is based on feeding or trophic structure, we shall not discuss nonfeeding symbioses in depth. However, nonfeeding symbiotic behavior can be extremely important to the structuring and efficient operation of model ecosystems. It is worthwhile for the aquarist to collect as much of the natural history information as possible for each species of moderate to larger size so that such adaptations can be allowed. Also, tremendously varied basic feeding symbiotic relations are described in the literature and would be of interest in specific modeling situations. The many cases of anemones and hydrozoans that inhabit the surfaces of snail shells are an example. However, our primary interest in this chapter, because they are so directly important to living ecosystem modeling, is plant–animal mutualism directly involving photosynthesis and also those involving parasitism.

Zooxanthellae and Their Animal Hosts

Cells from a number of major algal groups are known to occur symbiotically in the tissues of animal hosts. By far the most common are dinoflagellates, which occur in a wide variety of invertebrate phyla in the sea. The alga involved has generally been placed in a "super genus and species" *Symbiodinium microadriaticum*, though probably several taxa are involved. Cultured algal cells taken from their host have been repeatedly seen to develop the unique dinoflagellate, double flagella, and girdle-like sulcus form that unmistakeably place them in this group. Presumably it is also in this form that they migrate from host to host, although most vegetative biomass production occurs in the host by cell division. Flagellated zooxanthellae obtain access to their host through the mouth and stomach cavity, from which they enter the tissue lining the stomach. From there, they move to the cells of the tissues allocated for photosynthesis, dividing vegetatively until they reach a density tolerated by the host (Talbot, 1984) (Fig. 2). The host is capable of ejecting the algae in some situations, usually those involving stressful conditions.

A major part of the chemical production of zooxanthellae photosynthesis, often greater than 50%, is released to the host typically in the

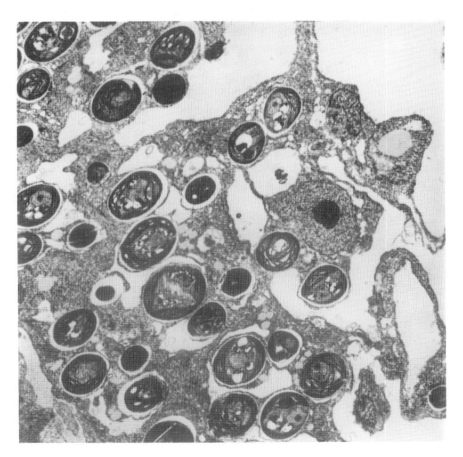

Figure 2 Algal cells (*Chlorella*) growing within the cells of the fresh water hydrozoan *Hydra*. From Smith and Douglas (1987).

form of glycerol (Spencer-Davies, 1984) (Fig. 3). It has been shown that this same release of glycerol to the medium can be also achieved in algal culture, but only when some of the host's tissues are present (Bold and Wynne, 1985). The proportion of energy aquired by the host from their zooxanthellae varies widely. In some cases the mutualism pair is functioning basically as a plant with some small but critical predation providing nutrient needs, somewhat like carnivorous land plants such as the Venus fly trap. In other cases the plant member of the pair may provide little in the way of energy support but instead maintains internal pH levels within the coral tissue that are conducive to calcium carbonate skeletal formation.

It is fairly clear what the invertebrate host gains from this arrangement: food, easier calcification, or both. The benefit to the alga is a little less obvious. Such relationships are considerably more common in nutrient-poor tropical seas, and the driving process from the algal point of view is generally ascribed to elevated nutrient supply, originally from animals captured and digested by the host. Enhanced protection from grazers is perhaps an additional incentive. While some scientists have suggested that these algae would be unable to make it in the typically nutrient-poor reef environment, it must be remembered that most coral reefs are rich in an algal turf that provides on the order of half to three-quarters of reef plant production (Table 1). Also, *Gambierdiscus toxicus*, an epiphytic dinoflagellate, is often abundant enough in reef environments to effectively poison the upper levels of the food chain with its toxins (Lobel *et al.* 1988). Perhaps the zooxanthellae can be regarded as being related to their hosts in the same way that cows, horses, and pigs relate to man. The price can be high for the individual, but as far as the genome is concerned reproduction is consistently successful. In the end that is the key element for all biological systems.

Biology and Ecology of Corals

Corals are coelenterates or cnidarians that are often predators, particularly on plankton and small fish and invertebrates. These animals were not

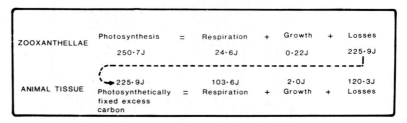

Figure 3 Twenty-four-hour energy budget of a stony coral; energy equivalents in joules (J). The dashed line represents transferring of chemical energy from algae to coral. After Spencer-Davies (1984).

	TABLE 1		
Tabulation of Data on Productivity of Primary Types of Plants Growing on Coral Reefs[a]			
Plant community	Productivity (g cm^{-2} day^{-1})	Range cover on reef communities (%)	Typical contribution to productivity (%)
Benthic macroalgae	0.1–4 (2)	0.1–5	4
Benthic turf algae	1–6 (5)	10–50	45
Zooxanthellae	0.6 (0.6)	10–50	4
Sand algae	0.1–0.5 (0.2)	10–50	0.7
Phytoplankton	0.1–0.5 (0.3)	10–50	0.5
Sea grasses	1–7 (5)	0–40	45
Total			99.2

[a]Note that on the average, the productivity of zooxanthellae in corals and other animals is not the major component, and turf algae and sea grasses are both dominant. After Larkum (1983).

discussed in Chapter 16 because so many have zooxanthellae and are thus a plant–animal symbiotic mix. Coelenterates are basically simple radially oriented animals with tentacles, a mouth, and only a single entrance to the gut (Fig. 4). They are also characterized by stinging cells, or cnidoblasts (Fig. 5), that are used to stun prey. Coelenterates include three major groups: the hydrozoans, the scyphozoans, and the anthozoans. The hydrozoans have only a few calcareous members that have zooxanthelleae, the fire coral, for example, and the scyphozoans are primarily jellyfish. It is

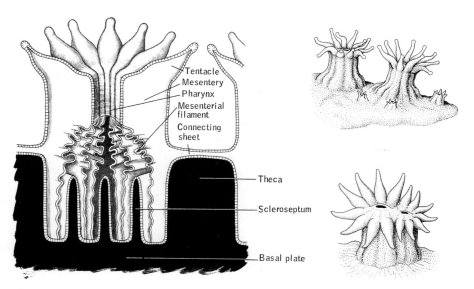

Figure 4 Longitudinal section of a stony coral showing the polyps, connecting tissue, and carbonate skeleton (black). After Barnes (1987).

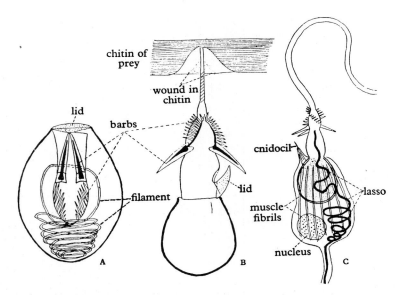

Figure 5 Nematocysts of coelenterates (in this case the fresh water hydroid, *Hydra*): (A) undischarged (within a coelenterate ectoderm); (B, C) discharged. After Borradaile and Potts (1958).

the last group that concerns us here as it includes sea anenomes, gorgonians, soft corals, and the stony corals, many of which have zooxanthellae. The basic structure of the gorgonians is shown in Figure 6. While these animals are mostly colonial and have carbonate and proteinaceous skeleta, the individual polyps are little different from their coral and anemone cousins.

Most stony corals require their zooxanthellae for continuous calcification, and it has been shown that calcification in sunlight exceeds that in the dark by 10 times. Indeed, it is well known that little coral growth and calcification occurs below 50 m in depth and peak growth occurs at less than 20 m depth, primarily due to the availability of light. Stony corals are major carbonate framework producers in reefs, with numerous other organisms providing the fill between the framework (see Smith, 1983). On the other hand, corals are probably only minor providers of photosynthetic energy to most reef ecosystems (Larkum, 1983; Adey and Steneck, 1985).

Stony or scleractinian corals and the reef structure they create require low-nutrient, low-turbidity, warm tropical seas for their success. Relatively small disturbances, whether by hurricanes or humans, provide sediment and nutrients that smother corals and allow the more productive algae to grow over them (Hughes, 1989). Scleractinian corals live a precarious existence in nutrient-poor seas, balanced between grazing levels that will keep their free-living algal competitors at low levels, and

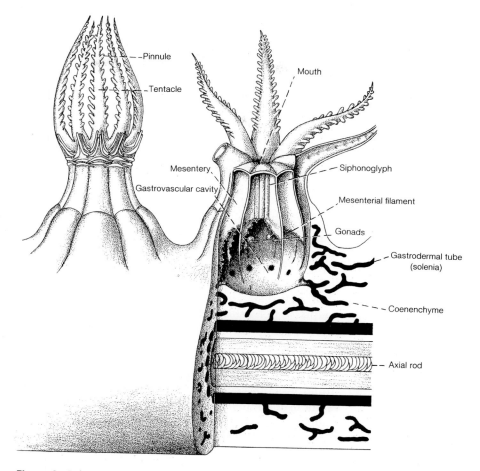

Figure 6 Polyps and flexible proteinaceous skeleton of a generalized octocoral. After Barnes (1987).

predation levels by reef fish and other invertebrates that are moderate. Stony corals have long been shown to be voracious predators themselves, particularly of plankton (Goreau *et al.*, 1979). Although in many cases their need for food in terms of volume is probably low, it seems likely that many species are highly specialized in their planktonic food requirements.

It is not only the stony corals (scleractinians) that harbor zooxanthellae and are important on coral reefs. Some clams, most notable the very large *Tridacna*, anemones, and a scattering of genera in many phyla, also keep algal "farms." Gorgonians and soft corals, perhaps with calcareous spines but with a largely proteinaceous skeleton, are especially crucial ecologically. While these octocorals provide little in the way of calcareous skeletal materials to a reef structure, they often provide a large element of surface heterogeneity and habitat for other organisms. Usually, octocorals are much more tolerant than stony corals of adverse environmental conditions, particularly including lower temperatures and light, increased turbidity, and nutrient levels. Also, most octocorals have developed chemical protection in the form of noxious compounds in their tissue, which make them unpalatable to all but the most specialized of animals, such as the flamingo tongue snail.

Anthozoans and Microcosms, Mesocosms, and Aquaria

The animals that we are considering here mostly have zooxanthellae and thus require light for photosynthesis. Kept in the dark, many corals and their cousins eject their zooxanthellae, turn white, and often do not survive. A species and even an individual is furthermore adapted to a certain light level in its home ecosystem. Changing that light level, even in the wild, can place an animal under considerable stress, at least until it re-adapts. When establishing a living ecosystem, it is important to know the approximate light level from which a colony was derived so that equivalent values can be simulated. Generally, if it is not possible to determine this, light-colored corals are usually from high light situations, less than 20 feet and greater than 700 $\mu E/m^2/s$. Darker corals are usually found under less intense light. The spectrum of the light in most cases probably is not crucial, just as long as it is a reasonably broad band. The accessory pigments of the zooxanthellae allow a broad range of light use, though certainly light in the yellow-orange band is of limited use.

Some anemones are characteristic of highly turbid, high-nutrient waters and provide little management difficulty in that respect. On the other hand, most gorgonians, soft corals, and of course stony corals require high-quality waters. It is our experience that gorgonians and soft corals have only moderate requirements, and if water quality in terms of oxygen, pH, turbity, and nutrients are reasonably correct, little problem is found in maintenance. Some stony corals similarly have minimal requirements. For example, the small *Astrangia danae* occurs from Florida north on the east coast of North America to Cape Cod. It can even be found in the highly turbid lower reaches of Chesapeake Bay and is generally successful in our estuarine models. Although *Astrangia* calcifies and makes colonies as large as a football, it lacks zooxanthellae. On the other hand, most tropical stony corals are extremely sensitive to poor water quality in all respects, both in the wild and in the model ecosystems. One rule of thumb, assuming light and turbidity levels are reasonable, is that nutrient levels, as measured by nitrite plus nitrate, must be below 2 μM for some species and below 1 μM for many species. Whether this situation derives from stony corals not being competitive with free-living algae at higher nutrient concentrations or whether a deeper problem, perhaps related to zooxanthellae capture and maintenance, is involved is not known. It is known that very rich wild reefs strongly disturbed by a hurricane will became densely algal coated, losing much of their live coral surface at least for many years. Although the long-term, large-geographic-scale synoptic evidence is lacking, there is concern among reef scientists that the stony corals of many wild reefs, particularly those in the Caribbean, are in a pattern of general degradation. However, there is no consensus as to whether this results from slightly elevated temperatures, due to a greenhouse effect, greater ultraviolet, due to ozone degradation, lowered pH, due to increased carbon dioxide uptake from the atmosphere, or possibly extremely low levels of organic pollutants.

"Thin-skinned" corals, particularly the Acroporids, do not accommodate well to shipping. In these cases, the water to specimen ratio must be very high. If transport times exceed 5–8 h, then special steps such as a traveling algal scrubber or its equivalent must be taken to insure the highest water quality. Even after the stony corals are safely ensconced in a microcosm or aquarium with the highest water quality, great care must be taken to limit the predation by fish (e.g., parrots, butterflies) and some predacious snails (Coralliomorpha) and worms (e.g., *Hermodice*). Transporting seems to result in a lowered effectiveness of stinging cells, perhaps simply due to agitation and mass firing off in transit.

Even when all the above factors are positive for stony coral survival and growth, colonies may gradually shrink back and eventually die after many years in a system. Perhaps the same chronic "pollution" problem that appears to be operating in the world oceans is also a factor in the fresh waters used for aquarium establishment. On the other hand, it also seems clear that we are dealing with a feeding or trophic balance problem. Corals must obtain enough plankton food, typically at night and often of a very specialized type so that their growth can exceed the predation by the many organisms that are in turn feeding on the coral. Pumps that do not kill zooplankton are discussed in Chapter 4. Methods of artificially raising zooplankton densities are discussed in Chapter 17. Specific coral reef microcosms, mesocosms, and aquaria are discussed in the next chapter.

Growing stony corals are not required for most elements of a coral reef community to survive. Indeed, intense hurricanes sometime damage reefs, leaving little coral, and yet 10–20 years later the calcareous structure of the reef will be growing again, in large part due to stony corals. However, in the permanent absence of growing stony coral, the boring, rasping, and wave action processes that break down the irregular reef structure continue until a smooth pavement remains. This end point has a devastating effect on the reef community, just as removing the larger trees has a devastating effect on the remainder of the rain forest community. Coral growth adequate to build in the face of predators and borers is nearly as difficult to achieve in the modern-day wild world as it is in the mesocosm. Man's effects in raising sediment and nutrient levels, in harvesting higher predators thus freeing direct coral predators to greater activity, and in physically damaging reef structures with boats and diving activites have been pervasive. As we mentioned above, one is even led to suspect the effects of more global human activities. Many reefs live near their upper temperature limits at the end of the warm season. Again, although not proven, many scientists thought that a Caribbean-wide pattern of zooxanthellae rejection by stony coral in the late summer of 1988 might have been due to unusually high temperatures. It is generally thought that the ocean has absorbed roughly one-half of the carbon dioxide that man has released from fossil fuels. If this is correct, some depression of pH is likely. Since stony corals must rely on their algae, in part, to achieve sufficient calcification, this could be just another factor in the environmental degradation of the extremely sensitive reef community. It is very important that mesocosm research on reefs continue to much more sophisticated levels so that we can begin to understand the critical nature of multiple environmental effects on complex natural communities such as coral reefs.

Parasitism

It is of the nature of organic evolution that any opportunities to "make a living" will be followed. The potential availability of food energy in the form of a living organism will inevitably lead to predation and parasitism by other organisms, and in this discussion of parasitism in closed aquatic systems, we include pathogenic bacteria and viruses. In the broadest sense, this is certainly not a bad situation. When a particular organism makes an evolutionary advance that allows its populations to greatly increase, the increase in numbers (because it makes the infection step easier) also renders the organism more suseptible to a parasite explosion. Ecologically the alternative is control by starvation or a runaway population increase of prey or host, which in the end is destructive of stable community structure. This process we humans have seen many times as we have transported organisms to places lacking their natural controls, such as rabbits and cactus to Australia. In both cases parasites had to be imported to keep the organisms from massively disrupting the new environments. Indeed, with the advance of modern medicine, in part removing or avoiding human parasites, we are witnessing an uncontrolled explosion of populations of *Homo sapiens*, which, because of human technology and mobility, will be devasting to the entire earth. To further try to dispel a tendency of humans to regard parasites in a very negative light, perhaps even as "unnatural," we would point out that parasitism has played a very important role in evolution. Much of the complexity and abilities of higher plant and animal cells have derived from the "capture" of parasites, or a détente between parasite and host, and the use of these initially foreign bodies as organelles. Also, at the tissue and organism level, much of the complexity now seen has derived from adaptations and evolution in response to the incursions of parasites. One of the most brilliant of scholarly environmentalists, René Dubos, spent the early part of his career as a bacteriologist, searching for means of germ control and "tempt(ing) humanity with an edenic vision of a sanitized world" (Nash, 1989). Yet, he eventually concluded that "people and germs should coexist just like people and wolves."

Thus, predators and parasites alike are important to balanced organic evolution. Parasites, by their nature, are closely associated with their hosts. The concomitant to an organism taking up the parasitic life, particularly an internal parasite, is the difficulty of meeting a sexual partner and then passing the results of reproduction on to another host. Pure distance and relative isolation of host make this process of infection more difficult. Model systems, by their relatively small nature, increase the likelihood of infection and of epidemics of parasites once a parasite has gained access. On the other hand, most model ecosystems are isolated or semi-isolated from plant and animal populations of wild ecosystems. The operator potentially possesses the ability to prevent the initial access of parasites more definitively than would be possible in the wild. Thus, either as an important modeling concern to achieve natural levels of parasitic loading, or as an esthetic concern as parasitized organisms may look or act poorly or even die, control of parasites becomes an important endeavor of the manager of model ecosystems.

Whether human, animal, or plant health is at stake, avoidance of disease, which is in large part parasitism, is achieved by three general approaches:

1. Provision of a high-quality environment and balanced food that allow organisms to function in an unstressed state (remember that some diseases result from poor physical or chemical environment or lack of proper nutrition and do not involve other organisms). Even where a parasite is involved, the successful functioning of a potential host also relates to physical and chemical health and a fully intact immune system.

2. Prevention of disease transmission.

3. Cure of the disease.

Although we will discuss all three approaches to disease control, it is the first two that we will emphasize in this chapter.

Environment, General Health, and Disease

Organisms are adapted in various ways to a wide variety of physical, chemical, and biological factors. Generally speaking, these parameters can be determined by those of the wild environment within which the organism is successful, remembering that extremes are likely to be marginal for many species. Establishing an organism under conditions of relatively high or low temperature, salinity, pH, oxygen, etc., in unusual habitats, such as no cover or improper substrate, or under unusual biotic conditions, will place an organism under stress. Under those conditions, parasites that are only rarely successful in gaining entrance to the host, or perhaps are established but under equilibrium conditions with the host's defenses, will become dangerous and perhaps lethal to the host. Likewise, any organism, plant or animal, has certain food or nutrient requirements. Poor health can result from inadequate diet or nutrient and water supply for a plant and allow parasite success where normal defenses might otherwise be adequate. Providing an environment biotically and physically as close as possible to the wild will most likely avoid this problem as well (Yeaman, 1987). However, in small model systems, the imported productivity or food supply from adjacent simulated ecosystems in the wild will necessarily lead to the need for a careful consideration of adequate artificial food supply (see Chapter 17).

The above points are crucial to operating ecosystems that are not to be controlled by parasites and pathogens. In summary, it is essential to maintain an environment more than just adequate for the organisms. If oxygen levels for a particular community are supersaturated in the wild for 8 h per day, the bubbling of air in a model of that community cannot maintain a proper environment. While bubbling of oxygen might suffice in this case, controlled plant photosynthesis offers a logical solution to this need as well as several concomitant requirements. Our attempts to achieve environmental accuracy for a number of model ecosystems are described in Chapters 21–25.

Quarantine (Prevention of Transmission)

The transport and temporary storage of organisms destined for model ecosystems is highly stressful. If at all possible every effort should be taken to reduce the transportation stress by limiting damage in collection and by providing a proper environment en route. Nevertheless, the immune systems of all introduced organisms, particularly intermediate- and large-sized organisms, are frequently in the process of collapse. Control of parasite injection into a model ecosystem can be handled by quarantine of the organisms in a separate quarantine tank for several weeks until diseases can become apparent. Since the transport is environmentally such a crucial time, it is essential, however, that the quarantine system be operated at a water quality that equals or exceeds that in the full model being attempted. Otherwise the quarantine process becomes a severe limitation instead of an advantage.

We have found that no matter how high the water quality of freshwater systems, general quarantine is essential if outbreaks of disease are to be avoided. While this process is desirable for saltwater ecosystems, con-siderable experience, on the part of the authors, in situations where quarantine could not be used for practical reasons suggests that this step can be omitted for saltwater systems with greater impunity. The reason for the apparent relative lack of parasite and pathogen effects in saltwater models is not known.

Disease Treatment in Model Ecosystems

Serious or epidemic disease in model ecosystems is best avoided by main-taining high-quality environments and by appropriate quarantine. Non-epidemic disease probably cannot and should not be avoided. In some cases where considerable efforts to manage epidemic disease involving a particular species in an ecosystem model are not successful, it is best to allow both the species and its parasite to run to extinction. This happens in the wild and may be inevitable in a given environmental situation and ecological patch. In a properly designed and diverse ecosystem, this should cause little problem with regard to overall operation.

Diseases that do not result from deficiencies in the basic require-ments of an organism can result from the effects of bacteria, fungi, viruses, protozoans, certain worms, particularly flatworms, nematodes, and some arthropods, especially among the isopods, amphipods, and insects. Para-sites occur in virtually all animal and plant groups, though those men-tioned above have some taxa that are particularly successful in this mode of life. Transmission from host to host is a crucial problem for a parasite. In many cases, the host-to-host transmission is direct, with infectious stages air-borne or with swimming stages in water. Penetration of the host can be direct, through body openings, gills, feeding, reproductive tract, etc. In many cases the life cycle is more complex, involving intermediate hosts. Elaborate double- or triple-fixed life cycles involving transfer through several hosts also occur. These are usually tied to feeding pat-terns, with parasites such as mosquitoes introducing smaller parasites such as protozoans and worms that have already gone through a phase in the mosquito body. In many cases the infection is acquired through the digestive tract through feeding on an intermediate host such as a snail or worm. Almost all larger parasites can effectively and inadvertently intro-duce bacteria, fungi, and viruses to form secondary infections.

This is not the place for an in-depth treatment of disease in general; the interested reader is referred to O. Olsen (1974).

Diseases of Fish

Humans have kept fish in culture for many centuries, and methods of treatment have developed for some of the most serious problems that are likely to be encountered. Most quality aquarium books have a section on fish diseases, and we will not repeat the appropriate treatments here. We refer the interested reader to several books: Spotte (1973), Emmens (1985), and Hunnam (1981).

We add a note of caution to all planned treatment of organisms in living ecosystems. It may be extremely important to the aquarist to save a favored fish or group of fish that have contracted a disease. However, if treatment cannot be handled by isolation or physical removal, and chemical methods must by resorted to, it should be remembered that rarely is a chemical specific to a single species of parasite. Algae and other plants, in most cases, are critical elements of the model ecosystem and are likely to be negatively affected. Likewise, most bacteria, fungi, protozoans, worms, and small arthropods are essential to ecosystem function. Most chemical treaments especially antibotics, copper, and formaldehyde must be scrupulously kept from the ecosystem itself. We have been moderately successful in using quinine to control ick (a protozoan) in fresh water systems, without obvious ecosystem effects.

It is valuable to recall that the widespread human use of insecticides and herbicides in the wild enviroment has placed natural ecosystems and specific organisms in jeopardy numerous times. The classic story of DDT is a prime example, and it is suspected that we will see many more chronic examples of this kind of pollution of wild ecosystems. The recent world-wide concern for a potential catastrophic loss of amphibians probably lies with insecticides or herbicides. Where parasitism must be dealt with in model ecosystems, biological methods should be used wherever possible Use of cleaner fish or shrimp is helpful for ectoparasites in marine systems. Human removal of a pest by physical processes, using hand, needles, water jets, etc. can be valuable, though often time-consuming. As we will describe in future chapters, we have found that in wetlands ecosystems, control of insects on vegetation can be achieved by appropriate use of jets of water. It would appear that in the wild swamps, marshes, and mangroves of the Everglades or Chesapeake Bay, wind and rain are major limitors of insect depredations. The simple approach of simulating a thunderstorm every week or so avoids serious insect predation. This is all the more interesting in that in the production greenhouse adjacent to the Everglades mesocosm (described below) extensive and repeated applications of highly toxic insecticides are employed as a routine element of management. Although we have lost a few plant species in the adjacent Everglades greenhouse, over 95% survival, with a minimum effect of parasites on the remaining species, has been achieved without resort to insecticides.

References

Adey, W., and Steneck, R. 1985. Highly productive eastern Caribbean reefs: synergistic effects of biological, chemical, physical and geological factors. *The Ecology of Coral Reefs.* M. Reaka (Ed.). *NOAA Symp. Ser. Underwater Res.* **3**: 163–187.

Bold, H., and Wynne, M. 1985. *Introduction to the Algae.* 2nd Ed. Prentice Hall, Englewood Cliffs, New Jersey.

Barnes, R. 1987. *Invertebrate Zoology.* 5th Ed. Saunders College, Philadelphia.

Borradaile, L., and Potts, F. 1958. *The Invertebrata.* Cambridge University Press, Cambridge.

Emmens, C. 1985. *The Marine Aquarium in Theory and Practice.* 2nd. Ed. TFH Publications. Neptune City, N.J.

Goreau, T., Goreau, N., and Goreau, T. 1979. Corals and coral reefs. *Sci. Am.* **241:** 124–136.

Hughes, T. 1989. Community structure and diversity of coral reefs: The role of history. *Ecology* **70:** 275–279.

Hunnam, P. 1981. *The Living Aquarium.* Ward Lock Ltd., London.

Larkum, A. 1983. The primary productivity of plant communities on coral reefs. In *Perspectives on Coral Reefs.* D. Barnes (Ed.). Australian Institute of Marine Sciences, Manuka, Australia.

Lobel, P., Anderson, D., and Durand-Clement, M. 1988. Assessment of ciguatera dinoflagellate populations: Sample variablity and algal substrate selections. *Biol. Bull.* **175:** 94–101.

Nash, R. F. 1989. *The Rights of Nature.* The University of Wisconsin Press. Madison.

Olsen, O. W. 1974. *Animal Parasites, Their Life Cycles and Ecology.* University Park Press. Baltimore

Smith, D. C., and Douglas, A. E. 1987. *The Biology of Symbiosis.* Edward Arnold, Baltimore.

Smith, S. 1983. Coral reef calcification. In *Perspectives on Coral Reefs.* D. Barnes (Ed.) Australian Institute of Marine Sciences, Manuka, Australia.

Spencer-Davies, P. 1984. The role of zooxanthellae in the nutritional energy requirements of *Pocillopora eydouxi. Coral Reefs* **2:** 181–186.

Spotte, S. 1973. *Marine Aquarium Keeping, the Science, Animals and Art.* John Wiley and Sons, New York.

Talbot, F. 1984. *Reader's Digest Book of the Great Barrier Reef.* Reader's Digest Books, Sydney, Australia.

Thurman, H., and Webber, H. 1984. *Marine Biology.* Merrill Publishing, Columbus, Ohio.

Yeaman, M. 1987. Aquarium microbiology. *Freshwater and Marine Aquarium* **10**(9):37–43.

PART IV

Ecological Systems in Microcosms, Mesocosms, and Aquaria

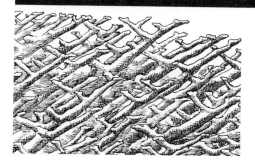

CHAPTER 21

MODELS OF CORAL REEF ECOSYSTEMS

Coral reefs are large calcium and magnesium carbonate structures formed by tropical, shallow-water communities of plants and animals occurring in clear, well-lit, high-quality coastal waters of moderate to intense wave action. They are dominated and built by stony corals and/or calcareous algae (see Kaplan, 1982; Talbot, 1984). Where they are algal-dominated and extend near or above the water surface, especially in strong wave-beaten areas, reefs are sometimes called algal ridges (Adey, 1978a,b). Large-scale reef structures such as atolls and barrier reefs are constructed in large part by species of the calcareous green alga *Halimeda* growing on back reefs and in lagoons. Wild coral reefs and their lagoons are typically highly productive communities due to the photosynthesis of abundant small algae (algal turfs), calcareous algae, sea grasses (in the lagoons), and the zooxanthellae algae of corals and other sedentary invertebrates (see Adey and Steneck, 1985).

Although long considered to be impossible to maintain in closed systems, considerable progress in maintaining some species characteristic of coral reefs has occurred in the past several decades. Bruce Carlson (1987) describes the history of the various technologies involved. In the more recent aquarium trade, however, a "reef system" has come to mean technically something rather specific. A typical "reef system" is a large, well-lighted, and fairly turbulent tank with abundant *Caulerpa* and *Bryopsis* algae, anemones, soft corals, and sometimes with a few of the resistant stony corals. Fish and the more mobile invertebrates are greatly limited. Strong lighting (metal halide or fluorescent lamps), often with actinic fluorescents to provide an esthetically desired blue color, is universal and the water-quality control apparatus is a complex array of filters, the exact number and type of which varies from system to system. Perhaps the best recent description of the modern "reef system" of the aquarium hobbyist is that of Moe (1989).

In this chapter we will describe three quite different model coral reefs that we have been involved in constructing and maintaining at various times over the past ten years. The first, a Caribbean reef ecosystem, is the 3000-gallon exhibit in the Smithsonian's Natural History Museum that followed a series of smaller prototypes, and is the system from which many of the basic modeling techniques that we have discussed have evolved (Walton, 1980; Adey, 1983). The second is a very large, 800,000-gallon, Indo-Pacific reef built in 1986–1988 at Townsville, Australia, for the Great Barrier Reef Marine Park Authority. Finally, we describe a 135-gallon generalized reef ecosystem that we have built and maintained in our home for several years.

Caribbean Coral Reef Microcosm

In this section we describe the 3000-gallon (7 kl) exhibit/research microcosm that was in operation at the Smithsonian Institution for about 8 years. It was the fourth closed reef system developed at the Marine Systems Laboratory. The earlier units were smaller prototypes. Its physical dimensions, arrangement, and morphology are shown in Figure 1. This system was taken over by the Exhibits group in 1988 and unfortunately the building heating/air conditioning system has been allowed to degenerate during the same time interval. The unit is still very much functioning as an ecosystem, but that of a highly temperature-stressed system such as one might see in a near shore community. In this discussion we treat the unit as it operated from 1980 to about 1988.

The shape and community structure of this microcosm reef were scaled after a typical Caribbean bank barrier reef, as were the light, wave energy, and current conditions. The critical physical–chemical variables and their relation to an actual reef in St. Croix, Virgin Islands, are discussed below. The reef substructure contained within the Smithsonian microcosm was built from dead reef carbonate material (derived from storm berms behind Caribbean reefs) to simulate characteristic barrier reef configuration. This "reef" structure is in deeper water in the forereef, crests near the water surface, and drops to a shallow sandy lagoon behind. Considerable fine carbonate sediment is produced by boring and rasping of organisms on this reef. Both the sea grass-rich lagoon and the interstices of the rubble-built reef itself continuously collect this sediment. Thus, these two areas act as geological storage systems (see Chapters 8 and 18)

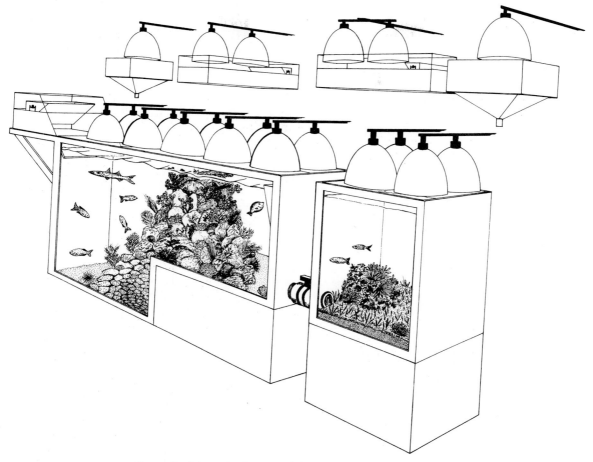

Figure 1 Schematic diagram of the Caribbean coral reef and lagoon model opened at the Smithsonian Institution in October 1980. Scrubbers and refugia overlie. Drawing by Charlotte Johnson.

for both carbonate sediment and entrained organic particulates. These processes are quite similar to those working on wild reefs and provide a basal structure to the model that is very much the same as that found in the wild (see Adey, 1978a). The rich benthic infauna occuring in this sediment (Chapter 18, Figs. 22 and 23) and the obvious bioturbation it carries out demonstrate the effectiveness of this sedimentation system. In this microcosm, the refugia (Figs. 1 and 2) also contain sediment traps. Exchange water is removed from the traps along with entrained sediment and this provides for "geological export" from the model.

In the prototypes, the sandy back reef area had developed the over-grazed and reworked aspect known as a reef halo. Thus, in the exhibit

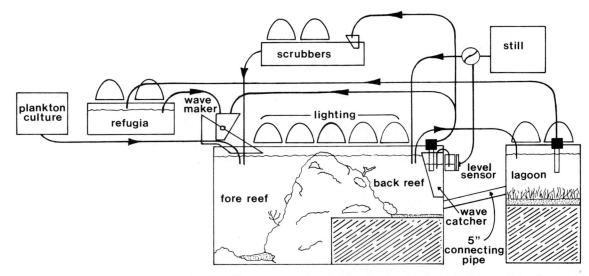

Figure 2 Plumbing/flow diagram of the Smithsonian coral reef model. This particular diagram is that of the prototype system. It is quite similar to the exhibit model except that the lagoon is connected to the reef proper by two 5-inch pipes and the replacement water supply (still) has been replaced by a reagent grade deionizer.

system described here, the lagoon was physically isolated by a pair of 5-inch pipes with $\frac{1}{2}$-inch mesh screen. This separation was designed to simulate distance from the reef, an approach that allows controlled grazing and successful development of a lagoon community while maintaining the normal water flow and sediment relations between the reef and its lagoon. The microcosm described here is more properly considered to be two attached ecosystems, reef and lagoon. The ocean or shelf ecosystem is not present but it is simulated by a wave generator and the bank of algal turf scrubbers as we discuss below.

Living material to be transplanted to this microcosm was collected in the field and returned to the laboratory by air in insulated containers containing aeration equipment. This operation averaged 6–10 h from ocean to tank. The water contained within the tank microcosm was initially prepared with artificial sea salts but has been mixed with natural sea water at a rate of about 2 gallons per day (2% per month). During most of the operating time of this system, the sea water used for exchange was obtained from the outer coast of Virginia or Delaware. It was coarse-filtered in batch culture, and an algal culture was allowed to bloom; after several days, each batch was diatom-filtered to scrub nutrients, but otherwise it was untreated. During later years, the tendency was to use commercial sea salts for the makeup media for logistic reasons. The primary reason for this small water exchange is to avoid drift of salt composition.

Salt loss occurs with evaporation, such salt being evident around the laboratory, particularly on the overhead beams.

The wave action and current flow necessary for reef function are induced by six centrifugal pumps totaling 140 gpm. These pumps remove water from the back reef and lagoon end of the tank and deliver it to a double-bucket "wave generator" in deeper water at the fore reef (Fig. 2). The wave generator overturns every 10–15 s, spilling the contents into the tank. This flow, combined with that of several additional pumps that do not cycle through the wave generator, produces a mean current across the reef flat of approximately 3 cm/s. This is low compared to the mean flow rate of 10 cm/s found on the windward reef flats off St. Croix, but it is within the range of the more protected sections of that reef. The wave-induced flow is irregular and surge-like, back and forth (Chapter 5, Fig.

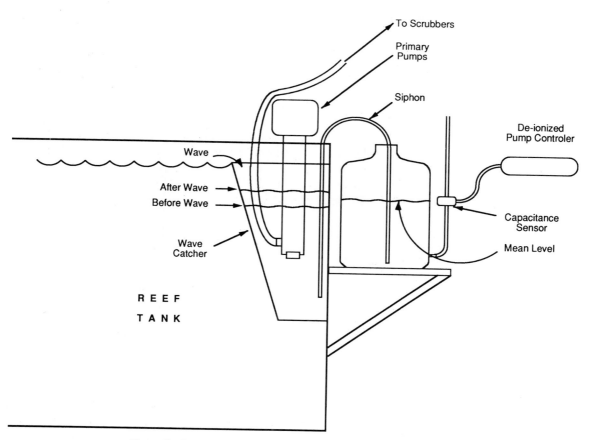

Figure 3 Diagram of wave catcher and water-level amplifying system on the Smithsonian coral reef.

11), and functions to enhance organism/water contact, much as happens in a scrubber (see Chapter 12).

Wave reflection at the downstream (back reef) end of the reef tank is prevented by a box-like "wave catcher" (as shown in Fig. 3). This is a separate reservoir containing the intakes for the pumps supplying the wave maker. As the water is pumped from this enclosure, the level drops, to be filled again as the next wave rolls across the tank and spills into the open top. This mechanism is also a critical element in salinity control, since water level in the catcher is amplified in proportion to tank levels. The wide level of fluctuation with wave fall in the catcher is smoothed by a small siphon/reservoir device and the level is monitored by an electronic capacitance sensor that activates a peristaltic pump delivering distilled water to the tank. With this device, detection and replacement of as little as 50–100 ml of evaporated water at one time is possible. The procedure mantains salinity over a very narrow range on an hour-to-hour basis. Generally, it has been maintained at 35.5–36.5 ppt, to match eastern Caribbean reef water, although on occasion it has been allowed to fall as low as 34.8 ppt or to climb as high as 37.0 ppt for several weeks. Approximately 40 gallons/day (80 l) is evaporated from the entire reef/lagoon system. Since such large volumes of replacement water are required, and the Washington, D.C., tap water ranges from 120 to 180 μM for dissolved nitrogen (nitrite plus nitrate), high-quality deionized water is used for evaporative replacement.

The fluorescent illumination commonly used for culture work on both terrestrial and aquatic organisms, even the more powerful VHO or HO lamps, cannot generate the intensity of natural solar radiation normally incident to shallow tropical reef areas in this size tank. Clear, metal-halide vapor lamps (General Electric and Sylvania) have proven satisfactory in intensity and spectral quality. Ten 1000-W lamps provide an irradiance over the 4.5-m^2 area of the reef tank, which is approximately that measured in the field (Fig. 4). In the back reef section of the microcosm, light energy measures 500–900 $\mu E/m^2/s$, as compared to 1100 $\mu E/m^2/s$ recorded in the late spring at a depth of 1 m on the reef flat at St. Croix. To compensate for the somewhat lower instantaneous energy, the light periods have been expanded to bring total light energy input to comparable levels in the microcosm (Fig. 5).

Lamps lighting the main reef tanks are controlled by five separate circuits, and the illumination periods vary for each. Activation of circuits is sequential, operating over approximately a 2-h period, simulating in step-like fashion a changing intensity from dawn to mid-morning and then again from late afternoon to dark. The crest area experiences the longest light period (16 h in summer), and the deepest areas experience the shortest (14 h in summer). This effectively increases the depth scaling factor.

A summer–winter temperature cycle approximating that on St. Croix reefs is followed partly by adjusting the level of air conditioning in

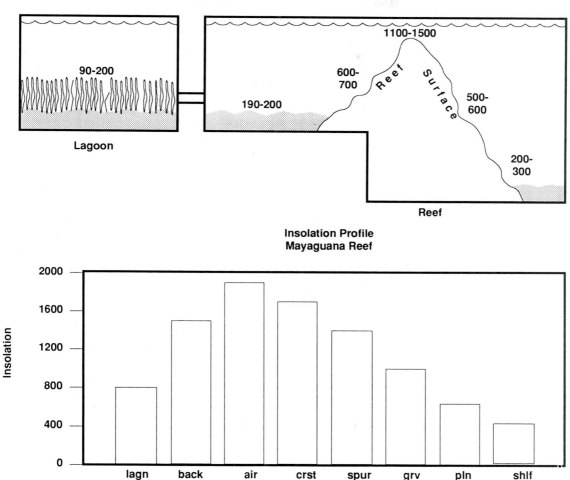

Figure 4 Light levels measured in μE/m²/s on the reef surface of the Smithsonian coral reef, as compared to light levels on a Bahamian reef at 1400 h on a clear day in May.

the tank room. A heat exchanger/refrigeration system is also used, particularly in late spring or during fluctuations in building air conditioning. Summer temperatures in the tank lie between 27 and 29°C, and usually follow a daily range of less than 1–1.5°C. The museum building is very old and the temperature control often malfunctions. In recent years this

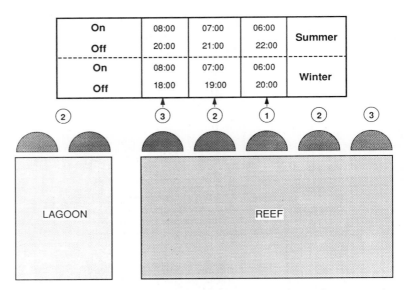

On	08:00	07:00	06:00	Summer
Off	20:00	21:00	22:00	
On	08:00	07:00	06:00	Winter
Off	18:00	19:00	20:00	

Figure 5   Typical timing of the main tank lighting system on the Smithsonian coral reef.

has led to occasional serious problems with high-temperature "spikes" (to 31°C), a wide range of water temperature (3–4°C(per day), and the loss of sensitive organisms. Winter temperatures are generally kept between 25 and 27°C, also with maximum daily range of less than 1.5°C, though in recent years, this too has tended to fluctuate more widely.

No mechanical or bacterial ("biological") filtration, air bubbling, or chemical conditioning is incorporated into this system, nor has it ever been necessary to use such methodology. During the daylight hours no conditioning is needed as extensive photosynthesis on the reef surface renders ammonia unmeasurable by standard techniques and maintains oxygen at supersaturated levels. A bank of two to eight "algal turf scrubbers" (see Chapter 12), maintains water quality during the dark hours. In this system, the number varies primarily in accordance with research projects that are underway.

Dissolved oxygen concentrations as measured on a Martek water-quality analyzer range from a minimum of 5.5–6.2 mg/l in early morning to a maximum of 7.5–8.3 mg/l in the late afternoon. This diurnal pattern is close to that measured on a St. Croix reef flat with moderate wave action (Chapter 10, Fig. 9). Since this microcosm is a closed system and the reef is physically very short, use of the upstream–downstream method to monitor primary productivity is impractical. The rate of change of oxygen concentration at morning and evening times of saturation is measured to determine utilization at night and production during full illumination. This avoids the complex problem of oxygen exchange with the atmosphere. This approach indicates a gross primary productivity of 32.5 g

$O_2/m^2/day$, which compares to 20–80 g $O_2/m^2/day$ for back reefs and 18–20 g $O_2/m^2/day$ for fore reefs as measured in St. Croix. Community respiration for the entire system is 0.9 g $O_2/m^2/day$. This compares to an average rate of 2.6g $O_2/m^2/day$ for St. Croix backreef zones, and 0.7 g $O_2/m^2/h$ for forereef. The tank measurements were taken without the algal turf scrubbers attached.

The pH is monitored primarily as a measure of the state of the carbonate system, particularly the concentration of CO_2 in the tank water. The relationship between pH, photosynthesis, and calcification in some degree affects all calcifying reef organisms and is critical in the microcosm. The microcosm range is between 8.05 and 8.25, in accordance with a daily cycle that peaks at about 1900 hours from a minimum at 0800 hours.

Nutrient levels in the tank vary according to the import–export regime employed (that is, the amount of algae removed as compared to feed added). The primary variables measured have been ammonia (NH_3), nitrite (NO_2^-), (nitrate NO_3^-), and phosphate (PO_4^{3-}), all of which show diurnal cycles when levels are relatively high. At low concentrations, only NO_2^- plus NO_3^- shows a clear diurnal cycle, correlated with primary production. Nutrient levels typically range from 0.6 μM (0.008 ppm) to 1.5 μM (0.02 ppm) for dissolved nitrogen as nitrite plus nitrate. These are slightly above the average found in St. Croix, although well within its upper limits. (see Chapter 12, Fig. 10). Ammonia has a mean concentration in the tank of about 0.2 μM. Comparable data are not available for the field, and this figure is likely high because of the museum environment.

A bank of five to eight algal turf scrubbers as described in detail in Chapter 12 are utilized to regulate nutrients, pH, and oxygen. During a period when the import/export ratio was experimentally maintained greater than one by greatly increasing animal (crab) biomass in a refugium tank, NO_2^- plus NO_3^- concentration remained at a constant 5–8 μM. Increasing the export rate eventually drove this measurement to values generally below 1.0 μM, other measured nutrients being proportional. When this decreasing concentration reached 2 μM, heterocysted blue-green algae, absent at higher levels, become important constituents in the algal turf community. This same phenomenon can be demonstrated in attached experimental algal turf scrubbers, and when concentrations of nitrate are below 1 μM, measurable atmospheric nitrogen is fixed by the algal turf (Lucid, 1989; also see Chapter 12).

Plankton supplied to a wild reef system from the open ocean can account for only a small part of the energy utilized in the reef food web. The estimate for a St. Croix reef is less than 1%. Nevertheless, it is probably an important component for some filter-feeding animals. Live brine shrimp (*Artemia salina*) and dried krill are added to the microcosm to simulate this input. Approximately 0.5 g $C/m^2/day$ (2 g dry weight) is introduced by this means as compared to about 17 g $C/m^2/day$ from primary productivity (plant photosynthesis).

Several additional sanctuary or refugium tanks are placed in the reef circulation (Fig. 2). Due to reduced levels of predation on some species within the refugia, species diversity can thus be increased to levels equivalent to much larger areas under natural conditions. A number of plants and animals that cannot currently be kept in the main tank in any numbers due to predation, function quite well in refugium tanks. The anemone *Aiptasia pallida* is a striking example, being virtually absent in the reef tank yet flourishing in the refugia.

Seasonal sampling in the fore and back reef areas in St. Croix has yielded from 35 to 50 larger benthic algae. The reef microcosm contains at least 35 species at any one time (Table 1). Most of the important turf species are prominent in both places, but several differences have been found: *Smithsoniella subterranea*, abundant in the Smithsonian tank, is found only occasionally in St. Croix. This species was discovered in the tank and subsequently located in the wild. Also, several important elements in the flora there, such as *Amphiroa fragilissima*, *Taenioma macrourum*, and *Asparagopsis taxiformis* (falkenbergia stage), have not been found in the microcosm. In spite of these deviations, the algal species diversity appears to approach natural levels, considering the small size of the reef microcosm. The algal turf species that characterize the algal scrubbers of this microcosm were shown in Chapter 12, Table 2.

The turtle grass *Thalassia testudinum* has been grown in the associated reef lagoon for nearly 10 years (Williams and Adey, 1983). Flowering was found toward the end of the first year. Mature blade length varies with time and the intensity of fish grazing and is generally well within the range encountered in the field. Under minimum grazing conditions, the blade length can greatly exceed that typical for field conditions. Blade length, life span, and colonization rate increase markedly when browsing fish are introduced to remove epiphytes.

The animal and protist community structure in the microcosm is certainly not as complex as that found on a large natural reef. However, many major elements are present in the microcosm. Those tallied during the past several years are given in Tables 2–4.

The vertebrate species have proven to be the easiest segment of the reef biota to manage. The problems of disease, common in many aquarium installations, are virtually nonexistent, and fish that have suffered injury through accident or territorial defense often return to health spontaneously if they do not undergo further predation. No quarantine procedures are used for the introduction of new fish, and on a number of occasions, fish apparently diseased when introduced have returned to good health without chemical medication. Behavior patterns are essentially normal. Feeding, territorial responses, and at least in some cases reproductive responses parallel those observed in the wild. The population includes common grazers, scavengers, and invertebrate predators. In the system described here, higher vertebrate predators have been excluded because of the depletion that would result in the absence of a large

---

TABLE 1

Plants Present in the Smithsonian Coral Reef and Lagoon Microcosm[a]

Kingdom Monera
  Division Cyanophycota
    Calothrix crustacea
    Oscillatoria submembranacea
    Schizothrix sp.
    Spirulina subsalsa
Kingdom Plantae
  Subkingdom Thallobionta
  Division Rhodophycota
    Asterocytis ramosa
    Goniotrichum alsidii
    Acanthophora spicifera
    Bryothamnion seaforthii
    Centroceras clavulatum
    Ceramium corniculatum
    Ceramium fastigiatum
    Heterosiphonia secunda
    Laurencia papillosa
    Polysiphonia havanensis
    Spyridia sp.
    Botryocladia occidentalis
    Coelothrix irregularis
    Halymenia floresia
    Halymenia pseudofloresia
    Porolithon pachydermum
    Neogoniolithan solubile
    Lithothamnium ruptile
    Jania adherens
    Peysonnellia rubra
    Gelidiella trinitatensis
    Gracilaria cylindrica
    Gracilaria domingensis
    Gracilaria sp.
    Hypnea spinella
    Neoagardhiella ramosissima
    Soleria tenera
  Division Chromophycota
    (browns)
    Ectocarpus elachistaeformis
    Giffordia rallsiae
    Sphacelaria tribuloides

Division Chromophycota continued
  (diatoms)
    Nitzschia closterum
    Nitzschia dissipata
    Nitzschia leutzingiana
    Nitzschia longissima
    Nitzschia 2 spp.
    Opephora schwartzii
    Plagiogramma 2 spp.
    Surirella fastuosa
    Amphora 5 spp.
    Cocconeis placentula
    Cocconeis trachyderma
    Cocconeis sp.
    Fragillaria sp.
    Navicula varians
    Striatella sp.
    (dinoflagellates)
    Gymnodinium sp.
    Polykrikos schwarzi
  Division Chlorophycota
    Chaetomorpha geniculata
    Chaetomorpha gracilis
    Chaetomorpha fascicularis
    Chaetomorpha fuliginosa
    Valonia ventricosa
    Enteromorpha prolifera
    Pringsheimiella sp.
    Penicillus capitatus
    Udotea flabellum
    Bryopsis hypnoides
    Caulerpa cupressoides
    Caulerpa mexicana
    Derbesia vaucheriaeformes
    Halimeda incrassata
    Halimeda monile
    Halimeda opuntia
    Halimeda tuna
  Subkingdom Eumetazoa
    Thalassia testudinum (turtle grass)

[a] This survey was undertaken during the years 1985 and 1986. The system was about 5 years old at that time.

TABLE 2

Benthic Protists Occurring in the Smithsonian Coral Reef

Subkingdom Protozoa
  Mastigophora (flagellates)
    Stephanopogon mobilensis
  Sarcodina (amoeboid types)
    Acanthamoeba sp.
    Allogromia sp.
    Gromia sp.
    Hyalodiscus sp.
    Mayorella sp.
    Actinophrys sp.
    Ameoba 2 spp.
    Hartmannela sp.
    Limax sp.
    Rotaliella sp.
  Ciliphora (ciliates)
    Acineta sp.
    Ascobius sp.
    Chilodonella sp.
    Cinetochilum marinum
    Condylosoma patens
    Diophyrus 3 spp.
    Euplotes 3 spp.
    Lagotia sp.
    Loxophylum sp.

Ciliphora (ciliates)
continued
    Pleuronema sp.
    Sonderia sp.
    Strombidium 2 spp.
    Tracheloraphis sp.
    Urosoma sp.
    Vorticella sp.
    Actinotricha 3 spp. etc.
    Aspidisca 3 spp.
    Ciliofaurea sp.
    Cohnilembus sp.
    Cyclidium sp.
    Dysteria 4 spp.
    Keranopsis rubra
    Lionotus 2 spp.
    Metafolliculina producta
    Prorodon sp.
    Stichotricha gracilis
    Trachelocerca sp.
    Uronychia transfuga
    Vorticella marina

ᵃThis survey was undertaken during the years 1985 and 1986. The system was about 5 years old at that time. Courtesy D. Spoon.

---

reef surface for feeding and an open ocean pool for recruitment. For over 2 years a 12- to 15-inch-long barracuda was maintained in the lagoon without significant predation on other fish. This higher predator was fed small goldfish daily. Since the goldfish show stress in the sea water, they are attacked instantly in preference to other fish.

Of particular interest is a school of striped parrot fish, *Scarus inserti*. Early in the development of the system, these fish were introduced as small juveniles and matured to assume the social structure characteristic of the species, including a single brightly colored terminal male and two heavily contrasted black-and-white dominant females, one dominating the forereef and one the backreef. Also, two species of damselfish, the beaugregory and the three-spot, have at times laid and fertilized egg clutches. These eggs are tended by the males and apparently hatch normally. After a clutch has hatched, the larval stages can be found in the tank plankton for 12–24 hours. To date, no attempt has been made to raise these larvae, and all have been lost to predation or the centrifugal pumps.

TABLE 3

Invertebrates Present in the Smithsonian Coral Reef and Lagoon Microcosm[a]

Kingdom Animalia
  Subkingdom Placozoa
    Trichoplax adhaerens
  Subkingdom Parazoa (porifera)
    Callyspongia sp.
    Chondrosia collectrix
    Clathria sp.
    Tethya sp.
    Chondrilla nucula
    Chondrosia sp.
    Monanchora barbadensis
  Subkingdom Eumetazoa
    Phylum Cnidaria (coelentrates)
      Class Anthozoa
        Aiptasia tagetes
        Condylactis gigantea
        Stoichactis helianthus
        Rhodactis sanctithomae
        Briareum abestinum
        Gorgonia flabellum
        Muricea sp.
        Plexaura flexuosa
        Pseudopterogorgia sp.
        Acropora palmata[b]
        Dendrogyra cylindricus
        Diploria clivosa
        Diploria strigosa
        Isophyllia sinuosa
        Montastrea annularis
        Mussa angulosa
        Porites asteroides
        Palythoa caribaeorum
        Bartholomea annulata
        Phymanthus crucifer
        Ricordea florida
        Eunicea mammosa
        Gorgonia ventalina
        Muriceopsis sp.
        Plexaura homomalla
        Colpophyllia natans
        Dichocoenia stokesi
        Diploria labyrinthiformis[b]
        Eusmilia fastigiata
        Madracis mirabilis
        Montastrea cavernosa
        Mycetophyllia lamarckiana[b]
        Porites porites
        Zoanthus sociatus
      Class Hydrozoa
        Staurocladia sp.
        Millepora alcicornis
        Millepora complanata

Phylum Platyhelminthes
  Anaperus sp.
  Plagiostomum sp.
Phylum Gastrotricha
  Macrodasys sp.
Phylum Mollusca
  Class Gastropoda
    Astraea tecta
    Cittarium pica
    Coralliophila caribaea
    Murex sp.
    Bulla occidentalis
    Coralliophila abbreviata
    Cyphoma gibbosum
    Strombus gigas
  Class Pelecypoda
    Diodora sp.
    Lima scabra
    Isognomon sp.
  Class Polyplacophora
    Chiton tuberculatus
Phylum Annelida
  Class Polychaeta
    Eupolymnia nebulosa
    Sebellastarte magnifica
    Hermodice carunculata
    Spirobranchus giganteus
Phylum Arthropoda
  Class Crustacea
    Artemia salinas
    unidentified demersal and harpacticoid
      sp. and larvae
    Amphithoe ramondi
    Alpheus armatus
    Mithrax sculptus
    Panulirus argus
    Stenopus hispidus
    unidentified gammaridean species
    Calcinus tibicen
    Mithrax spinosissimus
    Petrochirus diogenes
    Synalpheus brevicarpus
Phylum Echinodermata
  Diadema antillarum (long-spined urchin)
  Eucidaris tribuloides (slate pencil urchin)
  Ophiuroid spp.
  Echinometra lucunter (red rock urchin)
  Astichopus multifidus (furry Sea Cucumber)
Phylum Chordata
  Ascidiacean spp.

(continued)

TABLE 3
(Continued)

Planktonic invertebrates and protists	
Wave-breaking zone (fore reef)	
Zoomastiginids (flagellates)	Abundant
Hymenostomatid ciliates	Common
Vestibuliferid ciliates	Common
Cyclopoid copepods (some bearing eggs)	Occasional
Nematodes	Occasional
Copepod larvae (nauplii)	Occasional
Harpacticoid copepods	Rare
Over Thalassia bed (lagoon)	
Artemia larvae (nauplii)	Abundant
Nematodes (several spp.)	Abundant
Hymenostomatid ciliates (several spp.)	Common
Harpacticoid copepods	Occasional
Chaetonotid gastrotrichs	Rare
Foraminifera (several spp.)	Rare
Peritrich ciliates	Rare

aThis survey was undertaken during the years 1985 and 1986. The system was about 5 years old at that time.
bLong-term element not present at time of survey.

The most conspicuous sessile animals are stony corals, gorgonians, and anemones. Of the 40 to 45 common hermatypic scleractinia of the Caribbean-West Indian area, 24 have been introduced into the tank. Virtually all species survive for at least several months. Some species show little or no growth and eventually shrink marginally. The relatively high density of damselfish may place extra stress on the slower-growing species, which are often overgrown by dense filamentous algal growth propagated within the damsel territories.

Individual colonies of *Acropora palmata*, the dominant shallow-water reef builder in the eastern Caribbean, have been kept in the system for over 3 years at a time, growing at a rate of about 0.7 cm/month in spite of occasional predation by the crab *Mithrax spinosissimus*. This compares with 0.3–0.8 cm/month found by Gladfelter *et al.* in 1977. The morphology of the new growth, however, resembles the short, bushy branches of *Acropora prolifera* even though the original specimens had the typical bladelike appearance of *A. palmata*. Eventually, portions of colonies die suddenly after months or years of apparent good health, exhibiting symptoms of what has been called "white-band" disease (Gladfelter *et al.*, 1977). The true nature of this rapid death, a feature often present in natural reefs, is not known. In this microcosm, the occurrence of this "disease"

TABLE 4

Fish Typically Present in the Smithsonian Coral Reef
and Lagoon Microcosm[a]

Kingdom Animalia
  Subkingdom Eumetazoa
    Phylum Chordata
      Vertebrata
        Acanthurus bahianus (Ocean Surgeonfish)
        Acanthurus chirurgus (Doctorfish)
        Acanthurus coeruleus (Blue tang)
        Apogon maculatus (Flamefish)
        Chaetodon capistratus (Foureye Butterflyfish)
        Amblycirrhitus pinos (Red spotted Hawkfish)
        Thalassoma bifasciatum (Bluehead Wrasse)
        Acanthostracion polygonius (Honeycomb Cowfish)
        Holocanthus bermudensis (Blue angelfish)
        Holocanthus ciliaris (Queen Angelfish)
        Chromis cyanea (Blue chromis)
        Microspathodon chrysurus (Yellowtail Damselfish)
        Stegastes dorsopunicans (Dusky Damselfish)
        Stegastes planifrons (Yellow Damselfish)
        Eupomacentrus leucostictus (Beau Gregory)
        Haemulon flavolineatum (French Grunt)
        Haemulon sciurus (Bluestriped Grunt)
        Sparisoma viride (Spotlight Parrotfish)
        Pareques acuminatus (Cubbyu)

---

[a]This survey was undertaken during the years 1985 and
1986. The system was about 5 years old at that time.

---

is particularly marked during periods of relatively high nutrient levels, that is, greater than $1-2$ μM nitrogen as $NO_2^- + NO_3^-$ hr. In the reef model environment, human effects such as temperatures exceeding 29°C and perturbations resulting from the experimental manipulation of organisms are certainly a factor in coral health. Also, damselfish, butterfly fish, bristle worm (*Hermodice carunculata*), and snail (*Coralliophila*, Brawley and Adey, 1982) predation with resultant algal colonization is a recurrent problem for the coral in the tank as well as in the wild (see also discussion below on 130-gallon aquarium). The grazing activities of *Diadema antillarum* often result in the overturning of coral colonies. If this is not noticed the next day and corrected, the results are devastating. Several species of the hydrozoan *Millepora* are present and are generally long-lived, with considerable but sporadic growth. Octocorals, including species of *Gorgonia*, *Eunicea*, *Pseudopterogorgia*, *Plexaura*, and *Briareum*, as well as the Zoanthid soft corals *Zoanthus* and *Palythoa*, generally do well in the system and show moderate growth. Many filter feeders and detritivores, including numerous sponges, mysid shrimps, amphipods, foraminifera, and worms, live cryptically within the reef.

Larger invertebrates that have been successful in the microcosm are the lobster *Panulirus argus* and a variety of crabs, including *Mithrax* spp., *Calcinus tibicen*, *Petrochirus diogenes*, and the cleaner shrimp *Stenopus hispidus*. Young specimens of *Panulirus* grow rapidly and after molting four to five times eventually cause physical damage, particularly in the "uprooting" of uncemented carbonate. They are typically introduced as small specimens and then removed as adults a year or so later. The stomotopod *Pseudosquilla* has been present since the earliest stages of tank development. Probably because of its predation, it has not been possible to keep limpets in the reef tank for periods longer than a few months. The grazing gastropods *Cittarium pica*, generally in high-energy situations, especially on the reef crest and in the wave box, and *Strombus gigas*, the queen conch, in the lagoon, are also particularly successful. Numerous echinoids, including *Eucidaris tribuloides*, the slate pencil urchin, and particularly *Diadema antillarum*, are conspicuous nocturnal grazers. Like the lobster, the *Diadema* are placed in the system as young and removed after 1–2 years when overgrazing tends to become a problem.

More recent study of the exhibit tank microfauna has demonstrated over 80 species of protozoans and micrometazoans. In addition, a wide variety of small annelids, copepods, isopods, mysid shrimp, and ophiuroid brittle stars can be found on close examination, particularly at night.

Fish diversity of the microcosm seems limited in relation to normal reef conditions—21 species as compared to a predicted 200 for the 23-mile-long St. Croix reef. However, in a study of patch reefs behind the Buck Island barrier in St. Croix, only 22 species were found to be consistently present. Chief elements of fish community structure in our microcosms are the same as those on the patch reefs. The primary missing elements of the consistent fish fauna are the higher predators, which normally patrol larger territories than contained within the tank microcosm. Although figures as to the size of the patch reefs are not available, they are probably considerably larger than the microcosm surface.

In this microcosm many algae and invertebrates are successful in terms of reproduction, settling, and growth to maturity. A few fish spawn successfully but hatchlings fail to reach maturity, probably as the result of predation and cycling through centrifugal pumps. However, it should be expected that a few individuals would have developed to maturity. Most reef fish have planktonic larvae that escape into the open ocean, returning to coastal reefs at a later stage of development. Inclusion of a simulated open-ocean environment such as an attached saltwater swimming pool in the microcosm system would more closely duplicate natural conditions in this respect. Also, this reef and its prototypes have all been operated with centrifugal pumps. Since these pumps are prone to the destruction of many plankters (see Chapters 5 and 18), efforts are under way to develop reef models that do not rely on pumps that are damaging to the plankton.

Great Barrier Reef Mesocosm

The Australian Great Barrier Reef is the largest single coral reef system in the world ocean. Lying near the center of maximum shallow-water diversity in the East Indian Archipelago, the Indo-Pacific Biogeographic Province, the Barrier Reef also has a very high diversity of organisms. In the 1970s, the Australian Government designated the entire Barrier Reef Province as a national park. Later it was also designated a World Heritage Site. Most of the Great Barrier Reef is 10–50 miles offshore and often difficult to reach from the Australian mainland. Thus, the Marine Park Authority, needing a public display and educational tool to implement park policies, decided in the early 1980s to establish a large aquarium system on the mainland. The Smithsonian's Marine Systems Laboratory technology was chosen for this aquarium, and the authors of this book worked with the Park Authority and its chairman Graeme Kelleher for several years to design an appropriate mesocosm-sized coral reef. In June 1987, at Townsville, Australia, the system was opened to the public.

The "Great Barrier Reef Wonderland" aquarium is basically a 250 times larger version of the Smithsonian model reef. It is considerably more sophisticated in design, and along with its larger size is a major improvement in reef simulation over earlier systems. The aquarium director, in a paper written about a year after opening, said, "The performance of the algal turf system has exceeded original expectations. This simple system is capable of delivering a low nutrient regime and removing toxic metals together with maintaining seawater pH and oxygen levels and providing some degree of ultraviolet sterilization" (Jones, 1988).

The following discussion of the Great Barrier Reef Aquarium does not include the predator tank, a separate, smaller, and adjacent unit also run with algal scrubbers but at higher nutrient levels. This separate system is for maintainence of large sharks and other reef predators, which, even at the large scale of the Australian reef model, could not be included in the main reef.

The Townsville reef aquarium is 38 m long and 17 m wide (Color Plate 28). Its depth varies with the tide from 3.8 to 4.5 m. A separate unlighted tidal reservoir with 350,000 liters at high tide in the reef tank and 750,000 liters at low tide handles the tidal volume (Fig. 6). While the total volume of the reef proper is over 700,000 gallons, for comparison with other microcosm and mesocosm systems discussed in this book the tank includes 200 m^3 of coral sand and 700 metric tons of carbonate rock. The nominal water volume of this large model reef is 2,500,000 liters (625,000 gallons). Natural sunlight is used for the Great Barrier Reef

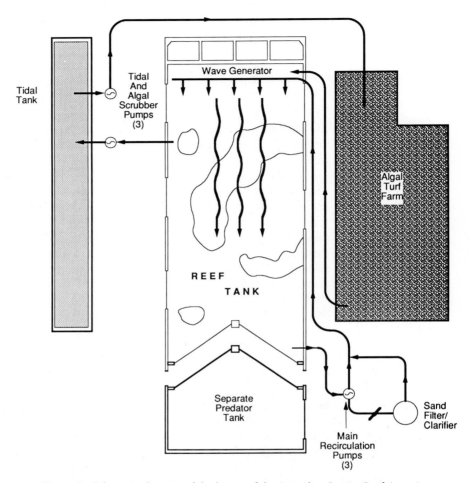

Figure 6 Schematic diagram of the layout of the Australian Barrier Reef Aquarium.

Aquarium. To provide a greater light range in summer and reduce heat loading some shading (shade cloth–plastic screen) is applied for several months in mid-summer. On the other hand, shading on the sides of the reef from the aquarium walls is such that some species appropriate for the actual depths in that area cannot be maintained there.

The Barrier Reef Aquarium tank uses three large centrifugal pumps that remove 1,000,000 liters/hour (4200 gpm) from the downstream (lagoon) end of the tank and return it to the forereef end. In addition to the current created by this flow, an air pressure-driven wave generator (see Chapter 5, Fig. 13) develops a 0.5- to 0.75-m wave every 2–8 seconds to create a marked wave surge motion on the surface of the reef.

Water quality in this reef model is managed primarily by a bank of 40 algal scrubbers (80 m²). These scrubbers (see Chapter 12, Fig. 8) are driven largely by natural light, though 6 hours of night lighting with 1000-W metal halide lamps is also provided. When stabilized and harvested every 10–14 days, these scrubbers have produced 5–7 g (dry)/m²/day (Morrissey et al., 1988). This value is generally lower than the 8–12 g (dry)/m²/day produced in other low-nutrient saltwater scrubbers (see Chapter 12). The difference is perhaps to be ascribed to the largely daytime operation, which puts the scrubber algae in competition with the reef algae for nutrients and carbon dioxide. Nutrient levels in the reef tank are shown in Figure 7.

Salinity in the Barrier World aquarium is maintained at about 34.5–35.5 ppt to match that characteristic of the Barrier Reef Province. Temperature is maintained at 20–29°C. Some cooling is required at the upper end of the range, since it is well established in the wild that coral bleaching and other negative effects are initiated at about 30°C. It is to be noted

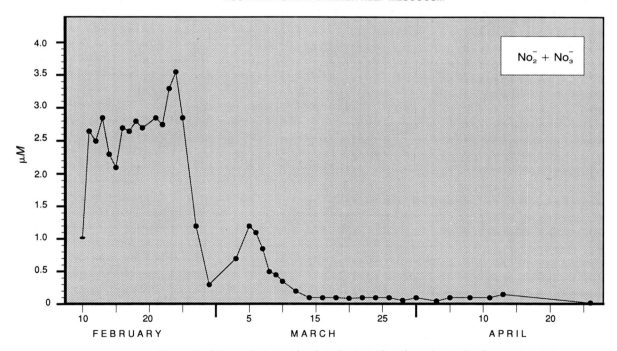

Figure 7 Dissolved nitrogen levels in the Australian Great Barrier Reef mesocosm during start up of the model ecosystem. Although the system was filled with very low offshore water delivered by tanker, early nitrogen levels were moderate due to sediment outwelling and beginning stocking losses. Once the scrubbers were fully operational, dissolved nitrogen fell below 1μM and remained there.

here as in most ecosystems that optimum temperature levels at the upper end of the range are often close to lethal values. Operational care is needed in most models in this regard, and in the case of the Great Barrier Reef Aquarium, some mortality has resulted from inadvertant short periods of operation above 30°C. In the Smithsonian model described above, this problem has become serious.

After about 1 year of operation of the Barrier Reef Aquarium, it was noticed that stony corals receiving partial shade from the tank walls (largely plastic-coated concrete) were doing poorly. The long axis of the tank is oriented north–south. These corals were removed and replaced with more shade-tolerant soft corals in those zones. This is a reminder that although ultraviolet can be harmful to corals, most ultraviolet is blocked by the intervening water and virtually all visible light is usable by the corals. This situation is even more marked for the algal turfs that occur on the reef (see Chapter 7).

In the Townsville model reef, several metals, mostly stainless steel but some zinc, were scattered throughout the system's plumbing. In the early months of operation, rising levels of dissolved heavy metals were threatening to be a serious problem. Most of this metal was eventually replaced by plastics. Heavy-metal concentrations in the water after about a year were moderate but within acceptable limits. It has been demonstrated that the algal scrubbers gradually removed these metal contaminants once the basic source of the metals was removed (Morrissey and Jones, 1988).

The original design of this very large reef system included sediment traps on the downstream end of each algal scrubber. For a variety of reasons these were omitted. Although not noticeably affecting the reef community, the net result of moderate turbidity for the long public viewing line across the aquarium was felt to be unacceptable. In spite of having to accept some plankton losses, this situation was brought under control with the partial operation of sand filters. The slight yellow coloration perceived (Gelbstoff) was managed by partial charcoal filtration. Neither of these methods is desirable. In future systems, the authors recommend careful attention to passive sedimentation devices to maintain turbidities at acceptable levels in very large systems. Many aquatic ecosystems have the equivalent of geological export and/or storage (see Chapter 8), and this must be allowed for in ecosystem models where appropriate.

Species counts have not been extensively developed for this system, although they are clearly far above the smaller models that we have developed in the past. Perhaps the most spectacular event indicating the relative veracity of this reef model is the general spawning of stony corals in November and December (spring) each year since 1987. Spawning took place within the same few days, controlled by rising temperature, the moon and tide (3–6 days after full moon), and hours of darkness, that it took place in the wild Great Barrier Reef (Wallace et al., 1986; Harriott, 1991).

A 130-Gallon Home Reef Aquarium

In October 1987, the 130-gallon home reef aquarium system shown in Figure 8 and Color Plate 29 was placed in operation. This relatively small home reef system is illuminated by six 72-inch VHO daylight fluorescent lamps having a total of 960 W or 1129 W/m². This is slightly lower than the Smithsonian reef exhibit described in the earlier part of this chapter at 1280 W/m² and is probably limiting for some sun-adapted corals. This lighting provides daytime oxygen levels of 7.0 mg/l, well above saturation for the operational temperature and salinity. In spite of the intense pumping and wave mixing, early in the afternoon the algal mats that coat the reef surface become densely covered with oxygen bubbles that break off and float to the surface.

The scrubber system as shown in Figure 8 is of the screen-bucket type and utilizes six 36-inch HO lamps having a combined power of 300 W. On a 7-day harvest schedule this provides 9–12 g (dry)/m²/day of export and maintains minimum night levels of dissolved oxygen in the system at about 5.5. mg/l. The scrubber algae are dominated by *Oscillatoria*, *Polysiphonia*, *Cladophora* and *Derbesia* (Table 5). Blue-greens are abundant, as one might expect in such a low-nutrient reef model (see

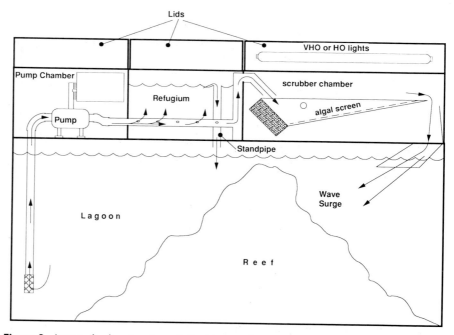

Figure 8 Longitudinal section through the home reef aquarium system shown in Color Plate 29. Not shown are the VHO tank lights. The lights indicated are the scrubber lights.

TABLE 5

Algae Typically Occurring in the Scrubber
of the 130-Gallon Home Coral Reef Aquarium
Shown in Figure 8[a]

Kingdom Monera
   Division Cyanophycota (blue greens)
     *_Oscillatoria limosa_
     Oscillatoris spp.
     Gloeocapsa sp.
Kingdom Plantae
   Subkingdom Thallobionta
     Division Rhodophycota (reds)
       *_Polysiphonia subtilissima_
       †_Callithamnion halliae_
       †_Erythrotrichea carnea_
       Goniotrichum alcidii
     Division Chromophycota (browns)
       †_Ectocarpus dasycarpus_ (diatoms)
       †several species, centric types dominate
     Division Chlorophycota (greens)
       *_Cladophora glaucescens_
       *_Derbesia lamourouxii_
       †_Enteromorpha lingulata_

    [a]This survey was taken after approximately three years of tank operation. This particular screen was about two years old and had been harvested approximately 100 times.
    *, dominant species; †, subdominant species.

below). When both scrubber and tank lights are in operation, oxygen concentration peaks out at a little over 7.5 mg/l. The range of oxygen concentration produced is well within the envelope for typical wild back-reef communities (Adey and Steneck, 1985).

Under stable conditions the algal scrubber maintains nutrient levels (measured as nitrite plus nitrate) at 0.5–1.0 μM in this small tank. Since the tank is so small and fish and invertebrate biomass are relatively high, 0.15 g (dry) of standard flake food is added to the system each day. About one-third of this addition occurs at 2200–2400 hs each night and is specifically intended for coral feeding. However, many motile invertebrates and a squirrel fish may consume a major part of the night feeding. While this may seem an exceptionally small amount of feed to be adding to the system per square meter of surface area, areally it is about the same as that fed to the Smithsonian exhibit reef.

The pH of this tank ranges from a minimum of 8.1 in the early morning to a maximum of 8.4 in the afternoon. Values of 8.3 occur through most of the middle of the day and allow particularly high calcification rates in this model, especially in the calcareous algae *Halimeda* and *Amphiroa* Although the scrubber area employed is only 0.18 m², per unit volume (3.8 cm²/l, scrubber area to tank volume) and wattage (1670 W/m²), this is one of the most effective scrubber units that we have employed.

The pumping utilized in this home reef system is that of two bellows pumps. These modified, fully plastic, hand bilge pumps, converted to geared electric motor drive, provide a flow rate of about 6 gpm for a tank overturn period of about 30 minutes. The bellows pumps were utilized in an attempt to reduce plankton mortality, although extensive comparison with centrifugal units has yet to be carried out. As shown in Figure 8, the scrubber bucket also provides for wave surge, producing a wave every 8– 12 seconds. This is quite adequate for the upstream third of the tank. It provides a quiet, more lagoon-like environment for the lower two-thirds of the tank. However, if rougher reef conditions were desired for most of such a tank, an additional surge device would have to be added.

Salinity is maintained in this system between 34.5 and 36.0 ppt. An automatic water leveling system is used. While this tends to maintain very stable salinity levels ±0.2 ppt on a day-to-day basis, gradual salt loss to the atmosphere and the sides of the tank and scrubber unit (which is wiped off periodically when the unit is cleaned) causes a very slow downward drift of salinity. About 1 gallon per month of 20–25% sea water is added to maintain salinity levels. An additional exchange of about 2 gallons per month of sea water is employed to avoid salt drift.

Temperature in the 135-gallon tank is maintained from 25 to 29°C. At the lower end, no additional heating over a room temperature of 65–70°F is required to keep a stable level in winter. At the upper end in summer, room temperature must be kept below about 74°F. From experience with later units, simple cooling fans on the tank lights allow room temperatures as high as about 80°F, although this results in greater evaporation rates. Above that level, a cooler and heat exchanger must be employed.

Although species diversity is not as high as it is in the larger reef tanks, it is nevertheless significant (Table 6). The three species of stony coral (including a hydrozoan) originally stocked in the system, *Montastrea cavernosa*, *Plerogyra sinuosa*, and *Millepora alcicornis*, remain present after nearly 3 years. *Millepora alcicornis* has shown considerable apical growth without appreciable lateral die back since the original stocking. On the other hand, a single small colony of *Acropora palmata*, placed several months after startup, slowly died, up from the base, after about 6 months, after showing about 1 cm of apical growth. In this case, perhaps the moderate light levels, for this "sun species," were responsible, preventing growth from keeping up with dieback from the base. All long-term successful stony corals in this model are under moderate wave action and all are moderate to severe stingers. It seems likely that overpredation is a major factor in small models lacking top predators acting

TABLE 6

Species List for a 130-Gallon Home Reef Aquarium
Operated with Algal Turf Scrubbers for 3 Years[a]

Kingdom Monera
  Division Cyanopycota (blue greens)
    Oscillatoria submembranacea
    Oscillatoria limosa
    Oscillatoria spp.
    Chroococcus spp.
    Gloeocapsa sp.
Kingdom Plantae
  Division Rhodophycota (reds)
    Erythrotrichia carnea
    Goniotrichum alcidii
    Hypnea musciformis
    Calithamnion halliae
    Digenia simplex
    Herposiphonia secunda
    Polysiphonia subtilissima
    Amphiroa fragillisima
    Porolithon pachydermum
    Neogoniolithon solubile
    Neogoniolithon spp.
    Mesophyllum mesomorphum
    Sporolithon dimotum
    Peysonnellia rubrum
  Division Chromophycota (browns and
    diatoms)
    Ectocarpus dasycarpus
    Sargassum sp.
    Nitzschia sp.
    numerous diatom species
  Division Chlorophycota (greens)
    Chaetomorpha minima
    Cladophora glaucescens
    Gomontia polyrhiza
    Valonia macrophysa
    Anadyomene stellata
    Halimeda opuntia
    Halimeda tuna
    Derbesia lamourouxii
    Enteromorpha lingulata
Kingdom Animalia
  Subkingdom Protozoa
    Homotrema rubrum
    Gypsina sp.
    Quingueloculina sp.
    Ciliates: *Hypotrich* sp.
      Oligotrichs (3 spp.)
      Vestibulates (3 spp.)
      Heterotrichs (3 spp.)

Kingdom Animalia
continued
  Subkingdom Parazoa (sponges)
    Tethya sp.
    Haliclona rubens (red finger
      sponge)
    Ulosa hispida (orange encrusting
      sponge)
    Monanchora barbadensis (red
      encrusting sponge)
    Siphonodictyon coralliophagum
      (yellow horn sponge)
    Chondrilla sp. (chicken liver
      sponge)
  Subkingdom Eumetazoa
    Phylum Cnidaria (coelentrates)
      Aiptasia tagetes (pale anemone)
      Stoichactis helianthus (sun
        anemone)
      Zoanthus sociatus (green colonial
        anemone)
      Palythoa sp. (encrusting anemone)
      Rhodoctis sanctithomae
      Coralliomorpharians
        (several spp.)
      Millepora alcicornis (fire coral)
      Eunicea sp. (knobby candelabra)
      Plexaura homomalla (bushy soft
        coral)
      Sarcophyton trocheliophorum
        (soft coral)
      Nephthya sp. (soft coral)
      Pleurogyra sinuosa (grape coral)
      Montastrea cavernosa (stone
        coral)
      Colpophyllia natans (large
        grooved brain)
      Mycetophyllia sp. (cactus coral)
    Phylum Platyhelminthes
      Acoel Turbellarian sp.
    Phylum Rotifera
      Notommata (?) sp.
    Phylum Gastrotricha
      Chaetonotid sp.
    Phylum Nematoda
      4 unidentified species
    Phylum Annelida
      Sabella melanostigma (fan worm)

(continued)

TABLE 6
(Continued)

Phylum Annelida	Phylum Arthropoda
continued	continued
Pomatostegus stellatus (red fan worm)	*Jaera* sp. (isopod)
Spirobranchus sp. (black christmas tree worm)	*Acrothoracid barnacle*
Schistomeringos rudolphi	*Pagurid* sp. (hermit crab)
Spirorbis sp.	*Enoplometopus occidentalis* (rare reef lobster)
Eusyllis sp.	Phylum Chordata
Capitallid sp.	*Dascyllus aranus* (humbug)
Nematonereis unicornis	*Amphiprion frenatus* (anemone fish)
Amphinomid sp.	*Pomacentrus caeruleana* (blue damselfish)
Phylum Mollusca	*Zebrasoma flavescens* (yellow tang)
Diadora listeri (listers keyhole limpet)	*Holacanthus ciliarus* (queen angelfish)
Anachis translirata (well-ribbed dove shell)	*Centropyge bicolor* (yellow and black angelfish)
Solariella sp. (top shell)	*Canthigaster margaritutus* (ocellated toby)
Petalochonchus erectus (vermetid)	*Pseudochromis dutoiti* (neon arabian)
Pinctada imbricata (scaly pearl oyster)	*Flammeo scammera* (blood spot squirrel fish)
Spondylus sp. (spiny oyster)	Phylum Echnodermata
Tridacna maxima (giant clam)	*Eucidaris tribuloides* (state pencil urchin)
Lucina sp. (lucine)	*Lytechnis* sp. (urchin)
Phylum Arthropoda	
Class Crustacea	
Harpacticoid copepods (2 spp.)	
Gammarid amphipods (2 spp.)	
Podocope ostracods (2 spp.)	
Pontocypris sp. (ostracod)	

ᵃBacteria, microprotozoa, and phytoplankton have not been studied in this system. The list includes only species that have become long-term elements (over 2 years).

continuously. It is instructive to follow the actions of angel and butterfly fish when provided with corals lacking such strong protection. When such specimens are moved in an attempt to confuse the fish, they immediately set up a search pattern until the potential food is relocated. When the specimen is placed in wave action, the feeding is slowed as the fish must now carefully negotiate to avoid injury.

Besides the stony corals, numerous alcyonarian soft corals and gorgonians as well as coralliomorphs and anemones are long-term residents of this home aquarium. The two gorgonians, *Eunicea* sp. and *Plexaura homomalla,* date from the original stocking. A number of small crustaceans, amphipods, isopods, ostracods, and copepods manage to avoid the constant fish predation pressure through burrows, tubes, or nocturnal activity.

The abundance of filter feeders, in spite of the small size of this system, is conspicuous. As can be seen from Table 6, numerous sponges, of a wide variety of genera, and filter-feeding annelid worms are long-term residents. The sponge *Tethya* is particularly abundant in spite of the presence of a large angelfish. This genus, common in Caribbean reefs, is known for its toxicity and is avoided by predators. A striking reproductive success in this group is *Petaloconchus*, a reef-forming vermetid gastropod that "spins a web" and then eats the entrapped plankton and organic particulates. We attribute the abundance of filter feeders to the presence of bellows pumps. This tank has never been operated with centrifugal pumps.

Finally, probably partly for the same reasons, the meiofaunal and larger microfaunal elements are particularly abundant. Twenty-one species of protozoans were tallied in a short search, and these were accompanied by numerous rotifers and gastrotrichs. Both common reef-encrusting foraminifera *Homotrema rubrum* and *Gypsina* sp. are also successful in this tank.

One of the most striking features of this miniature reef system is the small reef lobster (*Enoplometopus*) added when the tank was initiated. This primarily nocturnal animal functions as a scavenger but also acts as the highest predator. In 3 years it has not been necessary for us to function as the top predator. Older established fish have been present from the earliest days of stocking. However, fish newly added to the fully stocked community have a chance of about 1 in 10 of surviving the first few nights. Harrassment by the established fish generally results in cowering by new fish and apparently an inability to maintain protective behavior against the lobster at night. Nevertheless, even when no new fish are added for many months, the lobster continues its 6–10 weeks molting cycle (Fig. 9). Although we have had to suffer an occasional fish loss, it has been on a very long time frame, probably appropriate with the natural longevity of the fish involved. The lobster serves as a control to balance both added fish (the equivalent of fish dropping from the plankton in the wild) and the oldest, and failing, fish to maintain a stable population.

As a closing note, this tank has a very high grazer biomass, with numerous damsels and a yellow tang constantly browsing at the larger scale and abundant isopods and amphipods at the small scale, particularly nocturnally. Although the scrubber maintains an abundant turf and small macroalgae, on the reef itself the conspicuous plant elements either are calcified (*Halimeda*, *Amphiroa*, and *Mesophyllum*) or are very tough encrustors (*Valonia*). The daily supersaturated oxygen levels, the high pH values, and the strong calcification occuring in many phyla through the system all demonstrate that this miniature reef is strongly autotrophic

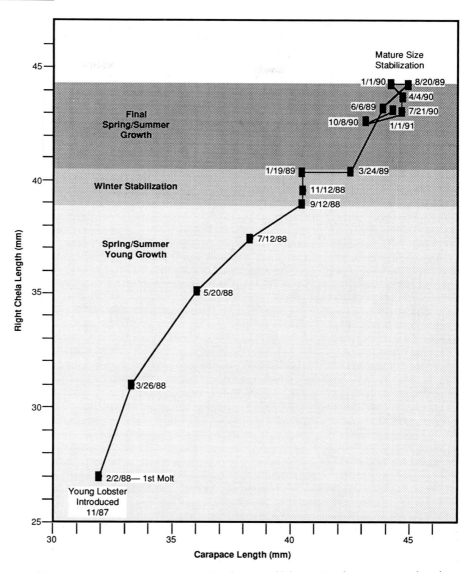

Figure 9 Molt dimensions and growth of rare reef lobster, *Enoplometopus occidentalis*.

(photosynthetic) in its overall balance. It is likely that over 95% of the energy, and therefore the feeding, that goes into this miniature ecosystem is derived from VHO lights. This may seem extraordinary, but it is the only rational way to avoid an impossibly complex problem of nutritious feed to what is likely over 100 species from dozens of major animal groups. See also Delbeek (1990) for relevance to home systems in general.

References

Adey, W. 1978a. Coral reef morphogenesis: A multidimensional model. *Science* **202**: 831–837.

Adey, W. 1978b. Algal ridges in the Caribbean Sea and West Indies. *Phycologia* **17**: 361–367.

Adey, W. 1983. The microcosm: A new tool for reef research. *Coral Reefs* **1**: 193–201.

Adey, W., and Steneck, R. 1985. Highly productive eastern Caribbean reefs: Synergistic effects of biological, chemical, physical and geological factors. In M. Reaka (Ed.). *NOAA Symp. Ser. on Underwater Research* **3**: 163–187.

Brawley, S., and Adey, W. 1982. *Coralliophila abbreviata:* A significant corallivore. *Bull. Marine Sci.* **32**: 595–599.

Carlson, B. 1987. Aquarium systems for living corals. *Int. Zoo. Yb.* **26**: 1–9.

Delbeek, J. C. 1990. Reef aquariums: nutrition. *Aquarium Fish Mag.* **2**(6): 26–35.

Gladfelter, W., Gladfelter, E., Monahan, E., Ogden, J., Dill, R. 1977. Environmental Studies of Buck Island Reef National Monument, St. Croix, U.S.V.I. West Indies Laboratory, St. Croix.

Harriott, V. 1991. Macro-Reef System? *Freshwater and Marine Aquarium* **14**(5): 8–10.

Jones, M. 1988. The Great Barrier Reef Aquarium—A matter of scale. Northern Regional Eng. Conf. (Australia), Townsville, June 10–13, pp. 1–10.

Kaplan, E. 1982. *A Field Guide to the Coral Reefs of the Caribbean and Florida.* Houghton Mifflin, Boston.

Lucid, D. 1989. Affects of dissolved inorganic nitrogen concentration on primary productivity, nitrogen fixation, and community composition of coral reef algal turf: A microcosm study. University of Maryland, M.S. thesis.

Moe, M. 1989. The Marine Aquarium Reference. Green Turtle Publications. Plantation, Fla. 507 pp.

Morrissey, J., and Jones, M. 1988. Water-clean, clear and warm. *Austr. Sci. Mag.* **3**: 33–41.

Morrissey, J., Jones, M., and Harriott, V. 1988. Nutrient cycling in the Great Barrier Reef Aquarium. *Proc. 6th Int. Coral Reef Symp.* (in press).

Talbot, F. 1984. *Reader's Digest Book of the Great Barrier Reef.* Reader's Digest Books, Sydney, Australia.

Wallace, C., Babcock, R., Harrison, P., Oliver, J., and Willis, B. 1986. Sex on the reef: Mass spawning of corals. *Oceanus* **29**: 38–42.

Walton, S. 1980. The reefs tale. *Bioscience* **30**: 805.

Williams, S., and Adey, W. 1983. *Thalassia testudinum* Banks ex Konig. Seedling success in a coral reef microcosm. *Aquatic Botany* **16**: 181–188.

CHAPTER 22

A SUBARCTIC SHORE
The Maine Coast

Cold-water, rocky, wave-beaten and tidal shores along with their accompanying protected mud flats and marshes have long fascinated scientists and the general public. They have also provided food for many small coastal human communities, and indeed, in some cases rocky coasts have provided sizable commercial fisheries with lobsters, clams, mussels, worms, Irish moss, sea urchins, etc. In nineteenth century Europe, naturalist coast walks and lectures became a routine for the educated public. In that same tradition in North America, the pre-World War II *Between Pacific Tides* by Ricketts and Calvin has held the interest of generations of Californians through many editions (Ricketts *et al.*, 1985). Yet, except for scattered coastal public aquaria and most recently one of the finest of modern aquaria in Monterey near Ricketts' old laboratory, little scientific or hobbyist interest in these shores has been transferred to experimental microcosms or home aquaria. There is much to be gained from increased public interest and understanding of these communities, many of which are under the increasing stress of population and fishing pressures.

In this chapter we describe a 2700-gallon model of a typical set of shore ecosystems from the Maine coast. This microcosm, consisting of a rocky shore, mud flat, and salt marsh, was completed in early 1985 and is still in operation. It remains on permanent display at the Smithsonian's National Museum of Natural History. Construction of this cold-water system followed 2 years of operation of two considerably smaller prototypes of 250 gallons each. The exhibit was first described in the literature at the time of its opening (Tangley, 1985).

Although numerous popular and scientific treatments of the flora, fauna, and ecology of Maine's coastal waters have been published, the authors suggest the six-volume work by Fefer and Schettig (1980) as general background. For organism identification, Boschung *et al.* (1983), Robins *et al.* (1986), and Gosner (1978) will be helpful. Most cold-water shores in populated areas can claim similar field guides.

Physical Layout

The Maine shore microcosm (Fig. 1) is laid out very much like the coral reef exhibit model in the sense that two partially separated tanks repre-

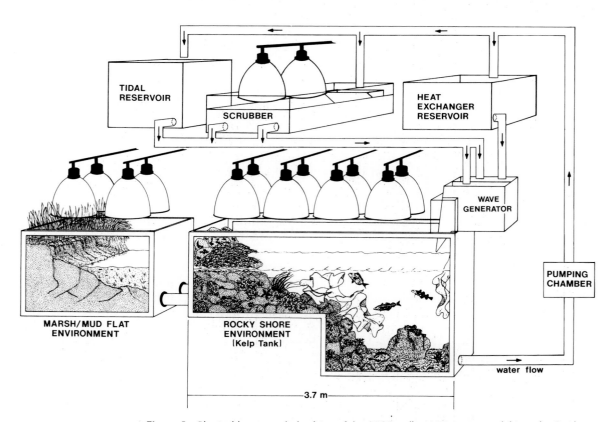

Figure 1 Physical layout and plumbing of the 2700-gallon Maine coast exhibit at the Smithsonian's Museum of Natural History.

senting exposed and protected shores are used. In both systems these include a larger stepped tank with wave action, in this case a rocky shore, and a smaller, quieter tank connected by piping. The mud flat and marsh are housed in the smaller tank, where significant wave action is lacking. In concept, these are two different model ecosystems linked together to provide the primary interactions that occur between the two in the wild.

The internal structure of the rocky shore was built like a stone wall with granitic or metamorphic cobbles and boulders. As mentioned in Chapter 2, some basal boulders were constructed of pumice stone, fiberglassed, to reduce weight. This structure (and the available wave energy in the microcosm) is typical for the shores of large, semiprotected bays in Maine, where the fine to pebble-sized fraction of the glacial till has been worked out by wave action leaving a rubble shore of great spatial heterogeneity. The protected basal sections along with the attached mud flat/marsh tanks serve as sediment traps and "geological storage" for fine organic particulates. As we discuss below, the upper foot or so of the rubble shore was constructed by block transfer of ledge, cobbles and boulders with encrusting living communities. The basal section of the mud flat and marsh unit were taken as sediment blocks from Maine mud flats and transported in coolers to the microcosm in Washington.

Also, like the coral reef, the Maine coast microcosm has intense metal halide lighting to simulate summer sun as described below. Algal scrubbers are likewise employed to simulate the larger offshore body of water that is present in the wild. In addition, being in a public environment with temperatures ranging from about 70 to 80°F, this mini ecosystem also has an extensive cooling system. There is provision for a tide, albeit much smaller in amplitude than that in the wild. The plumbing system employed to provide waves, current, cooling, and tide is shown in Figure 1.

Environmental Parameters

The Maine microcosm has a maximum water temperature of about 15°C (60°F) and a minimum of about 4°C (40°F). This provides a rather warm winter for the eastern Maine coast, which would typically be 1–2°C on outer shores in February. Otherwise, temperatures are close to those experienced in the wild. The cooling system with its heat exchanger differs only slightly from that employed during the spring and summer temperature maxima on the coral reef. It consists of a glass-tubed heat exchanger, in a fiberglassed box with a light brine and three 1-ton, immersion-type cooling units (Frigid Unit). The ecosystem's salt water is passed through the chilled brine in a set of 1-inch-diameter glass tubes. During the winter minimum, several additional cooling units are used to achieve the desired low temperatures for periods of up to about 8 weeks. Since this is an exhibit and summer humidity levels are relatively high in the public viewing areas, especially when crowds are present, the tank and its scrubbers are also contained in an air-conditioned room with acrylic viewing panels.

Beyond temperature, the distinctive feature of the Maine coast that differentiates it from many coastal environments is a large tide range, roughly 8–20 feet at spring tides, depending upon location. The range used in the microcosm is only one foot at spring tides and eight inches at neaps. The same basic method described in Chapter 6 is used to produce the pure semidiurnal spring/neap tides. An insulated fiberglass box placed on the next floor above the system serves as a tidal reservoir. Water is pumped to the reservoir and returns to the main tank by gravity through a level-controlled hose. Two geared and timed stepping motors, one rotating 360 degrees every 12 hours 20 minutes, the other rotating 360 degrees every 14 days, set the outflow hose level. As described below, although telescoped, the resulting community distribution in the intertidal is similar to that in the wild.

Salinity is maintained between 31 and 34 ppt in the Maine system on a seasonal basis, low in spring and high in autumn. Since a moderate amount of salinity variation occurs in the natural environment, a simple "top-up" method, to a mark, is used to replace evaporated water, rather than the more sophisticated control system used for reef models. Such control systems could, however, be used to reduce labor. The yearly cycling can be provided by manual adjustments on a seasonal basis.

The metal halide tank lighting, with ten 400-W units in this case, is basically the same as that described for the coral reef. Time clocks are used to set the day length as well as dawn and dusk times. Of course, the light cycle in winter is greatly reduced. Also, winter light intensities are reduced by raising the metal halide lights on vertical slides with small boat winches. Light levels on the surface of the rocky shore and on the mud flat and salt marsh for summer and winter are shown in Table 1. The irradiance levels at midday on the microcosm are 100–700 $\mu E/m^2$/second in summer and 30–450 $\mu E/m^2$/second in winter. These compare to 130–625 $\mu E/m^2$/second measured in July in outer Gouldsboro Bay, Maine, at 2.5–5 m and 45–280 measured at the same depths and conditions in March and April. Due to the lack of cloudiness on the model, total light received on the microcosm is probably somewhat higher than in the wild.

Wave action in this microcosm is created by a pair of dump buckets of 24 and 18 liters, respectively, and through which most of the pumped water is recycled. Depending upon the wave action desired, seven 10-gpm pumps are used to provide a wave period of 5–10 seconds. The current velocity half-way between the dump buckets and the rocky shore reaches 10–19 cm/second.

TABLE 1

Dimensions and Physical Parameters of the Maine Coast Microcosm

Community	Tank dimensions			Tank volume (l)	Substrate surface area
	L	W	D		
Rocky shore	3.65	× 1.21	× 1.82m	9100	3.98 m²
Marshland mud flat	1.21	× 1.21	× 1.21m	1800	2.00 m²
Total				10,900	5.98 m²

Principle operating characteristics	Summer	Winter
Lighting (metal halides); 12–400 W	700–100 µE/m²/second surface to 1.6 m (simulated depth 10 m)	450–30 µE/m²/second surface to 1.6 m (simulated depth 10 m)
Photoperiod	14 hours (maximum)	8 hours (minimum)
Temperature	15° C (maximum)	4° C (minimum)
Tide Semidiurnal	Spring 38 cm	Neap 20 cm
Wave action	Current velocity 10–19 cm/second; irregular with two dump buckets of 24 and 18 l driven by 70–100 gpm of centrifugal pumps, seasonal	

Principal Chemical Factors

The chemical environment in this cold-water ecosystem is generated as a balance between the functioning of the ecosystem itself and a bank of algal turf scrubbers as shown in Figure 1.

The scrubbers are two relatively small 0.5-m standard units as described in Chapters 12 and 21. Each scrubber has a single 400-W metal halide lamp. The scrubber lights are operated for 8–14 hours inversely to the ecosystem lights, thus simulating the night buffering effect of offshore water entering the immediate coast environment. These scrubber units are quite sufficient to drive nutrient levels in the system to less than the normal minimum for the wild system, which is about 1 μM nitrogen as $NO_2^- + NO_3^-$ (0.014 ppm). Through disturbance of the mud flat or excess feeding, it is possible to elevate nutrient levels to greater than 15–20 μM (nitrogen as $NO_2^- + NO_3^-$, 0.21–0.28 ppm). This is near maximum for the wild environment. Except for periods when experimental work is being carried out, the wild pattern of low nutrients (and high visibility) during

the winter and high nutrients (and low visibility) in the summer is followed. In Gouldsboro Bay, eastern Maine, dissolved inorganic nitrogen
ranges from about 5 μM offshore in late winter to 16 μM on the inner bay
in late summer. Although the experimental range used on this tank has
been wide-ranging, from 1–40 μM (NO_2^- + NO_3^-), typical microcosm
operation has ranged from about 3 to 10 μM. While this might be a little on
the low side for Gouldsboro Bay conditions today, it is probably close to
prehistoric values.

Scrubber harvest intervals have varied from 7 to 14 days and have
produced from 2 to 18 g dry weight of algae per day. The mean rate of
production under normal operation and over a 6-month period was found
to be 12 g (dry) per m² per day. A wide variety of algae can be found on the
scrubbers, including young sporophytic kelp fronds. However, the dominant genera are the greens *Enteromorpha* and *Cladophora*, the brown
Ectocarpus, and the reds *Polysiphonia* and *Porphyra*. These algae, which
are relatively small at reproductive maturity, are the plants that would be
expected in the wild in routinely heavily grazed environments.

The basic chemical environment developed in the water column of
this microcosm is shown in Table 2.

The Organisms

The Maine coast exhibit microcosm was established in large part by the
process of ecosystem block transfer described earlier. In the rocky-shore
community the "blocks" were individual rocks (pebbles, cobbles, boulders) and fragments of ledge chiseled from the wild. The primary organic
structuring elements were algae (blue-greens, rockweeds, Irish moss,
kelps, small reds) or in some cases barnacles or mussels attached to the
rocks. These rock/algae invertebrate units were moved in coolers as
rapidly as possible from the Maine coast to the microcosm in Washington,
D.C. They were initially placed in their equivalent microcosm zones with
plants and whatever encrusting attached fauna that were present. Many
smaller mobile molluscs, worms, and crustaceans also accompanied the
blocks. However, following an initial mass transport, relatively small
"block injections" were repeated over the next 2 years. Fish and larger
invertebrates were trapped or netted and introduced separately. On the
mud flat, marsh, and subtidal mud bottoms, the "blocks" consisted of
small ecosystem units of mud substrate with their marsh grass, sea grass,
or algal communities intact, carefully shoveled into coolers, and similarly
introduced. Since the original stocking, small collections, primarily of
species that have suffered overpredation, have been added about once per
year.

On the Maine coast, except for deeper, subtidal communities and

TABLE 2

Basic Operational Chemical Parameters
of the Maine Coast Microcosm

	Minimum	Maximum
Salinity	31 ppt (spring)	34 ppt (fall)
Dissolved oxygen	8.5 mg/l	9.5 mg/l
Dissolved nitrogen (NO_2^- + NO_3^-)	1 μM N[a]	10 (40) μM N[b]
Dissolved phosphorus	0.15 μM P	0.57 μM P

Ocean (coastal) simulation
    Algal turf scrubber of 1 m² lighted nightly with (2)400 W metal halides for 8–14 hours depending on season scrubber productivity rate: 1.9–11.4 g (dry)/m²/day, seasonal

Principal algal species:
    Ectocarpus, Enteromorpha, Cladophora, Polysiphonia, Porphyra

[a]Typical 1–10 μM N.
[b]For research purposes.

deeper muddy bottoms in bays, plants are the primary determiners of community structure. The basic plant zonation on soft bottoms includes *Spartina patens* (high marsh), *Spartina alterniflora* (low marsh) (Color Plate 31), *Zostera marina* (shallow muddy bottoms). On rocky bottoms blue-green algae (rocky supertidal), rockweeds (*Fucus* and *Ascophyllum*) (intertidal) (Color Plate 32), Irish moss (*Chondrus*) (lower intertidal and uppermost subtidal) (Color Plates 33, 34), kelp (*Laminaria, Saccorhiza,* and *Agarum*) (upper subtidal) (Color Plate 35 and Fig. 2), and red algae (*Phycodrys, Ptilotia,* and *Phyllophora*) (lower subtidal) (Color Plate 36). These plants with their associated communities of organisms were established initially in the microcosms and have been naturally maintained. It is interesting to note that although *Fucus* and *Ascophyllum* have shown good growth in this system and have repeatedly reproduced, the original adult plants established from the wild have shown a very slow process of dieback. Young plants of both genera have gradually established in the system, and we are watching closely for the development of the long-term zonation pattern. *Chrondrus crispus* (Irish moss) has done extraordinarily well in this model, generally creeping to shallower levels in the intertidal than it is found, in abundance, in the wild. This probably relates to the high humidity and moderate temperature levels of the microcosm. In the wild, Irish moss is likely sharply limited by high temperatures and relative dryness in summer as well as very cold temperatures and icing conditions in winter.

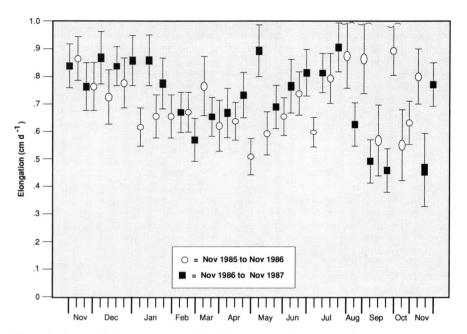

Figure 2 Seasonal elongation rates of the kelp *Laminaria longicruris* with constant nutrient and wave generation levels. These data are part of a several year study of kelp growth using the microcosm as an experimental tool and comparing results with those in the wild. Data and drawing from Brittsan, 1989.

Although the dominant plant species that characterize the Maine coast are present and easily maintained, as long as urchin populations *Strongylocentrotus drobachiensis* are controlled, a few species that are important in the wild are conspicuously absent or poorly developed in this system. *Alaria esculenta* is the most conspicuous of these. Normally restricted to the lowest intertidal of the most wave-beaten shores, one can assume that the microcosm environment lacks sufficient wave action to maintain this species. Whether this control is exerted directly on the plant itself or as a control for some predator is not known.

The muddy subtidal of many Maine bays has a richly developed eelgrass community (*Zostera marina*). While established in the muddy subtidal of the model several times, *Zostera* gradually dies out over 6–8 months. The intensive burrowing activities of green crabs (*Carcinas maenas*) may be responsible in this case. These crabs do exceptionally well in this system and without larger fish predators may require frequent human intervention. At this writing, we have removed all crabs from the mud flat and muddy subtidal and have again introduced young plants of *Zostera*. After nearly a year, this *Zostera* community has remained healthy.

The plants that are permanent long-term community elements of the Maine coast microcosm are given in Table 3. Although not specifically mentioned above, it should be noted that the understory of red, pink, and yellow calcified coralline algae prevalent in the wild, including *Clathromorphum*, *Phymatolithon*, and *Lithothamnium* species, are also characteristic understory elements in the model.

The encrusting and attached fauna so characteristic of the rocky Maine shore is also present in the Maine microcosm. Barnacles, blue mussels, horse mussels (Color Plate 36), and the calcified worm *Spirorbis* (Color Plate 37), are subject to heavy predation by *Thais* (dogwinkle) and *Buccinum* (waved welk) as well as by crabs, *Cancer* (rock crab), and *Carcinus* (green crab). Nevertheless, all of these attached animals manage to survive over the long term in the upper intertidal or in crevices where they cannot be easily reached.

The current list for the Maine coast system includes over 200 species. The animals are listed in Tables 4–6. However, some of the plankton and a number of minor worm groups have not been tallied. It is to be expected that the diversity of this tank complex is close to 250 species, about two-thirds of the similarly sized coral reef microcosm.

The "open water" volume available for mid-water fish in the Maine shore microcosm is relatively small. Thus, we have emphasized bottom or near-bottom fish and have kept biomass small. Of the larger variety of fish placed in the system, relatively few (Table 5) became permanently established. These fish tend to grow very large over several years, finally developing characteristics of old age. After several years, they are removed and replaced with younger animals. Although invertebrate and plant reproduction is abundant and successful for many species, in this very limited volume, fish reproduction has not been successful to date.

The Maine coast microcosm does not have a refugium from larger fish and invertebrates that has the same basic characteristics of the main system. The tidal tank is effectively a large refugium for some species. However, it is intertidal, with a muddy bottom and hard walls, and yet totally dark. It is certainly a peculiar environment, perhaps the equivalent of a long intertidal cave in the wild. Nevertheless, in the Maine coast system as a whole, the plants and the smaller encrusting invertebrates appear to be quite successful while the fish are relatively limited. Probably the greatest need of the microcosm as it stands is simply a larger volume of open water treated as a refugium.

Operation and Maintenance

The Maine coast microcosm is rather easy to operate mechanically. Evaporation rates are low and the requirements of make up water quality are not as sensitive as they are for the coral reef. The plumbing and piping

---

TABLE 3

Plants Occurring as Long-Term Residents of the Smithsonian Maine Coast Microcosm

Kingdom Plantae
  Subkingdom Thallobionta
    Division Rhodophycota (red algae)
      Porphyra umbilicalis
      Palmaria palmata
      Chondrus crispus
      Rhodophyllis dichotoma
      Euthora cristata
      Hildenbrandtia prototypus
      Corallina officinalis
      Clathromorphum compactum
      Lithothamnium glaciale
      Lithothamnium lemoineae
      Phycodrys rubens
      Ptilotia serrata
      Antithamnionella floccosa
      Phyllophora truncata
      Gigartina stellata
      Halosaccion ramentaceum
      Petrocelis middendorfii
      Phymatolithon laevigatum
      Phymatolithon rugulosum
      Phymatolithon lenormandi
      Polysiphonia sp.
      Ceramium sp.
    Division Chromophycophyta (brown
        algae, golden brown algae)
      (browns)
      Ectocarpus fasciculatus
      Punctaria sp.
      Petalonia sp.
      Laminaria longicruris
      Agarum cribosum
      Pylaiella curta
      Chordaria flagelliformis
      Ralfsia verrucosa
      Fucus vesiculosus
      Fucus spiralis
      Ascophyllum nodosum

Division Chromophycophyta *continued*
    (diatoms)
      Rhizosolenia sp.
      Cyclotella sp.
      Diatoma sp.
      Nitzschia sp.
      Licmophora sp.
      Fragillaria sp.
      Navicula sp.
      Pleurosigma sp.
    Division Chlorophycota (green algae)
      Chaetomorpha linum
      Rhizoclonium riparium
      Ulothrix sp.
      Spongomorpha sp.
      Ulva lactuca
      Enteromorpha linza
      Derbesia sp.
      Chaetomorpha melagonium
      Cladophora rupestris
      Cladophora serica
      Monostroma sp.
      Enteromorpha intestinalis
  Subkingdom Embryobionta
    Division Magnoliophyta (flowering plants)
      Spartina alterniflora (smooth cordgrass)
      Spartina patens (salt hay)
      Juncus gerardi (black rush)
      Suaeda linearis (sea blight)
      Solidago sempervirens (goldenrod)
      Salicornia virginica (saltwort)
      Limonium carolinianum (sea lavender)
      Atriplex patula (spearscale)
Kingdom Monera
  Division Cyanophycota (blue green algae)
      Lyngbya sp.
      Oscillatoria spp.
      3 spp. unicells
  Division Bacteria
      many species not tallied

---

concerns are little different from those in any model ecosystem. However, the unique operational aspect is temperature. A sudden breakdown of cooling apparatus, particularly in summer, could be quickly disastrous for this system. Because of this we maintain several stand-by cooling units.

The Maine coast is dynamic, but events tend to be much slower

TABLE 4

Macro Invertebrates Occurring as Long-Term Elements of the Smithsonian Maine Coast Microcosm

Kingdom Animalia
  Subkingdom Protozoa (see Table 21-6)
  Subkingdom Parazoa (sponges)
    Halichondria panicea (crumb of bread sponge)
    Leucosolenia sp. (organ pipe sponge)
  Subkingdom Eumetazoa
    Phylum Cnidaria (coelenterates)
      Bunadactis stella (silver-spotted anemone)
      Metridium senile (frilled anemone)
      Tealia felina northern red anemone)
      Hydroid sp.
    Phylum Mollusca
      Ishnochiton ruber (red chiton)
      Acmaea testudinalis (tortoise shell limpet)
      Buccinum undatum (waved whelk)
      Crepidula fornicata (common slipper)
      Nassarius obsoletus (mad dog whelk)
      Neptunea decemcostata (ten-ridged periwinkle)
      Thais lapillus (dogwinkle)
      Aequipecten irradians (bay scallop)
      Nucula delphinodonta (nutshell)
      Nucula proxima (nutshell)
      Placopecten magellanicus (deep sea scallop)
      Yoldia limatula (yoldia)
      Hydrobia minuta (swamp hydrobia)
      Littorina littorea (common periwinkle)
      Littorina obtusata (smooth periwinkle)
      Littorina saxatilis (rough periwinkle)
      Lacuna vincta (chink whelk)
      Coryphella sp. (nudibranch)
      Macoma balthica (baltic macoma)
      Modiolus modiolus (horse mussel)
      Mya arenaria (soft clam)
      Mytilis edulis (blue mussel)
    Phylum Annelida
      Cirratulus sp. (fringed worm)
      Haploscoloplos fragilis (obiniid worm)
      Lepidonotus sp. (12-scale worm)

Phylum Annelida *continued*
  Nereis sp. (clam worm)
  Ninoe sp. (thread worm)
  Polydora ligni (mud worm)
  Spirorbis sp. (hard tube worm)
  Thelepus sp. (terebellid worm)
Phylum Bryozoa (moss animals)
  Alcyonidium sp.
Phylum Platyhelminthes (flatworms)
  Macrostomum sp.
  Plagiostomum sp.
Phylum Arthropoda
  Class Crustacea
  (barnacles)
    Balanus balanoides (northern rock barnacle)
    Balanus balanus (rough barnacle)
  (amphipods)
    Ampelisca abdita
    Caprella sp.
    Corophium volutator
    Gammarus oceanicus
    Gammarus sp.
    Orchestia sp.
  (copepods)
    Ameira longipes
    Tisbe sp.
  (isopod)
    Idotea sp.
  (decapods)
    Cancer borealis (jonah crab)
    Cancer irroratus (rock crab)
    Homarus americanus (northern lobster)
    Crangon septemspinosa (sand shrimp)
    Carcinus maenas (green crab)
    Hyas areneus (toad crab)
    Pagurus sp. (hermit crab)
Phylum Echinodermata
  Asterius vulgaris (sea star)
  Ophiopholis aculeata (daisy brittlestar)
  Strongylocentrotus drobachiensis (green sea urchin)
  Henricia sp. (bloodstar)
  Ophiuroid spp. (brittle stars)
  Echinarachnius parma (sand dollar)

TABLE 5	
Protists Identified in the Smithsonian Maine Coast Microcosm[a]	

Kingdom Animalia
  Subkingdom Protozoa

Class Mastigophora (flagellates)	Class Ciliophora *continued*
Bodo saltans	*Aspidisca crenata*
Monosiga sp.	*Aspidisca* 5 spp.
Eutreptia sp.	*Diophrys appendiculata*
Euglena sulcata	*Cothurnia* sp.
Class Sarcodina (amoeboid types)	*Epistylis* sp.
Amoeba 8 spp.	*Vaginicola* sp.
Flabellula sp.	*Vorticella marina*
Hyalodiscus sp.	*Vorticella microstoma*
Limah hartmanella	*Zoothamnium* sp.
Mayorella sp.	*Loxophyllum* sp.
Pontifeh maximus	*Mesodinium* sp.
Rhixoamoeba sp.	*Nassula* sp.
Class Ciliophora (ciliates)	*Protocruzia* sp.
Amphileptus sp.	*Scuticociliate* sp.
Cinetochilum marinum	*Trachelophyllum* sp.
Coleps sp.	*Trochilia* sp.
Dysteria sp.	*Euplotes* spp.
Hemiophrys sp.	*Deranopsis* sp.
Lacrymaria sp.	*Uroleptus* 2 spp.
Lembus sp.	*Acineta* sp.
Lionotus spp.	*Metacineta* sp.
Actinotricha sp.	*Podophyra* sp.

[a]Courtesy D. Spoon.

paced than they are in the warmer microcosms and mesocosms. Without top fish predators to manage urchin, crab, and lobster numbers naturally, human intervention is needed to help smooth predator–prey cycles. However, the intervention process is relatively easy to accomplish. For example, impending problems with green crab or urchin predation can normally be detected many months before they are serious. Such "over-predation" is quite common in the wild on the Maine coast. However, with many hundreds of miles of irregular coast, the net result is a community structure that is patchy rather than depauperate of many species.

The rocky Maine shore microcosm has a relatively large wave generator, the effects of which have been the object of some research (see below). There is little question but that wave action is as crucial to plant production in the microcosm as it is in the wild ecosystem (see Leigh *et al.*, 1987). Although difficult to research in the sense of a comparison between wild and model, wave action is certainly a very major factor in determining the character of the entire model community.

A phycologist or ecologist would not tend to think of the Maine

TABLE 6

Fish and Lower Chordates Occurring as Long-Term Elements
in the Smithsonian Maine Coast Microcosom
1 Year After Establishment in 1984

Kingdom Animalia
  Phylum Chordata
    Vertebrata
      Class Osteicthyes (bony fish)
        Menidia menidia (Atlantic silversides)
        Myoxocephalus octodecemspinosus (longhorn sculpin)
        Fundulus heteroclitus (mummichog)
        Pollachius virens (pollock)
        Pungitius pungitius (ninespine stickleback)
        Tautogolabrus adspersus (cunner)
        Pseudopleuronectes americanus (winter flounder)
        Stenotemus chrysops (scup)
        Sygnathus fuscus (northern pipefish)
        Tunicata
        Botryllus schlosseri (golden star tunicate)
        Halocynthia pyriformis (sea peach)

shore as one with significant algal turf species. The large macroalgae, kelp, rockweeds, and even the smaller red "brush" zones are quite conspicuous. However, rich underlying and encrusting stories of filamentous algae are also present, and these are responsible for the well-developed algal turfs that form on the scrubbers. When algal substrates are constantly harvested, the larger algal species that do not reproduce until they have achieved many months of growth are unable to compete with the small, rapidly reproducing turf and smaller macroalgae. Unlike the slow development of a turf community that often characterizes freshwater systems, the subarctic scrubber turfs form quickly and effectively in this system.

Unfortunately, the design of the microcosm as it is currently laid out, with access between open coast and bay waters lying through the "rock mesh" of the lower rocky shore (see Fig. 1), does not allow larger rocky-shore fish to extend to the mud flat. *Fundulus* (killifish), *Crangon septemspinosa* (sand shrimp), and blue crabs work the very limited surface of muddy shore. It would be far more effective if a larger mud surface were available and the fish of the rocky shore could also browse in that habitat. Predation by larger fish might also naturally assist the establishment of a subtidal *Zostera marina* (eelgrass) community.

Research on the effects of temperature, light, nutrients, water movement, epiphytes, and grazing on kelp growth (*Laminaria longicruris*) has been carried out in this model and compared with that in the wild (Brittsan, 1989). Depending on simulated environmental parameters, kelp elongation in the model ranged from 0.02 to 2.1 cm/day (means 0.5–0.95 cm/day; Fig. 2). It is particularly interesting to note that the mean kelp elongation rates obtained in the model over 2 years of testing, 0.71 cm/day, were not significantly different from those found in the wild (0.72 cm/day). It is also of strong interest to those carrying out modeling efforts that with increased light levels, kelp growth in the model extended well beyond saturation levels found from previous research. It would appear that communities of aquatic plants generally use all available light, even though efficiency is relatively low.

Perhaps because of the necessarily lowered wave energies in the microcosm, urchin and snail grazing of kelp (44% of production for urchins) exceeded that typically found in the field (Adey, 1982; Vadas, 1977). Nevertheless, and unlike the reef communities discussed above, it would appear that on cold-water rocky shores, the amount of plant production consumed by herbivores is low to moderate (10–40%). In the wild, the remaining production is removed by wave action and ice and is deposited at the high tide line as shore drift. As detritus, this plant biomass is reduced to organic fragments by small crustaceans (particularly the amphipods *Orchestia* and *Talorchestia*, bacteria, and fungi). Returned to the water column on spring tides, heavy rains and in storms, this organic detritus becomes part of the tremendous suspended organic load in the water that is available to detritivores and filter feeders, particularly on quiet muddy bottoms. In the microcosm some algae break free in the microcosm due to snail and urchin grazing. Also, when plant growth is excessive, human intervention is used to remove plants, thus simulating storms. This "drift" algae is placed in the microcosm supratidal where it breaks down and returns to the system, much as in the wild. This is another example of the need for creative human intervention to simulate required physical or biotic processes that would be impractical to simulate precisely in a model.

Much more remains to be done in comparing a model ecosystem, such as this Maine coast exhibit with wild ecosystems. The larger the model, the closer it can be brought to reality. Nevertheless, with the use of algal turf scrubbing to simulate the chemical effects of the inevitably lacking larger body of water in closed systems, it seems clear that reasonably stable simulations of closed cold-water ecosystems can generally be developed.

Cold-Water Rocky Shores in Aquaria

We have not had the opportunity to develop a small Maine coast system in an aquarium. However, the prototype for the exhibit was quite successful at 250 gallons, and our experience with the Chesapeake model discussed in the next chapter suggests that smaller aquarium would be relatively easy to accomplish. We strongly encourage aquarists situated on boreal and subarctic rocky shores to experiment with those ecosystems in aquaria. The tidal unit discussed for the Chesapeake Bay would probably apply equally well to a Maine coast aquarium, although the separation of mud, flat, and marsh, if included, would be the best accomplished in a separate small tank as in the Smithsonian unit.

References

Adey, W. 1982. A Resource assessment of Gouldsboro Bay, Maine. Report to NOAA Marine Sanctuary Program. Final Report, Grant NA 81AA-DCZo76.

Boschung, H., Williams, J., Gotshall, D., and Caldwell, D. 1983. *Field Guide to North American Fishes, Whales and Dolphins.* The Audubon Society. Knopf, New York.

Brittsan, J. M. 1989. Regulation of kelp (*Laminaria longicruris*) growth in a subarctic marine microcosm and the rocky coast of Maine, U.S.A. M.S. thesis, University of Maryland.

Fefer, S., and Schettig, P. 1980. *An Ecological Characterization of Coastal Maine.* 5 vols. Department of the Interior, Northeast Region, Newton Corner, Massachusetts.

Gosner, K. 1978. *A Field Guide to the Atlantic Seashore.* The Peterson Field Guide Series. Houghton Mifflin, Boston.

Leigh, E., Paine, R., Quinn, J., and Suchanek, T. 1987. Wave energy and intertidal productivity. *Proc. Natl. Acad. Sci. USA* **84:** 1314–1318.

Ricketts, E., Calvin, J., and Hedgepeth, J. 1985. *Between Pacific Tides.* 5th Ed. Stanford University Press, Stanford, California.

Robins, C., Ray, R., and Douglass, J. 1986. *A Field Guide to Atlantic Coast Fishes of North America.* Houghton Mifflin, Boston.

Tangley, L. 1985. And live from the East Coast, a miniature Maine ecosystem. *Bioscience* **35:** 618–619.

Vadas, R. 1977. Preferential feeding: An optimization strategy in sea urchins. *Ecol. Monogr.* **47:** 337–371.

CHAPTER 23

ESTUARIES
Ecosystem Modeling Where Fresh and Salt Waters Interact

An estuary is a partly enclosed body of water in which fresh water coming off the land mixes with ocean water and provides more-or-less stable zones of intermediate salinities. During the past 10 thousand years, the earth's coastal waters have become relatively rich in large estuaries. The earth's climate is at a warm point in the glacial–interglacial cycle, and sea level is high. As a result, the lowermost reaches of many river valleys are flooded, creating elongate bodies of water grading from salt to fresh.

Estuaries tend to be very rich in organisms. Rivers, before they are diluted by the enormous body of ocean water, have generally high concentrations of many chemical elements needed by plants and animals to build their tissues. Organic particulates draining from the land tend to be sedimented-out in the estuary. Their breakdown on the often muddy bottoms tends to recycle needed elements to the estuarine communities of organisms. Ocean tides in the narrow confines of an estuary provide for relatively strong currents. As we pointed out in Chapters 6 and 19, coastal waters tend to be considerably more productive than fresh waters. Tidal currents, with their mixing action, help to drive this production. Also, we have described the process of evolution as an "arms race." Prey and predator evolve together, each developing more effective ways to defend themselves or to outsmart their prey. Accommodating to higher and, more particularly, lower salinities is a difficult step for many aquatic organisms. A marine organism, for example the oyster, having developed the ability to function at salinities much lower than ocean water, achieves considerable respite from its many predators. In many cases such an animal becomes highly abundant. Young marine fish of many species are hatched in the upper reaches of estuaries where they find not only abundant food for rapid growth but also a measure of protection from voracious larger fish that find the very shallow waters difficult to negotiate. In tight quarters, the latter are also exposed to a higher risk of predation from large animals such as birds and otters.

The rich food source provided by large numbers of fish, shrimp, crabs, bivalves, and aquatic birds has made estuaries a favored place of habitation for humans. In addition, estuaries have provided a haven for boats of commerce plying ocean waters. Through the estuarine rivers, these same boats or their coastal counterparts have found easy access to the interiors of continents. Many large cities have developed on estuaries where they function as the gateways to continents. Thus, in modern times these highly productive bodies of water have become the focus of human disturbance and pollution. We badly need a greatly extended understanding and appreciation of the ecological nature of estuaries or we are in great danger of creating extensive abiotic zones of the most useful and beautiful of coastal waters.

Stable, well-developed estuaries typically range from tens to hundreds of miles in length. Simulating the dynamics of estuaries would seem an almost impossible task in the confines of microcosms and mesocosms. In this chapter, we would like to describe our experiences over the past several years in simulating such systems.

In the pages that follow, we will describe a mesocosm of a very large temperate estuary, the Chesapeake Bay, followed by that of a complex of much smaller subtropical estuaries. The fresh waters of the Florida Everglades where they meet the Gulf of Mexico produce the subtropical estuarine complex that we have used as our wild analog. Finally, we will describe a small home aquarium system operated at a single, brackish level of salinity.

Day *et al.* (1989) provide a modern treatment of the ecology of estuaries. This work can provide the interested reader a general scientific background to the subject of this chapter. Most of the organisms specifically discussed in this chapter can be located in the following field guides: Audubon, 1983; Godfrey and Wooten, 1979; Gosner, 1978; Kaplan, 1988; Odum *et al.*, 1984. Local area citations are given below in the appropriate sections.

Chesapeake Bay in Mesocosm

About 180 miles long and 5–30 miles wide, Chesapeake Bay is one of the largest estuaries in the world. It is a relatively shallow body of water, the average depth, including tributaries, being only about 22 feet. However, in restricted areas of strong current, depths can reach over 150 feet. The freshwater flow of Chesapeake Bay, which is received from a number of rivers, the largest being the Susquehanna, the Potomac, and the James, derives from a watershed of about 64,000 square miles. Although there is a peak of flow in the spring and a minimum in the autumn, freshwater input and tidal exchange into this large bay are consistent enough to maintain salinity levels at any one point in open water at a yearly range of usually less than 4–7 ppt (Fig. 1). Tides in Chesapeake Bay are relatively small, 1–4 feet maximum. However, the currents created by this tide in the long, narrow confines of the bay are sufficient to provide a moderate amount of mixing. Thus, even though saltier water from the ocean does tend to work its way up the bay on the bottom as a heavier "salt wedge," and lighter fresh water from the rivers tends to flow over the top, mixing keeps surface to bottom salinity differences to less than 2–3 ppt.

A very large body of scientific literature exists for the Chesapeake Bay. A description of the bay and its organisms can be obtained from: Humm, 1979; Lippson et al. 1979a,b; and Wass, 1972. A particularly good introductory reference for the bay is *Life in the Chesapeake Bay* by Alice Jane Lippson and Robert L. Lippson (1984).

The first requirement for simulating an entire estuary in mesocosm is a means of maintaining connected bay segments in such a way that reasonably stable salinities are maintained in each segment. The Chesapeake Bay mesocosm model built at the Smithsonian Institution's Museum of Natural History in 1986 was designed first and foremost to accomplish this requirement for stable salinity (Figs. 2 and Color Plate 38). The model consists of a large, relatively flat and stepped fiberglass tank 40 feet by 12 feet by 5 feet with a total volume of about 15,000 gallons. Fiberglass and acrylic walls separate this large tank into eight segments of about 1900 gallons each. Each section has separate salinities about 4.5 ppt higher or lower than the adjacent section. The salinity separation and gradient is maintained dynamically with free access to organisms, as we shall describe below.

Each section of this estuarine model is connected to the adjacent section by hinged, circular, 4-inch-diameter flapper valves. These pneumatically operated gates or flapper valves (Fig. 3) are computer controlled to allow fresh water to move on the surface from the freshwater segments to the adjacent more saline segments and to allow the saltier water to flow in the other direction in the deeper parts of the system. Conductivity-type

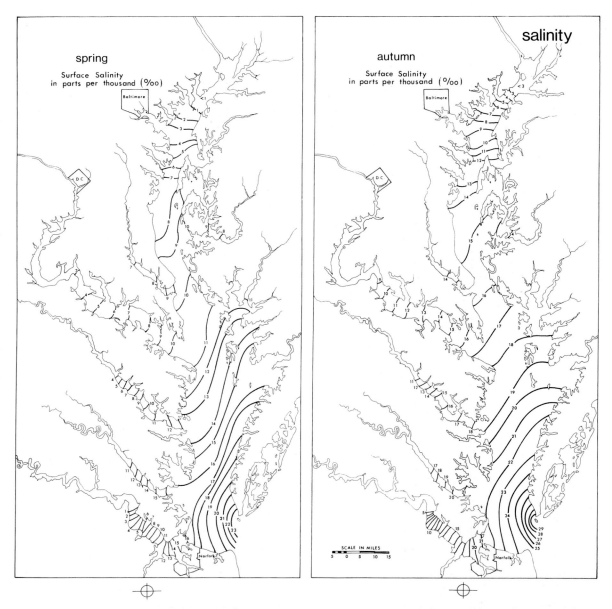

Figure 1 Typical surface salinity distribution in Chesapeake Bay during spring minimum and fall maximum. After Lippson *et al.* (1973a).

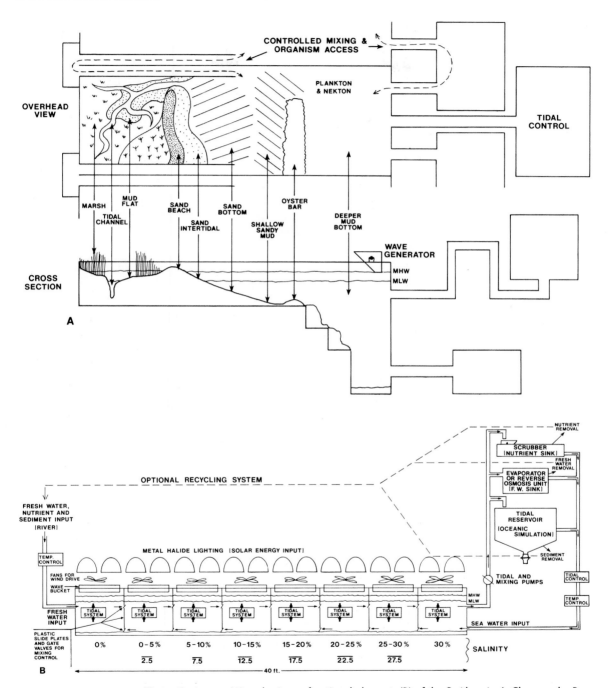

Figure 2 Layout (A) and primary functional elements (B) of the Smithsonian's Chesapeake Bay mesocosm. Individual tidal reservoirs not shown in (B).

Figure 3 Photograph of salinity control gate. In the initial version the flapper valve in the mixing chamber was 4 inches in diameter. On being activated by the computer, a solenoid valve delivers compressed air to the pneumatic piston, which drives the gate through plastic linkages. In the new and tighter unit pictured the gate slides between teflon bearings. Photo by Nick Caloyianis.

sensors are used to provide the computer with the salinity-level information needed for the decision to open the gate. Under computer operation, every 16 minutes each salinity sensor is queried as to whether or not the downstream tank has salinities less than a preselected value (shallow gate) or the upstream tank has salinities higher than a preselected value (deep gate). When the answer is positive, the associated gate is opened for 1 minute and then closed. Since the surface water is heated by the lamps and the water injected into the base of the higher, salinity tank is chilled (see below), the denser bottom water mixes relatively slowly with the surface and gradually moves through the deep gates, up the system toward fresh water. The net result is a flow and salinity pattern as shown in Fig. 4. Normal drift in sensor settings provides a variation in each tank of about ±2 ppt. While the computer control maintains narrow salinity ranges, it is

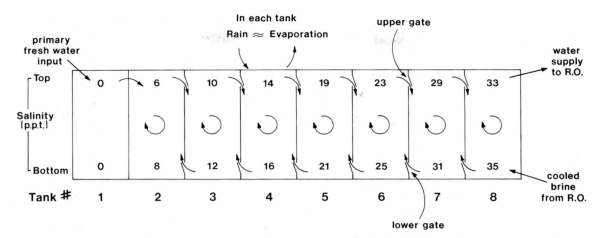

Figure 4 Flow and mean salinity distribution through the Chesapeake mesocosm.

hardly necessary. As configured, the Chesapeake mesocosm maintains stable salinity levels when the gates are operated manually several times a day. Perhaps the only problem with manual operation is that the frequency and random access of organisms from one salinity to the next is more difficult to maintain.

Since the salinity gates have an aperture of 4 inches and remain open for 1 minute whenever required by the salinity sensors and computer, mobile organisms such as crabs and fish can easily more throughout the mesocosm. Generally, this salinity gradient system has been quite successful. However, since rapid closure of the gates occasionally traps and kills fish and sometimes tears up the plastic fittings, a slower-acting system would be preferable. Also, in practice the lowest salinity gate on the shallow side between tanks one and two is not used since salt finds its way into the deeper parts of the tidal fresh tank and can only with difficulty be flushed out. To solve this problem, the freshwater tank is kept at a slightly higher level than the number 2 tank, and downstream flow occurs through a ½-inch hole between the two tanks. A shallow sluiceway like a dam fish ladder would allow greater access of estuarine organisms to fully fresh water in this system.

The gates described maintain stable salinity and allow organism movement throughout the mesocosm, effectively simulating a distance of almost 200 miles in the wild. However, the ends of the system must be held at freshwater and coastal saltwater salinities respectively, or the entire complex would level out to a single mid-level salinity within about 2 weeks. Technically, if the system were tight, the gates would cease to open if the ends were not held at 0 ppt and 30+ ppt, and no drift would occur. While the model estuary was not designed for fully tight gates, these

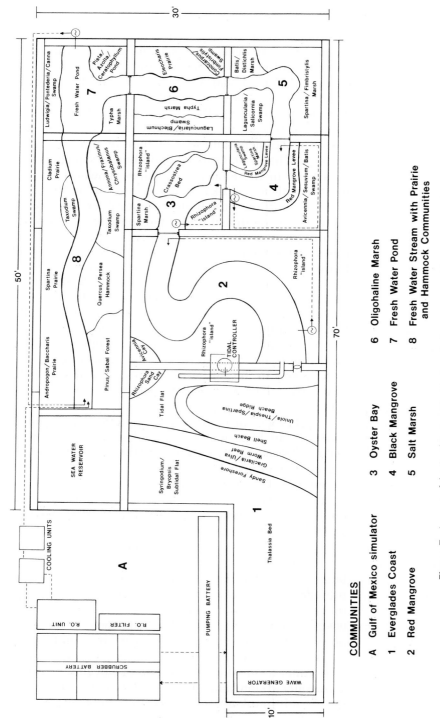

Figure 7 Layout of the Smithsonian Florida Everglades mesocosm. The community designations are based on the dominant plants that in most cases provide the structuring elements. The pumping battery of piston pumps was later changed to an Archimedes Screw (Color Plate 4 and Fig. 8).

COMMUNITIES
<u>COMMUNITIES</u>

A   Gulf of Mexico simulator

1   Everglades Coast

2   Red Mangrove

3   Oyster Bay

4   Black Mangrove

5   Salt Marsh

6   Oligohaline Marsh

7   Fresh Water Pond

8   Fresh Water Stream with Prairie and Hammock Communities

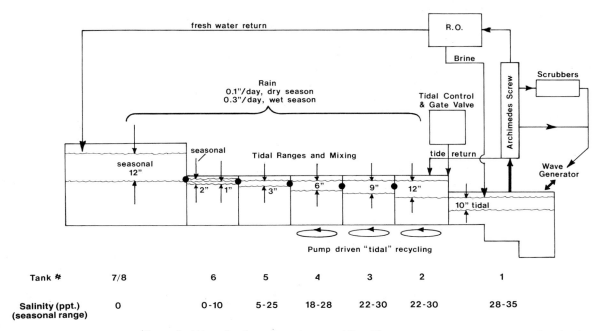

Figure 8 Water levels, weir-gate system (♦), and water management patterns in the Florida Everglades mesocosm.

fresh water from the Everglades coast tank and delivering it to the upper estuary is also an important element in simulating the evaporation–precipitation cycle and the salinity gradient. However, since this wind/rain pattern is quite mild relative to the typical intense thunderstorm of the wild Everglades, occasionally a hose is used to spray directly on the vegetation with some force. We have found this to be a major control device for insect pests, and it is the only control method that we have found to be both ecologically sound and effective.

The ocean simulator process and unit size employed on the Everglades tank in the sytem are the same as we used on the Chesapeake mesocosm. We will not repeat the discussion in detail except to note a few critical points. We first tried to use natural light for the scrubbers. This was not adequate, since intense cloudy intervals and particularly the fall and early winter dark periods are the times when scrubbing or ocean buffering is most important. Artificial light is necessary to provide this critical element of control. Also, on these and the Chesapeake scrubbers, while scrubbing rates of 15–18 g (dry algae)/m²/day are regularly produced, in month-to-month practice with levels of maintenance varying considerably, only 10–12 g (dry)/m²/day is routinely achieved. Micrograzer

TABLE 4

Primary Physical–Chemical Parameters of the Smithsonian Florida Everglades Mesocosm

Community	Tank dimensions (inches, feet)			Tank volume (l)	Surface area, ft² (m²)		Normal salinity range %
	D	W	L				
Marine (Gulf of Mexico) sea-grass bed, sandy beach, and beach ridge	43″	× 18′9″	× 36′	22,6.00 (1)	450	(42.1)	28–35
Red mangrove and tidal channel	34″	× 13′4″	× 18′9″	21,200	250	(23.4)	26–32
Oyster bay and mangrove	22″	× 8′9″	× 9′3″	4,440	81	(7.6)	24–30
Black mangrove and tidal channel	24″	× 9′4″	× 9′	5,165	86	(8.0)	18–28
Salt marsh	19″	× 9′4″	× 9′	3,978	84	(7.9)	5–25
Oligohaline marsh	22″	× 8′9″	× 9′	4,320	79	(7.4)	0–10
Freshwater Pond	40″	× 9′4″	× 9′	8,375	84	(7.9)	0
Savanna with hammocks and stream	26″	× 9′4″	× 30′	18,146	280	(26.2)	0
Total				(22,000 gal) 88,224 1	1394 ft² (130.3m²)		

Principle operating characteristics

Seasonal	Summer	Winter
Lighting	In greenhouse, natural light, Washington, D.C.	
Temperature		
Air	20–40°C	10–20°C
Water	25–29°C	15–20°C
Rain	Wet season 0.3 inch/day	Dry season 0.1 inch/day
Daily		
Wind	Afternoon, to 10 knots	
Tide	Mixed diurnal/semidiurnal, 1 ft marine and red mangrove, decreasing to 3 inch salt marsh	
Wave action	3–4 inch wave height, marine only	

(amphipod) grazing can be intense on these scrubbers. If amphipod populations in the absence of predation are allowed to explode, this can seriously reduce the efficiency of the scrubbers. Routine scraping at 7–12 days, with attention to the cleaning off of amphipods in the trays, is essential to adequate nutrient control. Note that here, as well as on the Chesapeake scrubbers, internal recycling of 10 gpm is used on each scrub-

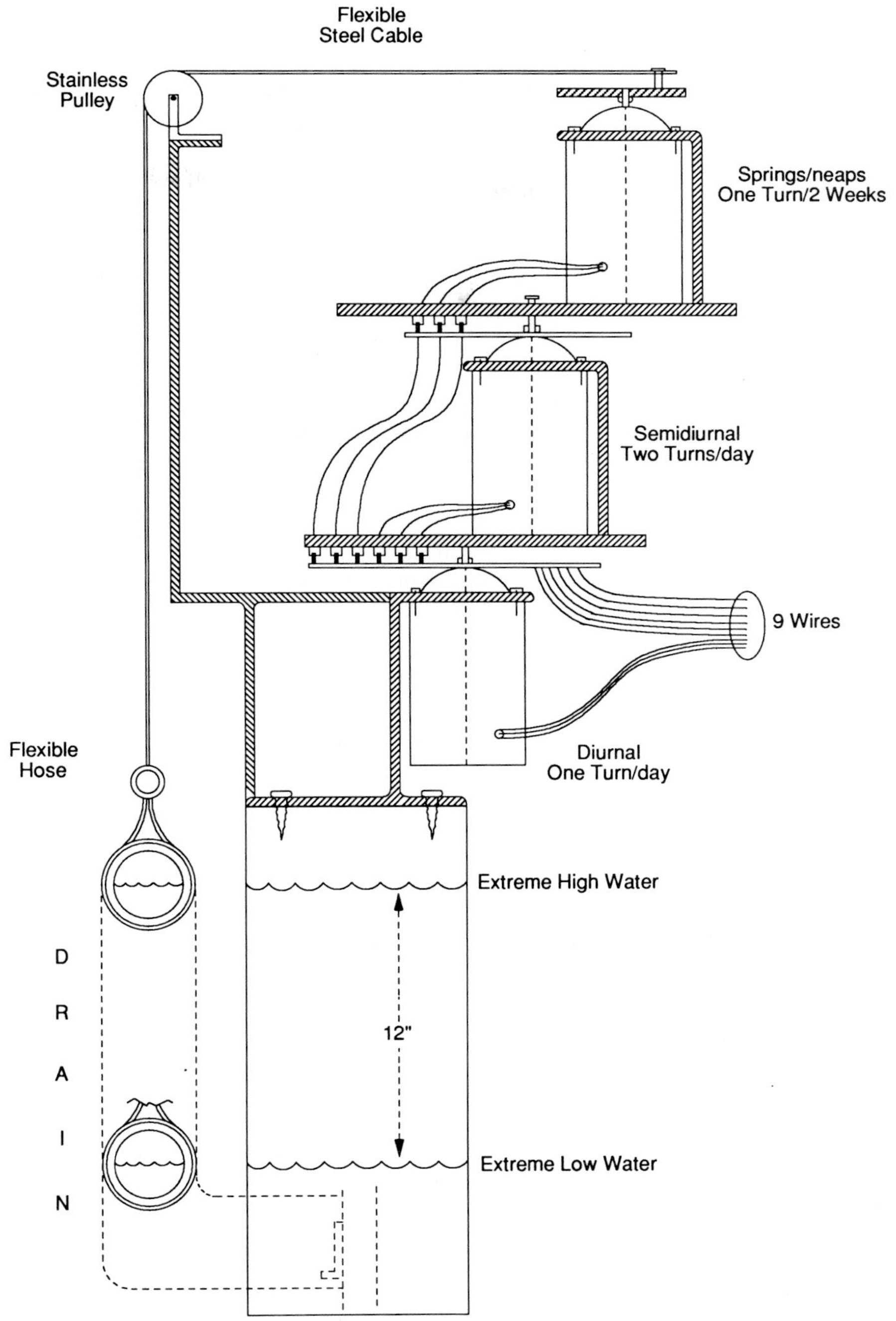

Figure 9 Tidal control system for the Everglades Estuary. The bottom stepping motor has a diurnal cycle (one rotation per 24 hours and 40 minutes), the middle motor has a semidiurnal cycle (12 hours 20 minutes), and the topmost motor simulating neaps and springs rotates once every 2 weeks. At every spring tide the system is readjusted to match the times of spring highs on the southwest outer coast of Florida.

ber to achieve adequate scrubber flow and surge while minimizing turn-over volume from the Everglades coast tank. Generally, 2–4 hours of R.O. operation per day (averaging 80–160 gallons of fresh water per day) along with 50–150 gallons of rain (replacing evaporation) is adequate to maintain the desired salinity gradient.

Water and nutrient dynamics are critical and mutually interactive components of both estuarine models and wild estuaries. Since both the Chesapeake and Everglades mesocosms operate in similar fashion in this respect, a separate section is necessary to properly treat this critical aspect. A nutrient dynamics section follows our discussion of the biota of the Everglades model.

Although incidental to system function, given proper climate, temperature control for a subtropical system is a matter of some concern in a greenhouse environment in Washington, D.C. Three propane heaters with a total capacity of 385,000 Btu are used to keep the Everglades greenhouse air at a minimum of 45°F on the coldest winter nights. Two sets of 1000-W water heaters are used to keep the Everglades coast and savanna streams at greater than 18°C under the same conditions. The estuary is allowed to follow, being heated at both ends, and has never gone below 15°C under this regime. A similar practice is followed in summer for cooling. The greenhouse atmosphere, with large wall-mounted fans pulling outside air through the house, is allowed to follow ambient temperature plus greenhouse effect. Occasionally the air in the greenhouse reaches 40°C (110°F). On the other hand, the waters of the saltier and fresher ends of the model are chilled using titanium heat exchangers, allowing a maximum temperature of 30°C in the coast and savanna stream waters. The estuary waters seldom rise significantly above this.

Wave action is not used on the Everglades estuary and stream. In the wild, these areas generally consist of narrow channels and only rarely are significant waves developed. The Everglades coast, on the other hand, while generally having only moderate wave energy in the wild, required simulation of wave action in the model. This was achieved with a single large dump bucket (Fig. 7).

An intense effort was made in this model to avoid centrifugal pumps. Slow-turning, gear motor-driven piston pumps were initially used. However, these caused significant maintenance problems. Eventually we installed a single, 200-gpm Archimedes screw on the Everglades coast tank (see Chapter 5, Fig. 18). This provides feed water for the algal scrubbers, the wave generator, the R.O., and the tidal lift. Because of the small size of the entire system, direct tide driven currents in the model are small. Thus, we have installed 10-gpm pumps for internal "tidal" circulation in the three highest-salinity estuary tanks. Temporarily these are standard centrifugal pumps. However, it is our intention to replace these with large bellows pumps.

The Biota of the Everglades Mesocosm

The Everglades mesocosm was stocked with substrate and organisms as described for the Chesapeake Bay model. The biota is listed in Tables 5 and 6. The species count in this system is currently over 500 species. Since neither plankton nor benthic protist species have been tabulated at this writing, it is to be expected that the stable (1–3 years) species count of the model is over 600 species.

A brief description of the benthic communities is given below, and is followed by a description of the fish and plankton that are present. Beginning in the Everglades coast tank, subtidally, a rich grass bed is developed. While the bed is dominated by *Thalassia testudinum* (turtle grass), the round-bladed grass *Syringodium* (manatee grass) and the smaller *Halodule* (Cuban shoal grass) are also present. Shoreward, in the zone between spring and neap low tides, a "worm reef" extends across the tank, in about the same position that it occupies in the more exposed localities in the wild. This low carbonate structure is richly endowed with sponges, small snails, porcelain and mud crabs, and sea squirts. The surface of the "worm reef" has a dense bed of algae with a wide variety of species, especially greens and reds. The higher predator, the stone crab, *Menippe mercenaria,* has been exceptionally successful in this system and seems to be able to find enough small "producer" species of crabs, snails, and clams to maintain a breeding population, at least on the scale of 2 years. Not shown in Fig. 7 is a series of four 200-gallon, natural-lighted refugia for soft bottom invertebrates attached to tank 1 in the lower lefthand corner of the diagram.

Color Plate 46 shows the "worm reef," shell beach, and the vegetated sand dune behind. To the left in the photograph is a small *Rhizophora* (red mangrove) sand key. These are abundant on the flanks of the coastline islands in the wild. In the model this little "key" is populated by large numbers of fiddler crabs (*Uca mordax*) and the sea roach (*Ligia*). The dune is unfortunately inactive since we have neither a continuous sand supply, a sand beach, nor the amount of wind that would be required to keep it in an active state. Nevertheless, the typical species such as sea grape, bay cedar, sea oats, finger grass, spanish bayonet, seaside mahoe, morning glories, alligator weed, and many others are present. If it became necessary to stop a succession from developing as supplied from the hammock community in tank number 8, we could probably achieve that need manually either through disturbance or actual removal of hammock species.

Across the tidal control wall and into the mangrove communities, the red mangrove above the water and the oyster below are the dominant elements. While the red mangroves are growing rapidly and have flowered and set seed in the system, the oyster barely holds its own, as we

TABLE 5

Plants Occurring in the Smithsonian Florida Everglades Mesocosm
about 2 Years After Stocking[a]

Kingdom Monera
  Division Cyanophyta (blue-greens)
    Gloeocapsa sp. (6)
    Anacystis sp. (3,6)
    Entophysalis conferva (2)
    Spirulina subsalsa (1,4,5,6)
    Schizothrix sp. (1,3,6)
    Schizothrix mexicana (2)
    Microcoleus sp. (5)
    Arthrospira sp. (4,5)
    Calothrix crustacea (1)
    Johannesbaptistia pellucida (6)
    Agmenellum thermale (4)
    Oscillatoria submembranacea (2)
    Oscillatoria sp. (3,5,6)
    Porphyrosiphon sp. (4)
    Scytonema sp. (2,4,5)
    Anabaina sp. (4,5,6)
Kingdom Plantae
  Subkingdom Thallobionta
  Division Rhodophycota (red algae)
    Acanthophora muscoides (1)
    Callithamnion cordatum (1)
    Spyridia filamentosa (2)
    Centroceras clavulatum (1)
    Polysiphonia gorgoniae (1,2)
    Murreyella sp. (2)
    Caloglossa leprieurii (2)
    Gracilaria sp. (1)
    Gigartina sp. (1)
    Heterosiphonia gibbesii (1)
    Pterosiphonia pennata (1)
    Ceramium nitens (1,2)
    Bostrychia sp. (2,3,4)
  Division Chromophycota
  (brown algae)
    Ectocarpus confervoides (1)
  (Diatoms)
    Fragilaria sp. (1,2)
    Licmophora sp. (1,6)
    Licmophora paradoxa (1)
    Licmophora longipes (1)
    Amphora 3 spp. (1,5)
    Auliscus coelatus (1)
    Actinoptychus undulatus (1)
    Ceratulus sp. (1)

Division Chromophycota
continued
(Diatoms)
continued
  Navicula sp. (5)
  Navicula clavata (1)
  Navicula granulata (1)
  Navicula peregrina (1)
  Navicula vividula (5)
  Rhaphoneis sp. (1)
  Rhopalodia musculus (1)
  Triceratium grande (1)
  Triceratium sp. (1)
  Tropidoneis lepidoptera (1)
  Amphiprova sulcata (5)
  Cocconeis sp. (5)
  Biddulphia pulchella (1)
  Biddulphia laevis (1,2)
  Biddulphia sp. (1)
  Tabellaria fenestrata (1)
  Achnanthes longipes (1)
  Campylodiscus daemelianus (1)
  Melosira (2) spp. (1)
  Melosira sulcata (1,5)
  Melosira numuloides (1)
  Nitzschia obtusa (5)
  Nitzschia lonenziana (5)
  Nitzschia longissima (1)
  Nitzschia paradoxa (5)
  Pleurosigma (2) spp. (1)
  Synedra crystallina (1)
  Synedra formosa (1)
  Synedra henneydyana (1)
  Synedra sp. (1)
  Capsulina sp. (2)
  Caloneis amphisbaena (5)
  Gyrosigma sp. (5)
Division Clorophyta (green algae)
  Oedogonium sp. (6)
  Cladophora sp. (2,3)
  Rhizoclonium hookeri (3)
  Bryopsis ramulosa (2)
  Chara sp. (6)
  Spirogyra sp. (6)
  Chaetomorpha sp. (2,3,4,5)
  Enteromorpha prolifera (4,5)

(continued)

TABLE 5

(Continued)

Division Clorophyta (green algae) *continued*
    Scenedesmus sp. (5)
Subkingdom Embyobionta
Division Filicophyta (ferns)
    Acrostichum aureum (leatherfern) (6)
    Azolla carolinianum (mosquito fern) (6)
Division Magnoliophyta (flowering plants)
    (Monocots)
    Hymenocallis crassifolia (spider lily) (1)
    Yucca aloifolia (spanish bayonet) (1)
    Typha domingensis (southern cat-tail) (6)
    Cladium jamaicensis (sawgrass) (6)
    Eleocharis robbinsii (spike rush) (6)
    Cyperus odoratus (umbrella plant) (1)
    Uniola paniculata (sea oats) (1)
    Spartina patens (salt hay) (1)
    Spartina alterniflora (smooth cordgrass) (5)
    Spartina spartinae (gulf cord-grass) (5)
    Distichlis spicata (salt grass) (4,5,6)
    Andropogon glaucopsis (6)
    Encyclia tampensis (butterfly orchid) (6)
    Thalassia testudinum (turtle grass) (1)
    Syringodium filiformis (manatee grass) (1)
    Halodule wrightii (cuban shoal grass) (1)
    Fimbristylis spathaceae (cone grass) (6)
    Frimbristylis castanea (cone grass) (5)
    Sporobolus virginicus (coastal dropseed) (1)
    Chloris petraea (finger grass) (1)

Division Magnoliophyta (flowering plants)
continued
    (Monocots) *continued*
    Chloris sp. (finger grass) (1)
    Cenchrus incertus (sandspur) (1)
    Setaria verticillata (bristle grass) (6)
    (Dicots)
    Coccoloba uvifera (sea grape) (1)
    Suriana maritima (bay cedar) (1)
    Thespia populnea (seaside mahoe) (1)
    Opuntia compressa (prickly pear) (1)
    Vaccinium stamineum (deer berry) (1)
    Chamaesyce mesembryanthemi-folia (spurge) (1)
    Limonium nashii (marsh rosemary) (4)
    Lycium carolinianum (christmas berry) (4,5)
    Philoxerus vermicularis (silver-head) (4,5)
    Ampelopsis arborea (pepper vine) (6)
    Myrica cerifera (wax myrtle) (6)
    Centella asiatica (centella) (6)
    Boehmeria cylindrica (false nettle) (6)
    Rhizophora mangle (red mangrove) (1,2,3,4,5)
    Conocarpus erectus (buttonwood) (5)
    Bidens pilosa (spanish needles) (1)
    Borrichia frutescens (sea daisy) (4,6)
    Mikania skandens (climbing hempweed) (6)
    Ipomoea sigittata (marsh creeper) (6)
    Ipomoea pes-capre (beach morning glory) (1)
    Ipomoea alba (moonflower) (1)
    Ipomoea indica (morning glory) (1)

(continued)

TABLE 6
(*Continued*)

Phylum Arthropoda
continued
  (Crustacea)
  continued
    Caprella equilibra (skeleton shrimp) (1,2)
    Leptochelia savignyi (4)
    Balanus amphitrite (striped barnacle) (1,3)
    Balanus eburneus (ivory barnacle) (1,2)
    Porcellana sayana (says porcellan) (1,2,3)
    Panopeus herbstii (herbs panopeus) (1,2,3)
    Petrolisthes armatus (porcelain crab) (1,2,3)
    Petrolisthes politus (porcelain crab) (1)
    Sesarma curacaoense (2)
    Sesarma cinera (4)
    Megalobrachium poeyi (2)
    Uca mordax (fiddler) (5)
    Penaeus duorarum (pink shrimp) (1)
  (Insecta)
    Paracoenia bisectosa (1)
    Orthocladine sp. (2,4,5)
    Microvelia hinei (6)
    Aedes taeniorhynchus (mosquito) (1,5)
    Erythemin simplicicollis (6)

Phylum Arthropoda
continued
  (Arachnida)
    Dolomedes triton (fishing spider) (1,5)
Phylum Chordata
  (Vertebrata)
    Floridicthys carpio (gold spot killifish) (1)
    Fundulus grandis (gulf killifish) (2,3,4)
    Cyprinodon variegatus (sheepshead minnow) (2,3,4,5)
    Eucinostomus gula (silver jenny) (1,2)
    Eucinostomus argenteus (spotfin moharra) (1)
    Serraniculus pumilio (pygmy sea bass) (1)
    Opsanus beta (gulf toadfish) (1)
    Micropogonias undulatus (atlantic croaker) (2)
    Achirus lineatus (lined sole) (1)
    Haemulon plumieri (white grunt) (2)
    Lagodon rhomboides (pinfish) (1,2,3,4)
    Mugil cephalus (striped mullet) (2)
    Poecilia latipinna (sailfin molly) (1,2,3,4,5,6)
    Gambusia affinis (mosquito fish) (2,3,4,5,6)

[a]The numbers following each species indicate the salinity levels in which it occurs (see Fig. 7; 1, Gulf of Mexico, full salinity; 6, oligohaline marsh). Plankton, protists, microscopic invertebrates, terrestrial insects, soil microfauna, and bacteria had not been fully studied at the time this table was prepared.

soggy, heavily rooted soil at low tide. The black mangroves were difficult to transport because of their extensive shallow cable roots. However, having become established in the model they have developed abundant new cable roots with the aerial pneumatophores that they require to survive in the anaerobic muds. Unlike in a large wild mangrove community, a number of lower-story plants densely appear in the model. These include sea purslane, saltwort, glasswort, and sea daisy. A number of small invertebrates, including several fiddler crabs, the coffee snail, the sea roach, and several copepods, are also abundant here, although at spring high tides, the killifish roam across the flooded muds seeking out small invertebrates.

Crossing the next weir brings us into the salt-marsh community (Color Plate 47). Another mangrove, the white, is highly successful here along with several salt grasses, rich growths of the sedge *Fimbristylis* and the conspicuous Christmas berry, a carpet of sea purslane, and a few ferns that can withstand moderate salt concentrations. Several snails, insects, and a spider are also abundant on this tiny marsh, although mosquitoes, so characteristic of the wild, are absent. We attribute this to the depredations of the mosquito fish that do so well in this part of the Everglades model. The mangrove trees, woody plants, from a wide variety of families, that survive in a salt water or a brackish environment are the basic structuring elements of much of the tidal Everglades. Only four species occur in the western Atlantic. All four of those species occur in this model ecosystem and have repeatedly flowered. All species except the Buttonwood have produced and set seed.

Passing the next weir brings us to the first zone that is seasonally fresh water, the oligohaline marsh (Color Plate 48). The last mangrove, the buttonwood, does well here, but the conspicuous species, both here and in the wild, are the very large cattail *Typha domingensis* and the thick carpet of the spikerush *Eleocharis robbinsii*. Also, for the first time in our transect, the namesake of the "River of Grass," saw grass, *Cladium jamaicensis*, appears in small numbers. Relatively few small animals occur in this zone as in the wild, including several small crustaceans and insects. Again the omnipresent mosquito of the wild systems has great difficult surviving with few refugia from mosquito fish.

In the Everglades greenhouse, we have attached to the upper part of the estuary the primary freshwater communities of the Florida Everglades, a small stream and prairie of varying hydroperiods with a variety of wooded hammocks. These we discuss in Chapter 24.

Returning to the fish community of the Everglades model estuary, it is apparent that species of two major groups, the Cyprinodonts and the Poecilids, are by far the most successful. The Poecilids, including the mollies, mosquito fish, and the guppies, are live bearers. In a system with minimum planktonic volume and space this is an obvious key to success. On the other hand, the killifish, including *Cyprinodon variegatus* (the sheepshead minnow), *Floridicthys carpio* (the gold-spotted killi), and several *Fundulus* species, are well adapted to marshes, flooded mangrove forests, and seasonal channels and would be expected to adjust well to this system. It is interesting that mosquito larvae are probably a major source of food for these fish in the wild. Mosquito larvae are virtually wiped out of the Everglades mesocosm; thus these highly successful and continuously breeding fish must have an alternate source of food, probably in the numerous small crustaceans, especially amphipods and copepods, that abound in the system.

Generally, the highest-salinity sections of the Everglades mesocosm are richest in fish species, with numerous mid-level predators, such as jennies and moharras (*Eucinostomous*), grunts, tomtate, pompano, tonguefish, and the eel (*Anguilla*) growing well over a long term of at least several years. Presumably much of the food of these fish is derived from the numerous fry of the killifish and the poecilids. Unfortunately, we have not yet seen reproduction in any of these mid-level fish. Perhaps the populations are too small for breeding or lack sufficient open-water habitat for the planktonic larvae.

Nutrient Dynamics in Estuarine Models

A summary of the distribution of nitrogen (nitrite plus nitrate) concentrations in these estuarine systems is given in Figure 10. Both systems show very similar patterns of nutrient distribution in spite of their great differences climatically and biotically. Through the middle salinities of both

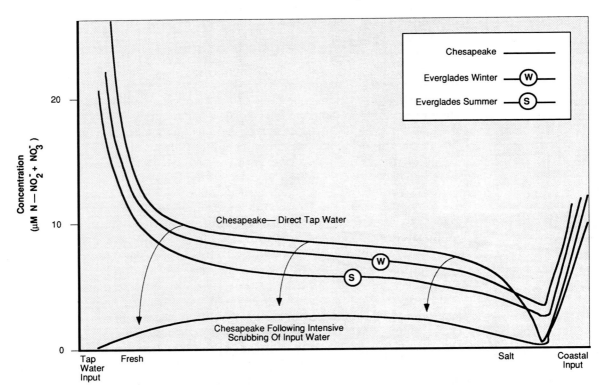

Figure 10 Summary of principal nutrient patterns in the Smithsonian estuarine mesocosms (see text).

estuaries, dissolved nitrogen is running 3–10 μM (μg-at/l) (0.04–0.14 ppm). In the Everglades system there is a significant summer–winter difference of about 3–4 μM due to high productivities and standing crops in summer. This seasonal pattern probably also applies to the Chesapeake model, but our long-term efforts to examine the effects of reducing nitrogen input on concentrations through the middle of the estuary have masked any seaonal differences. These features would be normal for large, wild, and undisturbed estuaries (Day et al., 1989). The pattern matches most wild systems where both ocean and mid estuary itself are sinking nitrogen and producing a concave downwards distribution rather than a straight line form river input to ocean. Also, it is plain that the scrubbers at the high-salinity end are "pulling" nutrients through the system and acting in the "sink" role that they were meant to play. Unfortunately, nitrogen levels are lower in the highest-salinity tanks of the Chesapeake system ($<$ 1 μM) than would be desired in a truly accurate model. This problem could be circumvented without losing the required amount of sinking by making the high-salinity tank somewhat larger, as has been done in the Everglades model.

The interesting part of both estuarine models is the nutrient dynamics at the low-salinity end. Both are characterized by the typical human-influenced high-nutrient input, the tap water from the Potomac River. Tap water typically ranges from 100 to 180 μM for dissolved inorganic nitrogen. We have added algal scrubbers and reverse osmosis units to the freshwater side and have been gradually lowering inputs. Rapid denitrification is probably responsible for much of the sharp drop in the early runs on the first two tanks. It seems likely that below about 10 μM nitrogen as nitrite plus nitrate, little denitrification occurs and virtually all injected nitrogen is passed to the ocean simulator.

Following the placement on the input tap water of a laboratory grade reverse osmosis unit plus a scrubber system for nutrient "polishing" of river input, dissolved nitrogen was reduced to less than 1.0 μM. Under those conditions nitrogen levels throughout the Chesapeake mesocosm fell sharply as shown in Fig. 10. The convex upwards curve that resulted indicates outwelling of stored nitrogen from the system. When concentrations throughout the model are consistently below 5μM, we will raise river input to the 5μM range (which we estimate to be prehistoric levels) and allow the system to come to equilibrium.

We have only limited dissolved phosporus information on these systems. Unfortunately early in the development of the Chesapeake model, we used a commercial filtering device for tap water that was reconstituted using phosphoric acid. This approach introduced considerable phosphorus into the model, and early measurements showed phosphate levels throughout mid salinities at about 10μM. Recent data under development for the Chesapeake Mesocosm shows current phosphate concentrations to be about the same as nitrate (Fig. 10), ranging up to 3–4 μM. While these values are normal to low for many estuaries, the nitrogen to phosphorus ratio is much too low. It is typically 4 or more in most estuaries today. On the other hand, considering that algal turf uptake ratios for N : P are likely to be in the 10–15 range, it is not unreasonable that high phosphorus levels should drop only slowly considering the past management history of the model. While it might be argued that low oxygen in the sediments or continued disturbance of the system by research activity is responsible for relatively high phosphorus, this seems unlikely, or at least secondary to the original source from the filtration process.

A 130-Gallon Brackish-Water Aquarium

We have developed a simple, single-salinity, brackish-water aquarium in our home near Chesapeake Bay (Color Plate 49). The tank lighting is

TABLE 7	
Organisms Occurring as Long-Term Elements of a 130-gallon Chesapeake Bay Aquarium[a]	

Kingdom Monera	Kingdom Animalia
Division Cyanophycota (blue-greens)	Subkingdom Protozoa
Microcoleus lyngbyaceous	Hypostomate ciliate, 2spp.
M. vaginatus	Peritrich ciliate, several spp.
Anacystis sp.	Vestibulate ciliate
Spirulina subsalsa	Subkindom Parazoa (sponges)
Kingdom Plantae	*Halichondria bowerbanki*
Division Rhodophycota (reds)	*Haliclona loosanoffi*
Hypnea musciformis	*Microciona prolifera*
Polysiphonia harveyii	Subkingdom Eumetazoa
Callithamnian byssoides	Phylum Cnidaria (coelentrates)
Division Chlorophycota (greens)	*Diadumene leucolena* (ghost
Ulva lactuca (sea lettuce)	anemone)
Enteromorpha intestinalis	*Haliphanella luciea*
Codium isthmocladum	*Astrangia danae* (northern cup
Caulerpa verticillata	coral)
Division Chromophyta (browns and	*Leptogorgia virgulata* (sea whip)
others)	*Hydractinia echinata* (snail fur)
Ectocarpus sp.	*Sertularia argentea* (hydroid)
Many diatoms and few	Subkingdom Eumetazoa
dinoflagellates	Phylum Platyhelminthes
Division Magnoliophyta (higher plants)	Turbellarian sp. (flatworm)
Monocots	Phylum Nematoda
Spartina patens (salt hay)	Two spp.
Spartina alterniflora (cord grass)	Phylum Rotifera
Distichlis spicata (salt grass)	*Ploima sp.*

(continued)

TABLE 7
(Continued)

Division Magnoliophyta (higher plants) continued	Subkingdom Eumetazoa continued
Monocots	Phylum Annelida
continued	Sabella micropthalma
Juncus roemerianus (black needle rush)	Polydora ligni
	Nereis succinea
Dicots	Hydroides dianthus
Aster tenuifolius (seaside aster)	Phylum Arthropoda
Iva frutescens (marsh elder)	Crustacea
Limonium caroliniana (seaside lavender)	Chthamalus fragilis (fragile star barnacle)
Salicornia virginica (glasswort)	Balanus eburneus (ivory barnacle)
	Cyclopoid copepod
	Harpacticoid copepod
	Ampithoe sp. (amphipod)
	Talorchestia longicornis (beach hopper)
	Cymadusa compta (amiphipod)
	Sphaeroma quadidentatum (isopod)
	Leptochelia savigni (tanaid)
	Neopanopeus sayi (mud crab)
	Sesarma cinerum (marsh crab)
	Uca pugnax (fiddler crab)
	Pagurus longicarpus (hermit crab)
	Eurypanopeus depressus (mud crab)
	Phylum Mollusca
	Crepidula fornicata (deck shell)
	Melampus bidentatus (marsh snail)
	Littorina irrorata (periwinkle)
	Nassarius obsoletus (dog whelk)
	Doris verrucosa (nudibranch)
	Ischadium recurvus (recurved mussel)
	Crassostrea virginica (oyster)
	Phylum Chordata
	Vertebrates (fish)
	Fundulus heteroclitus (mummichog)
	Menidia menidia (atlantic silversides)
	Prionotus carolinus (northern sea robin)
	Hypsoblennius hentzi (feather blenny)
	Gobiesox strumosus (skilletfish)
	Tunicata (sea squirts)
	Molgula manhattensi

ᵃBacteria, diatoms, protists, and plankton are largely not identified.

achieved by six 48-inch VHO flourescent lamps, and the scrubber unit is a smaller version of the home reef system described above. The unique feature of this tank is a 6-inch tide with a built-in combination tidal reservoir and refugium. In spite of its small size, this aquarium has about the same diversity as a single-salinity segment of the much larger Chesapeake mesocosm.

The Chesapeake aquarium was designed for a marsh characteristic of high brackish salinities (primarily *Spartina* grasses but some *Juncus roemerianus*). The marsh itself is also richly occupied by animals, with *Littorina* and *Melampus* snails, marsh and fiddler crabs, and marsh mussels dominating. Subtidally an oyster bed is occupied by abundant oysters as well as a variety of coelenterates, calcareous worms, the bent mussel, and the very striking gorgonion *Leptogorgia virgulata*. The small crustacean and protist fauna as well as the diatoms are equally rich, probably because of the algae refugium and the bellows pumps. Certainly this tank has fewer species than the much larger Chesapeake system described above. However, it is easily operated and highly stable with a moderate species diversity (Table 7). On a small scale, the whole Chesapeake estuary could be simulated as a home system with a series of such tanks having valves to allow water movement from one salinity to the next.

References

Audubon Society. 1983. *Field Guide to North American Fishes. Whales and Dolphins.* Knopf, New York.

Day, J., Hall, C., Kemp, W., and Yunez-Arancibia, A. 1989. *Estuarine Ecology.* Wiley and Sons, New York.

Douglass, M. 1988. *The Everglades.* Pineapple Press, Sarasota, Florida.

Drew, R., and Schomer, N. 1984. An ecological characterization of the Caloosahatchee River/Big Cypress watershed. Fish and Wildlife Service FWS/OBS-82/58.2.

Godfrey, R., and Wooten, J. 1979, 1981. *Aquatic and Wetland Plants of the Southeastern United States. Monocotyledons,* 1979. *Dicotyledons,* 1981. University of Georgia Press, Athens, Georgia.

Gosner, K. 1978. *A Field Guide to the Atlantic Seashore,* Peterson Field Guide Series. Houghton Mifflin, Boston.

Humm, H. 1979. *The Marine Algae of Virginia.* University of Virginia Press, Charlottesville, Virginia.

Kaplan, E. 1988. *A Field Guide to Southeastern and Caribbean Seashores.* Petersen Field Guide Series. Houghton Mifflin, Boston.

Lippson, A., and Lippson, R. 1984. *Life in the Chesapeake Bay.* Johns Hopkins University Press, Baltimore.

Lippson, A., (Ed.) 1979a. *The Chesapeake Bay in Maryland.* Johns Hopkins University Press, Baltimore.

Lippson, A., Haire, M., Holland, A. F., Jacobs, F., Jensen, J., Moran-Johnson, R. L., Polgar, T., and Richkus, W. 1979b. *Environmental Atlas of the Potomac Estuary,* Martin Marietta Corp., Environmental Center, Baltimore.

Odum, W., Smith, T. III, Hoover, J., and McIvor, C. 1984. The ecology of tidal freshwater marshes of the United States East Coast: A community Profile. U.S. Fish and Wildlife Service. FWS/OBS-83/17.

Schomer, N., and Drew, R. 1982. An ecological characterization of the lower Everglades, Florida Bay and the Florida Keys. U.S. Fish and Wildlife Service. FWS/OBS-82/58.1.

Wass, M. 1972. A check list of the biota of lower Chesapeake Bay. Virginia Institute of Marine Science Special Science Reprint 65. Gloucester Point, Virginia.

CHAPTER 24

FRESHWATER ECOSYSTEM MODELS

Freshwater bodies are generally smaller, shallower, and subject to greater variations in temperature, pH, and mineral content than are marine waters. The organisms that have evolved in fresh waters are adapted to wider-ranging conditions both in space and time. Thus, freshwater aquaria are usually simpler to set up and operate. The "original," simple balanced aquarium, with several dozen species of plants and animals, was surely a nonmarine tank or pool of tens of gallons. However, fresh waters can also be low-nutrient mountain streams or lakes or large, moderate-nutrient, high-surface-oxygen ponds and lakes. Also, freshwater systems for recreation or aquaculture in which food is added and a high biomass of organisms maintained can also be desired. In these latter cases, the techniques described in this book are helpful to achieving the model ecosystem that is desired and in reducing loss of fish and other organisms. In this chapter, as examples we describe several freshwater systems that we have established in recent years. These show a range of sizes and have quite different biological communities. The principles of establishment, control, and operation are basically the same as those we have discussed above for a variety of marine and estuarine ecosystem models.

For general background reading on the subject of this chapter, the authors suggest the following books: Burgis and Morris (1987), Moss (1988), Pennak (1953; 1989), Rataj and Horeman (1977), and Riehl and Baensch (1986). The earlier chapters in this book provide further background and more in-depth references to fresh water ecology.

A Florida Everglades Stream and Wetland

As part of the pilot system of the Florida Everglades being developed for Biosphere II, the Smithsonian's Marine Systems Laboratory built a 40-foot-long by 9-foot-wide stream and small pond flanked on one side by a series of "woodlands" or hammocks of different hydroperiods and on the other by a wet to dry prairie (Fig. 1). The reader is referred to the last chapter and to Drew and Schomer (1984) and Duever *et al.* (1986) for background information.

As described in Chapters 2 and 23, the tank itself was constructed of cement block and was lined with a nontoxic butyl rubber sealant. The critical bedrock geology of the Everglades in relation to the water table was desired. However, funds were inadequate to transport Florida limestone. To partially overcome this difficulty, an uncoated cement slab and block platform was constructed within the tank at about the level of seasonal low water table (Fig. 2). The "groundwater" is turned over about once per day (from downstream to upstream) with a 3-gpm centrifugal pump. The approximate water volumes of this fresh water microcosm segment of the larger estuarine mesocosm are 2500 gallons during the wet season and 4000 gallons in the dry season. As with the estuary that this stream and wet savanna accompany, light is ambient (a greenhouse in Washington, D.C.), and maximum and minimum temperatures are 10–35°C (air) and 18–29°C (water). Immersion heaters have to be used in winter to generally keep stream temperatures above 18°C. Wet-season (June–November) rainfall amounts to 0.3 inches per day; dry-season (January–May) rainfall is 0.1 inches per day. Freshwater supply to the stream, in addition to the rainfall, is derived from reverse osmosis and algal scrubbing of fresh water from the Gulf of Mexico tank at 30–34 ppt.

This Everglades wetlands system was designed with floating and

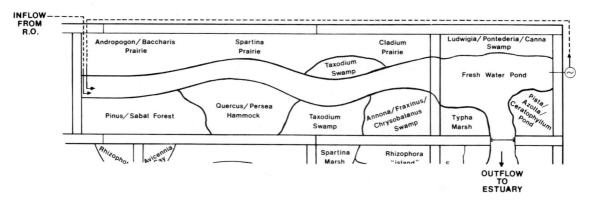

Figure 1 Diagram of the freshwater Florida Everglades model built by the Smithsonian's Marine Systems Laboratory. The communities established by block transfer are indicated.

Figure 2 Florida Everglades stream mesocosm (left) during construction. The double course of cement block defining the stream itself is placed on a mosaic of 4-inch thick cement slabs that separate "groundwater" from surface water.

emergent aquatic plants fringing the pond (Color Plate 50), and with hydroperiods (or water-saturated surface soils) ranging from about 10 months to 2 months in the hammocks and prairie (Color Plate 51). Lists of plant and animal species in the system are given in Table 1. Except for fish and a few large invertebrates, the model was stocked over roughly a 6-month period using block transfer methods. The primary trees in the hammock community were purchased from native plant greenhouses in Florida. In the early stages of operation, a few trees and bushes were lost to insects, primarily aphids and red spider mites. Further losses have been greatly reduced by periodically using a strong spray from a garden hose (thereby simulating the occasional heavy wind and rain that characterize the wild). We strongly recommend that commercial greenhouses not be the source of terrestrial plant communities for mesocosms because of pests usually associated with this source. If it is necessary to utilize commercial greenhouse material, quarantine and total elimination of pests is probably desirable, preferably by biological methods.

Stream flow in this mesocosm was achieved by recycling with a 50-gpm Flo-Tec centrifugal pump. While this is slower than some streams in southwest Florida, particularly in wet season or approaching tidal influence, it has the typical sluggish character of many streams in the area. We strongly feel that increased flow particularly using a small Archimedes screw that would not damage plankton and larvae would greatly increase small invertebrate and fish diversity. We have been unable to make this improvement for financial reasons. However, the Biosphere II system will use high-volume, vacuum-lift pumps to overcome the problem of plankton degradation.

Most crucial in the management of this model is that it is a "balanced" one in most respects. No feed is added, and all deciduous and cropped vegetation is allowed to rot within the system. Water is added directly and by rain to overcome evaporation (about 50 gallons per day). Approximately 200 gallons per day is also removed from the estuary below by a reverse osmosis machine and passed through the stream to the upper reaches of the estuary. The water used for evaporative makeup is Washington, D.C., water (which is extremely eutrophic: nitrogen as NO_2^- + NO_3^- equals 130–200 μM). However, this input water is constantly scrubbed to 50–80 μM dissolved nitrogen using a double 1.5-m-square algal turf system. One unit of this scrubber is lighted 12 h per day with eight 48-inch VHO lamps; the other uses natural light (greenhouse). Actual reverse osmosis injection levels are at 5–10 μM for nitrogen as NO_2^- + NO_3^- and 1–3 μM for phosphorus as PO_4^{3-}. The throughput of water to and from the estuary and its resulting flushing action have been sufficient to stabilize the open water of the stream model at 1–5 μM for dissolved nitrogen. Thus, effectively the Gulf of Mexico saltwater scrubbers described in Chapter 23 are providing low-nutrient source waters for the Everglades stream, and the rainwater is moderately elevated in nutrients. However, the stream and its surrounding prairie and hammocks are even further reducing dissolved nitrogen. While this description covers a roughly 6-month period of operation, there has been a considerable increase of plant biomass during this time. Once biomass increase can no longer occur, a further adjustment to nutrient balance (biomass removal or further nutrient reduction) may have to be undertaken. At that point scrubbers will be placed on the stream itself or the rain water (tap water) will be further scrubbed. Under full balance conditions if scrubbers are still necessary, they will be used only to buffer seasonal productivity. Nutrients would be removed in the autumn and replaced in the spring.

As can be seen in Table 2, this stream model is taxonomically quite diverse, considering its small size, and many of the small invertebrates and insects are reproducing in spite of heavy pressure from small

TABLE 1

Plants and Animals Occurring in the Smithsonian Fresh Water Florida Everglades Mesocosm After 2 Years of Operation[a]

Kingdom Monera
- Division Cyanophycota
  - *Microcoleus* sp.
  - *Oscillatoria* sp.

Kingdom Plantae
- Division Chromophycota
  - *Amphora* spp. (6)
  - *Fragillaria* sp.
  - *Thalasassionema nitzschiodes*
- Division Chlorophyta
  - *Oedogonium* sp.
- Division Filicophyta
  - *Azolla caroliniana* (mosquito fern)
  - *Salvinia rotundifolia* (water velvet)
  - *Blechnum* sp. (hammock fern)
  - *Pteris* sp. (brake fern)
  - *Thelypteris kunthi* (shield fern)
- Division Lycopodiophyta
  - *Selaginella floridana* (spike moss)
- Division Bryophyta
  - *Riccia fluitans* (thallose liverwort)
- Division Pinophyta
  - *Pinus elliotii* (slash pine)
  - *Taxodium distichum* (bald cypress)
- Division Magnoliophyta
  - monocots
    - *Cladium jamaicensis* (sawgrass)
    - *Carex lupuliformis* (sedge)
    - *Dichromena latifolia* (white bract sedge)
    - *Juncus* sp. (rush)
    - *Typha domingensis* (cattail)
    - *Crinum americanum* (swamp lily)
    - *Allium* sp. (wild onion)
    - *Thalia geniculata* (arrowroot)
    - *Pistia stratiotes* (water lettuce)
    - *Colocasia esculentum* (taro)
    - *Najas guadalupensis* (naiad)
    - *Tillandsia setacea* (wild pine)
    - *Tillandsia usneoides* (spanish moss)

Division Magnoliophyta
continued
monocots
continued
- *Billbergia pyramidalis* (bromeliad)
- *Canna generalis* (indian shot)
- *Zizania aquatica* (wild rice)
- *Chloris glauca* (finger grass)
- *Panicum dichotomum* (panic grass)
- *Erianthus brevibarbus* (plume grass)
- *Setaria* sp. (bristle grass)
- *Rhynchospora innundata* (beak rush)
- *Fimbristylis* sp. (cone grass)
- *Eleocharis robbinsii* (spike rush)
- *Pontederia cordata* (pickerel weed)
- *Eichhornia crassipes* (water hyacinth)
- *Alisma* sp. (water plantain)
- *Sysyrinchium atlanticum* (blue-eyed grass)
- *Lemna minor* (duckweed)
- *Hydrilla verticillata* (hydrilla)
- *Helicona latispatha* (lobster claw)
- *Pandanus veitchii* (screw pine)
- *Acoelorhaphe wrightii* (everglades palm)
- *Roystonea elata* (royal palm)
- *Sabal palmetta* (cabbage palm)
- *Andropogon capillipes* (broomsedge)
- *Andropogon virginicus* (broomsedge)
- *Andropogon glaucopsis* (broomsedge)
- *Sporobolus indicus* (smut grass)
dicots
- *Hyptis alata* (musky mint)
- *Hydrocotyle verticillata* (water pennywort)

(continued)

TABLE 1

(Continued)

Division Magnoliophyta
continued
 dicots
  continued
   Centella asiatica (centella)
   Polygonum hydropiperoides (smart weed)
   Rumex sp. (dock)
   Rhexia lutea (yellow rhexia)
   Sabatia sp. (sabatia)
   Agalinis fasciculata (false fox-glove)
   Solanum pseudogracile (black nightshade)
   Ceratophyllum submersum (coon-tail)
   Bacopa monnieri (water hyssop)
   Gratiola sp. (hedge hyssop)
   Bacopa sp.
   Chamaesyce hypercifolia (sponge)
   Cardamine hirsuta (bittercress)
   Vicia acutifolia (vetch)
   Zebrina pendula (wandering jew)
   Plantago aristata (buckhorn plantain)
   Lobelia glandulosa (glades lobelia)
   Ampelopsis arborea (pepper vine)
   Parthenocissus quinquefolia (woodbine)
   Vitus vulpina (winter grape)
   Ludwigia leptocarpa (ludwigia)
   Ludwigia sp.
   Cornus foemina (swamp dog-wood)
   Ilex cassine (dalhoon holly)
   Ilex glabra (gall berry)
   Swietenia mahogani (maho-gany)
   Acer rubrum (red maple)
   Myrsine guianensis (myrsine)
   Persea borbonea (red bay)
   Cassytha filiformis (love vine)

Division Magnoliophyta
continued
 dicots
  continued
   Cocoloba diversifolia (pigeon-plum)
   Eclipta alba (eclipta)
   Mikania scandens (climbing hemp weed)
   Solidago fistulosa (golden rod)
   Wedelia trilobata (creeping oxye)
   Cirsium horridulum (plume thistle)
   Hieracium sp. (hawk weed)
   Rudbechia sp. (black-eyed susan)
   Emilia fogbergii (tassel flower)
   Hydrolea corymbosa (sky flower)
   Cuscuta gronovii (dodder)
   Ipomoea sagittata (morning glory)
   Nerium oleander (oleander)
   Rubus cuneifolius (sand black-berry)
   Chrysobalanus icaco (coco-plumb)
   Fraxinus caroliniana (pop ash)
   Magnolia virginiana (sweetbay)
   Myrica cerifera (wax myrtle)
   Bursera simaruba (gumbo limbo)
   Annona glabra (pond apple)
   Dodonaea viscosa (varnish leaf)
   Quercus laurifolia (laurel oak)
   Salix caroliniana (willow)
   Baccharis glomeruliflora (groundsel)
   Baccharis halimifolia (salt brush)
   Aster sp.
   Erigeron sp. (flea bane)
   Bidens pilosa (spanish needles)
   Eupatorium incarnatum (agera-tum)
Kingdom Animalia
 Subkingdom Eumetazoa
  Phylum Cnidaria (coelentrates)
   Hydra sp.

(continued)

TABLE 1
(Continued)

Phylum Annelida	Phylum Arthropoda
Oligochaeta spp.	continued
Phylum Nemata	(Crustacea)
Choanolaimid sp.	continued
Chromadorid sp.	Palaemonetes pallidosa (shore
Linhomoeus sp.	shrimp)
Theristus sp.	Procambarus alleni (crayfish)
Phylum Mollusca	Cyclopoid copepods (2 spp.)
Gyraulus parvus (ash gyro)	Hyallela azteca (amphipod)
Melanoides tuberculata (red	(Insects) (ant species not tallied)
rimmed melania)	Caenis sp.
Planorbella duryi (seminole	Rheumatobates tenuipes
rams horn)	Limnoporus canaliculatus
Planorbella scalaris (mesa rams	Belostoma testaceum
horn)	Erythemis simplicicollis
Planorbella trivolvis (mesh rams	Ishnura vamburi
horn)	Ishnura posita
Viviparus georgianus (banded	Laccophilus sp.
mystery snail)	Mesovelia mulsanti
Micramenetus dilatatus	Hydrometia australis
Sphaerium sp.	Buenoa sp.
Polygyra uvifera	Telebasis byersi
Physella hendersoni (bayou	Enallagma geminata
physa)	Phylum Chordata
Physa sp.	(Vertebrata)
Pomacea palidosa (apple snail)	Fundulus chrysotus (golden top-
Tarebia granifera (guilted	minnow)
melania)	Lucania goodei (bluefin killifish)
Pseudosuccinea columnella	Heterandria formosa (least killi-
(mimic pond snail)	fish)
Ferrissia hendersoni	Poecilia latipinna (sailfin molly)
Laevapex peninsularae	Jordanella floridae (florida
Phylum Arthropoda	flagfish)
(Crustacea)	Lepomis macrochirus (blue gill)
Cypridopsis vidua (ostracod)	Gambusia affinis (mosquitofish)

ᵃThis system has not been systematically sampled for microflora and fauna, nor has the soil microfauna been studied.

invertebrate-eating fish. Two particularly interesting species that we have been following are the apple snail and a crayfish.

Pomacea palidosa, the apple snail, is among the largest of the North American freshwater gastropods. It is particularly unusual in having both gills and lungs, which impart amphibious capabilities. Under normal circumstances, apple snails can be extremely abundant, particularly on rotting vegetation, along the banks of small streams and pools. It is probably a detritivore on rotting vegetation and its bacterial and protozoan flora and fauna. During the dry season, *Pomacea palidosa* crawls out of the water on stalks of emergent vegetation (particularly cattails) and lays a cluster of large white eggs. After several months, these hatch and the young return to the water. The primary predator of the apple snail in the wild is the Everglades kite, which is especially adapted for capturing and removing the snails from their shells. Although many egg clusters have been laid in our stream model and young have successfully hatched, we have not yet had a population explosion of adults. Such an explosion might result after several seasons from the lack of the kite in the model. If this eventually occurs, the human operators of the ecosystem will have to play the role of the higher predator.

About a dozen adults of *Procambarus alleni*, a southeastern North American crayfish, were introduced into the model. After about a year, in late summer, young, independent animals began to appear. Crayfish are omnivorous to scavengers on dead animals. Since the model stream sides are made from open cement blocks, burrows are numerous. With abundant food and no predator for the adults, these animals may also have to be cropped by the system operators.

Freshwater marshes and wetlands in general can be strongly affected by birds and mammals (e.g., Weller, 1987). Except for the largest of mesocosms, these animals are not likely to be included in ecosystem models. In general, management by the operators with the intent of simulating low-level disturbance is necessary. If the ecosystem in question is to be maintained as the equivalent of an "average" wild community, the operators will have to be familiar enough with the wild, either through the literature or direct experience, to make the necessary judgements as to level and frequency of disturbance. Such management would have to include the crushing or burning in place of some plants along with removal of critical, highly productive animals. In the prairie community described above, we have not yet faced up to this need. If the wild analogs are any indication, we have another 3–6 years before this will be critical (Drew and Schomer, 1984).

Low-Nutrient Tropical Stream

This 2500-gallon, plywood–fiberglass tank is C-shaped with the ends of the C being about 2.5 feet deep and the center section about 6 feet deep. A 20-gpm standard centrifugal pump provides a continuously cycling flow that is quite strong in the effluent arm. The tank holds a composite community of Venezuelan black water stream leaf litter, Georgia blackwater stream plants, and South American river fish from the aquarium trade. The effect desired was a very low-nutrient system that would be the equivalent of a near-neutral pH jungle stream. The original sediments were pure silica sand with coolers of Venezuelan blackwater stream leaf litter transported damp by air freight. Lighting is ambient greenhouse in Washington, D.C. The tank is, however, strongly shaded by an overhanging growth of a variety of potted plants dominated by *Canna* sp. Summer water temperature reaches nearly 30°C and winter temperature is kept at a minimum of 22°C with submersible heaters. About 30 gallons per day is injected into the tank to replace water lost by evaporation. The source of the evaporated water is treated tap water, the same as that described earlier for the Florida Everglades stream.

A list of species of plants and fish in the blackwater model is given in Table 2. The invertebrates, as well as the plankton and protozoa, have not been fully tallied.

No feed is added to this system. A relatively small standing stock of emergent plants and benthic algae accompanied by breakdown of the original leaf litter is the only source of energy. An algal scrubber of 1.5 m provides nutrient levels of between 0.5 and 1.0 μM NO_2^- plus NO_3^-. The rich scrubber growth is typically a brilliant green upper story of long, filamentous *Spirogyra aequinoctiales*, embedded in a basal mat of unicells and small, filamentous diatoms, greens, and blue-greens. The pH is close to neutral. Thus, the blackwater model is as low in nutrients as open tropical seas, mountain streams, and a typical undisturbed tropical blackwater stream. It is not as acid as many blackwater streams but certainly lies within the range of such ecosystems.

This stream system, while it is biotically totally artificial, demonstrates feasibility in a freshwater model, using algal scrubbers to maintain extremely low nutrient levels, even with a large leaf litter load. It would be most desirable, given the opportunity to design and build a low-nutrient system, particularly with noncentrifugal pumps, to attempt to match an existing wild analog.

Black Water Home Aquarium

This 70-gallon, home aquarium system (Color Plate 52) is made up of the same basic species components as the 2500-gallon model described above. The substrate was initially established with pure silica sand, several inches thick, equivalent to the typical nature of small, tropical, black water tributary streams. This provides a silica-rich, poorly-buffered environment, which when coupled with nutrient scrubbing, give rise to a diatom-rich, poorly-productive benthic community. The operational parameters are also quite similar, though natural lighting in a shaded sunroom is supplemented by two 48-inch VHO fluorescent bulbs. Make up water as well as subsequent evaporative replacement water were both black water derived without filtering, primarily from sour gum/red maple swamps on the Virginia Coastal plain. Unlike the larger system described

TABLE 2

Macro Plants and Animals Occurring in the Smithsonian Black Water Stream Model[a]

Kingdom Monera
  Division Cyanophycota
    Chroococcus pallidus
    Gloeothece linearis
    Cylindrospermum sp.
    Oscillatoria 2 spp.
    Chamaesiphon incrustans
Kingdom Plantae
  Subkingdom Thallobionta
    Division Chlorophycota
      Ulothrix aegualis
      Mougeotia 2 spp.
      Spirogyra aequinoctiales
      Cladophora crispata
    Division Chromophycota
      numerous diatoms of many species
      present
  Subkingdom Embryobionta
    Division Magnoliophyta
      Family Araceae
        Pistia stratiotes (water lettuce)
      Family Alismaceae
        Sagittaria lancifolia (duck potato)
      Family Marantaceae
        Thalia geniculata (arrow root)
      Family Cyperaceae
        Cyperus spp. (sedges)
      Family Poaceae
        Panicum agrostoides (panic
        grass)
      Family Verbenaceae
        Phyla nodiflora (capeweed)
      Family Acanthaceae
        Justica ovata (water willow)
      Family Compositae
        Solidago altissima (golden rod)
        Mikania scandens (climbing
        hempweed)
        Aster sp. (aster)
      Family Onagraceae
        Ludwigia palustris (water
        primrose)
      Family Umbelliferae
        Centella asiatica (cypress
        creeper)

Kingdom Animalia
  Subkingdom Eumetazoa
    Phylum Arthropoda
      Procambarus alleni (crayfish)
      Palaeomonetes pallidosus (shore
      shrimp)
    Phylum Mollusca
      Melanoides tuberculata (red
      rimmed malcuria)
      Planorbrella diryi (seminole rams
      horn)
      Physella cubensis (carib physa)
    Phylum Chordata
      Family Characidae
        Cheirodon axelrodi (cardinal
        tetra)
        Hemigrammus ocellifer (head
        and tail light tetra)
        Nematobrycon palmeri (emporer
        tetra)
        Hyphessobrycon serpae (serpae
        tetra)
        Megalamphodus megalopterus
        (black phantom tetra)
      Family Serrasalmidae
        Metynnis argenteus (silver dollar)
      Family Gasteropelecidae
        Gasteropelecus levis (silver
        hatchet fish)
      Family Pimelodidae
        Pimelodus maculatus (spotted
        catfish)
      Family Callichthidae
        Corydoras schultzi (cory cat)
        Corydoras agassizi (cory cat)
      Family Loricaridae
        Rhineloricaria basemani (whip
        tail catfish)
        Otocinclus arnoldi (arnolds
        sucker catfish)
        Hypostomus plecostomus
        (plecostomus)
      Family Poeciliidae
        Poecilia velifera (sailfin molly)
        Xiphophorus helleri (swordtail)
        Xiphophorus variatus (rainbow
        platy)
        Gambusia affinis (mosquito fish)
      Family Cichlidae
        Apistogramma ramirez (butterfly
        ram cichlid)

[a] Except for some algae, organisms smaller than about 5mm have not been tallied in this system.

above, the home aquarium uses a single bellows pump for recycling and scrubbers and has never been operated with a centrifugal pump or filtration of any kind. Its stability after only one year of operation is quite remarkable for such a small system (Table 3).

Black water streams and swamps typically derive their coloration (which is usually reddish rather than blackish) from the presence of tannins leached from soil and leaf litter. The tannis are very resistant to breakdown and are in large part responsible for the "gelbstuff" or reddish/

TABLE 3

Organisms Occurring in a 70-Gallon, "Black Water" Home Aquarium Emphasizing South American Flora and Fauna

Kingdom Monera
  Division Cyanophycota
    Oscillatoria rubescens
    Oscillatoria sp.
Kingdom Plantae
  Subkingdom Thallobionta
    Divison Chlorophycota
      Spirogyra aequinoctiales
      Mougeotia sp.
      Ulothrix aequalis
      Cladophora glomerata
    Division Chromophycota
      numerous diatom spp.
  Subkingdom Embryobionta
    Division Magnoliophyta
      Lemna minor (duck weed)
      Echinodorus paniculatus
        (amazon sword plant)
      Potamogeten pusillus
        (pond weed)
      Heteranthera dubyi
        (water star grass)
      Nymphaeae elegans
        (blue water lily)
      Justicia americana
        (water willow)
      Ludwigia palustris
        (water primrose)
    Division Bryophyta
      Riccia fluitans
    Division Filicophyta
      Salvinia rotundifolia

Kingdom Animalia
  Subkingdom Eumetazoa
    Phylum Mollusca
      Melanoides tuberculata
        (red-rimmed melania)
      Planorbella duryi
        (seminole ramshorn)
      Physella cubanensis-peninsularis
        (Carib. Physa)
      Lampsilis sp. (freshwater mussel)
    Phylum Chordata (fish)
      Family Characidae
        Gymnocorymbus tennetzi
          (black tetra)
        Hyphessobrycon ersythrostigma
          (bleeding heart tetra)
        Pseudocoryinfopoma doriae
          (dragon fin tetra)
      Family Gasteropelecidae
        Gasteropelecus levis
          (silver hatchet fish)
      Family Callichthyidae
        Corydoras metae (coryoras)
      Family Poecilidae
        Xiphosphorus helleri
          (sword tail)
        Poecilia reticulata (guppy)
        Gambusia affinis
          (mosquito fish)
      Family Cichlidae
        Pterophyllum altum
          (deep angelfish)

ᵃAlgae, higher plants, macro invertebrates, and fish are included, while diatoms, plankton, and microinvertebrates are not cataloged.

yellow coloration of estuaries and coastal waters. Although all evaporative replacement water (about 5 gallons per week) in this tank is replaced by the strongly pink, source water described above, the water of the system itself has no visible tannin coloratin. Due to moderate scrubbing and intense submerged aquatic plant biomass production within the tank, the essentially neutral input water is raised to a pH of about 7.8 in the tank itself. Perhaps the relatively high pH is responsible for precipitation of the tannin, and if a more acid water were desired, a reduction of plant production in the tank would be required.

Typically operated at below 1–3 μM N-NO$_2^-$ + NO$_3^-$, a rough analysis of nitrogen import/export is useful for this small tank. A standard flake food is fed to the tank at the rate of 1.5 g (dry) per week. Estimating 10% nitrogen for the flake food, 6% for scrubber algae and 2% for higher plant fragment removal, the algal scrubber of 0.072 m^2 removes about one half of introduced nitrogen; the removal of old degenerating leaves of the submerged aquatics exports the other half. In the 1–3 μM N range, there is probably little nitrogen fixation or denitrification. The scrubber, lit by two 24-inch VHO bulbs, produces a typical basal mat of *Oscillatoria* spp. with some filamentous greens such as *Ulothrix aequalis* and *Mougeotia* spp. and abundant diatoms, but with a dark green upper story of *Cladophora glomerata*. This extensively branched and very coarse species has basal cells that appear to extend through the basal mat and attach directly to the plactic screen.

An "African" Pond

This home model of 70 gallons (Fig. 3) was designed and operated for entirely different reasons than any of the previously described systems. The purpose was to build up to the maximum biomass of behaviorally compatible fish that could be maintained using a freshwater scrubber. As a general reference to the fishes discussed, reference can be made to Fryer and Iles (1972).

This tank has a moderate abundance of submerged aquatic vegetation, most notably two *Aponogeton* species. Because of the constant grazing of the *Tilapia*, it would appear to be virtually devoid of algae. However, the abundant rocks in the system have a felt-like, greenish black coating. Similar to a constantly-scraped algal turf scrubber, this is a diatom and blue green algal mat with short tufts of *Cladophora* and *Oedogonium*.

At this writing, the "African" tank has been operating for about 3 years. It was started with approximately 25 inches of cichlids and 4 inches of catfish (Table 4). It now has approximately 40 inches of cichlids and 8 inches of catfish. Two species, *Melanochromis auratus* and *Pseudotrophius tropheops*, have repeatedly reproduced, though only a

Figure 3 Photograph of the 70-gallon high-biomass "African" aquarium showing the lighting and scrubbing control units.

very few fry reach adulthood from each hatch because of predation. *Tilapia marae*, although one of the most aggressive species in the tank and successful in an adjacent community aquarium, has not raised viable young to date in this tank. A single individual of *Melanochromis auratus* had its tail bitten off during the first year and survived for another year before succumbing to aggression of the other fish. Two adult fish and many young have been lost to predation in this tank. It seems clear in this intensely competitive environment that only mouth breeders hatching in burrows can successfully produce adult fish.

Approximately 0.3 g of dried commercial flake food is fed to the tank each day. The system is scrubbed with a 0.085 m² algal scrubber having two VHO lamps of 148 W total. This provides an algal production rate of about 5.8 g (dry)/m²/day. Temperature ranges from about 22°C (72°F) to 26°C (80°F) seasonally. The lowest oxygen levels are 7.6 mg/l, they peak out at about 8.5 mg/l, and are accompanied by a regular daily range of pH from 8.8 to 9.0. This balance has translated into an approximate tripling of the fish biomass. This in turn has led to relatively high nutrient levels (10–40 μM N as $NO_2^- + NO_3^-$) for a scrubber-based system, and the first planktonic algal blooms have begun to appear as this text is written. Nevertheless, the tank is quite stable as described. No disease has appeared in this system. This is likely partly due to complete isolation.

TABLE 4

Organisms Occurring in "African pond," 70-Gallon Tank on Long-Term Basis[a]

Kingdom Monera
  Division Cyanophycota
    Oscillatoria spp. (2)
    Schizothrix sp.
Kingdom Plantae
  Division Chromophycota
    Pinnularia sp.
    Anomoe sp.
    Stephanodiscus
  Division Magnoliophyta
    Vallisneria americana
    (tape grass)
    Cryptocoryne sp.
    Anubias sp.
    Aponogeton fenestralis
    Aponogeton crispus
  Division Chlorophycota
    Cladophora glomerata
    Scenedesmus sp.
    Closteriopsis sp.
    Ulothrix zonata
    Microspora sp.
    Coccochloris sp.
    Volvulina sp.
    Oedogonium sp.

Kingdom Animalia
  Subkingdom Protozoa
    Difflugia sp.
    Thecacineta sp.
    Homalozoan sp.
    Carchesium sp.
    Oileptus sp.
    Epistylis sp.
    Vorticella sp.
  Subkingdom Eumetazoa
    Phylum Nematoda
      2 sp nematodes
    Phylum Rotifera
      3 sp. rotifers
    Phylum Mollusca
      Melanoides tuberculata
      (red-rimmed melania)
      Planorbella duryi
      (seminole rams horn)
      Lampsilis sp. (freshwater mussel)
    Phylum Arthropoda
      Cyclopoid copepod
    Phylum Bryozoa
      lumatella repens
    Phylum Chordata
      Vertebrata (fish)
        Tilapia marae (Tilapia)
        Melanochromis auratus
        (Malawi golden cichlid)
        Pseudotrophius tropheops
        (golden tropheops)
        Arius berneyi (catfish)

[a] Bacteria and plankton have not been assessed.

However, along with the continuous reproduction and the general good health of the fish, it also suggests that given a compatible community even with high biomass and moderate nutrient levels, when accompanied by relatively high oxygen levels, a considerable stability can be achieved in a freshwater model ecosystem.

References

Axelrod, H., Burgess, W., Pronok, N., and Walls, J. 1986. *Dr. Axelrod's Atlas of Freshwater Aquarium Fishes*. 2nd Ed. T.F.H., Neptune City, New Jersey.

Burgis, M., and Morris, P. 1987. *The Natural History of Lakes.* Cambridge University Press, Cambridge.

Drew, R., and Schomer, N. S. 1984. An Ecological Characterization of the Caloosahatchee River/Big Cypress Watershed. Fish and Wildlife Service FWS/OBS-82/58.2. Tallahasse, Florida.

Duever, M. J., Carlson, J. E., Meeder, J. F., Duever, L. C., Gunderson, L. H., Riopelle, L. A., Alexander, T. R., Myers, R. L., and Spangler, D. P. 1986. *The Big Cypress National Preserve.* Research Report No. 8 of the National Audubon Society. New York.

Fryer, G., and Iles, T. 1972. *The Cichlid Fishes of the Great Lakes of Africa, Their Biology and Evolution.* T.F.H., Neptune City, New Jersey.

Moss, B. 1988. *Ecology of Fresh Waters.* 2nd Ed. Blackwell, Oxford.

Pennak, R. 1953. *Fresh Water Invertebrates of the United States.* Ronald Press, New York.

Pennak, R. 1989. *Fresh Water Inverftebrates of the United States,* 3rd Ed. Wiley, New York.

Rataj, K., and Horeman, T. 1977. *Aquarium Plants.* T.F.H., Neptune City, New Jersey.

Riehl, R., and Baensch, H. 1986. *Aquarium Atlas.* Baensch, Melle, W. Germany.

Weller, M. 1987. *Freshwater Marshes, Ecology and Wildlife Management.* 2nd Ed. University of Minnesota Press, Minneapolis.

PART V

Summary

CHAPTER 25

MICROCOSMS, MESOCOSMS, AND AQUARIA AS ECOSYSTEM MODELS
A Synthesis

By the present millenium, some fairly complex human civilizations had achieved at least a temporary balance with nature. Many cultures referred to "Mother Earth" with appropriate reverence and behavior. North American Indian tribes regarded bears and salmon, for example, as societies of "people"—to be utilized for certain but with great regard for their rights and role in the larger scale of things. While it is generally accepted that Stone Age man was responsible for the extinction of a number of large Pleistocene mammals and birds and extensive habitat modification through the use of fire, human effects on the biosphere as a whole were minimal prior to the industrial revolution. From the dawn of our modern civilizations up to the present decades, human relations to the remainder of the living world have increasingly consisted of a desire for total control of the natural world. Western civilizations and religions have generally regarded the extensive use of all nature as a right, perhaps even a duty. Depending upon the nature of the organism or wild community involved, this control has ranged from eradication—for disease or pest organisms, their vectors and habitats—and displacement (for space competitors), through management, for plants and animals of food, transportation, or recreational value. Where active development interests have yet to be viable, most other creatures and their habitats are given over to benign neglect. Certainly consideration for rare plants and animal rights and eventually conservation movements have developed and prospered, largely defended on the basis of ultimate human needs. However, human populations have grown at even faster rates, and species extinctions have followed equivalently, partly as a result of direct utilization under a regime of poor or little management. Today most extinctions result from habitat destruction and water and atmospheric pollution.

It is becoming increasingly clear that total human domination over a major part of the biosphere, at least as civilization is now structured, will lead to loss of many of the critical life support elements required by higher plants and animals, including man. In the spectrum of human control over the living world, it is apparent that human populations must fall and for an increasingly large wilderness area, benign neglect must change to a determined non-effect. Even at the other extreme of human desire for control, parasites and pests, we must learn to control only when we know that the effects of our efforts are limited tightly to the goals. The biosphere cannot withstand many DDT's, at least not at the complexity needed to support human beings, and except for human viruses, which are probably out-of-control fragments of the human genome, eradication of any species, even a pest, deserves very careful international consideration.

Although ecological modeling, mathematical or living, must deal with all interactions between humans and the remainder of the biosphere, in this book we are specifically focusing on the middle part of the spectrum of interactions, those organisms, habitats, and ecosystems that we wish to better understand or become intimate with for research, recreational, and esthetic reasons. In the past, in a direct management context, regarding food plants and animals, the goal was maximum productivity with minimum human effort. However, even in the growing and handling of food, this relationship must be reconsidered. Too great an area under cultivation with extensive fertilization, pesticide use, and fossil fuel consumption for cultivation, preparation, and transportation, as well as the ultimate disposal of the wastes, is costing us no less than the loss of our biosphere as we know it and of our civilization. Whether we achieve an ecological perspective in population and food management in the near future or barely in time to save ourselves, when it comes to the management of organisms for recreation, education, and research, we must think in an ecological context. We can no longer afford any other approach just because it is esthetic, provides interactive enjoyment, makes trade and income, or achieves limited education or scientific accomplishment within our budget. If we cannot reach a reasonably deep ecological view in these realms, we surely will not attain it in the rough-and-tumble world of economics and bare survival.

This summary chapter is written with these thoughts in mind. Where we manage plants and animals in captivity, for recreational, education, and research reasons, it must be in an ecological frame that we constantly strive to better understand and more accurately achieve.

Principles for Ecological Modeling

Over the past 14 years, we and our colleagues at the Smithsonian's Marine Systems Laboratory have designed, constructed, and operated aquatic and wetlands ecosystems in microcosms, mesocosms, and aquaria for a total of over 40 ecosystem years. Some of these systems have been surprisingly accurate reflections of the wild ecosystems they were meant to model. Others had considerably less of a relation to their wild counterparts. Our funding, while generous in some contexts, has been minimal in the perspective of our goals. Very little of the support could be given to research in the strict academic sense, although much could have been called research and development in the industrial sense. Often the support equipment, backups to mechanical and electrical failure, and sensory and human monitoring apparatus have been minimal or absent. Many physical "disasters" have occured. Nevertheless, all of these miniature ecosystems have suffered engineering failures and affronts with surprising stability to their basic character and ecosystem function. Not a single system has ever crashed, in the sense of most of the higher organisms dying. "Disasters" usually have led to no more than population shifts. Complex ecological systems that have not had key species or a number of secondary species "chipped away" and have not been placed under chronic physical or chemical stress have a surprising resistance to perturbation.

Here we present a list of loose "development and operational principles" that we feel are important to successful ecological simulation, at least in the aquatic and wetlands realm. They would probably mostly apply as well to the field of terrestrial ecosystem restoration. These are presented as a list below and then are briefly discussed. The in-depth discussion for each appears in the preceding 24 chapters.

Before initiating a modeling effort, it is assumed that the modeler or aquarist has become aquainted with the analog, the wild ecosystem to be modeled, through study of the ecological and systematic literature and preferably first-hand with the ecosystem itself. A period of mental interaction with the realities of the wild ecosystem, the modeling principles to be discussed below, and the realities of facilities and resources available is essential. While modifications and compromises will be necessary and a very important part of the process of synthetic ecology, a strong inital understanding of the nature of the ecosystem is crucial to success.

nutrient levels, the same basic problem that is destroying many of our wild rivers, lakes, and estuaries. In their particulate-capturing role, filters also do not distinguish between the organic particles and most plankton. Filters act like a giant filter feeder in the aquarium system and generally degrade water quality accordingly. Of the argument that "extra" microbes are required to recycle elements, one simply needs to point out that in the biosphere, most carbon is not rapidly recycled but is in storage. Indeed, the use and release of that storage is a major factor in the degeneration of the earth's biosphere as a whole. Stability requires locking up or removal of nutrient reactants. A nutrient-poor aquarium or mesocosm is not generally any aquarist's problem.

Of the modern systems, the addition of the trickle or spray part of the wet–dry system has been of great advantage in oxygenating and accelerating denitrification from otherwise eutrophic or nutrient-rich aquaria. However, nitrogen levels must be much higher than they should be for ecosystem health before significant denitrification can occur. Also, phosphorus levels and those of other nutrients and inevitable pollutants remain unaffected by wet-dry approaches. Plankters are likewise removed by both the attached physical filter and the trickle or spray unit. Of the remaining modern aquarium practices, ozonation, intense ultraviolet, and protein skimming have little to offer maintenance of whole ecosystems. They simply poison or remove organic elements that are crucial to a healthy ecosystem.

Modern aquarium techniques have improved tremendously in the past 10 years relative to maintaining some healthy animals species. However, as we discussed in the introduction to this chapter, the maintaining of a few healthy animals should be the sole goal of few aquarists. If we are to play a significant role in the restructuring of our relationship to our biosphere, as professionals or hobbyists, we must rapidly shift our focus to viewing all life as part of the ecosystems from which they are derived. We must not ask any esthetic, educational, or research animal to live outside the ecosystem or its model for which it evolved. At the ecosystem level, plants and their storage products are the "purifying" elements in the biosphere. The algal turf scrubber described in Chapter 12 and used in all the microcosms, mesocosms, and aquaria described in this book is one way to use plants in a controllable way that does not generally capture plankton or alternately degrade the environment. Algal scrubbers also provide the needed simulation of some adjacent ecosystems or, so often in much modeling work, simply a larger open body of water.

The Structuring Elements

Some communities of organisms are structured by physical elements—a sandy beach, for example. However, in the terrestrial environment and in most shallow aquatic environments, plants are the structuring elements.

They not only provide the food and a necessary water and atmospheric chemistry that make higher animal life possible, but also greatly increase surfaces for attachment and cover. In general, plants also provide a spatial heterogeneity that does not exist in the bare physical world. Particularly in the marine environment, where calcification is enhanced, many animals (the most dominant with plants in their tissues—for good chemical reasons) join plants to provide a community structure that consists of reef or shell framework. This framework is calcium carbonate instead of or along with cellulose.

In constructing any living model, it is essential that these structuring elements be added soon after the physical environment is formed. When they are plants, plant control of the chemical environment is a concomitant to structuring. On the other hand, if the structure is an animal community such as corals or oysters, it may be necessary to add the dead carbonate framework, "live rock" or oyster-free empty shells first. If it is part of the plan for the system in question, the needed planktonic food supply plus a scrubber to control water quality (the effects of the adjacent ecosystem) must then be added.

Community Blocks

Ecosystems can be functional units of any size. Normally, they are simply the units that the ecologist chooses to work with, usually physically defined (a river bank, a mountain range, a bay, etc.). Small ecosystems such as a mud flat, an oyster bar, or a *Spartina alterniflora* (cord grass) marsh can be defined, however, and within these units, subunits of one to several square feet function with all of the smaller components (plants, invertebrates, protozoa, algae, bacteria, and fungi) intact. It would be virtually impossible to extract the tens of species amounting to hundreds to thousands of small individuals that occur in these subunits. Many would be killed or damaged simply by the process of removal. Thus, the ecosystem of a model should be constructed by installing convenient blocks of wild ecosystem with all of the microspecies and their relationships intact. In part, this is what is done in modern "reef" systems by the addition of "live rock." If the physical environment is reasonably accurate, the first community blocks include the structuring elements, and if transport time is minimal, taking into account critical temperature, oxygen, and humidity factors, little problem should be encountered in installing these miniunits in the larger model.

For the typical benthic system, repeated efforts must eventually be taken to install "planktonic blocks." These are simply large water samples that are aerated and transported in minimum time to the model. These injections should be periodically carried out during system stocking, and at completion of benthic development they should be followed up by several final injections.

Multiple Injections

The process of cutting out or otherwise extracting an ecological block or ecosystem subunit and transporting it to the waiting model can be quite stressful. Even in the model, the block meets conditions that at least initially consist of the raw physical–chemical environment unameliorated by the effects of a functioning community of organisms. The first block injections are likely to lose many species. However, as time goes on and more collections are added, the model's mini-ecosystem becomes more like that of the wild counterpart. Thus, with each addition, the stable species diversity increases. Generally, depending on the size and nature of the model, tens of injections over 6 months to a year are required to bring the microcosm or mesocosm system to the point where it is complex and stable, with the larger and more mobile organisms a more-or-less continuous problem (see below).

Species Selection

All ecological communities are patchy. An island coral reef, a large salt marsh, a mud flat, all differ from place to place. Chance factors of organism settlement, negative and positive interactions between species, the local effect of environment (e.g., a log bouncing across a marsh with the tide), and real differences of environment (wave exposure, current, etc.) all lead to patchiness within a community. The model itself, no matter how accurate, is a patch, hopefully something that represents a "mean" of most wild patches. There is, therefore no patch quite like it in the wild; thus, it is extremely difficult to decide what species will be successful in the model. Surely the structuring species elements and the dominants in the wild should be included or at least attempted. However, after that, as far as possible the entire pool of available species from the type community (except rare or obviously difficult species) should be given a chance at immigration into the model. Large and mobile species are discussed below. Since the scientific information needed to determine the nature of the "model patch" will generally be lacking, the model community should be left free, as far as possible, to determine its own fate. This can never be strictly true, since the aquarist will determine the size and sequence of block injections. Nevertheless, we have found this a valuable principle for modeling success. In the form of the genetic codes of its constituent species the model ecosystem carries a tremendous quantity of information with regard to its structure and function. Particularly since we know and understand only a small part of this information, we should be loath to subvert ecosystem self development.

Large and Mobile Species

Once the microorganisms, plants and small invertebrates are in place and successfully functioning, and the water turbidity and nutrients are normal, the small to mid-size more mobile fauna can be added. Surprisingly, some of these, mostly juvenile fish and decapod crustaceans in fully aquatic environments, will already have been injected with the ecological blocks. At this point, decisions as to size and numbers of these usually netted or trapped animals become crucial. As a rule of thumb, younger members of a species usually transport and adapt better. Also, a thorough familiarity with size and numbers of each species per unit area or volume in the wild is very helpful. At least initially, adhering to wild density levels will help to prevent overstocking, overgrazing, and overpredation until the model is better understood. Also, species guilds should be considered, where they are known. A number of individuals of a single species performing a function, such as the browsing of sedentary invertebrates, sponges, worms, etc., is to be preferred over one to two individuals of many species of the same basic feeding habit. In general, the larger the population, the more likely that breeding success will be achieved. Higher predators are quickly recognized as a basic problem for a small model where insufficient feeding territory may be available. However, larger grazers can provide an even more critical problem. Particularly when freed from predation pressure, some herbivores can quickly overgraze and devastate an ecosystem model. Lacking higher predators, the aquarist must become the educated cropper and balancing element.

Top Predators

Depending on the size of the model, top or even mid-level predators must typically be omitted, with the aquarist then performing the required ecological role. To some extent, in microcosms and mesocosms, this can be overcome by the addition of fodder organisms and the replacement of the occasional loss of hopefully permanent ecosystem members. However, if this practice is followed, sufficient scrubbing capability must be included to handle the additional load and required export from the system. The difference between growth biomass of the predator as compared to feed added and immediate respiration losses is not easy to calculate. However, careful monitoring of nutrient levels or signs of increasing nutrient concentrations can avoid serious problems.

Problems to Solve

Virtually all aspects of the modeling techniques that we describe here can be improved. However, one major requirement, that of the movement of water, needs to be heavily researched and developed. Although we are just beginning to obtain hard and abundant experimental evidence, there is little question that the standard centrifugal pumps destroy many plankters, particularly larger zooplankton and swimming invertebrate larvae. As described in the pages above, we have taken several approaches to this problem, particularly slow-moving piston pumps, membrane pumps, and an Archimedes screw. All of these devices work well, though relative performance needs to be quantified. However, their chief failing lies in longevity, moderate maintenance, an the relatively high cost of individual construction. The pumps for the Biosphere II project are planned as vacuum-driven suction/valve units. For a very large multipump system, this approach has the advantage of economy in a bank of large, low-pressure vacuum pumps. Whether it can be effective for relatively small exhibit, research, or home systems remains to be seen. Once the aquarium manufacturing industry seriously addresses this issue, it will likely be solved with inexpensive and reliable units in a wide variety of sizes.

A variant of the pump–plankton problem lies in the reverse osmosis units that are required to remove fresh water in fully functional estuarine models. Evaporation units could solve this requirement if sufficient space were available, though efficiency would normally require flash evaporation, which would be equally damaging. An in-line, low-speed, continuous centrifuge to gently remove most plankton, especially larger varieties, could perhaps reduce the problem. Our approach has been to limit overturn in the tank connected to R.O. units to several days or more. The resulting slow loss of plankton can perhaps be equated to tidal flushing from the estuary. We have not tried to quantify this in our estuary models, but have simply lived with the requirement for freshwater sinking and the results.

Ecosystems in Home Aquaria

The basic principles of creating marine and freshwater ecosystems in home- or office-sized aquaria are no different from those behind the development of the microcosms and mesocosms that we have tended to emphasize in the previous four chapters of this book. The design techniques required to prevent structure and engineering from overwhelming the biological aspects of the tank in an esthetic sense are certainly more difficult, but well within existing technology. The principles emphasized in this book are relatively uncomplicated and require less space than more traditional technologies. The greatest difficulty in miniaturizing "captured" ecosystems is ecological and lies in the scaling down to tens of gallons or a few square feet an ecosystem that typically functions on the scale of multimillions of gallons over many acres of area. Some environmental parameters are certainly involved here. Primarily these are light, water quality, and waves and tides. However, mostly we are dealing with a problem of larger animals. There is a tendency to think of the problem as one only of the higher predators. To some extent that is true, since top predators tend to be large. Nevertheless, to point out how it can be otherwise, dugongs, or sea cows, are fully herbivorous and yet will never be placed in any but the very largest of mesocosms. On the other hand, some very agressive protozoan predators would never be limited by the minimum size of the system being constructed. Nevertheless, the smaller the microcosm or mesocosm, the more difficult it is to simulate a wild ecosystem and the more important will be the role of the aquarist in simulating the effects of larger plants and animals.

Generally speaking, mesocosms are over 10,000 gallons. In most cases they provide limitations on only the larger animals, primarily large fish and mammals. Larger microcosms in the 1000–10,000 gallon range also provide limited scale problems, although many predators and even larger algae such as the kelps could become limited in this sized range. In the 70–1000 gallon range, much depends on the type of community as to just how severe the scaling problems are becoming. Generally, for the hobbyist, a reasonable facsimile of most marine or aquatic communities can be developed in this size range. Systems below 70 gallons are severely limiting in their potential for simulating marine ecosystems, though they may still be adequate for some freshwater units.

We will first discuss the general engineering principles that we have developed to handle the miniaturizing of mesocosms and microcosms to aquarium size units. Generally, we will start with the most critical elements and gradually go on to those of lesser importance.

Metal halide gas lamps are capable of simulating the levels of solar radiation found in the wild. The light spectrum is also a reasonable facsimile of sunlight, though with a spike of yellow giving an undesirable hue for some uses. Unfortunately, while metal halides are among the most efficient of light sources (in terms of light output relative to electrical power used), they are relatively expensive. Equally important, they are inappropriately shaped to minimize the size of structure overlying an aquarium, though new technological developments will certainly overcome the latter problem. For shallow-depth aquaria of many kinds, very high output (VHO) or even high output (HO) fluorescent lights can provide an adequate light source that is easily managed. For these lamps, a wide range of spectral types or colors is available. Several approaches to utilizing VHO lamps were shown in Chapters 21–24. In general, we repeat that while natural photosynthetic systems usefully absorb only a small part of the incoming light (1–6%), relatively high intensities are required to achieve those rates of absorbtion. Also, while individual plant species and algal groups have specific absorbtion curves, an aquatic community with a range of species, particularly algae from different groups, absorbs nearly the entire spectrum of visible light. For a variety of reasons discussed in

Chapter 7, aquarists traditionally kept lighting at extremely low levels. Increasing interest in "reef" tanks is changing that view. The timing of day length is also crucial for most ecosystems. The management of day length as well as the developing of peak intensity is not different from that of larger systems, and a wide variety of inexpensive and quite adequate electrical control timers is now available from mass-market sources.

The next most important environmental element for developing benthic ecosystems in small aquaria is that of managing water quality. Water adjacent to wild benthic ecosystems with high biomass reaches the community from more open waters of rivers, lakes, or the ocean. These larger water bodies usually provide for the sinking of nutrients and also have excess oxygen for the use of a benthic community. In the larger microcosms and mesocosms described in this book these tasks were efficiently accomplished by algal scrubbers. For hobbyists, researchers, and exhibitors with adequate space the very same approaches are quite applicable. For placement in small aquarium environments, where size is crucial, several different variations on a common theme have been found to be successful. Several of these were described in Chapters 21–24. Typically the simple question to be answered in determining scrubber size is whether it will control nutrients and provide for long-term stable levels. The aquarium systems described will provide guidelines. However, much will also depend on whether or not the aquarist chooses to overload the ecosystem or, to put it a little differently, to "drive" the model faster than the simulated wild system. A 10–20% fish overload, for example, could easily be handled physiologically by increasing scrubber size. Ecologically this would generally be acceptable. On the other hand, in reef systems, for example, overloads greater than 10–20% would almost certainly result in excessive predation on corals or other invertebrates even if adequate water quality is provided by larger scrubbers. On the other hand, in Chapter 24, we discussed a small cichlid tank that was successfully driven to extreme overloading with a small scrubber.

It is quite possible that the home aquarist would wish to simulate a stagnant or nearly stagnant marine or freshwater community. On the other hand, the vast majority of ecosystems that are likely to be of interest to the aquarist have varying degrees of physical energy (waves, tides, currents). Some wild communities (coral reefs and rocky shores) have very high input levels of wave and current energy, and these cannot be ignored in any reasonable simulation. Current can usually be accomodated easily in the small microcosm by simply placing the pump intake at one end of the tank and the pump exhaust at the other. Pump size can then be adjusted to the amount of flow typically present in the wild system being modeled. Waves and tides in small microcosms can also be achieved just as in the mesocosms described above, assuming adequate space is available and esthetic needs can be managed. Methods for miniaturizing waves and tides in standard aquaria have also been devised and tested and were also shown earlier. Generally speaking, depending upon frequency of monitoring, pumps should be doubled or tripled in all cases where pump failure might not be noticed and the pump replaced quickly. Depending on the time of day and the loading in the system, loss of pumping can create a variety of serious problems and lead to difficulties in miniature ecosystems within a few hours. Simple lift pumps (preferably not air lifts) will increase plankton diversity and generally improve the quality of any aquarium. Simple modifications of common pumps were described in earlier chapters.

Of secondary importance to direct operation of aquaria, but potentially of great importance, is the question of heat transfer. Depending on the community simulated, energy input into a properly managed aquarium is considerable and may have to be compensated for. With the highlighted reef designs described above for 70- and 130-gallon aquaria, a temperature differential of 3–8°F over ambient is typical. This may not matter during the winter or in an air-conditioned house or apartment during the summer. However, in non-air-conditioned houses or apartments in a hot climate, serious problems could result. Lighting on non-reef systems usually would not create high temperature problems except in extreme situations. For cold-water systems or reefs requiring cooling, solutions to such problems were discussed in depth in Chapter 3.

Sediment traps and refugia were described on a number of the larger microcosm and mesocosm systems described above. They are critical only on reef tanks, although water clarity and small invertebrate diversity in any system can be improved with traps. Simple trap/refugia units were described in Chapters 21–24. Much more complex units, particularly as they relate to plankton simulation, can be provided by concatenated aquaria.

Labor-saving devices such as auto-feeders and auto-water-top-up that incidentally provide for more uniformly operated ecosystems are available commercially or can be easily constructed. Many fish become adapted to human feeding very rapidly, and this element of their behavior becomes rather unnatural. The auto-feeder described allows one to stretch out and randomize feeding. Also, using these devices can allow the aquarist to be away from a tank for several days or even several weeks.

Ideally, the dynamic aquarium can be self operating. In practice it is not. Mechanical apparatus is the primary problem, and no matter what the design and construction, such apparatus can potentially malfunction. The system should be checked regularly. Ecosystem adjustments are generally considerably less critical. Nevertheless, they must be made periodically and a plant cropped or an ever-enlarging fish or invertebrate removed. Even using automatic apparatus to top up water, and in some cases to provide feed as the equivalent of a plankton or large water body source, the base resource must be added to the automatic equipment. A reasonable time allotment for the successful operation of the kinds of

70- to 130-gallon aquaria described above is a brief check each morning and evening and $\frac{1}{2}$–1 h of general maintenance once a week.

Although it might be expected that these mini-ecosystems would be unstable and prone to problems, they are not. Early in the development of reef microcosms in the late 1970s, a small tank was lost due to glass failure. Otherwise, in over 30 system years of larger microcosms and mesocosms and 10 system years of small aquaria, not a single ecosystem failure or crash has occured. Many of the larger units and all of the small microcosms and aquaria were experimental systems, and mechanical or electrical failures were more-or-less common. Yet, the ecological systems adjusted to all problems. This record attests to the ultimate stability of high-diversity ecosystem simulations.

To return to the basic ecological rather than environmental problem of miniaturization, namely, that of large size, as the larger animals grow bigger they must be cropped. In reef systems our attentions in this regard have been directed to lobsters, barracuda, jacks, and parrot fish. In the cold-water Maine coast tank it is sea urchins and lobsters, and in the Chesapeake Bay systems, blue crabs, perch, drums, and pickerel. In this case, the aquarist is functioning as the top predator to provide control on those predators in mid-size levels. When the mid-level predator reaches such a size that overgrazing of plants or overcropping of animals is becoming serious, the animals involved must be removed from the system. The time for such removal is usually noncritical in the scale of weeks or even months, but the aquarist must have enough familiarity with the wild system or the captured ecosystem itself to recognize overgrazing and/or overcropping. This may sound like a difficult problem for the beginner, but in practice it is not. Even though this constitutes the major area of concern for managing captive ecosystems, normally such systems are so stable and insensitive that there is plenty of room for learning. Indeed, this heuristic or self-teaching aspect is a universal and most desirable characteristic of dynamic aquaria.

INDEX